Lecture Notes in Computer Science 13288

More information about this series at https://link.springer.com/bookseries/558

Luca Bernardinello · Laure Petrucci (Eds.)

Application and Theory of Petri Nets and Concurrency

43rd International Conference, PETRI NETS 2022 Bergen, Norway, June 19–24, 2022 Proceedings

Editors
Luca Bernardinello
University of Milano-Bicocca
Milano, Italy

Laure Petrucci
Paris 13 University
Villetaneuse, France

ISSN 0302-9743 ISSN 1611-3349 (electronic)
Lecture Notes in Computer Science
ISBN 978-3-031-06652-8 ISBN 978-3-031-06653-5 (eBook)
https://doi.org/10.1007/978-3-031-06653-5

This Springer imprint is published by the registered company Springer Nature Switzerland AG
The registered company address is: Gewerbestrasse 11, 6330 Cham, Switzerland

Preface

This volume collects the proceedings of the 43rd International Conference on Application and Theory of Petri Nets and Concurrency (Petri Nets 2022). This series of conferences serves as an annual meeting place to discuss progress in the field of Petri nets and related models of concurrency. These conferences provide a forum for researchers to present and discuss both applications and theoretical developments in this area. Novel tools and substantial enhancements to existing tools can also be presented. Petri Nets 2022 included a session devoted to the Application of Concurrency to System Design (ACSD), which was in the past a separate event. The final selection for this track was made by Jörg Desel and Alex Yakovlev. The event was organized by the Software Engineering Research Group at the Western Norway University of Applied Sciences and took place at Campus Bergen, Norway.

After two editions entirely held online because of the SARS-Cov-2 pandemic, this year the conference took place in the traditional style, gathering people in Bergen, on the western coast of Norway. On behalf of the authors and participants, we express our deepest thanks to the Organizing Committee, chaired by Lars Michael Kristensen and Violet Ka I Pun, for the time and effort invested in the organization of this event. This year, 35 papers were submitted to Petri Nets 2022 by authors from 18 different countries. Each paper was reviewed by at least three reviewers. The discussion phase and final selection process by the Program Committee (PC) were supported by the EasyChair conference system. From regular papers and tool papers, the PC selected 19 papers for presentation: 16 regular papers (with three papers on the ACSD track) and three tool papers. After the conference, some of these authors were invited to submit an extended version of their contribution for consideration in a special issue of a journal.

We thank the PC members and other reviewers for their careful and timely evaluation of the submissions and the fruitful constructive discussions that resulted in the final selection of papers. The Springer LNCS team provided excellent and welcome support in the preparation of this volume. The keynote presentations were given by Paulo Esteves-Verissimo (Carl Adam Petri distinguished lecture) Volker Diekert, and Marieke Huisman.

Alongside Petri Nets 2022, the following workshops and events took place: the 12th edition of the Model Checking Contest (MCC 2022), Algorithms and Theories for the Analysis of Event Data (ATAED 2022), Petri Nets and Software Engineering (PNSE 2022), and the International Health Data Workshop (HEDA 2022).

We hope you enjoy reading the contributions in this LNCS volume.

June 2022

Laure Petrucci
Luca Bernardinello

Organization

Program Committee

Elvio Gilberto Amparore	University of Turin, Italy
Abel Armas Cervantes	University of Melbourne, Australia
Paolo Baldan	Università di Padova, Italy
Benoît Barbot	Université Paris-Est Créteil, France
Luca Bernardinello (Co-chair)	Università degli Studi di Milano-Bicocca, Italy
Didier Buchs	University of Geneva, Switzerland
Béatrice Bérard	LIP6, Sorbonne Université and CNRS, France
Josep Carmona	Universitat Politècnica de Catalunya, Spain
David de Frutos Escrig	Universidad Complutense de Madrid, Spain
Jörg Desel	FernUniversität in Hagen, Germany
Raymond Devillers	ULB, Belgium
Susanna Donatelli	Università di Torino, Italy
Javier Esparza	Technical University of Munich, Germany
Stefan Haar	Inria and LSV, ENS Paris-Saclay, France
Xudong He	Florida International University, USA
Loïc Helouet	Inria, France
Ryszard Janicki	McMaster University, Canada
Anna Kalenkova	University of Adelaide, Australia
Jörg Keller	FernUniversität in Hagen, Germany
Ekkart Kindler	Technical University of Denmark, Denmark
Michael Köhler-Bußmeier	University of Applied Science Hamburg, Germany
Irina Lomazova	National Research University Higher School of Economics, Russia
Robert Lorenz	University of Augsburg, Germany
Łukasz Mikulski	Nicolaus Copernicus University, Poland
Andrew Miner	Iowa State University, USA
Marco Montali	Free University of Bozen-Bolzano, Italy
Laure Petrucci (Co-chair)	Université Paris 13, France
Artem Polyvyanyy	University of Melbourne, Australia
Pierre-Alain Reynier	Aix-Marseille Université, France
Arnaud Sangnier	IRIF, Université Paris Cité and CNRS, France
Natalia Sidorova	Technische Universiteit Eindhoven, The Netherlands
Jaco van de Pol	Aarhus University, Denmark

Additional Reviewers

Adobbati, Federica
Balasubramanian, A. R.
Barylska, Kamila
Bashkin, Vladimir
Chatain, Thomas
Claeys, Emmanuelle
Coet, Aurélien
Czerner, Philipp
De Weerdt, Jochen
Igor, Khmelnitsky
Julvez, Jorge
Koutny, Maciej
Morard, Damien
Nesterov, Roman
Oualhadj, Youssouf
Padoan, Tommaso
Petrak, Lisa
Remke, Anne
Rivkin, Andrey
Rykaczewski, Krzysztof
Schalk, Patrizia
Sekerinski, Emil
Shershakov, Sergey
Verbeek, Eric

Invited Talks

Assumptions in Computer Science: Mere Mathematical Hypotheses, or Representations of the Physical World?

Paulo Esteves-Verissimo

KAUST, CEMSE, RC3 (Resilient Computing and Cybersecurity Center)
Thuwal, Kingdom of Saudi Arabia

The praxis of algorithm or mechanism design, especially in concurrent or distributed systems, mandates clear statements of the assumptions underlying the design, such as topology, dimension, synchrony, performance, threats, etc. The design—say, of an algorithm whose behaviour is defined by a set of properties (safety, liveness)—is then shown correct by demonstrating it does secure those properties, given the assumptions. From a mathematical viewpoint, we are done. We never ask: are the hypotheses valid?

If we wish the algorithm to have any real world impact, it might be advisable to define an *abstract system* credibly materializing the assumptions made, where the protocol implemented from the algorithm will run correctly. However, this apparent detail completely changes the perspective: in essence, the protocol correctness becomes conditional to the likelihood of the assumptions being met, in such a system.

Suddenly, we must look at our assumptions from a physics viewpoint: How do I achieve perfect failure detection in an asynchronous environment? Why would an attacker compromise certain units and not others in an otherwise arbitrary failure environment? We should not be mistaken about these being "implementation details". They can and should be addressed as *systems theory* problems, related to the *substance* and the *robustness* of assumptions—i.e., the coverage of the mapping of the *abstract system* onto the physical world.

The rise of malicious threats to systems has been showing the importance of this argumentation. The practice of accepted deviations of some assumptions from physical reality, in face of accidental (stochastic) threats, completely crumbled in face of malicious(intentional) threats, exposing their lack of substance and/or robustness. Paraphrasing and extending my colleague and friend Fred Schneider's quote some years ago: "Every [non-substantiated] assumption is a vulnerability".

In the talk, I will delve into manifestations of the problems above, and approaches to solve them in a satisfactory way. Closing these gaps implies effort on some angles, and I will single-out two: (i) system architecture and design; (ii) modeling and verification. I will be discussing: system awareness (topology-, context-, hybridisation-), logical vs. physical centralisation, trust vs. trustworthiness); system-level impossibility results, lower bounds and safety predicates (coverage-stability, no-contamination, exhaustion-safety) denying substance to some commonly made assumptions—or providing guidance to achieve it.

Petri Nets and Mazurkiewicz Traces Partnership When Honeymoon is Forgotten

Volker Diekert

University of Stuttgart, Germany

History. In the mid 1970-ties it became clear that free partially commutative monoids play a central role for the analysis and the understanding of concurrent systems. The foundations were developed in the 1973 JACM-paper by Robert Keller [1]. A few years later, in 1977 Antoni Mazurkiewicz published his seminal Aarhus technical report where he showed that, indeed, partial commutation describes the semantics of one-safe Petri nets perfectly [2]. He also introduced a graphical representation of elements in free partially commutative monoids. In his notation a "trace" is not a (firing) sequence of the net, but a labeled, directed and acyclic graph with an immediate visual understanding.

Honeymoon. A concurrent system has a visual representation (a one-safe Petri net), the executions have a visual representation as a trace. The set of all executions are recognized by an asynchronous automaton in the sense of Zielonka ([3]) with a purely algebraic semantics using syntactic congruences. Languages accepted by asynchronous automata have finite syntactic monoids.
The Eldorado!

Honeymoon is over. Partnership remains: trace theory as a basic algebraic concept should not be forgotten.

My lecture. In my talk I will discuss the basic mathematics and extensions to notions of a semi-trace and/or a partial trace. These concepts can be applied to more general types of Petri nets, still the algebra can be expressed within the theory of free partially commutative monoids. It is a versatile tool connecting semantics and combinatorial algebra.

Trace theory, may you stay forever young. Bob Dylan (1974)

References

1. Keller, R.M.: Parallel program schemata and maximal parallelism I. Fundamental results. J. ACM 20(3), 514–537 (1973)
2. Mazurkiewicz, A.: Concurrent program schemes and their interpretations. DAIMI Rep. PB 78, Aarhus University, Aarhus (1977)
3. Zielonka, W.: Notes on finite asynchronous automata. R.A.I.R.O.—Informatique Théorique et Appl. **21**, 99–135 (1987)

VerCors and Alpinist: Correctness of GPU Applications Throughout the Development Cycle

Marieke Huisman

University of Twente, The Netherlands

GPU programs are widely used in industry. They make use of special hardware, supporting massive parallelism. To avoid the introduction of errors, we can augment GPU programs with (pre- and postcondition-style) annotations to capture functional properties. The VerCors program verifier, a verifier for concurrent software, has been tailored to reason about such GPU programs, and in this talk I will explain how this is done, and discuss various GPU case studies that have been verified with VerCors.

However, to obtain the best performance, a typical development process of a GPU application involves the manual or semi-automatic application of optimizations prior to compiling the code. Keeping these annotations correct when optimizing GPU programs is labor-intensive and error-prone. Therefore, this talk also introduces Alpinist, an annotation-aware GPU program optimizer. It applies frequently-used GPU optimizations, but besides transforming code, it also transforms the annotations. We evaluate Alpinist, in combination with the VerCors program verifier, to automatically optimize a collection of verified programs and reverify them.

Contents

Applications

Synthesis

Petri Nets Architecture

Process Mining

Application of Concurrency to System Design

Towards the Application of Coloured Petri Nets for Design and Validation of Power Electronics Converter Systems

Vegard Steinsland(✉), Lars Michael Kristensen, and Shujun Zhang

Department of Computer Science, Electrical Engineering and Mathematical Sciences, Western Norway University of Applied Sciences, Bergen, Norway
{vste,lmkr,shz}@hvl.no

Abstract. We apply Coloured Petri Nets (CPNs) and the CPN Tools to develop a formal model of an embedded system consisting of a power converter and an associated controller. Matlab/Simulink is the de-facto tool for embedded control and system design. Moreover, Matlab/Simulink relies on informal semantics and has limited support for transparent and integrated specification and validation of both the power converter electronics, controller (hardware), and the control logic (software). The contribution of this paper is to develop a timed hierarchical CPN model that mitigates the shortcomings of Simulink by relying on a Petri net formalisation. We demonstrate the application of our approach by developing a fully integrated model of a buck power converter with controller in CPN Tools. Furthermore, we perform time-domain simulation to verify the capability of the controller to serve the control objectives. To validate the developed CPN model, we compare the simulation results obtained in an open-loop configuration with a corresponding implementation in Simulink. The experimental results show correspondence between the CPN model and the Simulink model. As our CPN model reflects the fully integrated system, we are able to compare CPN simulation results to measurements obtained with a corresponding implementation in real hardware/software and compare closed-loop with open-loop configuration. The results show alignment for the steady state while further refinement of the control algorithm and validation is required.

1 Introduction

The development of power electronics converter systems is a multi-disciplinary task where the immediate time-critical response is related to the integrated system of electrical, electronics, cybernetics and embedded control design with software forming a cyber-physical system (CPS) [18].

Modelling is a valuable tool for understanding power electronics converters system and to control their behaviour, and can aid the design process from initial evaluation of proposed circuit topologies through physical prototyping with integration of the controller, to iterative improvements in both the physical design and the control. Given the comprehensive challenges in the design phase,

L. Bernardinello and L. Petrucci (Eds.): PETRI NETS 2022, LNCS 13288, pp. 3–22, 2022.
https://doi.org/10.1007/978-3-031-06653-5_1

abstraction and simplification of the system parts are needed to obtain models at the appropriate abstraction level to achieve and validate the design objective(s).

In power electronics and embedded control system design, Matlab/Simulink tool [16] is one of the de-facto standards and is supported by an integrated tool chain. Within the tool chain of MathWorks, the Simulink tool provides predefined modelling objects with approximation of the physical systems (via differential equation descriptions), control objects, tools for embedded code generation, and tools for real-time hardware (HW) monitoring to support integrated design and rapid prototyping. However, Matlab/Simulink relies on informal semantics and several researchers have identified application shortcomings implied by the lack of tools for model checking and validation [2–4,14,19].

The lack of formal semantics in Matlab/Simulink also implies challenges when executing embedded code on real hardware, where the actual performance has timing properties relying on the integrated CPS. This makes it difficult to validate the integrated system performance based on other metrics than the physical response. This means that the approach is partly non-transparent and the designer is somehow blindfolded by the complex power electronics converter system behaviour due to the lack of tools in Matlab/Simulink to perform formal validation of the CPS. For simple power electronics converters such as the buck converter, the lack of formalization may not be a practical challenge. However, when developing systems with higher complexity such as a Modular Multilevel Converter (MMC) and systems of converters forming microgrids, the lack of formal semantics and transparency may lead to unexpected events in the control schemes as discussed in [8,12,18].

The contribution of this paper is to explore the application of Petri nets to address the shortcomings of Matlab/Simulink discussed above. Our aim is to provide an integrated modelling approach that provides transparency and traceability on both the hardware part and on the control logic part of the CPS. Being based on Petri nets, our approach provides a formal semantics that can serve as a basis for model validation when applied for the design and implementation of power electronics converter systems. We rely on high-level Petri nets in order to conveniently model data related to sensed values (physical response), calculations, signalling, and system parameters. We specifically apply Coloured Petri Nets (CPNs) [5] and CPN Tools [6] as it supports the modelling and simulation of time, and the construction of hierarchical and parametric models divided into several modules. The latter is important in order to ensure transparency and scalability to larger power electronics converter systems.

As a case study, we focus on modelling the dynamics of a buck power electronics converter integrated with an embedded controller, and we use simulation and physical models to evaluate and compare the system performance in the time-domain. First the CPN model of the buck converter dynamical response is compared with the Matlab/Simulink implementation with fixed point operation, thus reducing the complexity. This is followed by results obtained via the integrated CPN modelling of the buck converter and the controller aimed at assessing the objective as an integrated CPS with closed-loop control logic serving the control objective of fixed output voltage. To further validate the constructed CPN model, we validate the CPN simulation against measurements obtained with a corresponding implementation of the power converter system in real hardware. The rest of this paper is organised as follows. Section 2 gives an overview of our modelling approach and the power converter system comprised of the buck converter topology description and the controller. Section 3 concentrates on the modelling of the buck converter, and Sect. 4 presents the modelling of the embedded controller. Section 5 provides a brief description of the Matlab/Simulink model that we use for comparison. The experimental results are presented in Sect. 6. In Sect. 7, we present our conclusions, discuss related work, and suggest directions for future work to further develop our modelling framework in the domain of power electronics systems. We assume that the reader is familiar with the basic concepts of Petri Nets and high-level Petri Nets.

2 Modelling Approach and Overview of the CPN Model

The developed CPN model comprises the complete system and consists of the buck converter integrated with the controller. Our modelling approach allows for system parametrisation of the physical converter model, the system interfaces, and the controller properties. The objective of the integrated buck power converter system is to reduce the input voltage (V_i) to a given output voltage (V_o) based on a Pulse Width Modulation (PWM) control signal from the controller to the converter.

Figure 1 shows the actucal physical CPS being modelled which is also the system against which we validate the simulation results.

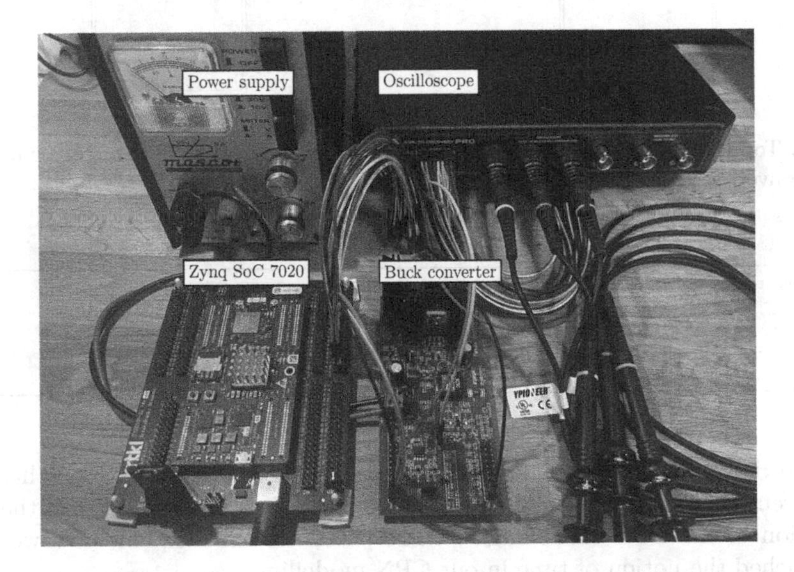

Fig. 1. The modelled CPS with KRTKL Snickerdoodle Black Zynq SoC 7020 controller and buck-converter. Connected to laboratory power supply and oscilloscope.

The controller is an embedded Xilinx Zynq System-on-Chip (SoC) 7020 hardware controller that consists of both an Advanced RISC Machines (ARM) processing system (PS) and a field-programmable gate array (FPGA) programmable logic (PL) in one single package. The main advantages of SoC with FPGA are the parallel task execution feature, very high processing speed, and the integrated ARM processor. These attractive features has resulted in the growing popularity of SoC being used in embedded controls of power electronics [11,19]. The complete CPN model is hierarchically organised into seven modules. The top-level module of the CPN model of the CPS is shown in Fig. 2 with the colour set definitions given in Listing 1.1. The substitution transition Buck Converter represents the physical electrical converter system and the substitution transition Controller represents the hardware and the control logic of the controller. The socket places Gate and Buck loop are used to pass the control signal to the converter and receive the acknowledgment of the converter state, respectively. The converter inductor current (i_L) and output voltage (v_o) measurements are interfaced with the controller and modelled by the socket place named iL vo.

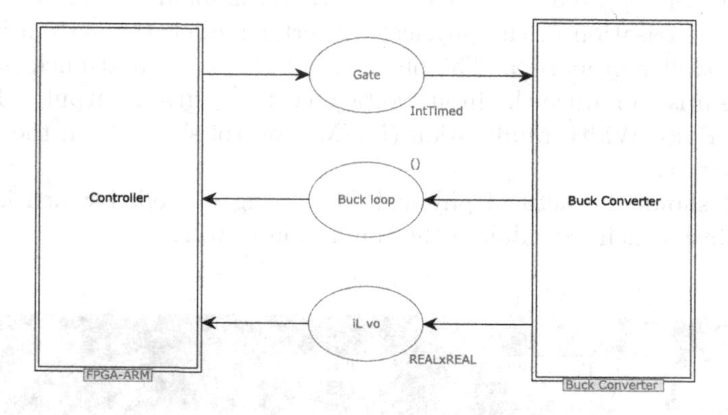

Fig. 2. Top-level CPN module of the CPS with integrated modelling of the controller, buck converter, and their interfaces.

```
colset IntTimed = int timed;

colset iL = real;
colset vo = real;
colset REALxREAL = product iL*vo;
```

Listing 1.1. Colour set definitions used in the top-level module.

The detailed CPN modelling of the buck converter and the controller will be presented in Sect. 3 and Sect. 4, respectively. Before proceeding with the presentation of the CPN modelling of these parts, we first explain how we have approached the notion of time in our CPN modelling.

The buck converter is a continuous-time dynamic system that evolves over time and the model description is implemented as a discrete event system (DES) together with the digital controller with real-time properties. For modelling the CPS as a real-time system, we must consider the continues-time dynamics of the modelled artifact and apply appropriate time discretisation to achieve the desired modelling accuracy. The physical behaviour of the buck converter over time is described through differential equations which we have implemented in CPN Tools with emphasis to clearly show graphically how each step of the calculation is performed (see Sect. 3). The reason for selecting this approach is to provide traceability in how the mathematical description is implemented to the explicit case of the buck converter. By default, the simulated timed properties in CPN Tools have no intrinsic relationship to real-time. To align the CPN Tools simulation time with the real-time physical properties, we introduce the simulated time iteration constant (k) and the discrete time-step (T_s) in the discretisation of both the buck converter and controller. It is the system under investigation that governs the required refinement of the simulation model discrete time-step (T_s). The buck converter holds switching states and therefore the discrete time-step (T_s) must be smaller than the cyclic period (T_p) to represent the dynamic response in the electric circuit in-between a switching event. The discrete time-step (T_s) by the simulator must be less than (T_p) to represent the circuit dynamics correctly. A drawback of making the discrete time-step (T_s) too small is its negative impact on the performance of the CPN Tools simulation engine. Analytical tools and domain-specific experience/experimental results may provide insights into determining a suitable discrete time-step (T_s).

The simulated time in the CPN model is represented via two functions: `Dynsys()` and `CPU_CLK()`. The task of the `Dynsys` function is to keep track on the time iteration of the dynamic system that forms the buck converter. As the dynamic system has the highest discretisation requirement in the simulation, the overall time reference is set as `Dynsys` and holds the iteration relation to the real-time trough the discrete time-step (T_s). The function `CPU_CLK` is used to describe the timing of the controller CPU cycle, and is aligning the cycle frequency of the controller to the real-time properties of `Dynsys`.

3 Buck Converter CPN Modelling

The buck converter is used to step down the input voltage (V_i) to the desired output voltage (V_o) The principal electrical topology of the buck converter is shown in Fig. 3 with the main components. The components introducing circuit states are the non-linear semiconductor elements with the active controlled transistor (u) and the passively controlled diode (D). In addition, the circuit contains linear elements in terms of inductance (L), capacitance (C) and resistance (R). All components values are derived from the International System of Units (SI), relating properties with respect to voltage (V) and current (I). In most basic circuit analysis (ignoring losses) the steady-state fundamental description of the relation between the input voltage (V_i) and the output voltage (V_o) may be described [10] with a factor of the duty-ratio (d):

$$V_o = V_i \cdot d \tag{1}$$

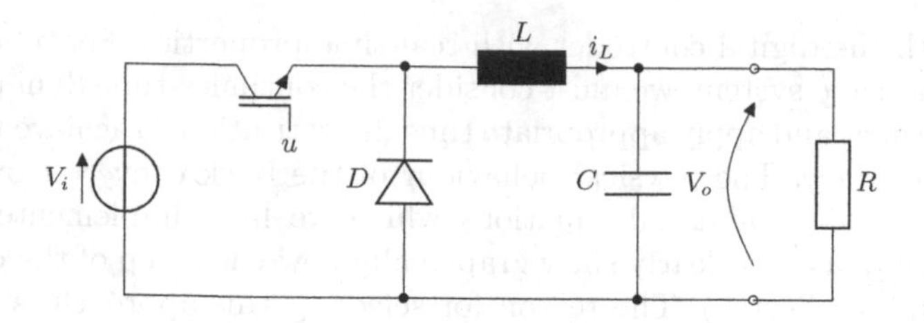

Fig. 3. Buck converter principal electrical topology.

The duty-ratio factor of the input/output voltage relation persists with limitation by the switching operation of the transistor element as long as the switching frequency (f_{sw}) is fixed and the circuit has a continuous inductor current (i_L) [10]. In the time-domain, the switching frequency (f_{sw}) may be expressed as the switching period (T_p).

Without further details of the design consideration, and the introduction of semiconductors, transistor and diodes, the buck converter circuit in Fig. 3 can be in two possible states. This is illustrated in Figs. 4a and 4b as equivalent circuits for the on- and off-state, respectively. As may be observed from Figs. 4a and 4b, the semiconductor elements is replaced by their feature in the switching state by ideal properties as either fully closed or opened.

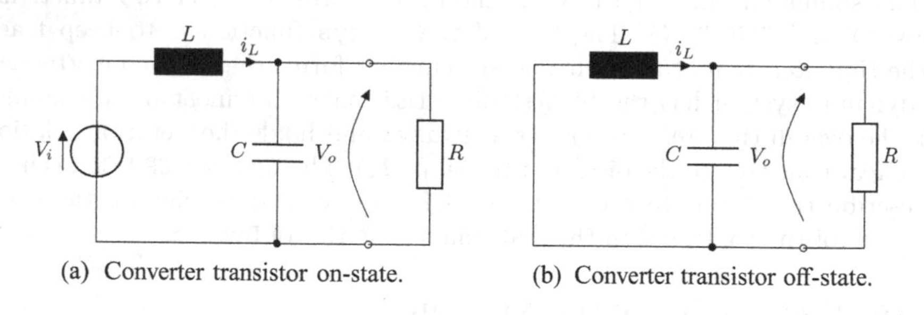

(a) Converter transistor on-state. (b) Converter transistor off-state.

Fig. 4. The circuit equivalents of Fig. 3 as determined by transistor state.

Based on the circuit states, the two differential equations [10] describing the circuit based on the change in the input inductor current (i_L) and output voltage (v_o) are given in Eq. (2a) and Eq. (2b):

$$\frac{di_L}{dt} = \frac{V_i}{L}(u) - \frac{v_o}{L} \tag{2a}$$

$$\frac{dv_o}{dt} = \frac{1}{C}i_L - \frac{1}{RC}v_o \tag{2b}$$

As can be seen, the transistor switching states only affects the $\frac{V_i}{L}(u)$ expression, where u is the switch state, zero (off) and one (on).

Figure 5 shows the buck converter submodule of the Buck Converter substitution transition found in the top-level model in Fig. 2. The port places Gate, Buck loop, and iL vo are associated with the accordingly named socket places in Fig. 2 and used to model the signalling between the buck converter and the controller. The submodules of the two substitutions transitions IL and Vo in Fig. 5 model the inductor current (I_L) and the output voltage (V_o) as described by the differential equations from Eq. (2). The socket places IL and Vo are used for modelling the interconnected behaviour of the two differential equation in Eq. (2), i.e., that v_o appears as a term in Eq. (2a) and that i_L appears as a term in Eq. (2b). Below we present in detail the modelling of the inductor current and the output voltage.

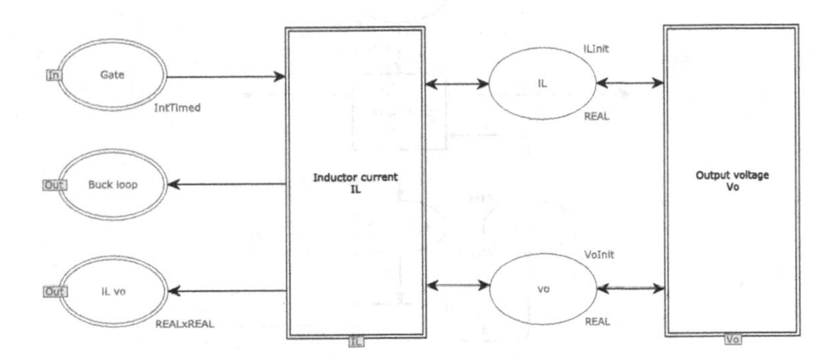

Fig. 5. Submodule of the Buck Converter with external interfaces to the controller (left) and calculation of I_L, Eq. (2a) and V_o, Eq. (2b) represented by substitutions transitions.

3.1 CPN Modelling of Inductor Current (IL)

Figure 6 shows the submodule of the Inductor Current IL substitution transition which implements the first part of the integrated behaviour of the buck converter. The colour set and function definitions are shown in Listing 1.2. The places iL and vo act as interface to the submodule Vo, and exchange the values for inductor current (i_L) and output voltage (v_o). As described by the differential equations in Eq. 2, the physical response between the modules is integrated.

The purpose of the IL module is to estimate the inductor current from Eq. (2a) based on the switch state and the circuit electrical properties (see Fig. 3) as it evolves over time. As may be observed, the modelling approach as represented by the submodule in Fig. 6 closely follows the buck converter description and behaviour as represented via the differential equations. This demonstrates how we have targeted a transparent modelling approach where the detailed modelling of the components and the steps involved in their execution are fully available for inspection and validation. This is in contrast to the black-box approach adopted by Matlab/Simulink.

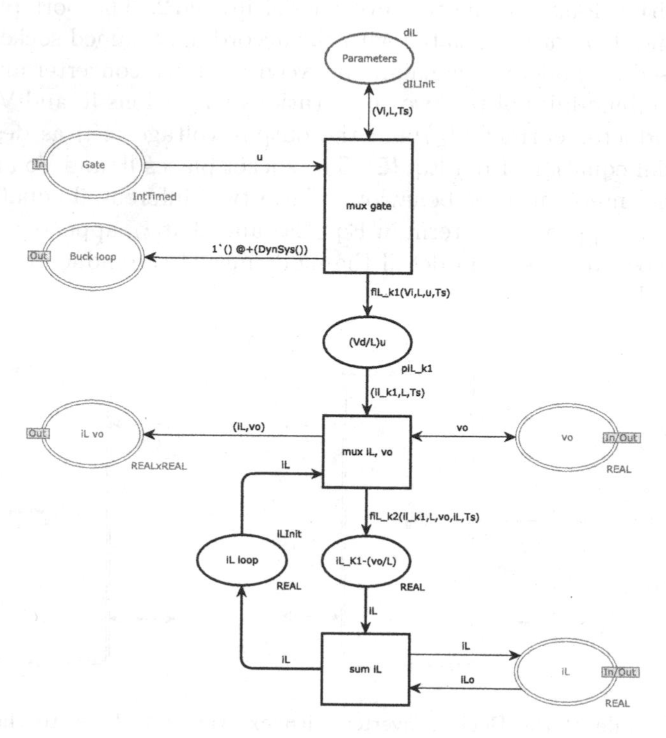

Fig. 6. Submodule IL estimating the inductor current (i_L) based on Eq. (3).

```
colset IntTimed  = int timed; (* Timed gate control signal *)
colset dILInit   = product Vi*L*Ts;
colset piL_k1    = product iL_k1*L*Ts;
colset REALxREAL = product iL*vo;

fun fiL_k1(Vi,L,u,Ts) = (if u=1 then  (Vi/L)*Ts else  0.0,L,Ts);

fun fiL_k2(il_k1,L,vo,iL,Ts) = (iL+(il_k1-(vo/L)*Ts));
```

Listing 1.2. Colour sets and functions used in the iL submodule.

Due to the requirement of time in the circuit analysis, the description of the inductor current found in Eq. (2a) must be adapted for discrete-event numerical simulation. As introduced at the end of Sect. 2, the modelling of time includes the iteration constant **k** and the real-time conversion constant **Ts** in the discretisation. Based on this discretisation, we obtain the description expressed by Eq. (3) which is what is implemented by the IL module in Fig. 6. The description is also dependent on the switching element u with either a dynamic or fixed switching frequency (f_{sw}) covering the corresponding switch period (T_p).

$$i_L((k+1)T_s) = T_s \left[\overbrace{\overbrace{\frac{V_i}{L}(u)(kT_s)}^{\texttt{fiL\_k1}} - \frac{v_o}{L}(kT_s)}^{\texttt{fiL\_k2}} \right] \tag{3}$$

The buck converter contains physical components each of which has a value, such as capacitance (C), inductance (L) and load resistanse (R). In addition, the considered operation of the converter is with a fixed value for the input voltage (V_i). The value of the components and input voltage forms the constant parameters defined at the place Parameters. The occurrence of the mux gate transition is triggered based on the received timed controller signal to the gate activation (u) from the place Gate. The function `fiL_k1` is used to estimate the first part of Eq. (3) based on the control signal state (u). The second part of Eq. (3) is estimated using the `fiL_k2` function as part of the firing of the mux Il,vO transition and summation of the resulting inductor current (i_L) from the first part. The arc expression from transition mux gate to place Buck loop is used to signal an acknowledgment which confirms when an iteration of the buck converter model has been executed and thereby allow the next token from the controller via the Gate place. Hence, the Buck loop interface place is purely a modelling artefact for executing the CPN model correctly wrt. time and does not have a physical counterpart in the real system.

3.2 CPN Modelling of Output Voltage (Vo)

Figure 7 shows the submodule of the output voltage Vo substitution transition implementing the second part of the integrated behaviour of the buck converter. The places iL and vo act as the interface to the submodule IL, and exchange the values for inductor current (i_L) and output voltage (v_o). The colour set definitions and the functions used in the arc expressions are listed in Listing 1.3. The output voltage (V_o) is dynamically connected to the inductor current (i_L) and the corresponding switch-state (u). The buck converter physical component characteristics are defined at the place `Parameters` with the load resistor (R) and capacitance (C).

The description found in Eq. (2b) is extended to include time properties of the discrete-event model using the same discretisation approach as for the inductor current (i_L). The derived equations are shown in Eq. (4).

$$v_o((k+1)T_s) = T_s \left[\overbrace{\overbrace{\frac{1}{C}i_L(kT_s)}^{\texttt{fvo\_k1}} - \frac{1}{RC}v_o(kT_s)}^{\texttt{fvo\_k2}} \right] \tag{4}$$

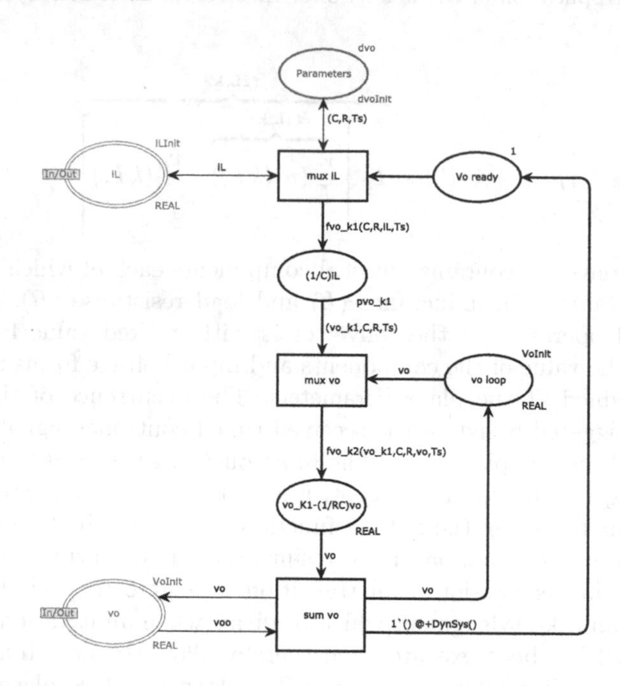

Fig. 7. Submodule Vo for estimating the output voltage (v_o) based on Eq. (4).

```
colset dvoInit = product C*R*Ts;
colset pvo_k1  = product vo_k1*C*R*Ts;

fun fvo_k1(C,R,iL,Ts) = (((iL/C)*Ts),C,R,Ts);

fun fvo_k2(vo_k1,C,R,vo,Ts) = (vo+(vo_k1-((Ts)*vo/(R*C))));
```

Listing 1.3. Definition of the color set related to substitution transition Vo.

The function `fvo_k1` estimates the first part of Eq. (4), and function `fvo_k2` is used for estimating the second part of Eq. (4) based on the firing of the transition, summation of the resulting voltage product from the first part, and the output load conditions.

4 Controller CPN Modelling

Figure 8 shows the FPGA-ARM Controller submodule of the FPGA-ARM Controller substitution transition in Fig. 2. The modelling of the FPGA-ARM controller in Fig. 8 is organised into two submodules: the XADC-Conditioning and Control Logic. The purpose of dividing controller in two parts (XADC-Conditioning and Control Logic) is to capture the data-flow and connected parsing delays in the Zynq SoC from sensing of input from the buck converter through execution of the control logic to the eventual effect on the controller output. The two socket places Control Vo and Control iL in Fig. 8 are used for modelling the interconnection of the two sub-modules. An advantage of having the two dedicated control tasks explicitly separated is that we can easily adopt more complex control logic into the model. The modelling of the control tasks is described in detail below.

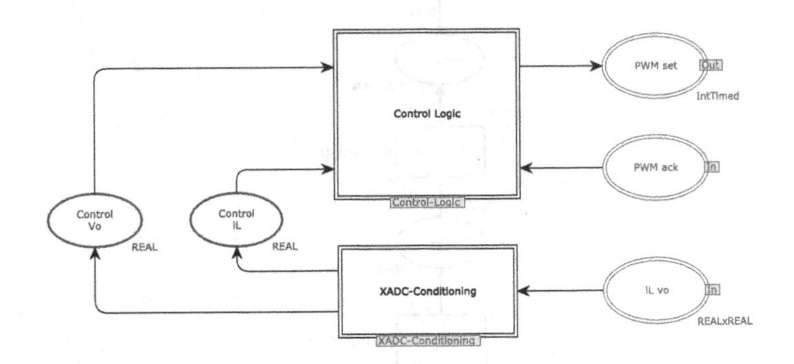

Fig. 8. Overview of FPGA-ARM modelling forming the Zynq SoC.

4.1 CPN Modelling of the XADC Conditioning

Figure 9 shows the submodule of the controller signal conditioning circuit associated with the XADC-Conditioning substitution transition from Fig. 8. It implements the required steps from parsing the signals from the input, through the analog to digital converter (ADC), and the preparation of the readings for use in control tasks. The input-pins are allocated to the FPGA and needs to be transferred to the integrated analog to digital converter (ADC) for signal conditioning in the ARM.

Each of the steps, moving the data between registers, requires FPGA and ARM processor execution implying a time delay. This is modelled by the aspect of `@+CPU_CLK()` time inscriptions placed next to the transitions. There is no time delay associated with the Input read as it does not involve any CPU processing for the sensed input on place iL vo to be available via the associated Register in.

4.2 CPN Modelling of the Control Logic

Figure 10 shows the submodule of the Controller Logic substitution transition. The associated colour set and control functions are defined in Listing 1.4. It models the execution of the control logic in the FPGA and the setting of the control output to the buck converter through the Control logic substitution transition in Fig. 8. The controller uses the power of the FPGA by executing all logic concurrently based on a single read from the bulk register with the input from the buck converter output voltage, `vo`, and inductor current, `iL` from the XADC-Conditioning. The time control is therefore determined by the XADC-Conditioning module, adding a `@+CPU_CLK()` for the pull request of the analog sensing values to the controller.

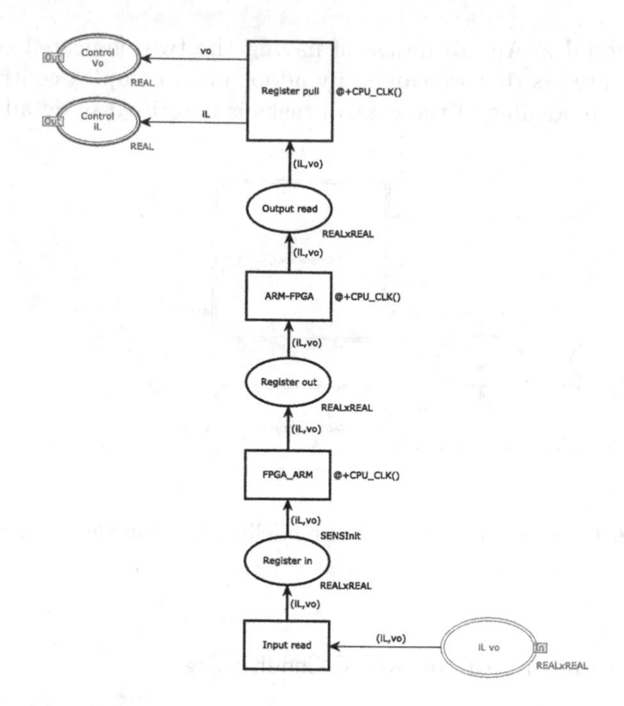

Fig. 9. The XADC-Conditioning submodule modelling of FPGA and the ARM interface

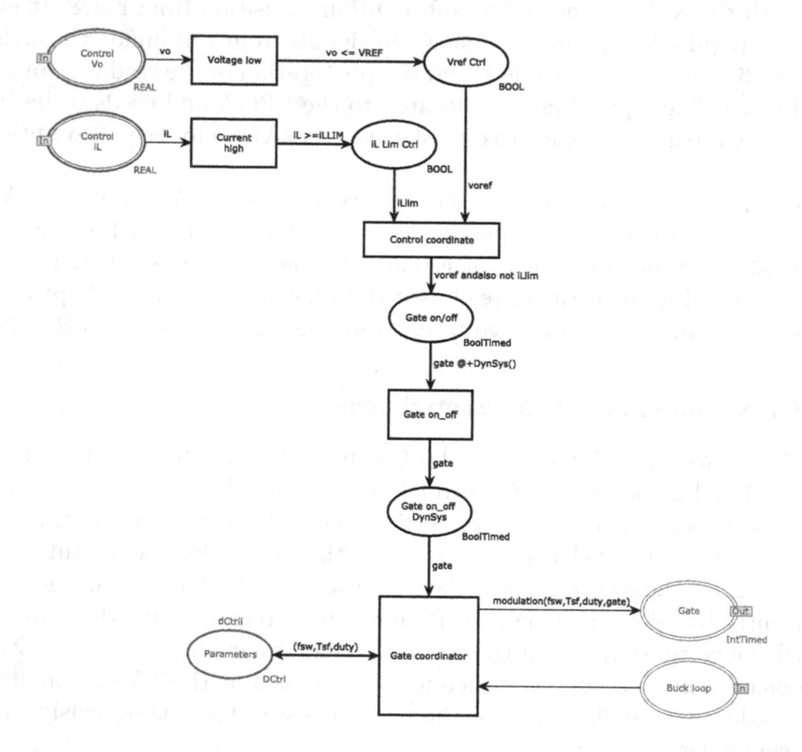

Fig. 10. The Control Logic submodule with open-loop and closed-loop controls.

```
colset DCtrl = product fsw * Tsf * duty;

fun modulation(fsw,Tsf,duty,gate) =
let
  val time_stamp = time()
  val time_stamp_real = ((IntInfToReal 2 time_stamp)*100.0)
  val Tp = 1.0/fsw
  val Ts = Tp/Tsf

  (* Sets the duty ratio relativ to the simulator timesteps *)
  val relative_duty = Tsf*duty

  (* Defines the count of cycles, Tp, relative to simulator timesteps *)
  val TpCount = Real.fromInt(round (((IntInfToReal 0 (time()))/Tsf)-0.50))
in
  if TpCount<1.0 (* As initial point the Tp counter is ignored*)
  then
    if relative_duty>time_stamp_real andalso gate then 1 else 0
  else
   if relative_duty>(time_stamp_real-(Tsf*TpCount)) andalso gate then 1
        else 0
end;
```

Listing 1.4. Definition of the colour set related to substitution transition Control logic.

Three modes of control are represented by the model of the controller logic:

1. The duty ratio control which ensures a gate switching at given duty ratio (d) of the switching frequency, f_{sw}. With this operation, the ideal properties of the converter input/output voltage relation is shown in Eq. (1).
2. Voltage reference control ensures that the gate is attempting to compensate a voltage deviation based on the sensing of the output voltage, `vo`, to the voltage reference `VREF` as closed-loop control.
3. Inductor current protection ensures that the gate is not overloading the inductor and converter by sensing the inductor current, `iL`, and are blocking the gate if the threshold current `iLLIM` is exceeded.

The gate signal (u) is modulated at the output stage between the controller and buck converter with the function `modulation`. Through the modulation function, the predefined switching frequency `fsw` and time-step discretisation factor `Tsf` is aligned with the real-time properties of time-step `Ts`. The predefined parameters are defined at the place Parameters. As the duty ratio, `duty`, is a fraction of the switching time periode `Tp` it also implies that the duty ratio is a fraction of the number of the time-step discretisation factor `Tsf`. In the `if`-statements, the relative duty ratio to the switching period `Tp` is compared to the real-time execution of the model. In addition, the gate output considers the `gate`-conditions from `VREF` and `iLLIM`.

5 Simulink Model of the Buck Converter

For validation and comparison, we have implemented the buck converter using the standard component library of Simscape (v 5.1) and Simscape Electrical (v 7.5) under Matlab/Simulink R2021a (Fig. 11).

The topology of Fig. 3 is adopted, adding voltage source, switching elements, passive components and measurements. All Simulink model components have simplified real characteristics which is defined by modifying the parameters of each individual part. In simulation, the discrete solver is selected to catch transients from the switching events consistently and the results are gathered from the scope measurement for output voltage (v_o) and inductor current (i_L). The model has simple operation of PWM fixed duty ratio control with open-loop feedback, controlled by the signal generator connected to the transistor gate.

As may be observed, the Simulink model is implemented with ease based on standard components from a library. However, the detailed modelling of each component is not openly accessible and how Matlab/Simulink implements the circuit approximation is not transparent in an approach to evaluate the model step execution. This gives the CPN Tools model a benefit by its transparent representation and the added value for the user can be examined by comparing the results from the Matlab/Simulink model. In addition, a Matlab/Simulink model that forms the physical controller hardware with its control dynamics is not found publicly available nor methods on how to implement such a system.

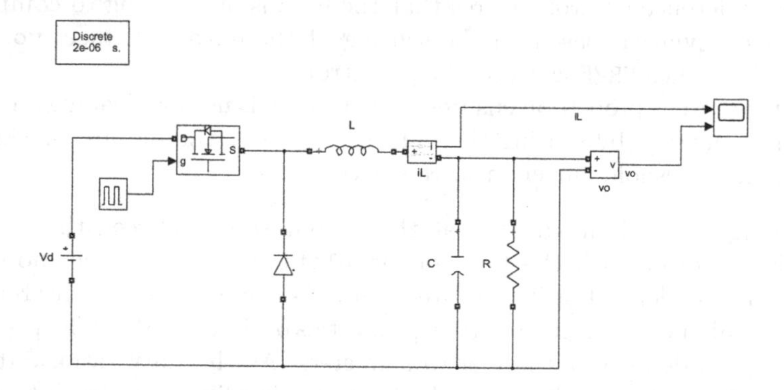

Fig. 11. Matlab/Simulink model of the buck converter without controller

6 Model Validation and Experimental Results

Our experimental model validation is organised in two parts. In the fist part, the open-loop response is simulated in both CPN Tools and Matlab/Simulink. The goal of this simulation is to ensure the correct implementation of the developed

CPN model of the buck converter by comparing the results in a controlled environment. The controlled measure is the scenario of fixed duty ratio (d) which draws the attention to the buck converter dynamical performance. In both models, we collect the measured dataset for inductor current (i_L), output voltage (v_o), and the gate signal (u). In CPN Tools, we obtain the simulation results using simulation monitors associated with the places corresponding to the measured values, and store the observed values in external data-files for post-processing and visualisation.

In the second part, the simulations are performed in CPN Tools and forms the integrated CPS by interfacing the controller and buck converter. The initial closed-loop response is shown based on the FPGA controller functions and the identified time delays added in the interfacing loops. The results are compared to the hardware setup of the buck converter and controller in an open-loop configuration. The results are found in Sects. 6.1 and 6.2, respectively and are based on the component and control parameters given in Table 1.

Table 1. Model parameters used in the simulation and physical experiments

Section	Symbol	Value	Description
6.1 and 6.2	f_{sw}	200 kHz	Switching frequency of the transistor
	T_s	0.05 µs	Discrete time-step
	V_i	12 V	Input voltage source
	L	9.5 mH	Inductor inductance
	C	20 µF	Capacitor capacitance at the output
	R	2.4 Ω	Resistor resistance as the load
6.1	d	50%	Duty ratio of switching frequency
6.2	d	10%	Duty ratio of switching frequency
	V_{REF}	1.2 V	Output voltage reference
	I_{LIM}	0.7 A	Inductor current limit

6.1 Open-Loop Response Evaluation

The simulation results in Fig. 12 shows the initialization and transient response of the buck converter in open-loop configuration at fixed duty ratio (d) of the switching frequency, $f_{sw} = 200$ kHz. In Fig. 12a the output voltage (v_o) and inductor current (i_L) are shown for both the CPN Tools and the Simulink model. The measurements are overlayed, where the CPN Tools measurements are represented with solid lines and the Simulink measurements are dashed. It can be seen that the transient response of the CPN Tools model correspond perfectly with the Simulink model for both the output voltage (v_o) and inductor current (i_L). The inductor properties are clearly interconnected correctly as can be seen from the relation between the gate triggering signal (u) in Fig. 12b and the corresponding response on the inductor current (i_L) in Fig. 12a.

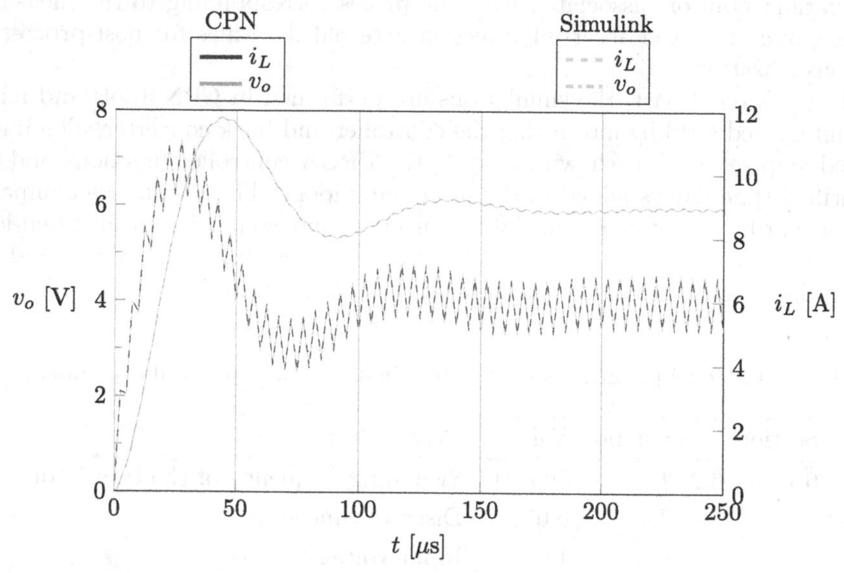

(a) Output voltage (v_o) and inductor current (i_L) for the CPN model and the Matlab/Simulink model.

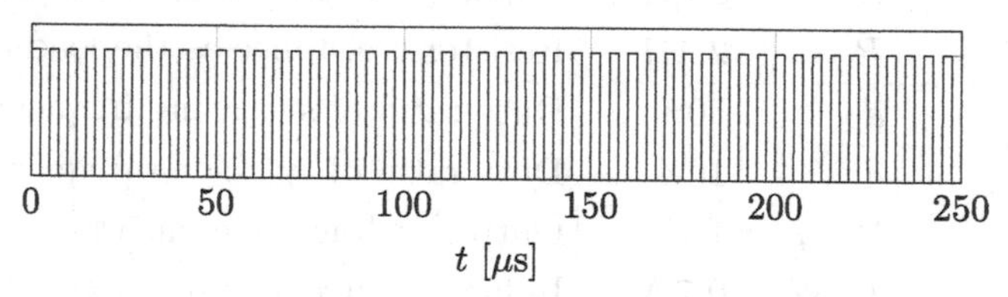

(b) The gate signal operating at duty ratio d=50 % of the switching frequency, fsw=200 kHz. The signal is collected from the CPN model.

Fig. 12. Comparing of open-loop response from initial zero state to duty radio $d = 50\%$ with the CPN model and the Matlab/Simulink model.

6.2 CPN Tools and Physical Response Evaluation

The results in Fig. 13 are comparing the closed-loop control features of the CPN Tools model and the physical measurements of the open-loop buck converter setup (see Fig. 1). In CPN Tools, the closed-loop control uses the output voltage (vo) measurements and continuously compare against the predefined reference value V_{REF}. In addition, the inductor current (i_L) protection function is used to limit the maximum current to I_{LIM}. The physical system operates at fixed duty ratio, $d = 10\%$, of the switching frequency, $f_{sw} = 200\,\text{kHz}$.

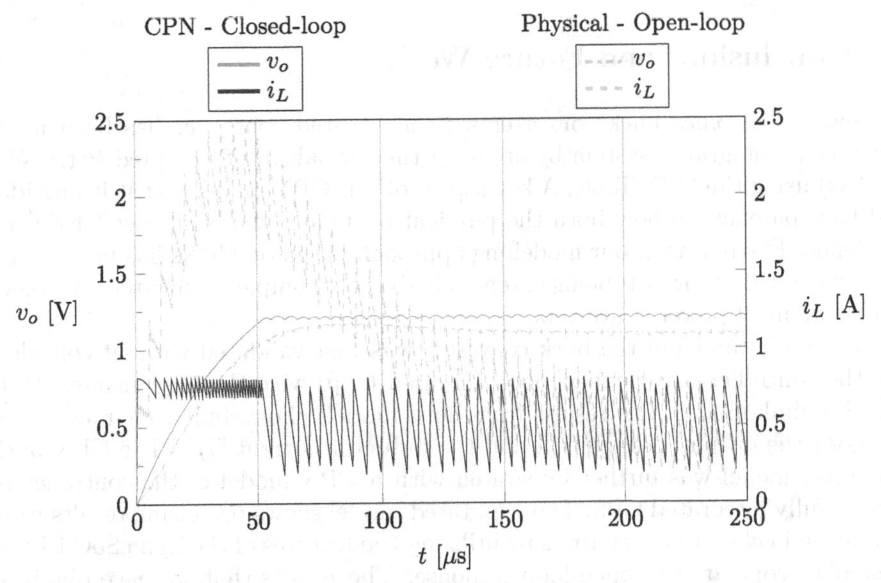

(a) Closed-loop response from initial zero state to a voltage reference set-point of 10 %. Compared with open-loop response of physical CPS setup.

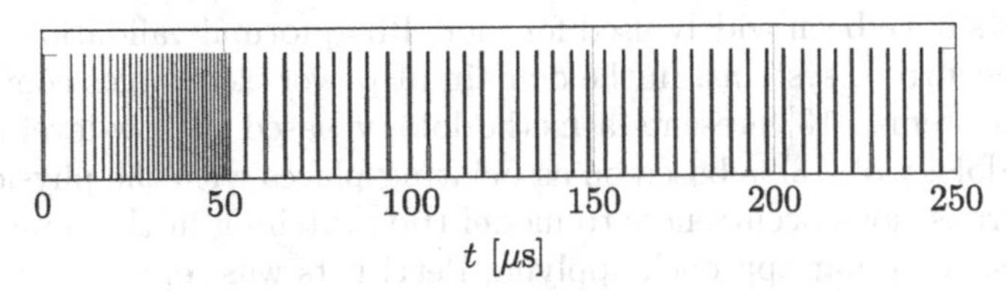

(b) The gate signal operating at dynamic duty based on the closed-loop V_{ref} CPN control.

Fig. 13. Comparison of the response from initial zero-state of the closed-loop CPN model, FPGA V_{ref}-controller and open-loop control physical setup with duty ratio $d = 10\%$

The transient response of the CPN model with closed-loop controls tends to give an even more rapid output increase, increased steady-state and a more critically damped output voltage (v_o) response than the open-loop configuration of the physical system. When evaluating the simple implementation of the controller this is expected as the gate is switched on as long as the output voltage (v_o) is below the set-point value, as a consequence of the control logic for V_{ref} shown in Fig. 10. The additional inductor current protection function is also acting in the initial phase, $t < 50\,\mu\text{s}$, and limiting ($i_L < I_{LIM}$) which is a favourable feature in a design. An important note with practical limitation for the CPN Tools controller is the high switching frequency observed in the initial phase ($t < 50\,\mu s$), which is observed in Fig. 13b. The physical transistor switching speed has some limitation which needs to be considered in converter design and also when considering limitation in control features.

7 Conclusions and Future Work

A power electronics buck converter with associated controller has been modelled as an integrated system by applying the formalism of Coloured Petri Nets (CPNs) using the CPN Tools. A key aspect of our CPN model is that it provides full transparency on how both the physical parts and the controller have been modelled. Furthermore, our modelling approach based on CPNs has been developed with a view towards being extendable to more complex converter topologies and systems of power converters.

The CPN model of the buck converter was first validated without considering the controller and feedback loop (in open-loop) with the implemented Matlab/Simulink model. The results from both models show similar characteristics for both the output voltage (v_o) and the inductor current (i_L). The CPN buck converter model was further integrated with a CPN model of the controller to form a fully integrated CPS. The integrated model shows promising results from operating in closed-loop controls by utilizing the features of the Zynq SoC FPGA controller compared to open-loop response. The results that we have obtained show a rapid voltage increase and good steady-state operation which corresponds to the physical system response and validates the implementation.

Petri nets have been widely used for modelling, formal validation, and verification of concurrent systems. In the domain of power electronics converters and embedded systems, [13] presents a methodology based on low-level Petri nets. The methodology of [13] is based on associating places with the physical sensing and using transitions occurrences to model the switching in the power electronics elements. A similar approach applying Petri nets was reported in [12,17,18]. There are also examples of extremely high frequency converters of small size, using asynchronous circuits, which applies an interpreter of Petri Nets with Signal Transition Graphs (STG) as a framework for the controller logic [15]. Further the approach of STG has been used to implement protection functions related to the transistor switching [7]. Reverse approaches can be found in [1,3,9], where the state- and control flow statements in Matlab/Simulink are used to encode Petri net constructs. As discussed in [3], Petri nets are well-suited for modelling power electronics systems as they can be represented as a DES whose state evolution depends on discrete events over time in the form of switching events. Still, as our literature study has shown, the application of Petri nets to power electronics converter systems is limited. Furthermore, the approaches and results reported in the literature have primarily served a design objective by applying Petri nets as a basic framework, and there is a lack of proven track record demonstrating validation capabilities with a Petri net-based approach.

There are several interesting directions to advance the modelling approach developed in this paper. The basic component description can be advanced to include unideal characteristics (parasitics effects) and further apply as a general component library. In addition, the convert system could be implemented as separate components of linearized models to form more complex power electronics converters and option to simulate the dynamics with systems of converters. This may allow introduction of more advanced classical analytical tools to assess system stability and frequency response. The currently implemented closed-loop control is simple, and many improvements are required to ensure safe operation. The potential for exploring such more advanced control schemes is easily achievable with our CPN modelling approach of the buck converter system. Another direction is in the detailed modelling of components and integration to benefit from the formalism in system design of the CPS and to conduct a more comprehensive validation against a real hardware implementation. With the higher level of details, more advanced and time-critical control algorithms may be explored and validated to ensure correct operation. As part of the current research trends and challenges in improving DC-microgrids, complex converter topologies such as MMC and systems of converters consist of time-critical control objectives where CPNs may be applicable. Investigating a top-down approach, considering the overall system design and improvements to the control architecture is also a possible direction for future work. With such an approach, the level of details may be added gradually and better support an iterative design approach. It may serve both in the initial design phase with model checking and verification of the structural properties to avoid formal errors and mistakes in prototyping or when implementing the higher-level control schemes.

References

1. Alcaraz, M., Campos-Rodriguez, R.: A framework based on matlab/simulink for the simulation of DES using petri net models. Int. J. Simul. Model. **18**, 420–431 (2019). https://doi.org/10.2507/IJSIMM18(3)479
2. Barbot, B., Bérard, B., Duplouy, Y., Haddad, S.: Integrating simulink models into the model checker cosmos. In: Khomenko, V., Roux, O.H. (eds.) PETRI NETS 2018. LNCS, vol. 10877, pp. 363–373. Springer, Cham (2018). https://doi.org/10.1007/978-3-319-91268-4_19
3. Bera, D., van Hee, K., Nijmeijer, H.: Relationship between simulink and petri nets. In: 2014 4th International Conference On Simulation And Modeling Methodologies, Technologies and Applications (SIMULTECH), pp. 12–23, August 2014. https://doi.org/10.5220/0005012000120023
4. Gutierrez, A., Bressan, M., Jimenez, J.F., Alonso, C.: Real-time emulation of boost inverter using the systems modeling language and petri nets. Math. Comput. Simul. **158**, 216–234 (2019). https://doi.org/10.1016/j.matcom.2018.07.011
5. Jensen, K.: Coloured Petri Nets: Basic Concepts, Analysis Methods and Practical Use. Springer, Heidelberg (1997). https://doi.org/10.1007/978-3-642-60794-3
6. Jensen, K., Kristensen, L.M., Wells, L.: Coloured petri nets and CPN tools for modelling and validation of concurrent systems. Int. J. Softw. Tools Technol. Transfer **9**(3), 213–254 (2007). https://doi.org/10.1007/s10009-007-0038-x
7. Li, D., Xia, F., Luo, J., Yakovlev, A.: Modelling reversion loss and shoot-through current in switched-capacitor DC-DC converters with petri nets. In: 2019 29th

International Symposium on Power and Timing Modeling, Optimization and Simulation (PATMOS), pp. 69–74, July 2019. https://doi.org/10.1109/PATMOS.2019.8862124
8. Mboup, A.B., Guerin, F., Ndiaye, P.A., Lefebvre, D.: Petri nets control design for hybrid electrical energy systems. In: 2009 American Control Conference, pp. 5012–5017, June 2009. https://doi.org/10.1109/ACC.2009.5159890
9. Mohamed Lamine, H., Denoun, H., Fekik, A., Benyahia, N., Zaouia, M., Benamrouche, N.: A Comparison Between PWM and Petri Nets Control Applied for Series Multi-cells Chopper, October 2017
10. Mohan, N., Undeland, T.M., Robbins, W.P.: Power Electronics: Converters, Applications, and Design, 2nd edn. Wiley, New York (1995)
11. Ormaetxea, E., et al.: Matrix converter protection and computational capabilities based on a system on chip design with an FPGA. IEEE Trans. Power Electron. **26**(1), 272–287 (2011). https://doi.org/10.1109/TPEL.2010.2062539
12. Ortiz, L., Gutiérrez, L.B., González, J.W., Águila, A.: A novel strategy for dynamic identification in AC/DC microgrids based on ARX and Petri Nets. Heliyon **6**(3), e03559 (2020). https://doi.org/10.1016/j.heliyon.2020.e03559
13. Salinas, F., González, M.A., Escalante, M.F., de León Morales, J.: Control design strategy for flying capacitor multilevel converters based on petri nets. IEEE Trans. Industr. Electron. **63**(3), 1728–1736 (2016). https://doi.org/10.1109/TIE.2015.2494535
14. Singh, N.K., Saha, I.: Specification-guided automated debugging of CPS models. IEEE Trans. Comput. Aided Des. Integr. Circuits Syst. **39**(11), 4142–4153 (2020). https://doi.org/10.1109/TCAD.2020.3012862
15. Sokolov, D., Khomenko, V., Mokhov, A., Dubikhin, V., Lloyd, D., Yakovlev, A.: Automating the design of asynchronous logic control for AMS electronics. IEEE Trans. Comput. Aided Des. Integr. Circuits Syst. **39**(5), 952–965 (2020). https://doi.org/10.1109/TCAD.2019.2907905
16. Szcześniak, P., Grobelna, I., Novak, M., Nyman, U.: Overview of control algorithm verification methods in power electronics systems. Energies **14**(14), 4360 (2021). https://doi.org/10.3390/en14144360
17. Wiśniewski, R.: Dynamic partial reconfiguration of concurrent control systems specified by petri nets and implemented in Xilinx FPGA devices. IEEE Access **6**, 32376–32391 (2018). https://doi.org/10.1109/ACCESS.2018.2836858
18. Wisniewski, R., Bazydło, G., Szcześniak, P., Grobelna, I., Wojnakowski, M.: Design and verification of cyber-physical systems specified by petri nets—a case study of a direct matrix converter. Mathematics **7**(9), 812 (2019). https://doi.org/10.3390/math7090812
19. Wiśniewski, R., Bazydło, G., Szcześniak, P., Wojnakowski, M.: Petri net-based specification of cyber-physical systems oriented to control direct matrix converters with space vector modulation. IEEE Access **7**, 23407–23420 (2019). https://doi.org/10.1109/ACCESS.2019.2899316

Discovering Unseen Behaviour from Event Logs

Abel Armas Cervantes(✉) and Farbod Taymouri

The University of Melbourne, Melbourne, Australia
abel.armas@unimelb.edu.au

Abstract. Process mining techniques aim to discover insights into the performance of a business process by analysing its event logs. These logs capture historical process executions as sequences of activity occurrences (events). Often, event logs capture only part of the possible process behaviour because the number of executions can be very large, particularly when many activities are executed concurrently. A highly incomplete event log is problematic because process mining techniques use the event log as a starting point. This paper proposes a technique to discover behaviour from an incomplete log. In order to do so, the presented technique builds distributive lattices from the executions captured in the log, which have well-defined notions of completeness and can be used to discover behaviour from few observations. The paper tests the presented approach in a set of real-life event logs and measures the amount of behaviour that can be discovered.

Keywords: Process mining · Distributive lattices · Partial orders · Concurrency detection

1 Introduction

Process mining analyses historical business process executions to help discover fact-based opportunities for process improvements [17]. These historical executions are captured as event logs (or simply *logs*), where process executions are recorded as sequences (*traces*) of activity instances (*events*). These traces of events describe the order in which the activities were executed, thus the concurrent execution of activities are captured as interleavings. These logs are used as the starting point for various process mining operations, but there are three main ones: automated process model discovery, conformance checking (checking the conformance between a model describing expected behaviour and the log describing the observed behaviour) and process enhancement.

Event logs can capture only a handful of possible executions of the underlying process [20]. As the complexity of the process increases, it becomes more difficult to observe all possible traces that a process can generate, which is infinite in cases when there is looping behaviour. However, even in the finite case, when there is a large amount of activities that can be executed concurrently, it will be nearly impossible to capture all possible interleavings. In the worst case, it is

L. Bernardinello and L. Petrucci (Eds.): PETRI NETS 2022, LNCS 13288, pp. 23–42, 2022.
https://doi.org/10.1007/978-3-031-06653-5_2

necessary to have $n!$ traces to represent all possible interleavings of n concurrent activities. The completeness of a log is a critical issue because the great majority of process mining operations use the log as a starting point, and if the log is highly incomplete, then the analysis results can be highly inaccurate. In practice, the process model describing the actual behavior of the process is not available, hence it is impossible to assess the completeness of the log. But by making some assumptions about the underlying process, it is possible to gain a rough idea of the behavior the process generates.

The seminal work of Winskel et al. [12] shows the relationship between a family of Petri nets (conflict-free 1-safe nets), event structures and distributive lattices. While Petri nets and event structures explicitly represent concurrency between events, lattices represent concurrency as interleavings, where pairs of concurrent events form squares in the lattices. Figure 1 shows an example of the transformations defined in [12], where (a) shows a 1-safe Petri net, (b) its event structure and (c) the distributive lattice [5,16] representing each of the execution states of the Petri net. Note that in (c), every node in the lattice represents an execution state where a set of events have taken place and are ordered by subset inclusion. This lattice represents the evolution of a process execution by means of the edges between nodes; for instance, the edge from $\{A\}$ to $\{A, B\}$ represents the occurrence of event B. In the lattices, the traces that the process can generate are represented as paths (chain of consecutive nodes in the lattice) from the empty execution state to the final execution state. In the example displayed in Fig. 1, any path from $\{\}$ to $\{A, B, C, D, E\}$ represents a possible trace of the process.

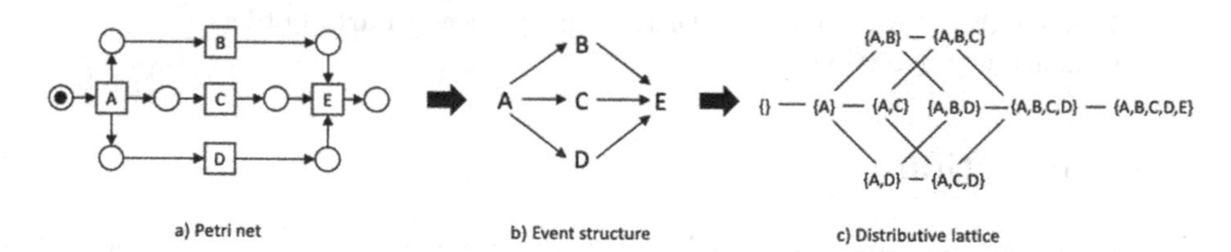

Fig. 1. From model to domain of configurations.

The aim of this paper is twofold. First, it aims to build distributive lattices from an event log. Each of these distributive lattices represents a concurrent execution that is derived from some traces in the log. If the lattices constructed from the log do not meet the distributivity property because not all interleavings were observed, then they are completed to be distributive, which will discover unseen behaviour. Second, by detecting missing behavior, a notion of event log completeness is defined, which can be used as a reference to assess the quality of a log. For the use of distributive lattices as a valid representation of an event log, two main assumptions are made: 1) the event log represents the behaviour of a 1-safe Petri net, so that each trace in the log represents the execution of the Petri net, and 2) there is no auto-concurrency in the underlying process (i.e., two activities with the same name cannot be concurrent).

The paper is structured as follows. Section 2 introduces the relevant definitions and notation for partial orders, lattices and event logs. Then, Sect. 3 presents the reconstruction of the distributive lattices from an event log, defines a notion of completeness based on the discovered behaviour, and presents a set of experiments using a set of real-life logs. Discussion and future work is presented in Sect. 4 and related work in Sect. 5. Finally, Sect. 6 concludes the paper.

2 Preliminaries

This section establishes the foundations for the rest of the paper. The first part introduces partial orders and lattices, and the second part introduces traces and event logs. Given that Petri nets are not a central element of this paper, we assume the reader is familiar with the basic notions of 1-safe Petri nets, see for example [15].

2.1 Partial Orders and Lattices

Let R be a binary relation over a set X, R is an ordering relation in X if it is reflexive ($(x, x) \in R$ for all $x \in X$), antisymmetric ($(x, y), (y, x) \in R$ implies $x = y$) and transitive ($(x, y), (y, z) \in R$ implies $(x, z) \in R$). The following definition presents reflexive partially ordered set, *poset* for short.

Definition 1 (Poset). *A* partially ordered set, *or simply* poset, *is a pair* $\langle X, \leq \rangle$, *where* X *is a set and* $\leq$ *is a reflexive, antisymmetric and transitive relation.*

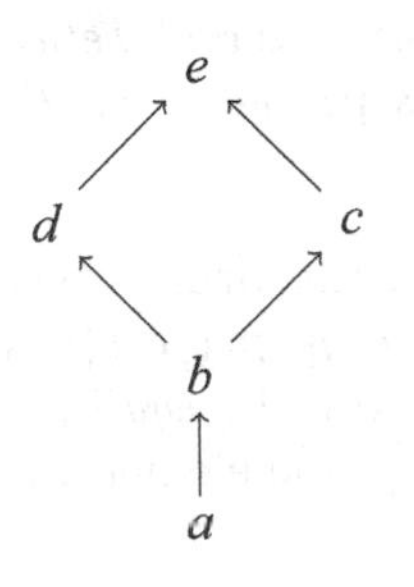

Fig. 2. Hasse diagram

Posets can be graphically represented as Hasse diagrams. These diagrams represent the *"cover" relation*, which is the transitive reduction of the graph representing the relation $\leq$ over the elements X. Given a pair of elements $x, y \in X$, y *covers* x, denoted as $x \prec y$, if $x < y$ and $\forall z \in X \; : \; x \leq z < y$ implies $x = z$. The Hasse diagram of a poset $\mathcal{P} = \langle X, \leq \rangle$ is shorthanded as $\mathcal{H}(\mathcal{P}) = \langle X, \prec \rangle$. Figure 2 shows a Hasse diagram where the cover relation is represented as arrows and the elements are represented by letters; e.g., b covers a in the diagram.

Meet and *join* are two common operators in a poset. Given a poset $\langle X, \leq \rangle$, $x \in X$ is the meet of a set $Y \subseteq X$, denoted $x = \bigsqcap Y$, iff (1) $\forall y \in Y \; : \; x \leq y$ and (2) $\forall z \in X : (\forall y \in Y : z \leq y) \Rightarrow z \leq x$. If Y contains only two elements, $X = \{a, b\}$, then the meet operation is written as $a \sqcap b$. Analogously, the join of $Y \subseteq X$ is an element x, denoted as $x = \bigsqcup Y$, iff (1) $\forall y \in Y \; : \; y \leq x$ and (2) $\forall z \in X : (\forall y \in Y : y \leq z) \Rightarrow x \leq z$. If $Y = \{a, b\}$, the join operation is written as $a \sqcup b$. The meet and join are also known as greatest lower bound (glb) and least upper bound (lub), respectively. For example, in the Hasse diagram in Fig. 2, $d \sqcup c = e$ and $d \sqcap c = b$.

A lattice is a special type of poset that contains a join and a meet for every pair of elements in the set X. One can easily check that the Hasse diagram in

Fig. 2 represents a lattice. Lattices are *distributive* [5] if they are distributive over $\sqcup$ and $\sqcap$. The following definition formalises lattices and the distributive property.

Definition 2 (Lattices and distributive lattices). *A* lattice *is a poset* $\langle X, \leq \rangle$ *where* $\forall x, y \in X : x \sqcup y$ *and* $x \sqcap y$ *exist. A lattice is* distributive *if* $\forall x, y, z \in X \ : \ x \sqcap (y \sqcup z) = (x \sqcap y) \sqcup (x \sqcap z)$.

Some elements in the lattice can be of one of three types: top element, bottom element and prime. An element x is the bottom element if $\forall y \in X : x \leq y$, and x is the top element if $\forall y \in X : y \leq x$. An element in a poset is *prime* if it is neither the bottom element nor the join of two other elements, see the formal definition below.

Definition 3 (Prime elements). *Let* $\mathcal{P} = \langle X, \leq \rangle$ *be a lattice. An element* $x \in X$ *is a complete prime (*prime *for short) iff for every* $Y \subseteq X$ *iff* $\bigsqcup Y$ *exists and* $x \leq \bigsqcup Y$*, then there exists* $y \in Y$ *such that* $x \leq y$*. The set of complete primes for* $\mathcal{P}$ *is denoted as* $\mathcal{C}_{\mathcal{P}}$.

In a distributive lattice, prime elements are those covering exactly one element [10]. The set of primes below an element $x \in X$ (w.r.t. $\leq$) are denoted as $\phi(x) = \{x' \in \mathcal{C}_{\mathcal{P}} \mid x' \leq x\}$. For example, in Fig. 2, the elements b, c and d are primes, and $\phi(d) = \{d, b, a\}$. All the elements between a pair of elements x and y define an *interval*, and it is called a *prime interval* if it contains only x and y.

Definition 4 (Prime interval). *Let* $\langle X, \leq \rangle$ *be a poset, an* interval *between* $x_1, x_2 \in X$ *is* $[x_1, x_2] = \{x_3 \in X \mid x_1 \leq x_3 \leq x_2\}$ *and it is* prime *iff* $x_1 \neq x_2$ *and* $[x_1, x_2] = \{x_1, x_2\}$.

Let $pr([x, y]) = \phi(y) \backslash \phi(x)$ be the set difference between the primes below x and the primes below y. In the case of a prime interval $[x, y]$, $pr([x, y])$ is a singleton. A pair of prime intervals $[x_1, x_2]$ and $[x_3, x_4]$ are said to be *equivalent*, denoted as $[x_1, x_2] \equiv [x_3, x_4]$ iff their difference is the same prime element, i.e., $pr([x_1, x_2]) = pr([x_3, x_4])$.

2.2 Distributive Lattices and Concurrency

Winskel et al. [12] shows the connection between conflict-free 1-safe Petri nets, elementary event structures and distributive lattices[1]. Elementary event structures are posets describing the execution of events by means of causality $\leq$, such that an event a has to occur before an event b when $a < b$. The authors showed that the execution states of the unfolding (elementary event structure) of a conflict-free 1-safe Petri net form a distributive lattice when ordered by set

[1] The connection defined in [12] considers a family of Petri nets called *causal nets* where there is not conflict – every place has at most one transition connected from and to it – and F^+ is irreflexive, where $F \subseteq (P \times T) \cup (T \times P)$ is the flow relation of the net with places P and transitions T.

inclusion; furthermore, the prime elements of the distributive lattice correspond to the events in the elementary event structure. The distributive lattice captures all possible interleavings of concurrent events, which can occur at different executions states, and the order $\leq=\subseteq$. Additionally, the meet ($\sqcap$) and join ($\sqcup$) operators correspond to the set intersection ($\cap$) and union ($\cup$) set operations, respectively. As shown in [5,16], a distributive lattice is a ring of sets where the union and intersection of every pair of elements are present in the lattice.

Figure 3 shows a conflict-free 1-safe Petri net N (Fig. 3a), the corresponding elementary event structure $\mathcal{P}$ (Fig. 3b) and the distributive lattice $\mathcal{H}(\mathcal{P})$ of its execution states ordered by $\subseteq$ (Fig. 3c). In the *elementary event structure*, the behavior is described by means of causality ($\leq$) and concurrency between events, such that events $x, y \in X$ are concurrent iff $\neg(x \leq y \ \vee \ y \leq x)$. The execution states, *a.k.a. configurations*, of an elementary event structure are left-closed subsets of events (i.e., $Y \subseteq X$ is left closed iff $x \in Y \wedge x' \leq x \Rightarrow x' \in Y$). In $\mathcal{H}(\mathcal{P})$, the nodes are the execution states, and the cover relation is denoted by a line, please disregard the different colors and line formats as they are explained later. The cover relation describes the evolution of a configuration and represents the occurrence of an event. For instance, in Fig. 3c, the cover relation between $\{a\}$ and $\{a, d\}$ represents the execution of the event d after a. The bottom and top element in this distributive lattice represents the state where no event has been executed $\{\}$ and the final state $\{a, b, c, d, e, f\}$ where all events have occurred.

In the distributive lattice, the concurrent executions of pairs of events form diamond-like shapes representing interleavings. For example, in Fig. 3c, the states

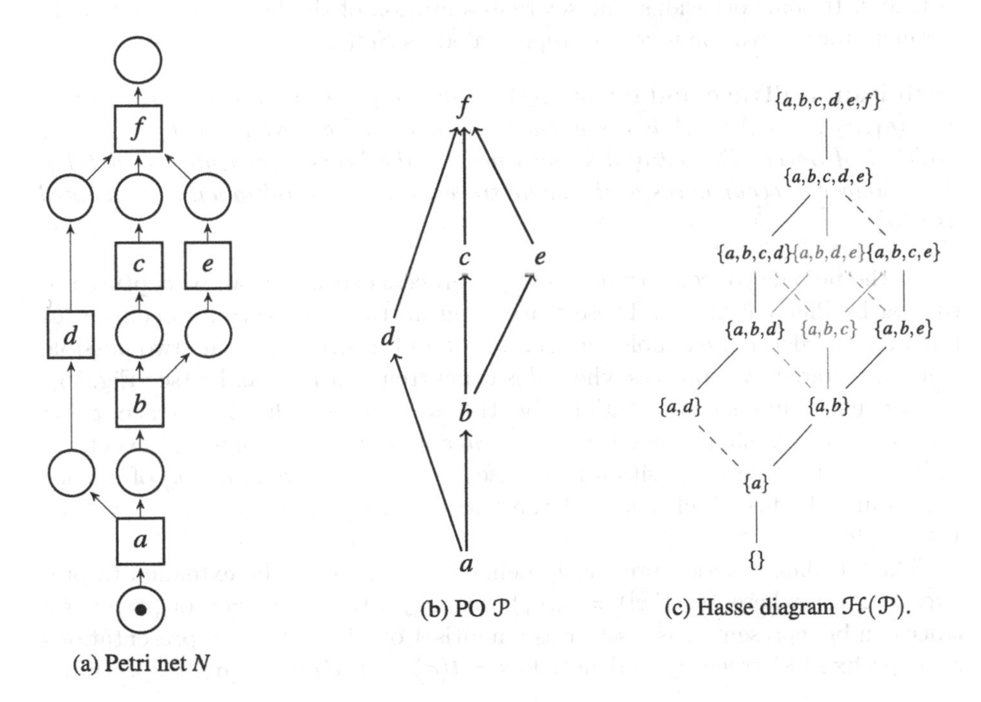

(a) Petri net N (b) PO $\mathcal{P}$ (c) Hasse diagram $\mathcal{H}(\mathcal{P})$.

Fig. 3. Petri net, partial order and Hasse diagram

$\{a\}, \{a, d\}, \{a, b\}$ and $\{a, b, d\}$ represent the concurrent execution of events b and d. Note, for instance, both intervals $[\{a\}, \{a, d\}]$ and $[\{a, b\}, \{a, b, d\}]$ represent the execution of the event d (i.e., $pr([\{a\}, \{a, d\}]) = pr([\{a, b\}, \{a, b, d\}])$ and thus $[\{a\}, \{a, d\}] \equiv [\{a, b\}, \{a, b, d\}]$). In fact, all dotted lines in Fig. 3c are an equivalence class because they represent occurrences of the event d at different execution states.

2.3 Traces and Event Logs

An *event log* (or simply *log*) captures historical executions of a process, where every execution of a process activity produces an event in the log. Thus, several events in the log may stem from the same activity. Hereinafter the activities of a process are represented as Σ and the events as E. An event can have different attributes, such as the name of the corresponding activity, resources or execution time. In this paper, we assume that the only available attribute of an event is the name of the corresponding activity. In order to relate activities to events, $\lambda : E \rightarrow \Sigma$ is a labeling function, such that the activity of an event e is denoted as $\lambda(e) = l$, where $l \in \Sigma$.

Process executions are captured in the log by means of traces. These traces are sequences of events ordered by their order of observation. A pair of traces are considered the same if they have the same number of events and those events are instances of the same activities executed in the same order. A log L can contain several occurrences of the same traces, thus a log is defined as a multiset of traces. In some occasions, the set representation of the log – containing only distinct traces – will be used and represented as $Set(L)$.

Definition 5 (Trace and event log). *Given a finite set of events E,* a trace *$\sigma = \langle e_1, e_2, \ldots, e_n \rangle \in E^*$ is a sequence of events and an* event log *$L \subseteq E^*$ is a multiset of traces. The set of distinct traces in the log is represented as $Set(L)$. The number of occurrences of the same trace in the log (multiplicity) is denoted as $\gamma(\sigma)$.*

In the presence of concurrency, a single process execution can be captured in the log by different traces. These traces contain the same activity occurrences but vary in order. For example, $\langle a, b, e, c, d, f \rangle$ and $\langle a, d, b, c, e, f \rangle$ are two possible traces generated by a process where d is concurrent with b, c and e (see Fig. 3b).

Let us define some notation for traces. The length of a trace $\sigma = \langle e_1, e_2, \ldots, e_n \rangle$ is shorthanded as $|\sigma| = n$ and it is the number of events in σ. The event at the ith position is accessed as $\sigma[i]$. The prefix $\sigma[1, k]$ of a trace σ contains the first k elements of the trace, i.e., $\sigma[1, k] = \langle e_1, e_1, \ldots, e_k \rangle$ for $1 \leq k \leq |\sigma|$.

The labelling function previously defined for events can be extended to prefixes of traces where $\lambda(\sigma[1, k]) = \langle \lambda(e_1), \lambda(e_2), \ldots, \lambda(e_k) \rangle$. A trace or prefix of a trace can be represented as a set or as a multiset of labels. These representations of a (prefix of a) trace σ are denoted as $Set(\sigma)$ and $MultiSet(\sigma)$, respectively.

The cardinality of a set and a multiset Z is denoted as $|Z|$ and refers to the number of elements in Z; note that in the case of multisets, the cardinality considers the multiplicities of the elements in Z.

The following section presents the main contribution of the paper, where traces are used as "seeds" to construct distributive lattices.

3 Distributive Lattices of an Event Log

This section presents the main contributions of the paper. The central idea is the reconstruction of distributive lattices representing execution states of a process execution. This reconstruction starts by merging groups of event log traces into lattices, such that prefixes of a trace represent an execution state, and then missing elements are added until the lattices are distributive.

The completion operation over the lattice can discover new behavior by introducing new elements (unseen execution states or event executions) until the distributivity property is met. However, in the presence of noise, this operation can introduce a large amount of new behavior that may be undesirable. Thus, Subsect. 3.1 presents the steps for the construction of the distributive lattice from a set of traces, and Subsect. 3.2 describes a way to tame the possibly large amount of behavior introduced during the completion operation.

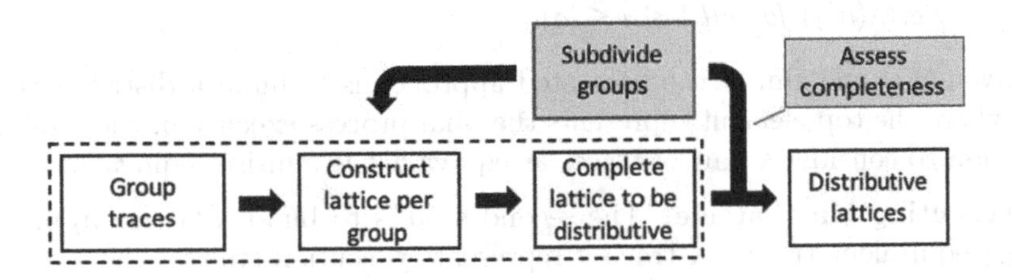

Fig. 4. Overview of the proposed approach.

3.1 Reconstruction of Distributive Lattice

The reconstruction of the distributive lattice is inspired by two main existing results: the behavior of a conflict-free 1-safe Petri net forms a distributive lattice [12] and a distributive lattice is a ring of sets [5,16]. Thus, the proposed approach consists of three central steps (dotted area in Fig. 4): 1. group traces representing the same process executions (interleavings of concurrent executions), 2. construct a lattice for each group of traces, and 3. complete the lattice until it is distributive (missing elements can be computed by using two set operations: union and intersection).

Before presenting the three steps of the reconstruction, let us define a special labelling for handling traces with events with the same name. In a trace, every event is unique but, when the process originating the log contains repeatable behavior, several events within a trace can stem from the same activity. In order to differentiate the events with the same label within a trace σ, $\bar{\lambda}$ is a special labelling function, such that $\bar{\lambda}(e) = \lambda(e)_{w(e,i)}$, where $w(e,i) = |\{e' = \sigma[j] \mid 1 \leq j \leq i \ \wedge \ \lambda(e') = \lambda(e)\}|$. The special labelling can be applied to traces, where $\bar{\lambda}(\sigma) = \langle \bar{\lambda}(e_1), \bar{\lambda}(e_2), \ldots \bar{\lambda}(e_{|\sigma|}) \rangle$. Intuitively, the special label of an event has a sub-index representing its number of occurrence within the trace. For instance, given a trace $\sigma = \langle a, b, c, c \ldots \rangle$, then $\bar{\lambda}(\sigma) = \langle a_1, b_1, c_1, c_2 \ldots \rangle$.

Grouping Traces. The first step is to define a notion of equivalence over the log traces, such that a pair of traces are equivalent if they represent the same process execution. A simple notion of equivalence is multiset equivalence, where traces are equivalent if they have events of the same activities that were executed the same number of times. For example, the traces $\langle a, b, e, c, d, f \rangle$ and $\langle a, d, b, c, e, f \rangle$ would be considered as multiset equivalent because they represent a single execution of the same activities. Note that if the special labelling is used, the multiset equivalence of a trace can be defined as a set equivalence as shown next.

Definition 6 (Set equivalent traces). *A pair of traces σ, σ' are set equivalent, denoted as $\sigma \sim_{set} \sigma'$, iff $Set(\bar{\lambda}(\sigma)) = Set(\bar{\lambda}(\sigma'))$. I.e., $|\sigma| = |\sigma'|$ and $\bar{\lambda}(\sigma[i]) \in Set(\bar{\lambda}(\sigma'))$ for all $1 \leq i \leq |\sigma|$.*

Given that the aim of the presented approach is to build a distributive lattice, where the top element represents the final process execution, the minimum condition to consider a pair of traces as equivalent is multiset equivalence.

Constructing the Lattice. The second step is to build a lattice from a set of $\sim_{set}$-equivalent traces. A trace σ represents execution states, where a prefix $\sigma[1, k]$ is the state where events $\langle e_1, e_1, \ldots, e_k \rangle$ have occurred. Indeed, each element in the lattice will represent a trace prefix. The following definition formally defines a prefix equivalence between traces.

Definition 7 (Equivalent prefixes). *Let σ_1, σ_2 be two traces. The prefixes $\sigma_1[1, k]$ and $\sigma_2[1, k]$ are equivalent iff $\sigma_1[1, k] \sim_{set} \sigma_2[1, k]$.*

Given a group of equivalent traces G, let $\sigma[1, k]_{\equiv} = \{\sigma'[1, k] \mid \sigma' \in G \wedge \sigma[1, k] \sim_{set} \sigma'[1, k]\}$ be the equivalence class for the trace prefix $\sigma[1, k]$. The lattice representing G is the pair $\langle X, \subseteq \rangle$, where X represents the equivalence classes for the prefixes of traces in G and the order between the elements in X is the subset containment between a pair of elements in the equivalence classes.

Definition 8 (Lattice of traces). *Given a log L, the lattice of a set of traces $T \subseteq L$ is a pair $\langle X, \subseteq \rangle$, where $X = \{\sigma[1, k]_{\equiv} \mid \sigma \in L \ \wedge \ 1 \leq k \leq |\sigma|\} \cup \epsilon$ and ϵ is a special state representing the bottom element, and $x \subseteq y$ for $x, y \in X$, if $\bar{\lambda}(\sigma') \subseteq \bar{\lambda}(\sigma'')$, such that $\sigma' \in x$ and $\sigma'' \in y$.*

Observe that the above definition constructs a valid lattice because there is a unique bottom element ϵ and a unique top element $Set(\sigma)$ that can be the meet and join elements, respectively, for any pair of elements. Please, observe that in the Hasse diagram we denote ϵ as the empty set $\{\}$.

Reconstructing Distributive Lattices. The completion of the lattice consists of introducing the relations and elements needed to transform any lattice into a distributive one. This completion operation is based on the fact that a distributive lattice is isomorphic to a ring of sets [5], where the union and intersection of every pair of elements is also an element in the lattice.

Definition 9 (Lattice completion). *Let $\mathcal{D} = \langle X, \leq \rangle$ be a lattice. The completion of $\mathcal{D}$ for a distributive lattice is a lattice $\mathcal{D}^* = \langle Y, \leq \rangle$ where $X \subseteq Y$ and $(x \cup y) \in Y$ and $(x \cap y) \in Y$ for all $x, y \in Y$.*

Consider the conflict-free 1-safe Petri nets shown in Fig. 5. The possible execution states of these nets can be represented as the distributive lattices displayed in Fig. 5. These lattices can be reconstructed from only two traces (represented by the black elements and relations). For example, traces $\langle a, b, c, d \rangle$ and $\langle b, c, d, a \rangle$ are the only traces necessary to reconstruct the lattice in Fig. 6a, where activity a can occur after $\{b\}$ and $\{b, c\}$; after completion, the lattice represents four traces. In Fig. 6b, traces $\langle a, b, c, d \rangle$ and $\langle c, d, a, b \rangle$ can be used to generate the distributive lattice where $\{a, c\}$ is the union (intersection) of $\{a\}$ and $\{c\}$ (resp. $\{a, b, c\}$ and $\{a, c, d\}$), which represents six traces. Finally, for the reconstruction of the lattice in Fig. 6c only traces $\langle a, b, c \rangle$ and $\langle c, b, a \rangle$ are necessary to obtain the lattice representing six traces.

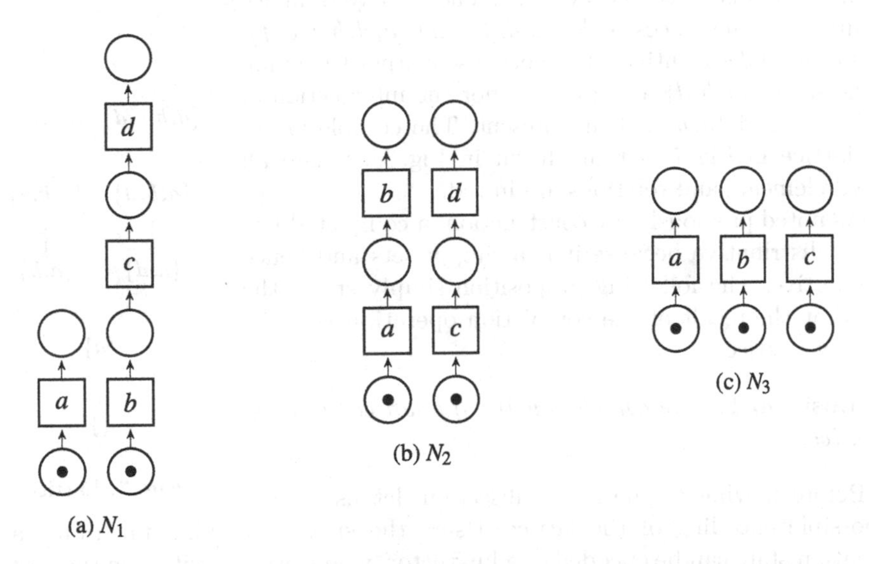

Fig. 5. Petri nets with computations represented in the distributive lattices in Fig. 6

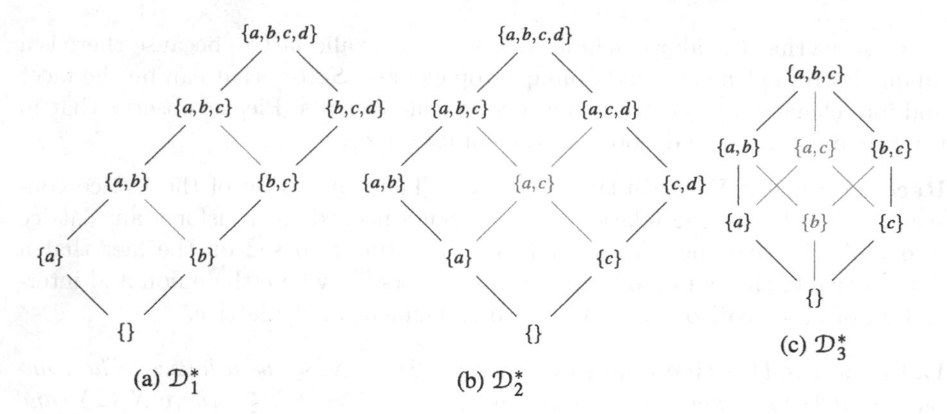

Fig. 6. Completed distributive lattices representing the execution states of the nets in Fig. 5

Note that other equivalences have been proposed in the context of process mining (see [19]) for constructing transition system-based representations of an event log, which are not necessarily lattices (e.g., some of the transitions systems can represent loops). [19] puts forward the idea of closing the diamonds (called *extend strategy*) in a transition system. This operation is able to handle cases as that in Fig. 6a, but it is limited to discover relations between events, and fails to discover missing elements (i.e., execution states). Thus, [19] would fail to reconstruct the lattices shown in Figs. 6b and 6c. This discussion is expanded in the related work in Sect. 5.

Another example of the possible lattices that can be reconstructed is presented in Fig. 7. The displayed lattice is built from the traces $\langle a, b, e, c, d, f \rangle$ and $\langle a, d, b, c, e, f \rangle$. This is not a distributive lattice because neither the union of the states $\{a, b, d\}$ and $\{a, b, e\}$, nor the intersection of $\{a, b, c, d\}$ and $\{a, b, c, e\}$ are present. The completion of the lattice in Fig. 7 is that shown in Fig. 3c, where all added elements and relations are in red.

As noted previously, by construction, a completed lattice is distributive because it is a ring of sets and hence distributive. The following proposition simply states the fact that the result of the completion operation is a distributive lattice.

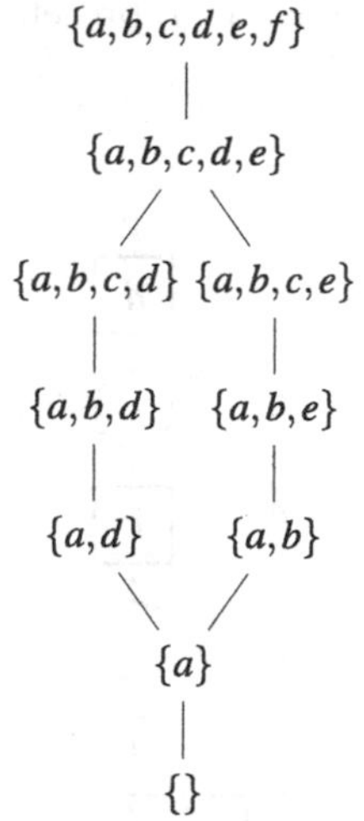

{a,b,c,d,e,f}

{a,b,c,d,e}

{a,b,c,d} {a,b,c,e}

{a,b,d} {a,b,e}

{a,d} {a,b}

{a}

{}

Fig. 7. Lattice

Proposition 1. *The completion $\mathcal{D}^*$ of a lattice $\mathcal{D}$ is distributive.*

Before moving to the next subsection, let us discuss a possible encoding of the lattice. Using the special labelling function, each execution state can be encoded as a bit-vector, where every position in the vector represents an event's special label. A value of 1 at a given position represents the occurrence of such an event. The bit-vector encoding will be particularly useful during the completion of the lattice when many union and intersection operations will be performed.

3.2 Containing the Amount of Behaviour Discovered During Completion

The completion operation, while simple, can be a double-edged sword. In the case of processes capable of generating large amounts of distinct traces (for example when there are many concurrent activities), even the majority of the behaviour can be reconstructed from a few traces acting as seeds. For instance, the lattices in Fig. 6 can be fully reconstructed from only two traces. However, this makes the technique too sensitive to noise or exceptional behavior that is not part of the "usual" process behaviour. In this case, a single (noisy) trace can lead to the insertion of a large amount of behaviour when computing the missing elements of a lattice. Note that, the filtering of noise in an event log is an orthogonal problem that has been studied independently (see for example the works in [6,8,22]), thus a noise-filtering technique can be applied to the log as a pre-processing step before the reconstruction of the lattices.

In order to control large numbers of behaviours that might be introduced during the completion operation, this subsection presents a strategy to, first, compute the completeness of a log, and then to control the amount of behavior introduced in the lattice.

Measuring Completeness. The completeness of a lattice is defined with respect of the number of traces represented after completion. Then, a Hasse diagram of a distributive lattice is seen as a graph where every path (sequence of contiguous edges) from the bottom to the top element represents a trace. Formally, a path p in a poset $\langle X, \subseteq \rangle$ is $p = \langle s_1, s_2, s_3, \ldots, s_n \rangle$, where $s_i \in X$, for $1 \leq i \leq n$, and $s_j \prec s_{j+1}$ for $1 \leq j < n$. Note that for extracting the traces from the paths, it is necessary to look at the events represented by the prime intervals of two consecutive nodes. Thus, the trace represented by a path p is $t(p) = \lambda(\langle pr([s_1, s_2]), pr([s_2, s_3]), \ldots, pr([s_{n-1}, s_n])\rangle)$. For instance, Fig. 7 has a path $p = \langle \{\}, \{a\}, \{a, b\}, \{a, b, e\}, \{a, b, c, e\}, \{a, b, c, d, e\}, \{a, b, c, d, e, f\} \rangle$, and the corresponding trace of p is $t(p) = \langle a, b, e, c, d, f \rangle$.

The measure of completeness Θ for a lattice is computed as follows. $\Theta(\mathcal{D})$ is defined as the ratio between the total number of paths in the completed lattice $\mathcal{D}^*$ and the number of paths in the lattice $\mathcal{D}$ prior completion, where $\mathcal{D}_{\#P}$ and $\mathcal{D}^*{}_{\#P}$ are the set of paths in the lattice and its completed version, respectively.

$$\Theta(\mathcal{D}) = \frac{|\mathcal{D}_{\#P}|}{|\mathcal{D}^*{}_{\#P}|} \tag{1}$$

A Hasse diagram is a directed acyclic graph, thus using dynamic programming, the number of paths can be computed in $\mathcal{O}(V+E)$ where V is the number of set elements and E is the number of cover relations.

The completeness measure can be used to control the amount of behavior introduced in the lattice during completion. For instance, the lattice of a set of traces G can be deemed as *valid* if $\Theta(\mathcal{D}) \geq \beta$, where β is a given threshold. Then, if $\Theta(\mathcal{D}) < \beta$ then G has to be refined into subgroups and the lattices for such subgroups have to be constructed independently. Intuitively, if the completion operation introduced more behavior than desired according to β, i.e. $\Theta(\mathcal{D}) < \beta$, it is necessary to subdivide the group of traces where the completion operation will introduce less new elements, relations and, as a consequence, fewer paths. The way to subdivide the groups of traces is left for future work. In the current tool implementation, a hierarchical clustering was used as a black box, such that the distance between σ and σ' is $|\{\sigma[i] \mid \lambda(\sigma[i]) \neq \lambda(\sigma'[i]) \text{ for } 1 \leq i \leq |\sigma|\}|$. In words, the distance between a pair of traces is the number of events that are not the same at a given position in both traces.

Consider the lattice in Fig. 6c. In this lattice, the black lines represent two traces $\langle a, b, c\rangle$ and $\langle c, b, a\rangle$, while the red elements are inserted. In this example, the completed lattice represents six traces, while only two were given as seeds, thus $\Theta(\mathcal{D}_3) = \frac{2}{6} = 0.333$. Then, if $\beta > 0.333$, then it would be necessary to build one lattice for the trace $\langle d, c, e\rangle$ and one for $\langle e, c, d\rangle$.

3.3 Experiments

In order to test how much behaviour we can discover in real-life event logs, the approach was implemented and tested using a set of publicly available logs. This section presents the results of the presented approach with a series of real-life event logs. For reproducibility purposes, the library, benchmark and results (lattices and elementary event structures were included for completeness) can be found in this link: *Latticer* at https://blogs.unimelb.edu.au/bpm/tools/.

Datasets. The experiments were conducted using 11 publicly available real-life logs obtained from the 4TU Centre for Research Data.[2] Table 1 shows the

Table 1. Event logs

Log name	#Events	#Distinct events	#Traces	#Distinct Traces	Trace length	
					Min.	Max.
$BPIC12$	262200	36	13087	4366	3	175
$BPIC13_{cp}$	6660	4	1487	183	1	35
$BPIC13_{inc}$	65533	4	7554	1511	1	123
$BPIC14_{f}$	369485	9	41353	14948	3	167
$BPIC15_{1f}$	21656	70	902	295	5	50
$BPIC15_{2f}$	24678	82	681	420	4	63
$BPIC15_{3f}$	43786	62	1369	826	4	54
$BPIC15_{4f}$	29403	65	860	451	5	54
$BPIC15_{5f}$	30030	74	975	446	4	61
$RTFMP$	561470	11	150370	231	2	20
$SEPSIPS$	15214	16	1050	846	3	185

[2] https://data.4tu.nl/Eindhoven_University_of_Technology/categories/Commerce_Management_Tourism_and_Services/13500.

characteristics of these logs including the number of (unique) events, (unique) traces, and the minimum and maximum length of traces for each event log.

Results. The distributive lattices for each of the event logs were computed with different thresholds β ranging from 0.0 to 1.0. Table 2 shows the number of lattices generated with different thresholds β. Intuitively, the lower β, the more behaviour is accepted when completing the lattice and potentially the fewer lattices.

The best result was obtained in the case of $BPIC13_i$, where the number of lattices decreased from 1394 to 535 for $\beta = 1.0$ and $\beta = 0.0$, respectively; whereas the smallest reduction in number of lattices was observed in the case of $BPIC15_5f$, where the number decreased from 295 to 264 for $\beta = 1.0$ and $\beta = 0.0$, respectively.

Table 2. Number of lattices for different β

	Threshold β										
Dataset	**1.0**	**0.9**	**0.8**	**0.7**	**0.6**	**0.5**	**0.4**	**0.3**	**0.2**	**0.1**	**0.0**
$BPIC12$	3921	3921	3914	3832	3778	3632	3606	3479	3390	3310	3159
$BPIC13_{cp}$	152	152	151	139	134	131	125	120	113	110	106
$BPIC13_i$	1394	1394	1382	1341	1292	1236	1204	1143	1040	949	535
$BPIC14_f$	13670	13656	13582	13398	13086	12559	12428	11843	11297	10511	7225
$BPIC15_{1f}$	295	295	295	295	295	295	295	288	278	277	264
$BPIC15_{2f}$	416	416	416	416	412	410	410	409	402	398	368
$BPIC15_{3f}$	785	785	785	775	750	738	736	708	702	682	598
$BPIC15_{4f}$	451	451	451	451	442	437	437	425	423	419	370
$BPIC15_{5f}$	446	446	446	446	436	436	436	436	434	434	424
$RTFMP$	152	152	151	137	123	108	106	94	91	86	85
$SEPSIS$	791	791	784	777	751	736	734	709	673	621	434

Even though the reduction of the number of traces can be considerable, the biggest impact of the technique is in the amount of behaviour introduced in the lattices. Table 3 shows the total number of paths represented by the constructed lattices. As shown in the last column, $\beta = 0.0$ can lead to a huge amount of paths. For example, in the case of $BPIC15_{4f}$, the number of paths increases from 451 when $\beta = 1.0$, which is the number of distinct traces, to more than 2 billion when $\beta = 0.0$. In this latter example, the great majority of the paths is extracted from a lattice that represents 2,087,976,600 paths and that is computed from 14 distinct traces. A reason for this can be that there is noise in the logs, or traces that are multiset equivalent but do not represent executions of a concurrent activities (e.g., the process can have activities with the same name that are simply performed in a specific order).

Table 3. Number of paths in the lattices for different β

	Threshold β										
Dataset	**1.0**	**0.9**	**0.8**	**0.7**	**0.6**	**0.5**	**0.4**	**0.3**	**0.2**	**0.1**	**0.0**
$BPIC12$	4366	4366	4368	4446	4507	4797	4877	5338	6003	7028	1128721
$BPIC13_{cp}$	183	183	184	191	197	202	213	234	286	319	461
$BPIC13_{i}$	1511	1511	1515	1540	1590	1670	1747	1967	2543	3565	95633528
$BPIC14_{f}$	14948	14951	14979	15132	15513	16596	16984	19269	22870	34609	628335832
$BPIC15_{1f}$	295	295	295	295	295	295	295	318	369	378	1454
$BPIC15_{2f}$	420	420	420	420	424	428	428	433	475	507	4308820
$BPIC15_{3f}$	826	826	826	836	863	889	899	1007	1052	1347	5416969
$BPIC15_{4f}$	451	451	451	451	460	470	470	506	517	571	2092799796
$BPIC15_{5f}$	446	446	446	446	456	456	456	456	470	470	5032
$RTFMP$	231	231	232	245	262	292	302	402	440	628	753
$SEPSIS$	846	846	848	855	882	912	928	1029	1254	1986	208796

Figures 8 and 9 show the lattices for each dataset, where every dot represents a lattice. Each graph shows the threshold which was used to create the lattice, the lattice's completeness $\Theta(\mathcal{D})$ and the number of paths it represents after completion. Please note that the dot colors are according to the thresholds β. In these graphs it is possible to observe that there are few lattices that contribute with the largest amount of paths when the threshold $\beta = 0.0$.

4 Discussion and Future Work

The main purpose of this paper is to present distributive lattices as a representation that can guide the discovery of unseen behavior from event logs. However, the advantages of using distributive lattices as a representation of the information of the event logs goes beyond the approach presented in this paper. Distributive lattices can be used as the bridge to go from traces representing interleavings of a concurrent process to models with true concurrency semantics, such as the elementary event structures (see [12]). This is particularly relevant because event structures have been proposed as suitable representations for the behaviour of process models and event logs [7].

The construction of distributive lattices from event logs has many potential uses and directions for our future work, we list some of them below.

Detection of duplicates: As a future work, we will explore the possibility to detect duplicates (activities which carry the same name but are executed in different contexts). In particular, we want to explore if the detection of these duplicates can improve the quality of the models that can be generated using automated discovery of process models.

Detection of undesirable behaviour: While this paper was focused on deriving behavior from observed traces, it may also represent behaviour that should be forbidden if the log is complete. In this case, any other trace inserted during

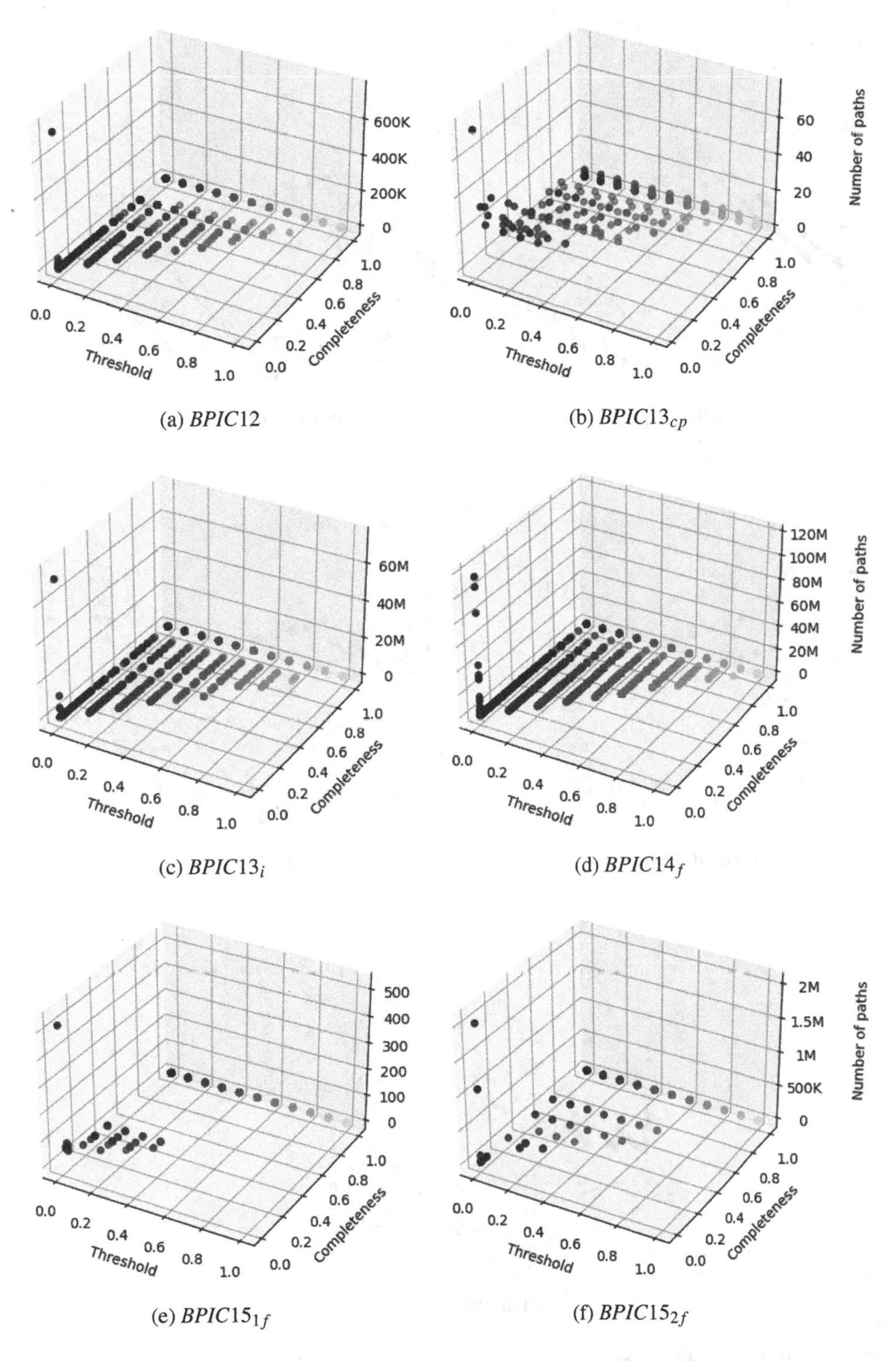

(a) $BPIC12$

(b) $BPIC13_{cp}$

(c) $BPIC13_{i}$

(d) $BPIC14_{f}$

(e) $BPIC15_{1f}$

(f) $BPIC15_{2f}$

Fig. 8. Thresholds, completeness and number of paths for the lattices per dataset

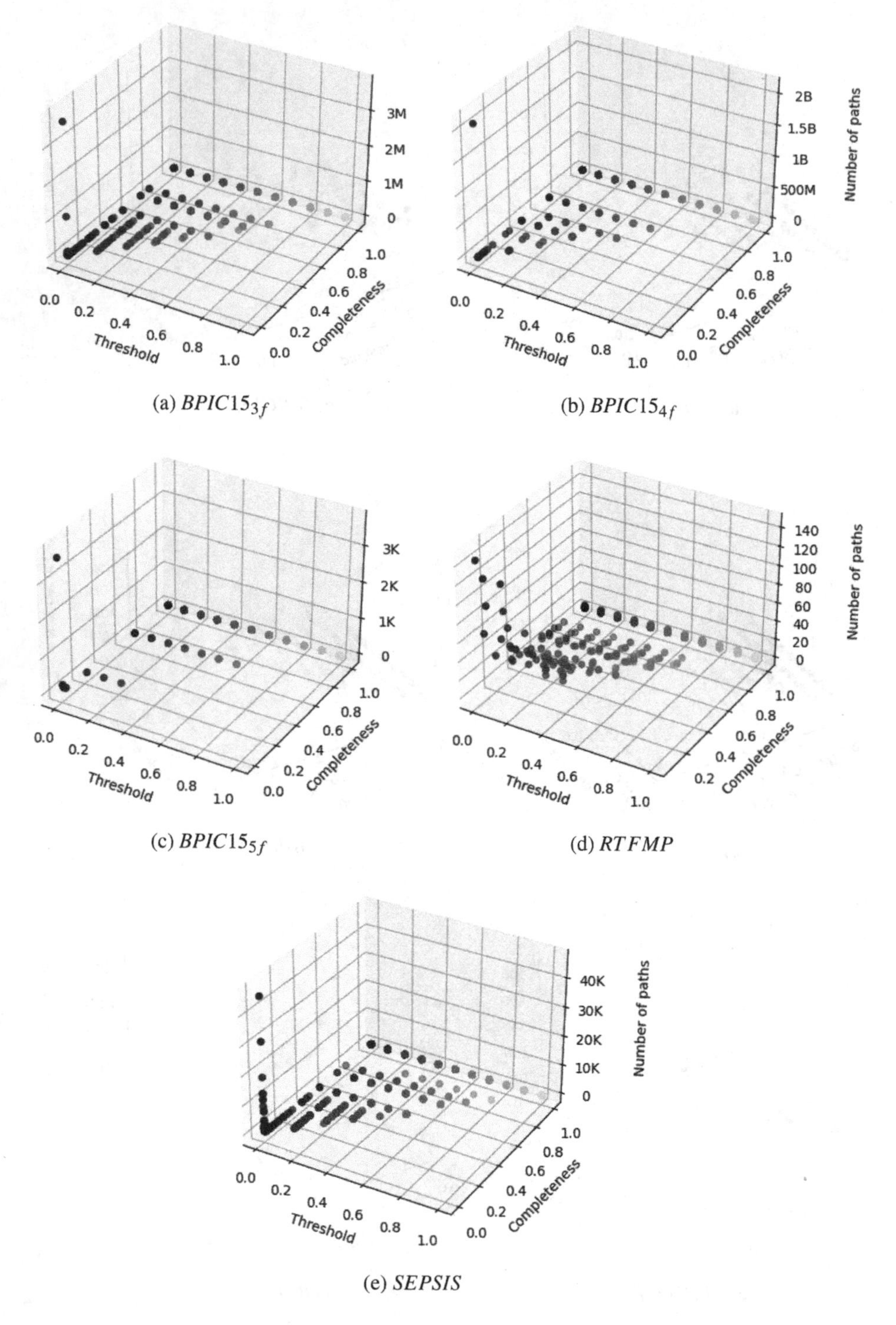

(a) $BPIC15_{3f}$

(b) $BPIC15_{4f}$

(c) $BPIC15_{5f}$

(d) $RTFMP$

(e) $SEPSIS$

Fig. 9. (2) Thresholds, completeness and number of paths for the lattices per dataset

the construction and completion of the distributive lattices can be a trace that should be forbidden by the process.

Process model discovery from partial orders: A promising direction for future work is to define automated model discovery techniques that take partial orders as input and generate process models with guarantees (e.g., free-choice sound workflow nets or models in BPMN notation). Some works moving in this direction are [3,4,14]. Currently, the majority of automated process model discovery techniques take event logs as input and derive concurrency relations over the activities, rather than over the events (e.g., [2,11]).

Generate unseen behavior as event log traces: Once the distributive lattices have been constructed, it is possible to compute all traces they represent. While the generation of the traces is trivial, in this paper we only consider control flow information and do not take into account other possible event attributes. As a future work, we will explore the possible attributes that can be derived/extrapolated during the generation of the traces from the distributive lattices.

Finally, another promising direction for future work is to consider other types of lattices (e.g., semi-modular lattices) and the Petri net classes to which they correspond.

5 Related Work

The closest related work is that in van der Aalst et al. [19], where different strategies for constructing transitions systems from event logs are presented. Once the transition systems are constructed, the authors put forward the idea of adding new edges between states in the transition systems (called "extend" strategy) as a way to discover behaviour that was not observed in the log but was likely to be present in the process. Our work differs from such approach in two ways. First, we adopt a well-known formalism, distributive lattices, with a well-defined notion of completeness. Thus, when introducing new behaviour in the distributive lattices, it is possible to determine when all missing behaviour has been added, which is not the case when using transition systems. Furthermore, the extend strategy in [19] can only add missing edges between pairs of states, but in the case of distributive lattices, it is possible to discover missing states as well as edges.

Another related work is that on concurrency oracles. [9] uses concurrency oracles to transform event log traces into elementary event structures. In such case, the amount of behaviour added when inserting concurrency will depend on the quality of the oracle. In particular, [9] uses the alpha relations [18] as oracle, which deems a pair of events as concurrent if the activities were ever observed in different orders in the log. For example, alpha relations deem (a, b) concurrent if a is executed before b in one trace, and b is executed before a in another trace. These relations can be spurious because it is possible that the order between a and b is particular to some executions (e.g., a is always executed before b in the trace $\langle c, a, b\rangle$; whereas b is always executed before a in the trace $\langle f, b, a\rangle$). In order to address that issue, [1] put forward the idea of local concurrency oracles that find concurrency relations that only apply to particular areas of the traces. Such technique uses two threshold, arbitrarily defined by the user, to control the sensitivity of the oracle. Different from the concurrency oracle approaches, the use of distributive lattices gives a reference as to what is the missing behaviour without having to rely on arbitrary thresholds or inserting concurrency derived from distinct computations.

The problem of measuring the completeness of an event log is not new. Probabilistic approaches to measure log completeness can be found in [13,21,23]. Different from these approaches, our aim is to discover the behaviour that is not observed but likely to be present in the process (under some assumptions). Instead, the probabilistic approaches aim at computing a lower bound representing the completeness of a log. While informative, it does not allow us to obtain the missing behavior.

While there are few techniques that compare directly to the approach presented in the paper, process mining operations, and in particular automated process discovery techniques, implicitly discover behavior when abstracting the behavior in a log (e.g., when creating models [2,11]). In fact, a way to assess the quality of a discovered process model is by measuring its generalization, behavior that is not observed in the log but likely to be part of the process [17]. The approach presented in this paper can be seen as a pre-processing step that discovers concurrency from groups of equivalent traces, which can then be inserted during the construction of other more sophisticated models, such as models in BPMN notation.

6 Conclusion

This paper presented an approach to discover unseen behavior from an event log. The approach is based on the reconstruction of distributive lattices, which represent execution states of process instances. This approach was inspired by the results presented in [12], where the relationship between a family of Petri nets (conflict-free 1-safe nets), event structures and lattices was shown. There, the authors showed that the lattices representing the execution states of the Petri nets were distributive when ordered by subset inclusion. The approach presented in this paper starts by computing lattices from groups of traces representing the same computation. Then, these lattices are completed by inserting missing elements until the distributivity property is fulfilled. Using the discovered behavior, a measure of completeness is defined to assess the volume of traces discovered during the completion operation. Finally, it was shown how this measure of completeness can be used to control the amount of behavior discovered.

In order to test the effect of the proposed approach in real-life logs, a set of experiments were run to measure the amount of behavior discovered in a collection of publicly available event logs. It was observed that in some event logs, the number of traces that could be computed from the distributive lattice could be very large. The latter may be due to noise in the log, which can lead to the discovery of too much behavior that may not be part of the "normal" process. Thus, while the completion operation is simple, it can be too sensitive to noise. Nonetheless, by using the measure of completeness as threshold, it is possible to control the amount of discovered behavior.

Acknowledgements. The authors would like to thank all the reviewers for their valuable comments. Special thanks to reviewer 2 for their insightful suggestions.

References

1. Armas-Cervantes, A., Dumas, M., Rosa, M.L., Maaradji, A.: Local concurrency detection in business process event logs. ACM Trans. Internet Technol. **19**(1), 1–23 (2019)
2. Augusto, A., Conforti, R., Rosa, M.L., Dumas, M.: Split miner: discovering accurate and simple business process models from event logs. In: 2017 IEEE International Conference on Data Mining (ICDM), pp. 1–10 (2017)
3. Bergenthum, R.: Prime miner - process discovery using prime event structures. In: 2019 International Conference on Process Mining (ICPM), pp. 41–48 (2019)
4. Bergenthum, R., Desel, J., Mauser, S., Lorenz, R.: Synthesis of petri nets from term based representations of infinite partial languages. Fundam. Inf. **95**(1), 187–217 (2009)
5. Birkhoff, G.: On the combination of subalgebras. Math. Proc. Cambridge Philos. Soc. **29**(4), 441–464 (1933)
6. Conforti, R., Rosa, M.L., ter Hofstede, A.H.M.: Filtering out infrequent behavior from business process event logs. IEEE Trans. Knowl. Data Eng. **29**(2), 300–314 (2017)
7. Dumas, M., García-Bañuelos, L.: Process mining reloaded: event structures as a unified representation of process models and event logs. In: Devillers, R., Valmari, A. (eds.) PETRI NETS 2015. LNCS, vol. 9115, pp. 33–48. Springer, Cham (2015). https://doi.org/10.1007/978-3-319-19488-2_2
8. Fani Sani, M., van Zelst, S.J., van der Aalst, W.M.P.: Repairing outlier behaviour in event logs. In: Abramowicz, W., Paschke, A. (eds.) BIS 2018. LNBIP, vol. 320, pp. 115–131. Springer, Cham (2018). https://doi.org/10.1007/978-3-319-93931-5_9
9. García-Bañuelos, L., van Beest, N.R.T.P., Dumas, M., Rosa, M.L., Mertens, W.: Complete and interpretable conformance checking of business processes. IEEE Trans. Softw. Eng. **44**(3), 262–290 (2018)
10. Habib, M., Nourine, L.: Tree structure for distributive lattices and its applications. Theoret. Comput. Sci. **165**(2), 391–405 (1996)
11. Leemans, S.J.J., Fahland, D., van der Aalst, W.M.P.: Discovering block-structured process models from event logs - a constructive approach. In: Colom, J.-M., Desel, J. (eds.) PETRI NETS 2013. LNCS, vol. 7927, pp. 311–329. Springer, Heidelberg (2013). https://doi.org/10.1007/978-3-642-38697-8_17
12. Nielsen, M., Plotkin, G.D., Winskel, G.: Petri nets, event structures and domains, Part I. Theor. Comput. Sci. **13**, 85–108 (1981)
13. Pei, J., Wen, L., Yang, H., Wang, J., Ye, X.: Estimating global completeness of event logs: a comparative study. IEEE Trans. Serv. Comput. **14**(2), 441–457 (2021)

14. Ponce-de-León, H., Rodríguez, C., Carmona, J., Heljanko, K., Haar, S.: Unfolding-based process discovery. In: Finkbeiner, B., Pu, G., Zhang, L. (eds.) ATVA 2015. LNCS, vol. 9364, pp. 31–47. Springer, Cham (2015). https://doi.org/10.1007/978-3-319-24953-7_4
15. Reisig, W.: Petri Nets: An Introduction. Springer, Heidelberg (1985)
16. Stone, M.H.: The theory of representation for boolean algebras. Trans. Am. Math. Soc. **40**(1), 37–111 (1936)
17. van der Aalst, W.: Process Mining: Data Science in Action, 2nd edn. Springer, Heidelberg (2016). https://doi.org/10.1007/978-3-662-49851-4
18. van der Aalst, W.M.P., Weijters, A.J.M.M., Maruster, L.: Workflow mining: discovering process models from event logs. IEEE Trans. Knowl. Data Eng. **16**(9), 1128–1142 (2004)
19. Van der Aalst, W.M.P., Rubin, V., Verbeek, H.M.W., van Dongen, B.F., Kindler, E., Günther, C.W.: Process mining: a two-step approach to balance between underfitting and overfitting. Softw. Syst. Model. **9**(1), 87–111 (2010)
20. van der Aalst, W.M.P., Weijters, A.J.M.M.: Process mining: a research agenda. Comput. Ind. **53**(3), 231–244 (2004)
21. van Hee, K.M., Liu, Z., Sidorova, N.: Is my event log complete? - a probabilistic approach to process mining. In: International Conference on Research Challenges in Information Science, pp. 1–12 (2011)
22. Wang, J., Song, S., Lin, X., Zhu, X., Pei, J.: Cleaning structured event logs: a graph repair approach. In: 2015 IEEE 31st International Conference on Data Engineering, pp. 30–41 (2015)
23. Yang, H., Ter Hofstede, A.H., Van Dongen, B.F., Wynn, M.T., Wang, J.: On global completeness of event logs. BPM Center Report BPM-10-09 (2010)

Layered Memory Automata: Recognizers for Quasi-Regular Languages with Unbounded Memory

Clément Bertrand[1(✉)], Hanna Klaudel[2], and Frédéric Peschanski[3]

[1] Scalian Digital Systems, Valbonne, France
clement.bertrand@scalian.com
[2] IBISC, Univ. Evry, Université Paris-Saclay, Évry, France
hanna.klaudel@univ-evry.fr
[3] LIP6, Sorbonne Université, CNRS UMR7606, Paris, France
frederic.peschanski@lip6.fr

Abstract. This paper presents the model of Layered Memory Automata (LaMA) to deal with languages involving infinite alphabets, with practical applications in the analysis of datastreams, or modeling complex resource usages in concurrent systems. The LaMA can be seen as an extension of the Finite Memory Automata (FMA) with memory layers and the capacity of dealing with an unbounded amount of memory. Despite the increased expressiveness, the LaMA preserve most of the "good" properties of the FMA, in particular the closure properties for the so-called quasi-regular constructions. Moreover, the layering of the memory enables particularly economical constructions, which is an important focus of our study. The capacity of dealing with an unbounded amount of memory brings the LaMA closer to more powerful automata models such as the history register automata (HRA), thus occupying an interesting position at the crossroad between the operational and the more abstract points of view over data-languages.

Keywords: data languages · memory automata · register automata · unbounded memory · quasi-regular languages

1 Introduction

Automata on *datawords*, involving infinite alphabets, represent an influential foundation for the analysis of datastreams [13]. Resource analysis frameworks for concurrent systems have also been investigated based on similar automata-theoretic foundations, e.g. in [1] or our own previous work [7]. Quoting [10]:

> *Actions of concurrent processes, when concurrency and communication are restricted to very simple patterns, are another possible interpretation of infinite alphabets.*

The classification of automata models for datawords can be roughly decomposed in two major families. The first family, pioneered by the *finite memory*

L. Bernardinello and L. Petrucci (Eds.): PETRI NETS 2022, LNCS 13288, pp. 43–63, 2022.
https://doi.org/10.1007/978-3-031-06653-5_3

automata (FMA) of [10] (colloquially known as *register automata*), adopts a mostly *operational* point of view similar to the classic *finite state automata* (FA). With FMA, letters (ranging over an infinite alphabet) can be temporarily or permanently stored in a *finite* amount of dedicated memory cells (or *registers*). They can then be compared with letters read at a later time during the recognition process. These models characterize an important notion of *freshness*: the property of a recorded letter to be unique among the ones already stored. In the FMA, this is obtained thanks to an *injectivity* constraint: the fact that the registers must hold distinct letters at any given time. The languages recognized by FMA are called *quasi-regular*, emphasizing their "classical" roots. In particular, they enjoy important closure properties, especially for the regular operators with the notable exception of *complementation*. Moreover, several important decision problems (e.g. emptiness checking) for FA remain decidable in FMA and related models. At the other end of the spectrum, the family related to *data automata* (DA) [5] adopts a more *abstract* point of view. They adopt principles, such as *guessing* that are very high level in comparison. Unsurprisingly, the decision problems are much harder for these models.

An important distinction can be made between these two *operational* vs. *abstract* families regarding the nature of the memory store. In the FMA and related models, the memory is finitely *bounded*. The automata cannot store more letters, in a given configuration, than the number of available registers. As a consequence, it is impossible expressing a language, which needs an unbounded number of different letters such as the language of words where each letter occurs at most once. This is a particularly strong constraint that one would like to lift in order to take more advantage of the infinite alphabets.

In this paper, we introduce an extension of the FMA, namely the model of *Layered Memory Automata* (LaMA), with both practical and theoretical benefits. Essentially, the intent is to establish a link between the *abstract* and *operational* families of automata. On the one side, the LaMA possess a strong operational nature in that they are a (conservative) extension of the FMA with the extra ability to handle an unbounded amount of memory. LaMA are non-deteministic finite state automata that have a finite number of variables, each of them able to store a finite set of letters. Upon reading a letter, a transition can test if the letter is already stored in a variable, can store the letter in a variable, or can reset a variable to emptyset. Like FMA, the variables of LaMA are under an injectivity constraint, which means that two variables cannot stores the same letter. This constraint is partially relaxed with the introduction of a finite set of *memory layers*. Variables are grouped into layers, and the injectivity constraint is only required between variables of a same layer.

Other works designed models extending FMA to manage an unbounded number of different letters. The FRA (*Fresh-Register Automata*) introduce the history, a memory cell able to store a set of letters from the infinite alphabet not restricted by injectivity constraint. It is used as well as the registers of FMA to express the notion of globally fresh letter, a letter never stored in a register before. However, this extension only slightly increases the expressivity of FMA as the set of languages recognized by FRA is not close to concatenation and Kleene star. The HRA (*History-Register Automata*) uses multiple of this histories instead of registers to store letter of the infinite alphabet. Without injectivity constraints, HRA uses a similar transition as M-FMA (*M-Automata* from [10]), the transition guard is satisfied when the input letter is stored in the exact set of histories annotating the transition. HRA transitions can clear a history and transfer the inputted letter among histories. Our model LaMA is similar to HRA, but without the possibility of performing *transfers* among its variables (histories) and preserving the injectivity constraint from FMA.

As a primary contribution, we argue that the LaMA provide a kind of a *sweet spot* between the "good" operational properties of the FMA, and (at least some of) the expressiveness of higher-level models with unbounded memory capabilities. A second contribution we defend in this paper is the *economical* nature of the proposed model regarding the (quasi-)regular constructions. The regular constructions proposed for FMA in [10,15] or [8] all yield automata of exponential sizes. Despite the fact that the LaMA strictly subsume the FMA (with unbounded memory), the constructions we propose for concatenation, disjunction and conjunction[1] of (the language recognized by) LaMA remain polynomial. Despite its simplicity, the idea of the memory layers plays here a crucial role. Unfortunately, the construction for the *Kleene star* remains exponential for LaMA. In [2] we introduce a variant of the LaMA with *transfer* capabilities that allows to obtain a polynomial construction. However, this variant only preserves the membership problem, and most other "good" properties are lost. Because of this and of space constraints, this variant will not be presented in detail in this paper.

The outline of the presentation is as follows. The LaMA model is presented in Sect. 2 with a discussion of related work, and its main closure properties are discussed in Sect. 3. Important language inclusion links between LaMA and other automata models are presented in Sect. 4. Finally, in Sect. 5 we discuss the important aspect of the sizes of the regular constructions in LaMA and related models.

2 Layered Memory Automata

We present in this section the model of layered memory automata (LaMA), an extension and improvement of the ν-automata presented in previous works [3]. The principle is to recognize datawords based on a countably infinite alphabet of letters, that we denote by $\mathcal{U}$. During the recognition process, the LaMA use *variables* to identify memory cells that can store sets of letters read as input.

The main specificity of this memory model is its structuring in *layers*. A memory context M corresponds to a memory divided in distinct layers. Each layer, identified by a natural number, can store a finite set of letters (over the infinite alphabet). Thus, for example, we can say that, in M, a variable X contains (is associated with) the finite set $E \subseteq \mathcal{U}$ at layer l, which will be denoted by $M(X^l) = E$. By a slight abuse of terminology, we will often write

[1] Without complementation, the conjunction operator becomes primitive in FMA and related models.

"the variable X^l" to in fact designate "the variable X at layer l". The formal definition is given below.

Definition 1 (Memory context). *Given a finite set of variables V, a finite set of layers L and an infinite alphabet $\mathcal{U}$, we define a memory context M as an association function whose signature is as follows: $M : V \times L \rightarrow 2^{\mathcal{U}}$ where $M(X^l) \subset \mathcal{U}$ is the finite set of letters* associated *with variable X^l.*

The most important feature of memory contexts is the following *injectivity* constraint.

Definition 2 (Injectivity of layers). *Let a memory context M be defined on the finite sets of variables V and layers L, the injectivity constraint is:*

$$\forall (X, Y) \in V \times V, \forall l \in L, X \neq Y \implies M(X^l) \cap M(Y^l) = \emptyset$$

In more informal terms: it is forbidden for a given letter to be stored in the memory corresponding to distinct variables at the same layer. If compared to FMA, we can say that each layer resembles the memory context of a FMA, but that distinct layers remain independent. The second, and fundamental difference with FMA is that the memory of LaMA is *unbounded*: each memory cell X^l can store an arbitrary number of letters.

Thanks to the injectivity constraint, we can define the notion of a *fresh letter at layer l*, i.e., a letter that is associated with no variable of the layer l. This subsumes the usual notion of a fresh letter, i.e., a letter being fresh at all layers.

We now explain the composition of a LaMA as *state-transition machines.*

Definition 3 (Layered Memory Automata). *Layered Memory Automata are defined with respect to an infinite alphabet $\mathcal{U}$ and are represented as tuples of the form $A = (Q, q_0, F, \Delta, V, L, M_0)$ where:*

- *Q is a finite set of states*[2]*,*
- *$q_0 \in Q$ and $F \subseteq Q$ are respectively the initial state and the set of accepting states,*
- *Δ is a finite set of transitions*[3]*, described below in Definition 4 and 5,*
- *V and L are respectively the finite set of variables and the finite set of layers, and*
- *$M_0 : V \times L \mapsto 2^{\mathcal{U}}$ is the initial memory context.*

The initial memory context M_0 indicates the letters initially associated with each variable. This makes it possible to define a finite alphabet of *constants,*

[2] The term *state* is rather connoted, being also used in e.g. "state-space" to designate "runtime" artifacts. We will use the term *configuration* to designate the notion of a "running state".

[3] For transitions, we will make the distinction between the transition itself, and its *firing*, i.e., the fact of effecting the transition on a previous configuration, to construct a next configuration, at "runtime".

similarly to FMA, with thus the possibility to simulate classical FA (a feature we will not take advantage of in this paper).

The set Δ of transitions encompasses two kinds of transitions: (1) the *observable transitions* that are fired when a letter is read in input, and which *consume* the letter, and (2) the ε-*transitions*, which are non-observable and thus can be fired at any time (without consuming the input).

Definition 4 (Observable transition). *The observable transitions are tuples of the form:* $\delta = (q, \nu, \alpha, \overline{\nu}, q') \in \Delta$ *where:*

- $q, q' \in Q$ *are the source and destination states of the transition,*
- $\nu \subseteq 2^{V \times L}$ *is the set of variables modifiable by the transition,*
- $\alpha : L \rightarrow V \cup \{\sharp\}$ *indicates for each layer the variable consulted by the transition,*
- $\overline{\nu} \subseteq 2^{V \times L}$ *is the set of variables which are reset by the transition.*

Input letters can only be consumed when firing such observable transitions. The precise definition of a transition firing is given below (cf. Definitions 7 and 8) but we summarize the informal intent now. The α function indicates the variables consulted by the transition, with the constraint that at most one variable can be consulted for each layer. The special symbol $\sharp$ is used to indicate that no variable is to be consulted for this layer when firing the transition. When reading a letter u, the transition may be fired if, for each variable X^l such that $\alpha(l) = X$:

- either X^l is not modifiable ($X^l \notin \nu$) and u is already associated with X^l;
- or, if X^l is modifiable ($X^l \in \nu$), then u is fresh for layer l (i.e., associated with no variable of layer l).

Because of the injectivity constraint, only a *fresh* letter can be associated with a variable X^l. That is, upon reading, the letter must not be associated with any variable in layer l, not even X^l.

Remark 1 (Universal transition). If no variable is consulted by a transition (i.e., $\forall l \in L, \alpha(l) = \sharp$), then the transition can be fired when reading any letter.

The set $\overline{\nu}$ is the set of variables that must be reset by the transition. No letter is associated with the variables of $\overline{\nu}$ in the configuration reached by the transition.

Definition 5 (ε-transition). *The non-observable ε-transitions are tuples of the form* $\delta_\varepsilon = (q, \overline{\nu}, q') \in \Delta$ *where:*

- $q, q' \in Q$ *are the source and destination states of* δ_ε*,*
- $\overline{\nu} \subseteq 2^{V \times L}$ *is the set of variables reset by the transition.*

We now turn to the dynamics of the model, describing the behavior of LaMA as language *recognizers*. We begin with the definition of a configuration, i.e., a "running state".

Definition 6 (configuration). *A configuration of a LaMA is a pair* (q, M) *of a state* q *and a memory context* M. *Given an automaton* $A = (Q, q_0, F, \Delta, V, L, M_0)$, *the initial configuration is* (q_0, M_0) *and an accepting configuration is a pair* (q_f, M), *for a reachable memory context* M *and an accepting, final state* $q_f \in F$.

A dataword belongs to the language of a LaMA if there is a (finite) sequence of firings of transitions going from the initial configuration to an accepting one. For observable transitions the question is the following: given a source configuration (q, M) and an input letter $u \in \mathcal{U}$, is there an observable transition $\delta \in \Delta$ which is *enabled* such that, as an *effect*, we can construct a destination configuration (q', M')? In such a case, the actual firing is denoted by $(q, M) \xrightarrow[\delta]{u} (q', M')$ (for observable transitions), or alternatively $(q, M) \xrightarrow[\delta]{\varepsilon} (q', M')$ (for ε-transitions).

Observable transitions, to be fired, must be *enabled*, under the following conditions.

Definition 7 (Enabling of an observable-transition). *For a configuration* (q, M) *and an input letter* $u \in \mathcal{U}$, *an observable transition* $(q, \nu, \alpha, \overline{\nu}, q') \in \Delta$ *is enabled if and only if for each layer* $l \in L$ *and variable* $X \in V$ *such that* $\alpha(l) = X \in V$:

- *if* X^l *is modifiable* ($X^l \in \nu$) *then no variable must be already associated with* u *in layer* l, *i.e.*, $\nexists Y \in V, u \in M(Y^l)$;
- *otherwise* ($X^l \notin \nu$), u *must be associated with* X *in layer* l, *i.e.*, $u \in M(X^l)$.

Informally, the role of the enabling conditions is: (1) to preserve the injectivity of each layer, and (2) to check the capability of consuming the input and store it in the required memory cells. The ε-transitions are enabled independently from the inputted letter. Once a transition is enabled, it can be non-deterministically *fired*, which produces as an *effect* a resulting configuration, as explained by the following definition.

Definition 8 (Effect of a transition firing). *For a source configuration* (q, M), *an input letter* $u \in \mathcal{U}$ *and an enabled transition* $\delta \in \Delta$, *the firing* $(q, M) \xrightarrow[\delta]{u} (q', M')$ *produces the configuration* (q', M') *constructed as follows:*

- *if* $\delta = (q, \overline{\nu}, q')$, *then* M' *consists of* M *where the variables in* $\overline{\nu}$ *are reset,* $M' = M[\overline{\nu} \to \emptyset]$,
- *if* $\delta = (q, \nu, \alpha, \overline{\nu}, q')$, *then in* M' *the modifiable variables are associated with* u *and the variables of* $\overline{\nu}$ *are reset,*

$$\text{i.e., for each } X^l,\quad M'(X^l) = \begin{cases} \emptyset & \text{if } X^l \in \overline{\nu} & (1) \\ M(X^l) \cup \{u\} & \text{if } \alpha(l) = X \wedge X^l \in \nu & (2) \\ M(X^l) & \text{if } X^l \notin \nu \vee \alpha(l) \neq X & (3) \end{cases}$$

The memory context M' produced by a transition firing is the result of a combination of three different cases of effects, denoted by (1) - (3) in the definition above. Case (1) corresponds to the reset of the memory cell X^l, which

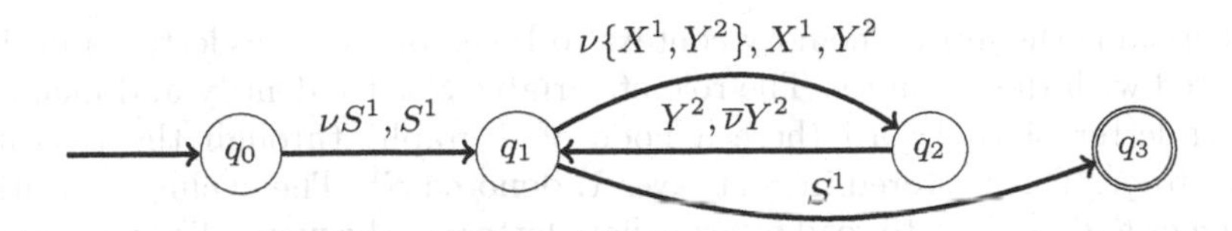

Fig. 1. A layered memory automaton recognizing words of the form *abbccdd*... *a*

is thus emptied. Case (2) corresponds to the actual consumption of the letter u, which is placed in all the required memory cells. Finally Case (3) aims at preserving the unchanged parts of the memory.

The language recognized by a LaMA is now naturally defined by sequences of firings from the initial configuration to accepting ones. To simplify the definition, we first introduce the notion of a *weak firing* that encompasses the firing of a single observable transition, surrounded by (possibly empty) sequences of ε-transitions.

We denote by $(q, M) \overset{u}{\underset{\delta}{\Rightarrow}} (q'', M'')$ a weak transition firing, corresponding to any firing sequence of the form: $(q, M) \overset{\varepsilon}{\underset{\gamma}{\rightarrow}}^* \cdots \overset{u}{\underset{\delta}{\rightarrow}} (q', M') \overset{\varepsilon}{\underset{\eta}{\rightarrow}}^* (q'', M'')$.

Definition 9 (Language of a LaMA). *Let A be a LaMA and $\mathbb{L}(A)$ the language it recognizes. A word $w = u_1\ u_2\ \ldots\ u_n \in \mathcal{U}^*$ belongs to $\mathbb{L}(A)$ iff there exists a sequence of weak transition firings: $(q_0, M_0) \overset{u_1}{\underset{\delta_1}{\Longrightarrow}} (q_1, M_1) \overset{u_2}{\underset{\delta_2}{\Longrightarrow}} \cdots \overset{u_n}{\underset{\delta_n}{\Longrightarrow}} (q_n, M_n)$ such that $q_n \in F$.*

We depict in Fig. 1 an example of a LaMA with 4 states and 5 transitions. The memory structure of the automaton involves the variables X, Y, S with two distinct layers 1 and 2.

For the sake of readability, we use a slightly simplified notation for the transition label. A transition labeled $\nu\{X^1, \ldots\}Y^1Z^2 \ldots \overline{\nu}\{U^1, \ldots\}$ in a diagram, from a state labeled q to a state labeled q', corresponds more formally to a transition $\delta = (q, \nu, \alpha, \overline{\nu}, q')$ such that $\nu = \{X^1, \ldots\}$, $\alpha = \{1 \mapsto Y, 2 \mapsto Z, \ldots\}$ and $\overline{\nu} = \{U^1, \ldots\}$. Also, we omit the brackets for singleton sets, and we also omit the empty sets and the epsilons. For example, in the diagram of Fig. 1, for the transition labeled $Y^2, \overline{\nu}Y^2$ we in fact mean $\nu = \emptyset$, $\alpha = \{1 \mapsto \sharp, 2 \mapsto Y\}$ and $\overline{\nu} = \{Y^2\}$. Thus, the transition labeled S^1 actually means $\nu = \overline{\nu} = \emptyset$ and $\alpha = \{1 \mapsto S, 2 \mapsto \sharp\}$.

Now that the simplified notation is in place, we can explain the behavior of the depicted automaton. The language it recognizes is the following one:

$$\{sx_0x_0x_2x_2 \ldots x_nx_ns \mid \forall i, j \in \mathbb{N}, s, x_i \in \mathcal{U}, i \neq j \implies x_i \neq x_j\}$$

This is an example inspired from [3] where we study the pattern recognition in dynamic graphs, with datawords representing sequences of edges established dynamically (so-called *link streams*). In this representation, the automaton characterizes a Hamiltonian circuit as a pattern.

We assume the initial memory context to be empty, i.e., no letter is initially associated with the variables. The role of variable S is to identify and memorize the first letter of the word (here a node of a graph) through the transition from q_0 to q_1. This is stored in S at layer 1, denoted S^1. The cycling transitions between q_1 and q_2 allow to read intermediate letters of the word. The variable X^1 memorizes these intermediate letters when ensuring that letters in even positions are all different from each other. Since X^1 belongs to the same layer as S^1, the injectivity constraint ensures that all letters are different from the first one. Then, the variable Y^2 ensures that the letters in odd positions are identical to the ones which immediately precede them. In the transition going from q_1 to q_2 the letter in even position is associated with Y^2. The only letter enabling the transition from q_2 to q_1 is the one previously associated with Y^2. Then, Y^2 is reset in order to track the next letter, and not confuse it with the one previously stored. Eventually, the last letter is read, which has to be in even position and to be the same letter as the one stored in S^1 to enable the transition from q_1 to the accepting state q_3.

3 Regular Constructions and Closure Properties

One of the most important properties of FMA, beyond their extended expressiveness, is the fact that they preserve most of the "good" properties of FA, especially closure properties for all the *regular constructions*, except for complement. This aspect is emphasized by the authors of [10] by defining the class of languages recognized by FMA as *quasi-regular.*

The LaMA we introduce in this paper correspond to a strict extension of the FMA (and in fact an extension of both the FRA and the GRA, as discussed in Sect. 4). But most importantly, we aim with the LaMA to an extension that is as *conservative* as possible, wrt. the "good" properties of FMA. In particular, the LaMA ensure the same closure properties as the FMA wrt. the regular constructions. In fact, most regular constructions are greatly facilitated by the availability of layers that allow to compose memory contexts without interference (e.g. composing two LaMA for concatenation). With the notable exception of the Kleene star, the proof schemes thus resemble the ones of FA. As such, we will only present proof sketches, the details being available in [2]. Note, also, that Sect. 5 discusses quantitative aspects related to these constructions.

Theorem 1 (Closure properties of basic operators). *Let the two LaMA $A_1 = (Q_1, q_1, F_1, \Delta_1, V_1, L_1, M_1)$ and $A_2 = (Q_2, q_2, F_2, \Delta_2, V_2, L_2, M_2)$, such that $L_1 \cap L_2 = \emptyset$, then:*

- *(Concatenation) there is a LaMA $A_{1\cdot 2}$ such that $\mathbb{L}(A_{1\cdot 2}) = \mathbb{L}(A_1) \cdot \mathbb{L}(A_2)$.*
- *(Union) there is a LaMA $A_{1\cup 2}$ such that $\mathbb{L}(A_{1\cup 2}) = \mathbb{L}(A_1) \cup \mathbb{L}(A_2)$.*
- *(Intersection) there is a LaMA $A_{1\cap 2}$ such that $\mathbb{L}(A_{1\cap 2}) = \mathbb{L}(A_1) \cap \mathbb{L}(A_2)$.*

Proof (Proof sketches). The assumption $L_1 \cap L_2 = \emptyset$ is without loss of generality because a trivial fact is that the injective renaming of the set of layers of a LaMA (with *fresh* layer identities) does not change the language it recognizes. Now, we consider the basic operators in turn.

Concatenation. It is possible to construct automaton $A_{1\cdot 2}$ following the classical construction of finite state automata, which consists in adding ε-transitions allowing to access the initial state of A_2 from each accepting state of A_1. As the layers of A_1 and A_2 are disjoint, their memories are actually put side by side and the variables of both automata do not interact together. Thus, we ensure that there is no side effect of A_1 on A_2 and the initial values of the variables of A_2 do not change when firing transitions in A_1.

Union. Similarly as above, the classical construction of FA applies here, which consists in adding a new initial state connected to the former initial states of A_1 and A_2 with ε-transitions (without reset). As for concatenation, the variables of both automata do not interact thus the initial context of the A_1 has no impact on the recognized language of A_2.

Intersection. As the memories of both automata are disjoint, it is possible to use the classical construction of a synchronized product of automata. The synchronization of two non-ε-transitions consists forming a transition labeled with the union of the sets ν, α and $\overline{\nu}$ of both transitions. Formally, the synchronization of observable transitions $(q_1, \nu_1, \alpha_1, \overline{\nu}_1, q_1') \in \Delta_1$ with $(q_2, \nu_2, \alpha_2, \overline{\nu}_2, q_2') \in \Delta_2$ is the transition : $((q_1, q_2), \nu_1 \cup \nu_2, \alpha_{1\cap 2}, \overline{\nu}_1 \cup \overline{\nu}_2, (q_1', q_2'))$ where $\forall i \in \{1, 2\}, l \in L_i, \alpha_{1\cap 2}(l) = \alpha_i(l)$. This construction is illustrated in Appendix A.

The case of iteration, or Kleene star, is a little bit less straightforward because during an iteration the memory context of the automaton may change, however such effect should be "canceled" for further iterations. Indeed, each (regular) iteration has to recognize exactly the same language, and not a language changed due to memory effects of previous iterations.

Theorem 2 (Closure property of Kleene star). *Let $A = (Q, q, F, \Delta, V, L, M)$ be a LaMA, then there is a LaMA A^* such that $\mathbb{L}(A^*) = \mathbb{L}(A)^*$.*

Proof (Proof sketch). The proposed construction is based on the classical one for FA which requires adding "ε-loops" from accepting states to the initial, thus allowing to iterate on the content of automaton A. As with all kinds of register automata, one difficulty with LaMA is that the language recognizable from a configuration depends on its memory context. And the latter can change at each iteration. In a way similar to what is done in the case of M-automata [10], the required "cancelling" of memory effects is realized thanks to a mechanism simulating a reset of the memory context to its initial value M_0. To do so, the principle is to duplicate the set of variables of layers in L on a set of "shadow" layers L_s. The variables of L are used to memorize the initial values of M_0 while the variables of L_s are used to store the fresh values recognized during the iterations. This way, in order to retrieve the initial values of the memory context, it is enough to remove at the end of each iteration all the letters stored in the variables of L_s.

If a transition is enabled in A when the letter read is associated with variable X^l, then this transition has to be duplicated in A^* such that it is possible to

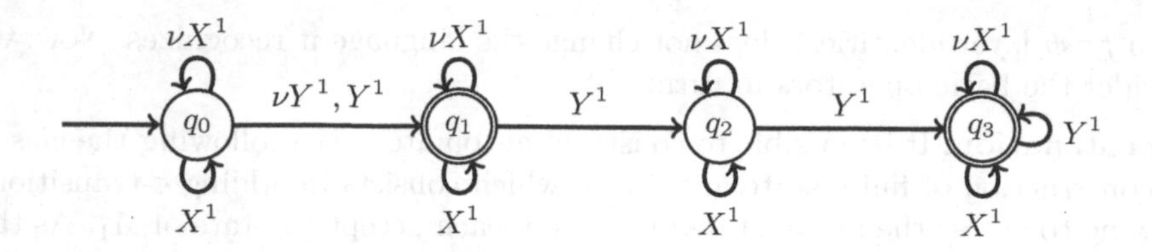

Fig. 2. LaMA accepting the language $\mathbb{L}_{\neq 2}$.

access either X^l (the initial values) or X^{l_s} (the possibly updated ones), with l_s the "shadow" layer corresponding to l. Moreover, if a transition in A has a guard referencing several variables, it is necessary to duplicate this transition in A^*. For example, a transition accessing the variables X^l, Y^k will be duplicated 4 times, once for each pair of : $(X^l, Y^k), (X^{l_s}, Y^k), (X^l, Y^{k_s}), (X^{l_s}, Y^{k_s})$. This duplication is required, in the absence of e.g. a transfer mechanism (cf. Sect. 5), because it is not effective to consult the variables in the layers of L and L_s simultaneously. Indeed, their sets of values are disjoint (e.g. initially the layer L_s is empty). In consequence, this construction leads to an exponential growth in terms of the number of transitions of the resulting automaton A^*. Moreover, it is also necessary to know which variables have been reset during each iteration, which is realized by duplicating states, implying also an exponential growth in terms of constructed states. These exponential growth phenomena are discussed further in Sect. 5.

The infinite nature of the alphabet manipulated by all the classes of memory automata (at least all the classes discussed in this paper) is in contradiction with the principle of complementation and determinism. Thus, unsurprisingly the following negative result also applies to LaMA.

Proposition 1 (Complement). *The set of languages recognized by LaMA is not closed under complement.*

Proof. The LaMA represented in Fig. 2 recognizes the language $\mathbb{L}_{\neq 2}$ of words containing at least one letter not appearing twice in all words. It does so by non-deterministicaly selecting a letter when it occurs for the first time, associating it to variable Y^1 and accepting the word only if this letter does not occur in the word exactly twice. The variable X^1 is used to store all the other letters and to never forget them, which ensures that the selection of a letter may only happen at its first occurrence.

The complement of $\mathbb{L}_{\neq 2}$ is the language $\mathbb{L}_{=2}$ containing *only* words with all their letters occurring exactly twice. In order to encode $\mathbb{L}_{=2}$, it is necessary to enumerate the occurrences of all the letters of words recognized by this language. An automaton recognizing this language would have to count an arbitrary number of occurrences of distinct letters. With a finite number of variables and states, such a construction is not possible with LaMA.

A deterministic LaMA is an automaton such that for all configurations, when reading any letter of $\mathcal{U}$, at most one transition can be fired. This restriction implies that when reading a globally fresh letter there is at each step only one way to identify it (associate it with a variable).

Proposition 2 (Determinism). *The set of languages recognized by deterministic LaMA is strictly included in the set of languages recognized by non-deterministic LaMA.*

Proof. The language $\mathbb{L}_{\neq 2}$ recognized by the non-deterministic LaMA from Fig. 2 cannot be recognized by a deterministic LaMA. To recognize this language, the automaton would have to "find" a letter that will not occur exactly twice. However, the words from this language are finite but may contain an arbitrary amount of different letters. Thus, it is not possible to track the number of occurrences of each of them with a finite amount of variables and layers.

4 A Classification of LaMA (Related Work)

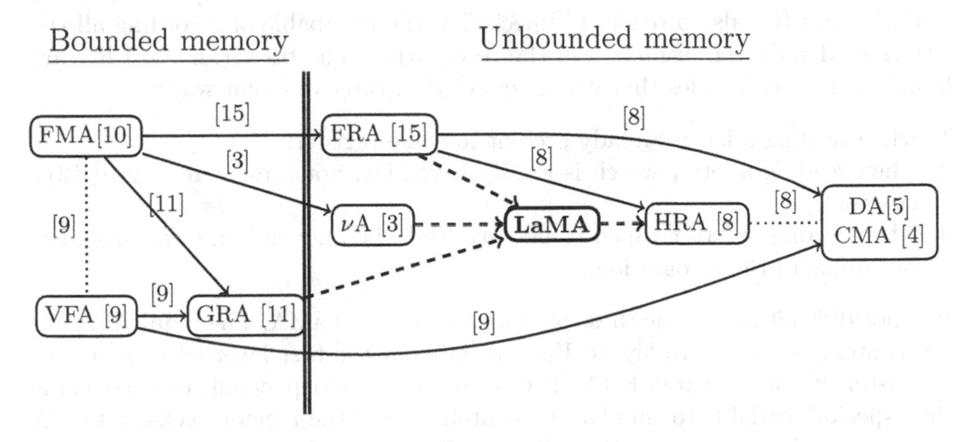

Fig. 3. A classification of automata over datawords, based on [12]. The arrows represent (strict) language inclusions, the dashed arrows are presented in Sect. 4, and dotted lines denote language incomparability.

Figure 3 represents most of the automata models we investigated while developing our proposition. The arrows on the figure are (strict) language inclusions. In this discussion, we denote by $A \sqsubset B$ the fact that the languages recognized by automata of model A strictly includes those of model B. For example, we know from [15] that the FRA (*fresh register automata*) can simulate the FMA, and thus FMA $\sqsubset$ FRA. The models related by dotted lines are knowingly *incomparable*. In this section we discuss the positioning of the LaMA in the family of data language recognizers. More precisely, we present the language inclusions depicted by dashed arrows on the figure. Since we cannot describe the related automata models with enough details in this paper, the discussion remains mostly informal, with the complete proof available in [2].

The LaMA were designed, broadly speaking, as a variant of FMA with unbounded memory capabilities. It is thus expected that LaMA are able to simulate FMA. Since the LaMA with one layer correspond exactly to ν-automata, we can reuse the result of [6] to show that LaMA are able to simulate the FMA.

Proposition 3. *FMA* $\sqsubset$ *LaMA*

However, in technical terms, it is interesting to compare the LaMA with other models proposed as extensions or variations of the FMA. FRA (*fresh register automata*) is a conservative extension of FMA capable of dealing with (a restricted kind of) unbounded memory. It is possible to simulate a FRA with a 2-layer LaMA, and thus to simulate a FMA by transitivity.

Proposition 4. *FRA* $\sqsubset$ *LaMA*

Proof (Proof sketch). The FRA model is based on a memory composed of a set of registers capable to memorize a unique letter, and constrained by injectivity. The model is thus quite similar to the FMA, however with a little but important "twist". An FRA also provides a "special" variable capable of recording all the letters read since the beginning of the recognition. The transitions of FRA are found in three categories that can be enabled in three different ways:

1. when reading a letter already present in some register;
2. when reading a letter which is locally fresh, i.e., not present in any register currently;
3. when reading a letter which is globally fresh, i.e., not encountered since the beginning of the recognition.

It is not difficult to provide these mechanisms with a LaMA. The required memory context contains two layers. Each variable of the first layer corresponds to a register of the simulated FRA. The second layer, independent, only concerns the "special" variable to simulate its content. Since the memory cells of LaMA are not bounded, we can say that all the variables of LaMA are "special", in the FRA understanding of the term. Put in other terms, the FRA can be seen as a special cases of LaMA with a FMA-like layer of bounded memory, and a unique variable of unbounded memory in a second layer.

The LaMA are also strictly more expressive than the FRA. One may observe, indeed, that FRA are not closed under concatenation. For example, the language $\mathbb{L}_{\neq}$ of words composed of all-distinct letters, may be recognized by both FRA or LaMA. But the language $\mathbb{L}_{\neq} \cdot \mathbb{L}_{\neq}$ is only recognized by LaMA.

The GRA (*guessing register automata*) model is an interesting variant of FMA using a non-deterministic assignation (*guessing*) principle. By proving, below, that LaMA are able to simulate GRA it emphasizes the fact that the LaMA are also capable of simulating its *guessing* principle, and not only the operational principles of the FMA. This establishes an interesting connection with the "logical" family that also rely on guessing features (note the inclusion link between VFA and DA in Fig. 3).

Proposition 5. *GRA* $\sqsubset$ *LaMA*

Proof. The GRA model is a variant of FMA with a modified variable assignment method. The memory of a GRA is composed of a finite set of registers, each containing at most one letter, together with an injectivity constraint. The transitions of GRA are found in two categories:

- the observable ones are annotated by the register containing the letter that has to be consumed to fire the transition;
- the ε-transitions are annotated with a register which is reassigned to a non-deterministically *guessed* letter.

The assigned letter will be decided when firing the next observable transition annotated with this register. However, if other registers are reassigned in the meantime, they cannot be assigned the same letter due to the injectivity constraint.

Given a GRA, it is possible to construct a LaMA which recognizes the same language. After the reassignment of a register r, an arbitrary letter of the infinite alphabet is non-deterministically assigned to it. To find out which letter was assigned to r, it is necessary to memorize all letters currently assigned to the other registers and those that will be assigned to them until an observable transition labeled with r is fired. This transition will be enabled by any letter not recorded since the reassignation.

Hence, for each register of a GRA, the simulating LaMA will use as many variables as necessary to memorize all the values stored by every other registers between its reassignment and the transition that will determine the guessed value. This way, when an observable transition allowing to determine the value of the input letter is enabled, the injectivity constraint ensures that the letter is different from those already associated with other registers. The actual construction is in consequence quite intricate, and we delegate to [2] for the formal details.

The inclusion is strict since it is known (from [11]) that there is no GRA that can recognize the language of words of any length with all letters occurring only once.

Perhaps the most interesting inclusion link is the one connecting the LaMA to the more expressive HRA (*history-register Automata*).

Proposition 6. *LaMA* $\sqsubset$ *HRA*

Proof (Proof sketch). The HRA memory is constituted of variables associated with histories that can store an unbounded amount of letters. This is very much like the ν-automata and thus the LaMA with a single memory layer. However, a very important difference is that the HRA histories are not restricted by an injectivity constraint. There are thus quite similar to the M-automata of [10], but with unbounded memory. The observable transitions are annotated with two sets of histories: R (read) and W (write). A transition is enabled when the input letter is exactly associated with all histories of R. After the firing, in the

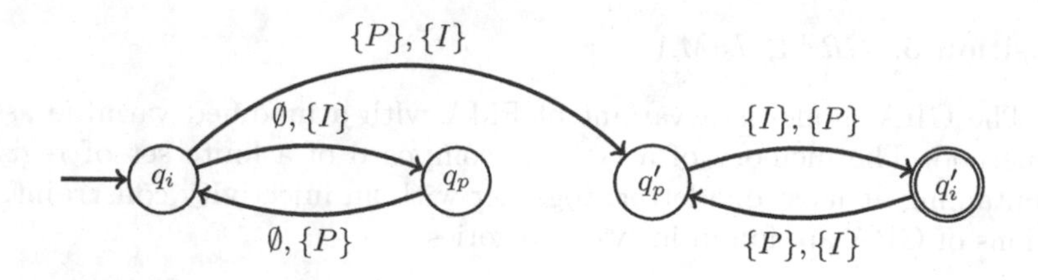

Fig. 4. HRA recognizing a language which is not recognized by a LaMA

resulting configuration, the letter is associated exactly to all histories of W. Thus, the letter can be transferred among the histories, or erased from them, in the resulting configuration. The ε-transitions are annotated with a set of histories C containing histories cleared (reset) in the resulting configuration.

It is possible to simulate a LaMA with a HRA by encoding the memory layers and the injectivity constraint. The simulating HRA has the same set of states, as well as a history for each variable of the original LaMA. Since the observable transitions of HRA cannot reset variables, they are split in two parts: (1) a transition for the enabling and firing, and (2) a transition for the reset. To simulate the enabling and firing of a LaMA transition, multiple observable transitions are needed in the HRA:

- for each variable X^l consulted in the LaMA transition, $\alpha(l) = X, X^l \notin \nu$, the matching history is part of both R and W;
- for each variable X^l modified in the LaMA transition, $\alpha(l) = X, X^l \in \nu$, the matching history is only part of W.

As R needs to encompass the histories containing the input letter in order to be enabled, when no variable is consulted for some layer, $\exists l, \alpha(l) = \sharp$, then the transition needs to be duplicated in the HRA to search if the value is present in one of the histories of this layer. If multiple layers are not consulted, then the transition is duplicated to search the letter in each combination of histories for those layers. To enforce the layer injectivity constraint, the construction is designed so that the transitions are never annotated by histories that simulate variables of the same layer. This way, during the recognition, it is not possible to reach a configuration in which the histories corresponding to the same layer contain a common letter.

The observable transitions can remove the input letter from the histories it is annotated with, when $R \backslash W \neq \emptyset$. It will thus be possible to delete a particular letter from a history, which is impossible for LaMA. Thus, it is rather easy to come up with a language recognizable by a HRA, and not recoginizable by a LaMA. For example, no LaMA can recognize the language of the HRA in Fig. 4, which is the language of words of the form $w = uv$ where:

- the prefix u is a word whose length is **even** and in which all letters are different;

- the suffix $v = v_1v_2v_3 \dots v_n$ is a word in which each letter v_i satisfies that if i is odd then the occurrence of v_i is in an even position in w, and if i is even then the previous occurrence of v_i is in an odd position in w.

It is known, from [8], that the HRA recognize languages that are incomparable with those of the CMA and DA (class memory automata and data automata). This is due to the capability of resetting histories in HRA, which cannot be simulated by a CMA/DA. We have not studied the problem finely, but, for the same reason, we expect the incomparability of LaMA vs. CMA/DA, although it is for now only a conjecture.

The connections we established with related automata models allow us to give some insight about the complexity (and decidability) of some decision problems concerning LaMA. First, the strict inclusion of FMA induces the undecidability of the same problems as FMA, in particular the *language inclusion* and the *universal language* problems (cf. [14]). The inclusion links discussed previously allow to establish the following:

Fact 1. *The emptiness checking and membership problems for LaMA are both NP-hard.*

Proof. The *emptiness checking* problem consists in detecting if the language of an automaton is empty. The problem is known to be NP-complete in the case of FMA [10,14]. Moreover, the same problem is known to be Ackermann-complete for HRA [8], thus trivially decidable for LaMA. The situation is in fact exactly the same for the *membership problem*: NP-complete for FMA and "at-most" Ackermann-complete for HRA. Indeed, the membership problem can be solved through emptiness, although for some automata model the membership problem can be solved by better, dedicated ways (starting with FA). It is unlikely that this would be the case for LaMA since it is already not the case for FMA (cf. [10,14]).

As a future work, we intend to study more finely the complexity of these two problems for LaMA. It would be interesting to see if the use of unbounded memory *without* a transfer mechanism simplifies the emptiness problem (put in other terms, do we reach the Ackermann bound?).

5 A Quantitative Point of View on Regular Constructions

Expressiveness is not the only important aspect to consider when comparing classes of automata. For example, many "regular"-expression packages (e.g. PCRE[4]) adopt the non-deterministic finite state automata (NFA) rather than the theoretically "more efficient" and equivalent determistic ones (DFA), because of the exponential growth when translating the former to the latter. In the same spirit, the prototype analysis tool we develop[5] requires the construction of an

[4] Perl compatible regular expressions, cf. https://www.pcre.org/.
[5] PaMaTina, cf. https://github.com/clementber/MaTiNA.

Table 1. Translation between FMA and M-FMA

	♯states	♯transitions	♯registers
FMA → M-FMA	$\lvert Q\rvert * (\lvert M\rvert !)$	$\lvert \Delta\rvert * (\lvert M\rvert !)$	$\lvert M\rvert + 1$
M-FMA → FMA	$\lvert Q\rvert * \lvert M\rvert^{\lvert M\rvert}$	$\lvert \Delta\rvert * \lvert M\rvert^{\lvert M\rvert}$	$\lvert M\rvert$

automaton, akin to a (timed variant of the) LaMA, from an extension of regular expressions (cf. [2,3]). In this compilation step, the size of the resulting automaton plays a significant role.

In this section we compare the sizes of the regular constructions for three models of automata: the LaMA, the FMA (taking the constructions proposed in [10]) and the HRA (taking those of [8]). Note that these sizes are not given in the aforementioned papers, and we established them while learning about those constructions. As a consequence, all encountered errors about these computations would be ours, not those of the original authors. We evaluated the sizes of the constructions of the FMA presented in the proof of Theorem 3 of [10]. For the HRA, we evaluated the sizes of the constructions presented in Sect. 3 of [8]. The GRA [11] and FRA [15] constructions are not studied here as they are based on the ones presented for FMA and HRA. For the sake of concision, we only consider the (most intricate) cases of concatenation and Kleene star in this paper (the other constructions being also detailed in [8,10]).

Most importantly, our intent is *not* to say that the construction we propose are "better", in any sense of the word, but instead: (1) to motivate the fact that reasoning about the size of the constructions is important, and (2) trying to find ways to make such construction as *compact* as possible. A positive point of view is that if we find compact constructions for LaMA, then they can also be used almost directly as compact constructions for FMA (by first translating FMA to LaMA, which is both straightforward and economical), and similarly for FRA, GRA and VFA.

To compare the constructions, the sizes we consider are the worst-case estimates of the automata, with respect to:

- the number of states in the automata, denoted by $|Q|$,
- the number of transitions, denoted by $|\Delta|$,
- and the number of memory identifiers, denoted by $|M|$.

What we call memory identifiers here are the registers in the FMA, the histories in the HRA and the variables $X^l \in V \times L$ in the LaMA. This quantification on the identifiers does not take into account the number of letters that may be stored in memory, simply because there is no bound in the case of LaMA and HRA. In the following tables we denote by $|\Sigma|$ the number of letters initially stored in the memory of an automaton and by $|L|$ the number of layers of the LaMA.

In [10], the regular constructions are not established directly for FMA but rather rely on the equivalent model of *M-Automata* (M-FMA). Thus, the FMA are first converted to M-FMA, which in fact already causes an exponential growth, as described on Table 1. The M-FMA resulting from the translations use approximately the same number of registers. However, the loss of the injectivity constraint in M-FMA causes an explosion in the number of states required to simulate the correct (i.e., injective) use of registers. The duplication of transitions follows from the duplication of states. Perhaps surprisingly the exponential growth is also present when translating back to FMA (which could perhaps be avoided by keeping a little bit more structural information in M-FMA). But as it is, none of the regular constructions proposed for FMA has polynomial size.

Concatenation. Table 2 represents the sizes of the automata constructed for concatenation. The constructions for the three models try to duplicate that of the finite state automata by keeping the structures of the two automata and by adding transitions allowing access to the initial state of the suffix automaton at the end of the prefix automaton path.

Table 2. Sizes of constructions for concatenation $\mathbb{L}(A_1) \cdot \mathbb{L}(A_2)$

	♯states	♯transitions
M-FMA	$\|Q_1\| + \|Q_2\|$	$\|\Delta_1\| * 2^{\|M_2\|} + \|\Delta_2\| * 2^{\|M_1\|}$
HRA	$(\|Q_1\| + \|Q_2\|) * 2^{\|\Sigma_2\| * \|M_2\|}$	$(\|\Delta_1\| + \|\Delta_2\|) * (\|\Sigma_2\| + 1) * 2^{\|\Sigma_2\| * \|M_2\|}$
LaMA	$\|Q_1\| + \|Q_2\|$	$2 * \|\Delta_1\| + \|\Delta_2\|$

	♯registers
M-FMA	$\|M_1\| + \|M_2\|$
HRA	$\max(\|M_1\|, \|M_2\|) + \|\Sigma_2\|$
LaMA	$\|M_1\| + \|M_2\|$

In M-FMAs, the constructed automaton uses all the registers of the two concatenated automata, as well as their initial valuations. However, due to the nature of the transitions, similar to that of the HRA, it becomes necessary to duplicate all the transitions for each subset of registers of the other automaton. Thus, this leads to a combinatorial explosion in the number of transitions in the automaton resulting from the construction.

In the HRA, before the construction is carried out, all the letters initially associated with the histories of the suffix automaton are extracted from the two automata. These letters are each associated with a new history. This preserves the initial value of the suffix automaton memory when transiting the prefix one. However, when these values are extracted, it is necessary to add transitions in order to preserve the language of the automaton. Thus, in the resulting automaton, the transitions leading to the initial position of the suffix automaton reset all the histories except those containing the extracted letters.

Kleene Star. Table 3 presents the sizes of constructions for Kleene star. The construction used in the LaMA is inspired from that of M-FMA. Thus, the sizes are of the same order.

Table 3. Sizes of constructions for the Kleene star $\mathbb{L}(A)^*$.

	$\sharp$states	$\sharp$transitions	$\sharp$registers
M-FMA	$\|Q\| * 2^{\|M\|}$	$\|\Delta\| * 2^{2*\|M\|}$	$2 * \|M\|$
HRA	$\|Q\| * 2^{\|\Sigma\|*\|M\|}$	$\|\Delta\| * 2^{\|\Sigma\|*\|M\|} * (\|\Sigma\| + 1)$	$\|M\| + \|\Sigma\|$
LaMA	$\|Q\| * 2^{\|M\|}$	$\|\Delta\| * 2^{2*\|M\|}$	$(2 * \|M\|) + \|L\|$

For the Kleene star, the construction in the HRA consists first of all in extracting all the letters initially stored in the histories and in storing them in new dedicated histories, as in the construction for concatenation. It is again necessary to duplicate the transitions and the states so that the automaton always recognizes the same language. So at the end of each iteration it suffices to reset all the other histories in order to reset the memory to its initial value.

As a summary, the constructions proposed for LaMA are in most cases more compact than the ones proposed for FMA (and M-FMA) and HRA. This is not shown here but the situation is the same for all the regular operators. In fact, all constructions are polynomial for LaMA with the notable exception of the Kleene star. To address this issue, we propose in [2] a variant of LaMA with a transfer mechanism that allows to copy all the letters associated with a variable from one layer to another layer. This allows to "dump" the memory from the layers in L to the layers in L_s in the final transitions of an iteration, enabling an exponential reduction in the number of required states and transitions. However this new mechanism is quite "powerfull", causing a loss of several "good" properties of the model (if only the closure properties). However, it is shown in [2] that this alternative model is conservative wrt. the membership problem, which explains why we use it in practice.

6 Conclusion

In this paper we introduced the model of LaMA, characterized by the layered structure of their memory, and the fact that this memory is not bounded. We mostly discussed the quasi-regular constructions (insisting on quantitative aspects) and language inclusion links with related models. Beyond such (important) theoretical considerations, we find important to emphasize the fact that the LaMA were also designed with practical applications in mind. This is the main reason why we emphasized so much the "compactness" of the quasi-regular constructions, the layered architecture playing a significant role here.

For future works, we intend to study two more aspects of the model. First, we know that the class of *deterministic* LaMA is strictly less expressive than the non-deterministic ones. However, this class is still worth studying given the fact that the membership problem becomes much easier in this case. Second, we would also like to investigate the relationship between subclasses of MSO and language classes recognizable by LaMA, or a restricted version (without reset for example) as it is done for DA wrt. $\exists$MSO.

A Examples of Regular Constructions (Complement to Sect. 3)

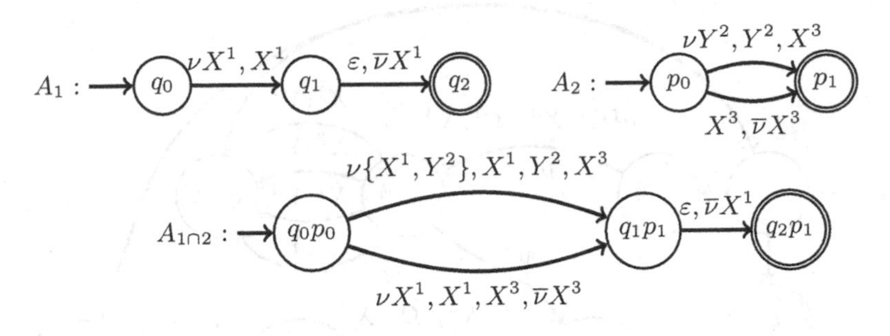

Fig. 5. Intersection construction

In this appendix, we give some more details about the constructions corresponding to the intersection and iteration of LaMa, as a complement to Sect. 3.

Intersection. Figure 5 illustrates the synchronized product of two LaMA on the left, A_1 and A_2, used to produce the LaMA recognizing the intersection of the languages of A_1 and A_2. The resulting LaMA $A_{1\cap 2}$, on the right of the Figure, contains only the states reachable by transitions from the initial states. The construction is thus quite similar to the usual construction for finite automata. One notable difference relates to then handling of observable transitions. In fact, only observable transitions are synchronized together, while non-observable ones are not. The reason is the firing of non-observable transition does not consume letters, and are thus "transparent" wrt. language intersection.

Iteration. Figure 6 illustrates the Kleene star construction, with on top a LaMA A, recognizing language $\mathbb{L}(A)$, and below the LaMA A^* constructed such that $\mathbb{L}(A^*) = \mathbb{L}(A)^*$.

The construction is in principle close to the equivalent construction for finite automata. However, the handling of memory layers requires some care. To illustrate this, the automaton A in the figure uses two layers, 1 and 2. To simulate the reset, two so-called "shadow layers", resp. 3 and 4, are added in A^*. A variable Ω is added on the layers 1 and 2 (even if not used on 2) to check the layer freshness without altering the values initially associated with their other variables.

The states of A are duplicated in A^* where they are annotated with the variables that were reset since the beginning of an iteration. These annotations are used in the construction to create the outgoing transitions. When the variable X^1 is consulted by a transition in A, the matching transitions in A^* are going to consult both X^1 and X^3 if X^1 was never reset before. However, if X^1 was reset, then only X^3 is consulted as the values associated with X^1 should have been deleted. When the variable X^1 is modified by a transition of A, the matching

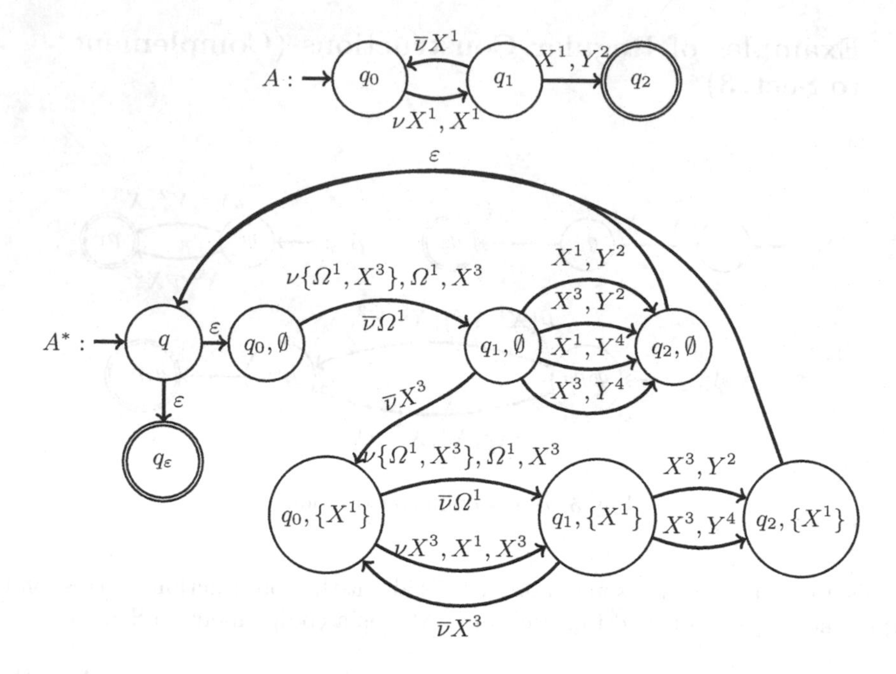

Fig. 6. Kleene star construction

transition in A^* will modify X^3 and it will also check if the value is fresh on layer 1, using Ω^1. However, if X^1 is supposed to have been reset earlier in the iteration, then the transition is duplicated to check if the letter is associated with X^1 instead, as the values it is associated with are supposed to be fresh.

References

1. Bartoletti, M.: Usage automata. In: Degano, P., Viganò, L. (eds.) ARSPA-WITS 2009. LNCS, vol. 5511, pp. 52–69. Springer, Heidelberg (2009). https://doi.org/10.1007/978-3-642-03459-6_4
2. Bertrand, C.: Reconnaissance de motifs dynamiques par automates temporisés à mémoire. (Matching of dynamic patterns with timed memory automata). Ph.D. thesis, University of Paris-Saclay, France (2020). https://tel.archives-ouvertes.fr/tel-03172600
3. Bertrand, C., Peschanski, F., Klaudel, H., Latapy, M.: Pattern matching in link streams: timed-automata with finite memory. Sci. Ann. Comput. Sci. **28**(2), 161–198 (2018). http://www.info.uaic.ro/bin/Annals/Article?v=XXVIII2&a=1
4. Björklund, H., Schwentick, T.: On notions of regularity for data languages. Theor. Comput. Sci. **411**(4), 702–715 (2010). https://doi.org/10.1016/j.tcs.2009.10.009. https://www.sciencedirect.com/science/article/pii/S0304397509007518. Fundamentals of Computation Theory
5. Bojanczyk, M., David, C., Muscholl, A., Schwentick, T., Segoufin, L.: Two-variable logic on data words. ACM Trans. Comput. Log. **12**(4), 27:1–27:26 (2011). https://doi.org/10.1145/1970398.1970403

6. Deharbe, A.: Analyse de ressources pour les systèmes concurrents dynamiques. Ph.D. thesis, Université Pierre et Marie Curie, France, September 2016. https://tel.archives-ouvertes.fr/tel-01523979
7. Deharbe, A., Peschanski, F.: The omniscient garbage collector: a resource analysis framework. In: 14th International Conference on Application of Concurrency to System Design, ACSD 2014, Tunis La Marsa, Tunisia, 23–27 June 2014, pp. 102–111. IEEE Computer Society (2014). https://doi.org/10.1109/ACSD.2014.18
8. Grigore, R., Tzevelekos, N.: History-register automata. Log. Methods Comput. Sci. **12**(1) (2016). https://doi.org/10.2168/LMCS-12(1:7)2016
9. Grumberg, O., Kupferman, O., Sheinvald, S.: Variable automata over infinite alphabets. In: Dediu, A.-H., Fernau, H., Martín-Vide, C. (eds.) LATA 2010. LNCS, vol. 6031, pp. 561–572. Springer, Heidelberg (2010). https://doi.org/10.1007/978-3-642-13089-2_47
10. Kaminski, M., Francez, N.: Finite-memory automata. Theor. Comput. Sci. **134**(2), 329–363 (1994). https://doi.org/10.1016/0304-3975(94)90242-9
11. Kaminski, M., Zeitlin, D.: Finite-memory automata with non-deterministic reassignment. Int. J. Found. Comput. Sci. **21**(5), 741–760 (2010). https://doi.org/10.1142/S0129054110007532
12. Kara, A.: Logics on data words: expressivity, satisfiability, model checking. Ph.D. thesis, Technical University of Dortmund, Germany (2016). http://hdl.handle.net/2003/35216
13. Libkin, L., Tan, T., Vrgoč, D.: Regular expressions for data words. J. Comput. Syst. Sci. **81**(7), 1278–1297 (2015). https://doi.org/10.1016/j.jcss.2015.03.005
14. Neven, F., Schwentick, T., Vianu, V.: Towards regular languages over infinite alphabets. In: Sgall, J., Pultr, A., Kolman, P. (eds.) MFCS 2001. LNCS, vol. 2136, pp. 560–572. Springer, Heidelberg (2001). https://doi.org/10.1007/3-540-44683-4_49
15. Tzevelekos, N.: Fresh-register automata. In: Ball, T., Sagiv, M. (eds.) Proceedings of the 38th ACM SIGPLAN-SIGACT Symposium on Principles of Programming Languages, POPL 2011, Austin, TX, USA, 26–28 January 2011, pp. 295–306. ACM (2011). https://doi.org/10.1145/1926385.1926420

Timed Models

Waiting Nets

Loïc Hélouët[1(✉)] and Pranay Agrawal[2]

[1] University Rennes, Inria, CNRS, IRISA, Rennes, France
loic.helouet@inria.fr
[2] ENS-Paris-Saclay, Paris-Saclay, France
pranay.agrawal@ens-paris-saclay.fr

Abstract. In Time Petri nets (TPNs), time and control are tightly connected: time measurement for a transition starts only when all resources needed to fire it are available. For many systems, one wants to start measuring time as soon as a part of the preset of a transition is filled, and fire it after some delay <u>and</u> when all needed resources are available. This paper considers an extension of TPN called *waiting nets* decoupling time measurement and control. Their semantics ignores clocks when upper bounds of intervals are reached but all resources needed to fire are not yet available. Firing of a transition is then allowed as soon as missing resources are available. It is known that extending bounded TPNs with stopwatches leads to undecidability. Our extension is weaker, and we show how to compute a finite state class graph for bounded waiting nets, yielding decidability of reachability and coverability. We then compare expressiveness of waiting nets with that of other models and show that they are strictly more expressive than TPNs.

1 Introduction

Time Petri nets (TPNs) are an interesting model to specify cyber-physical systems introduced in [22]. They allow for the specification of concurrent or sequential events, modeled as transitions occurrences, resources, time measurement, and urgency. In TPNs, time constraints are modeled by attaching an interval $[\alpha_t, \beta_t]$ to every transition t. If t has been enabled for at least α_t time units it *can* fire. If t has been enabled for β_t time units, it is *urgent*: time cannot elapse, and t *must* either fire or be disabled. Urgency is an important feature of TPNs, as it allows for the modeling of strict deadlines, but gives them a huge expressive power. In their full generality, TPNs are Turing powerful. A consequence is that most properties that are decidable for Petri Nets [15] (coverability [25], reachability [21], boundedness [25]...) are undecidable for TPNs. Yet, for the class of bounded TPNs, reachability [24] and coverability are decidable. The decision procedure relies on a symbolic representation of states with *state classes* and then on the definition of abstract runs as paths in a so-called state class graph [7,20].

There are many variants of Petri nets with time. An example is *timed Petri nets* (TaPN), where tokens have an age, and time constraints are attached to arcs

L. Bernardinello and L. Petrucci (Eds.): PETRI NETS 2022, LNCS 13288, pp. 67–89, 2022.
https://doi.org/10.1007/978-3-031-06653-5_4

Table 1. Decidability and complexity results for time(d) variants of Petri nets.

	Reachability	coverability	Boundedness
Time Petri Nets	Undecidable [19]	Undecidable [19]	Undecidable [19]
(bounded)	Decidable	Decidable	–
Timed Petri nets	Undecidable [27]	Decidable [1,16]	Decidable [16]
(bounded)	Decidable	Decidable	–
Restricted Urgency	Undecidable [2]	Decidable [2]	Decidable [2]
(bounded)	Decidable	Decidable	–
Stopwatch Petri nets	Undecidable [8]	Undecidable [8]	Undecidable [8]
(bounded)	Undecidable [8]	Undecidable [8]	–
TPNR	Undecidable [23]	Undecidable [23]	Undecidable [23]
(bounded)	Decidable [23]	Decidable [23]	–
Waiting Nets	Undecidable (Remark 1)	Undecidable (Remark 1)	Undecidable (Remark 1)
(bounded)	PSPACE-Complete (Theorem 2)	PSPACE-Complete (Theorem 2)	–

of the net. In TaPNs, a token whose age reaches the upper bound of constraints becomes useless. The semantics of TaPNs enjoys some monotonicity, and well-quasi-ordering techniques allow to solve coverability or boundedness problems [1, 26]. However, reachability remains undecidable [27]. We refer readers to [18] for a survey on TaPN and their verification. Without any notion of urgency, TaPN cannot model delay expiration. In [2], a model mixing TaPN and urgency is proposed, with decidable coverability, even for unbounded nets.

Working with bounded models is enough for many cyber-physical systems. However, bounded TPNs suffer another drawback: time measurement and control are too tightly connected. In TPNs, time is measured by starting a new clock for every transition that becomes enabled. By doing so, measuring a duration for a transition t starts only when all resources needed to fire t are available. Hence, one cannot stop and restart a clock, nor start measuring time while waiting for resources. To solve this problem, [8] equips bounded TPNs with stopwatches. Nets are extended with read arcs, and the understanding of a read arc from a place p to a transition t is that when p is filled, the clock attached to t is frozen. Extending bounded TPNs with stopwatches leads to undecidability of coverability, boundedness and reachability. This is not a surprise, as timed automata with stopwatches are already a highly undecidable model [10]. For similar reasons, time Petri nets with preemptable resources [9], where time progress depends on the availability of resources cannot be formally verified.

This paper considers *waiting nets*, a new extension of TPN that decouples time measurement and control. Waiting nets distinguish between enabling of a transition and enabling of its firing, which allows rules of the form "start measuring time for t as soon as p is filled, and fire t within $[\alpha, \beta]$ time units when p and q are filled". This model is strictly more expressive than TPN, as TPN are a simple syntactic restriction of waiting nets. Waiting nets allow clocks of enabled transitions to reach their upper bounds, and wait for missing control to fire. A former attempt called Timed Petri nets with Resets (TPNR) distinguishes some delayable transitions that can fire later than their upper bounds [23]. For bounded TPNR, reachability and TCTL model checking are decidable. However, delayable transitions are never urgent, and once delayed can only fire during a maximal step with another transition fired on time. Further, delayable transitions start measuring time as soon as their preset is filled, and hence do not allow decoupling of time and control as in waiting nets. As a second contribution, we show that the state class graphs of bounded waiting nets are finite, yielding decidability of reachability and coverability (which are PSPACE-complete). This is a particularly interesting result, as these properties are undecidable for *stopwatch Petri nets, even in the bounded case.* The Table 1 summarizes known decidability results for reachability, coverability and boundedness problems for time variants of Petri nets, including the new results for waiting nets proved in this paper. Our last contribution is a study of the expressiveness of waiting nets w.r.t timed language equivalence. Interestingly, the expressiveness of bounded waiting nets lays between that of bounded TPNs and timed automata. Due to space limitation, proofs in this paper are only sketched, but can be found in an extended version [17].

2 Preliminaries

We denote by $\mathbb{R}^{\geq 0}$ the set of non-negative real values, and by $\mathbb{Q}$ the set of rational numbers. A *rational interval* $[\alpha, \beta]$ is the set of values between a lower bound $\alpha \in \mathbb{Q}$ and an upper bound $\beta \in \mathbb{Q}$. We also consider intervals without upper bounds of the form $[\alpha, \infty)$, to define values that are greater than or equal to α.

A *clock* is a variable x taking values in $\mathbb{R}^{\geq 0}$. A variable x_t will be used to measure the time elapsed since transition t of a net was last newly enabled. Let X be a set of clocks. A *valuation* for X is a map $v : X \rightarrow \mathbb{R}^{\geq 0}$ that associates a positive or zero real value $v(x)$ to every variable $x \in X$. Intervals alone are not sufficient to define the domains of clock valuations met with TPNs and timed automata. An *atomic constraint* on X is an inequality of the form $a \leq x$, $x \leq b$, $a \leq x - y$ or $x - y \leq b$ where $a, b \in \mathbb{Q}$ and $x, y \in X$. A *constraint* is a conjunction of atomic constraints. We denote by $Cons(X)$ the set of constraints over clocks in X. We will say that a valuation v satisfies a constraint ϕ, and write $v \models \phi$ iff replacing x by $v(x)$ in ϕ yields a tautology. A constraint ϕ is *satisfiable* iff there exists a valuation v for X such that $v \models \phi$. Constraints over real-valued variables can be encoded with Difference bound Matrices (DBMs) and their satisfiability checked in $O(n^3)$ [14]. The *domain* specified by a constraint ϕ is the (possibly infinite) set of valuations that satisfy ϕ.

Given an alphabet Σ, a *timed word* is an element of $(\Sigma \times \mathbb{R}^+)^*$ of the form $w = (\sigma_1, d_1)(\sigma_2, d_2)\ldots$ such that $d_i \leq d_{i+1}$. A timed language is a set of timed words. Timed automata [4] are frequently used to recognize timed languages.

Definition 1 (Timed Automaton). *A* Timed Automaton $\mathcal{A}$ *is a tuple* $\mathcal{A} = (L, \ell_0, X, \Sigma, Inv, E, F)$, *where* L *is a set of locations,* $\ell_0 \in L$ *is the initial location,* X *is a set of clocks,* Σ *is an alphabet,* $Inv : L \rightarrow Cons(X)$ *is a map associating an invariant to every location. The set of states* $F \subseteq L$ *is a set of final locations, and* E *is a set of edges. Every edge is of the form* $(\ell, g, \sigma, R, \ell') \in L \times Cons(X) \times \Sigma \times 2^X \times L$.

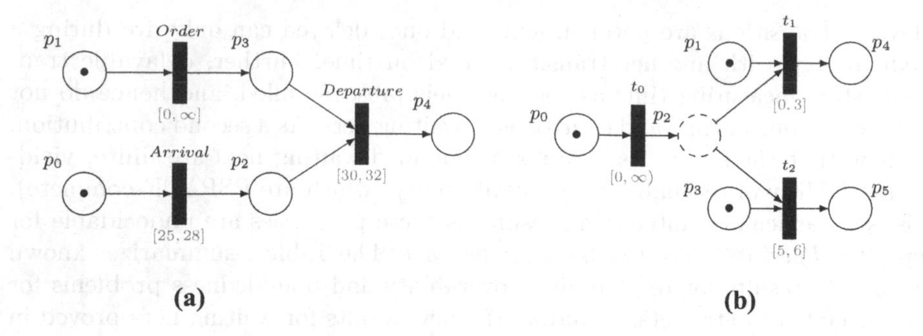

Fig. 1. A simple TPN *a*) and a simple waiting net *b*)

Intuitively, the semantics of a timed automaton allows elapsing time in a location ℓ (in which case clocks valuations grow uniformly), or firing a discrete transition $(\ell, g, \sigma, R, \ell')$ from location ℓ with clock valuation v if v satisfies guard g, and the valuation v' obtained by resetting all clocks in R to 0 satisfies $Inv(\ell')$. One can notice that invariants can prevent firing a transition. Every run of a timed automaton starts from (ℓ_0, v_0), where v_0 is the valuation that assigns value 0 to every clock in X. For completeness, we recall the semantics of timed automata in appendix. The timed language recognized by $\mathcal{A}$ is denoted $\mathcal{L}(\mathcal{A})$.

In the rest of the paper, we will denote by TA the class of timed automata. We will be in particular interested by the subclass $TA(\leq, \geq)$ in which guards are conjunctions of atomic constraints of the form $x \geq c$ and invariants are conjunctions of atomic constraints of the form $x \leq c$. Several translations from TPNs to TAs have been proposed, and in particular, the solution of [20] uses the state class graph of a TPN to build a time-bisimilar timed automaton in class $TA(\leq, \geq)$. This shows that one needs not the whole expressive power of timed automata to encode timed languages recognized by TPNs.

3 Waiting Nets

TPN are a powerful model: they can be used to encode a two-counter machine, and can hence simulate the semantics of many other formal models. A counterpart to this expressiveness is that most problems (reachability, coverability, verification of temporal logics...) are undecidable. Decidability is easily recovered when considering the class of bounded TPNs. Indeed, for bounded TPNs, one can compute a finite symbolic model called a state class graph, in which timing information is symbolically represented by firing domains. For many applications, working with bounded resources is sufficient. However, TPN do not distinguish between places that represent control (the "state" of a system), and those that represent resources: transitions are enabled when all places in their preset are filled. A consequence is that one cannot measure time spent in a control state, when some resources are missing.

Consider the example of Fig. 1, that represents an arrival of a train followed by a departure. The arrival in a station is modeled by transition *Arrival*, that should occur between 25 and 28 minutes after beginning of a run of the net. The station is modeled by place p_2, and the departure of the train by transition *Departure*. A train can leave a station only if a departure order has been sent, which is modeled by transition *Order*. The time constraint attached to *Departure* is an interval of the form $[30, 32]$. Assume that one wants to implement a scenario of the form "the train leaves the station between 30 and 32 min after its arrival if it has received a departure order". The TPN of Fig. 1-a) does not implement this scenario, but rather behaviors in which the train leaves the station between 30 and 32 min after the instant when it is in station **and** a departure order is received. This means that a train may spend more that 32 min in station, if the order is not released first. Similarly, Timed Petri nets, that do not have a notion of urgency, cannot encode this scenario where a transition has to fire after 32 time units.

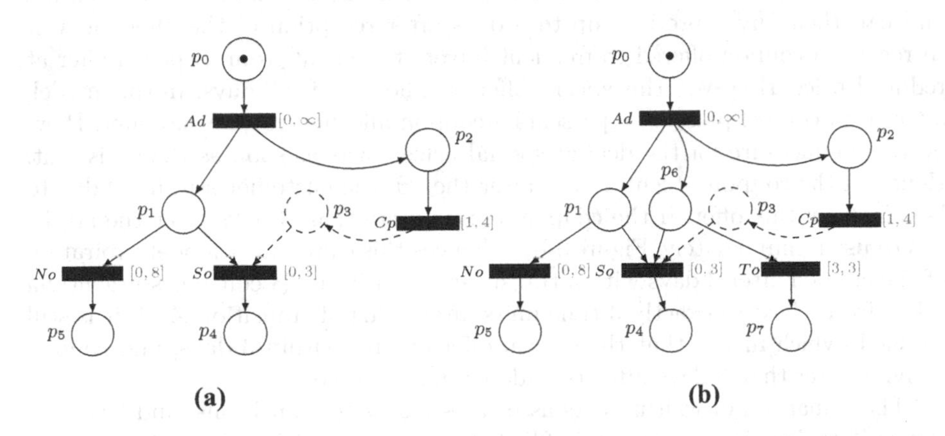

Fig. 2. a) Decoupled time and control in a waiting net. b) ... with a timeout transition.

We propose an extension of TPNs called *Waiting nets* (WTPN for short), that decouples control and resources during time measurement. We consider two types of places: *standard* places, and *control* places, with the following functions: Time measurement for a transition t starts as soon as t has enough tokens in the standard places of its preset. Then, t can fire if its clock value lays in its timing interval, and if it has enough tokens in the control places of its preset.

Definition 2. *A* waiting net *is a tuple* $\mathcal{W} = (P, C, T, {}^\bullet(), ()^\bullet, \alpha, \beta, \lambda, (M_0.N_0))$, *where*

- *P is a finite set of standard places, C is finite set of control places, such that* $P \cup C \neq \emptyset$ *and* $P \cap C = \emptyset$. *A* marking $M.N$ *is a pair of maps* $M : P \to \mathbb{N}$, $N : C \to \mathbb{N}$ *that associate an integral number of tokens respectively to standard and control places.*

- *T is a finite set of transitions. Every $t \in T$ has a label $\lambda(t)$,*
- *${}^\bullet() \in (\mathbb{N}^{P \cup C})^T$ is the backward incidence function, $()^\bullet \in (\mathbb{N}^{P \cup C})^T$ is the forward incidence function,*
- *$(M_0.N_0) \in \mathbb{N}^{P \cup C}$ is the initial marking of the net,*
- *$\alpha : T \to \mathbb{Q}^+$ and $\beta : T \to \mathbb{Q}^+ \cup \infty$ are functions giving for each transition respectively its earliest and latest firing times ($\alpha(t) \leq \beta(t)$).*

Labeling map λ can be injective or not. To differentiate standard and control places in the preset of a transition, we will denote by ${}^\circ(t)$ the restriction of ${}^\bullet(t)$ to standard places, and by ${}^c()$ the restriction of ${}^\bullet()$ to control places. We will write $M(p) = k$ (resp. $N(c) = k$) to denote the fact that standard place $p \in P$ (resp. control place $c \in C$) contains k tokens. Given two markings $M.N$ and $M'.N'$ we will say that $M.N$ is greater than $M'.N'$ and write $M.N \geq M'.N'$ iff $\forall p \in P, M(p) \geq M'(p)$ and $\forall c \in C, N(c) \geq N'(C)$.

Figure 2-a) is a waiting net modelling an online sale offer, with limited duration. Control places are represented with dashed lines. A client receives an ad, and can then buy a product up to 8 days after reception of the offer, or wait to receive a coupon offered to frequent buyers to benefit from a special offer at reduced price. However, this special offer is valid only for 3 days. In this model, a token in control place p_3 represents a coupon allowing the special offer. However, time measure for the deal at special price starts as soon as the ad is sent. Hence, if the coupon is sent 2 days after the ad, the customer still has 1 day to benefit from this offer. If the coupon arrives more than 3 days after the ad, he has to use it immediately. Figure 2-b) enhances this example to model expiration of the coupon after 3 days with a transition. Transition T_O consumes urgently a token from place p_6 exactly 3 time units after firing of transition *Ad* if it is still enabled, which means that the special offer expires within 3 days, and coupon arriving later that 3 days after the add cannot be used.

The semantics of waiting nets associates clocks to transitions, and lets time elapse if their standard preset is filled. It allows firing of a transition t if the standard and the control preset of t is filled.

Definition 3. *(Enabled, fully enabled, waiting transitions)*

- *A transition t is* enabled *in marking $M.N$ iff $M \geq {}^\circ(t)$ (for every standard place p in the preset of t, $M(p) \geq {}^\circ(t)(p)$). We denote by* $\mathsf{Enabled}(M)$ *the set of transitions which are enabled from marking M, i.e.* $\mathsf{Enabled}(M) := \{t \mid M \geq {}^\circ(t)\}$
- *A transition t is* fully enabled *in $M.N$ iff, for every place in the preset of t, $M.N(p) \geq {}^\bullet(t,p)$.* $\mathsf{FullyEnabled}(M.N)$ *is the set of transitions which are fully enabled in marking $M.N$, i.e.* $\mathsf{FullyEnabled}(M.N) := \{t \mid M.N \geq {}^\bullet t\}$
- *A transition t is* waiting *in $M.N$ iff $t \in$* $\mathsf{Enabled}(M) \setminus \mathsf{FullyEnabled}(M.N)$ *(t is enabled, but is still waiting for the control part of its preset). We denote by* $\mathsf{Waiting}(M.N)$ *the set of waiting transitions.*

Obviously, $\mathsf{FullyEnabled}(M.N) \subseteq \mathsf{Enabled}(M)$. For every enabled transition t, there is a clock x_t that measures for how long t has been enabled. For every fully enabled transition t, t can fire when $x_t \in [\alpha(t), \beta(t)]$. We adopt an *urgent* semantics, i.e. when a transition is fully enabled and $x_t = \beta(t)$, then this transition, or another one enabled at this precise instant *has to* fire without letting time elapse. Firing of a transition t from marking $M.N$ consumes tokens from all places in $^\bullet(t)$ and produces tokens in all places of $(t)^\bullet$. A consequence of this token movement is that some transitions are disabled, and some other transitions become enabled after firing of t.

Definition 4 (Transition Firing). *Firing of a transition t from marking $M.N$ is done in two steps. It first computes an intermediate marking $M''.N'' = M.N - {}^\bullet(t)$ obtained by removing tokens consumed by the transition from its preset. Then, a new marking $M'.N' = M''.N'' + (t)^\bullet$ is computed. We will write $M.N \xrightarrow{t} M'.N'$ whenever Firing of t from $M.N$ produces marking $M'.N'$ A transition t_i is* newly enabled *after firing of t from $M.N$ iff it is enabled in $M'.N'$, and either it is not enabled in $M''.N''$, or it is a new occurrence of t. We denote by $\uparrow \mathsf{enabled}(M.N, t)$ the set of transitions newly enabled after firing t from marking $M.N$.*

$$\uparrow \mathsf{enabled}(M.N,t) := \{t_i \in T \mid {}^\bullet(t_i) \leq M.N - {}^\bullet(t) + (t)^\bullet \wedge ((t_i = t) \vee ({}^\bullet(t_i) \geq M.N - {}^\bullet(t))\}$$

As explained informally with the examples of Fig. 2, the semantics of waiting nets allows transitions firing when some time constraints on the duration of enabling are met. Hence, a proper notion of state for a waiting net has to consider both place contents and time elapsed. This is captured by the notion of *configuration*. In configurations, time is measured by attaching a clock to every enabled transition. To simplify notations, we define valuations of clocks on a set $X_T = \{x_t \mid t \in T\}$ and write $x_t = \bot$ if $t \notin$ *enabled*(M). To be consistent, for every value $r \in \mathbb{R}$, we set $\bot + r := \bot$.

Definition 5 (Configuration). *A* Configuration *of a waiting net is a pair $(M.N, v)$ where $M.N$ is a marking and v is a valuation of clocks in X_T. The* initial configuration *of a net is a pair $(M_0.N_0, v_0)$, where $v_0(x_t) = 0$ if $t \in \mathsf{enabled}(M_0)$ and $v_0(x_t) = \bot$ otherwise. A transition t is* firable *from configuration $(M.N, v)$ iff it is fully enabled, and $v(x_t) \in [\alpha(t), \beta(t)]$.*

The semantics of waiting nets is defined in terms of *timed* or *discrete* moves from one configuration to the next one. Timed moves increase the value of clocks attached to enabled transitions (when time elapsing is allowed) while discrete moves are transitions firings that reset clocks of newly enabled transitions.

$$\frac{\begin{array}{l}\forall t \in \mathsf{Waiting}(M.N),\\ \quad v'(x_t) = \min(\beta(t), v(x_t) + d)\\ \forall t \in \mathsf{FullyEnabled}(M.N),\\ \quad v(x_t) + d \leq \beta(t)\\ \quad \text{and } v'(x_t) = v(x_t) + d\\ \forall t \in T \setminus \mathsf{enabled}(M), v'(x_t) = \bot\end{array}}{(M.N, v) \xrightarrow{d} (M.N, v')}
\qquad
\frac{\begin{array}{c}M.N \geq {}^\bullet(t)\\ M'.N' = M.N - {}^\bullet(t) + (t)^\bullet\\ \alpha(t) \leq v(t) \leq \beta(t)\\ \forall t_i \in T, v'(t_i) = \begin{cases}0 \text{ if } t_i \in \uparrow \mathsf{enabled}(M.N, t)\\ \bot \text{ if } t_i \notin \mathsf{enabled}(M)\\ v(t_i) \text{ otherwise}\end{cases}\end{array}}{(M.N, v) \xrightarrow{t} (M'.N', v')}$$

Timed moves let $d \in \mathbb{R}^{\geq 0}$ time units elapse, but leave markings unchanged. We adopt an *urgent semantics* that considers differently *fully enabled* transitions and *waiting* transitions. If t is a fully enabled transitions then, t *allows* elapsing of d time units from $(M.N, v)$ iff $v(t) + d \leq \beta(t)$. The new valuation reached after elapsing d time units is $v(t) + d$. If we already have $v(t) = \beta(t)$, then t does not allow time elapsing. We say that firing of t is *urgent*, that is t *has to* be fired or disabled by the firing of another transition before elapsing time. If $v(t) + d > \beta(t)$ then t becomes urgent before d time units, and letting a duration d elapse from $(M.N, v)$ is forbidden. Urgency does not apply to waiting transitions, which can let an arbitrary amount of time elapse when at least one control places in their preset is not filled. Now, as we model the fact that an event has been enabled for a sufficient duration, we let the value of clocks attached increase up to the upper bound allowed by their time interval, and then freeze these clocks. So, for a waiting transition, we have $v'(t) = \min(\beta(t), v(t) + d)$. We will write $v \oplus d$ to denote the valuation of clocks reached after elapsing d time units from valuation v. A timed move of duration d from configuration $(M.N, v)$ to $(M'.N', v')$ is denoted $(M.N, v) \xrightarrow{d} (M'.N', v')$. As one can expect, waiting nets enjoy time additivity (i.e. $(M.N, v) \xrightarrow{d_1} (M.N, v_1) \xrightarrow{d_2} (M.N, v_2)$ implies that $(M.N, v) \xrightarrow{d_1+d_2} (M.N, v_2)$, and continuity, i.e. if $(M.N, v) \xrightarrow{d} (M.N, v')$, then for every $d' < d$ $(M.N, v) \xrightarrow{d'} (M.N, v'')$.

Discrete moves fire transitions that meet their time constraints, and reset clocks attached to transitions newly enabled by token moves. A discrete move relation from configuration $(M.N, v)$ to $(M'.N', v')$ via transition $t_i \in T$ is denoted $(M.N, v) \xrightarrow{t_i} (M'.N', v')$. Overall, the semantics of a waiting net $\mathcal{W}$ is a timed transition system (TTS) with initial state $q_0 = (M_0.N_0, v_0)$ and which transition relation follows the time and discrete move semantics rules.

Definition 6. *A* run *of a Waiting net $\mathcal{W}$ from a configuration $(M.N, v)$ is a sequence $\rho = (M.N, v) \xrightarrow{e_1} (M_1.N_1, v_1) \xrightarrow{e_2} (M_2.N_2, v_2) \cdots \xrightarrow{e_k} (M_k.N_k, v_k)$, where every e_i is either a duration $d_i \in \mathbb{R}^{\geq 0}$, or a transition $t_i \in T$, and every $(M_{i-1}.N_{i-1}, v_{i-1}) \xrightarrow{e_i} (M_i.N_i, v_i)$ is a legal move of $\mathcal{W}$.*

We denote by $\mathsf{Runs}(\mathcal{W})$ the set of runs of $\mathcal{W}$. A marking $M.N$ is *reachable* iff there exists a run from $(M_0.N_0, v_0)$ to a configuration $(M.N, v)$ for some v. $M.N$ is *coverable* iff there exists a reachable marking $M'.N' \geq M.N$. We will say that a waiting net is *bounded* iff there exists an integer K such that, for every reachable marking $M.N$ and every place $p \in P$ and $p' \in C$, we have $M(p) \leq K$ and $N(p') \leq K$. Given two markings $M_0.N_0$ and $M.N$ the *reachability* problem asks whether $M.N$ is reachable from $(M_0.N_0, v_0)$, and the *coverability* problem whether there exists a marking $M'.N' \geq M.N$ reachable from $(M_0.N_0, v_0)$.

Remark 1. A waiting net with an empty set of control places is a TPN. Hence, waiting nets inherit all undecidability results of TPNs: reachability, coverability, and boundeness are undecidable in general for unbounded waiting nets.

Given a run $\rho = (M_0.N_0, v_0) \xrightarrow{e_1} (M_1.N_1, v_1) \xrightarrow{e_2} (M_2.N_2, v_2) \cdots$, the timed word associated with ρ is the word $w_\rho = (t_1, d_1) \cdot (t_2, d_2) \cdots$ where the sequence $t_1 \cdot t_2 \ldots$ is the projection of $e_1 \cdot e_2 \cdots$ on T, and for every (t_i, d_i) such that t_i appears on move $(M_{k-1}.N_{k-1}, v_{k-1}) \xrightarrow{e_k} (M_k.N_k, v_k)$, d_i is the sum of all durations in $e_1 \ldots e_{k-1}$. The sequence $t_1.t_2 \ldots$ is called the *untiming* of w_ρ. The *timed language* of a waiting net is the set of timed words $\mathcal{L}(\mathcal{W}) = \{w_\rho \mid \rho \in \mathsf{Runs}(\mathcal{W})\}$. Notice that unlike in timed automata and unlike in the models proposed in [6], we do not define accepting conditions for runs of timed words, and hence consider that the timed language of a net is prefix closed. The *untimed language* of a waiting net $\mathcal{W}$ is the language $\mathcal{L}^U(\mathcal{W}) = \{w \in T^* \mid \exists w_\rho \in \mathcal{L}(\mathcal{W}), w \text{ is the untiming of } w_\rho\}$. To simplify notations, we will consider runs alternating timed and discrete moves. This results in no loss of generality, since durations of consecutive timed moves can be summed up, and a sequence of two discrete move can be seen as a sequence of transitions with 0 delays between discrete moves. In the rest of the paper, we will write $(M.N, v) \xrightarrow{(d,t)} (M'.N', v')$ to denote the sequence of moves $(M.N, v) \xrightarrow{d} (M.N, v \oplus d) \xrightarrow{t} (M'.N', v')$.

Let us illustrate definitions with the example in Fig. 2-a). In this net, we have $P = \{p_0, p_1, p_2, p_4, p_5\}$, $C = \{p_3\}$, $T = \{Ad, No, So, Cp\}$, $\alpha(Ad) = \alpha(No) = \alpha(So) = 0$, $\alpha(Cp) = 1$, $\beta(Ad) = \infty$, $\beta(No) = 8$, $\beta(So) = 3$, $\beta(Cp) = 4$. We also have ${}^{\circ}(So) = p_1$ and ${}^{c}(So) = p_3$, $(So)^{\bullet} = p_4$ (we let the reader infer ${}^{\bullet}()$ and $()^{\bullet}$ for other transitions). The net starts in an initial configuration $(M_0.N_0, v_0)$ where $M_0(p_0) = 1$ and $M_0(p_i) = 0$ for all other places in P, $N_0(p_3) = 0$, $v_0(Ad) = 0$ and $v_0(t) = \perp$ for all other transitions in T. From this configuration, one can let an arbitrary duration d_0 elapse before firing transition Ad, leading to a configuration $M_1.N_0$ with $M_1(p_1) = M_1(p_2) = 1$, and $v_1(Cp) = v_1(No) = v_1(So) = 0$. Then, one can let a duration smaller than 4 elapse and fire No, or let a duration between 1 and 4 time units elapse and fire Cp. Notice that the net cannot let more than 4 time units elapse before taking a discrete move, as firing of Cp becomes urgent 4 time units after enabling of the transition. Let us assume that Cp is fired after elapsing 2.3 time units. This leads to a new configuration $(M_2.N_2, v_2)$ where $M_2(p_1) = M_2(p_2) = 1, N_2(p_3) = 1$, $v_2(No) = v_2(S_o) = 2.3$. In this net, firing of So can only occur after firing of Cp, but yet time measurement starts for So as soon as ${}^{\circ}(So)$ is filled, i.e. immediately after firing of Ad. This example is rather simple: the net is acyclic, and each transition is enabled/disabled only once. One can rapidly see that the only markings reachable are $M_0.N_0$, $M_1.N_0$, $M_2.N_2$ described above, plus two additional markings $M_3.N_0$ where $M_3(p_5) = 1$ and $M_4.N_0$ where $M_4(p4) = 1$. A normal order can be sent at most 8 time units after advertising, a special order must be sent at most 3 time units after advertising if a coupon was received, etc. We give a more complex example in [17].

4 Reachability

In a configuration $(M.N, v)$ of a waiting net $\mathcal{W}$, v assigns real values to clocks. The timed transition system giving the semantics of a waiting net is hence in general infinite, even when $\mathcal{W}$ is bounded. For TPNs, the set of reachable valuations can be abstracted to get a finite set of domains, to build a *state class graph* [7]. In this section, we build similar graphs for waiting nets. We also prove that the set of domains in these graphs is always finite, and use this result to show that reachability and coverability are decidable for bounded waiting nets.

Let t be a transition with $\alpha(t) = 3$ and $\beta(t) = 12$, and assume that t has been enabled for 1.6 time units. According to the semantics of WPNs, $v(x_t) = 1.6$, and t cannot fire yet, as $x_t < \alpha(t)$. Transition t can fire only after a certain duration θ_t such that $1.4 \leq \theta_t \leq 10.4$. Similar constraints hold for all enabled transitions. We will show later that these constraint are not only upper and lower bounds on $\theta_t's$, but also constraints of the form $\theta_i - \theta_j \leq c_{ij}$.

Definition 7 (State Class, Domain). *A* state class *of a waiting net $\mathcal{W}$ is a pair $(M.N, D)$, where $M.N$ is a marking of $\mathcal{W}$ and D is a set of inequalities called* firing domain. *The inequalities in D are of two types:*

$$\begin{cases} a_i \leq \theta_i \leq b_i, & \textit{where } a_i, b_i \in \mathbb{Q}^+ \textit{ and } t_i \in \mathsf{Enabled}(M) \\ \theta_j - \theta_k \leq c_{jk}. & \textit{where } \forall j, k\ j \neq k \textit{ and } t_j, t_k \in \mathsf{Enabled}(M). \end{cases}$$

A variable θ_i in a firing domain D over variables $\theta_1, \ldots, \theta_m$ represents the time that can elapse before firing transition t_i if t_i is fully enabled, and the time that can elapse before the clock attached to t_i reaches the upper bound $\beta(t_i)$ if t_i is waiting. Hence, if a transition is fully enabled, and $a_i \leq \theta_i \leq b_i$, then t_i cannot fire before a_i time units, and cannot let more than b_i time units elapse, because it becomes urgent and has to fire or be disabled before b_i time units. Now, maintaining an interval for values of $\theta_i's$ is not sufficient. Allowing a transition t_i to fire means that no other transition t_j becomes urgent before firing of t_i, i.e. that adding constraint $\theta_i \leq \theta_j$ for every fully enabled transition t_j still allows to find a possible value for θ_i. Then, assuming that t_i fires, the new firing domain D' over variables $\theta'_1, \ldots, \theta'_q$ will constrain the possible values of $\theta'_j s$ for all transitions t_j that remain enabled after firing of t_i. As time progresses, we have $\theta'_j = \theta_j - \theta_i$, which gives rise to diagonal constraint of the form $\theta'_j - \theta'_k \leq c_{jk}$ after elimination of variables appearing in D.

A firing domain D defines a set of possible values for $\theta'_i s$. We denote by $[\![D]\!]$ the set of solutions for a firing domain D. Now, the way to define a set of solutions is not unique. We will say that D_1, D_2 are equivalent, denoted $D_1 \equiv D_2$ iff $[\![D_1]\!] = [\![D_2]\!]$. A set of solutions $[\![D]\!]$ is hence not uniquely defined, but fortunately, a unique representation called a *canonical form* exists.

Definition 8 (Canonical Form). *The canonical form of a firing domain D is the unique domain* $D^* = \begin{cases} a^*_i \leq \theta_i \leq b^*_i \\ \theta_j - \theta_k \leq c^*_{jk}. \end{cases}$, *where* $a^*_i = Inf(\theta_i)$, $b^*_i = Sup(\theta_i)$, *and* $c^*_{jk} = Sup(\theta_j - \theta_k)$

The canonical form D^* is the minimal set of constraints defining $[\![D]\!]$. If two sets of constraints are equivalent then they have the same canonical form. The constraints we consider are of the form $K_1 \leq x \leq K_2$ and $K_1 \leq x - y \leq K_2$, where K_1, K_2 are rational values. This type of constraints can be easily encoded by *Difference Bound Matrices* [14]. Checking satisfiability of a domain D, or computing a canonical form D^* can be done in $O(n^3)$, where n is the number of variables (see [17] for details and [5] for a survey on DBMs). Syntactically, state classes and canonical forms of waiting nets have the same definition as those of TPNs: the fact that a transition is waiting or fully enabled does not affect the representation of constraints. Now, there is a major difference between state graphs of TPNs and those of waiting nets: for waiting nets, the maximal duration that can elapse in a state class in contrained by fully enabled transitions only. However, at the same time, when elapsing time, one has to adapt contraints attached to waiting transitions which clocks have reached their upper bound. In some sense, for waiting transitions, variable θ_t represents a *time to upper bound of intervals* rather than a *time to fire*. When computing the effect of firing a fully enabled transition, one has to consider which waiting transitions have reached their upper bounds. A consequence is that state class graphs of waiting nets are not deterministic, as a class has several successors via the same transition.

Following the semantics of Sect. 3, a transition t_i can fire from a domain D if one can find a value for θ_i that does not violate urgency of other fully enabled transitions. However, the upper bound of waiting transitions should not prevent t_i from firing. To get rid of this upper bound, we can use the notion of *projection*.

Definition 9 (Projection). *Let D be a firing domain with variables a_i, b_i, c_{jk} set as in Definition 7. The* projection *of D on its fully enabled transitions is a domain*

$$\begin{aligned} D_{|full} = &\{a_i \leq \theta_i \leq b_i \mid t_i \in \mathsf{FullyEnabled}(M.N)\} \\ &\cup \{a_i \leq \theta_i \leq \infty, \mid t_i \in \mathsf{Waiting}(M.N)\} \\ &\cup \{\theta_j - \theta_k \leq c_{jk} \in D \mid t_j, t_k \in \mathsf{FullyEnabled}(M.N)\}. \end{aligned}$$

A transition t_i can fire from a configuration $(M.N, v)$ iff it is fully enabled and $v(t_i) \in [\alpha(t_i), \beta(t_i)]$. Hence, from configuration $(M.N, v)$, firing of t_i is one of the next discrete moves iff there exists a duration θ_i such that t_i can fire from $(M.N, v + \theta_i)$, i.e., after letting duration θ_i elapse, and no other transition becomes urgent before θ_i time units. We say that t_i is *firable* from a state class $(M.N, D)$ iff $M.N \geq {}^\bullet(t_i)$ and $D_{|full} \cup \{\theta_i \leq \theta_j \mid t_j \in \mathsf{FullyEnabled}(M.N)\}$ is satisfiable. So, t_i can be the next transition fired iff there exists a value θ_j greater than or equal to θ_i that does not exceed b_j for every fully enabled transition t_j.

The construction of the set of reachable state classes of a waiting net is an inductive procedure. Originally, a waiting net starts in a configuration $(M_0.N_0, v_0)$, so the initial state class of our system is (M_0, D_0), where $D_0 = \{\alpha(t_i) \leq \theta_i \leq \beta(t_i) \mid t_i \in \mathsf{Enabled}(M_0.N_0)\}$. Then, for every state class $(M.N, D)$, and every transition t firable from $(M.N, D)$, we compute all possible successors $(M'.N', D')$ reachable after firing of t. Note that we only need to consider $t \in \mathsf{FullyEnabled}(M.N)$, as t can fire only when $N > {}^{\circ}(t)$. Computing $M'.N'$ follows the usual firing rule of a Petri net: $M'.N' = M.N - {}^\bullet(t) + (t)^\bullet$ and we can hence also compute $\uparrow \mathsf{enabled}(M.N, t)$, $\mathsf{enabled}(M'.N')$ and $\mathsf{FullyEnabled}(M'.N')$. It remains to show the effect of transitions firing on domains to compute all possible successors of a class. Firing a transition t from $(M.N, D)$ propagates constraints of the firing domain D on variables attached to transitions that remain enabled. Variables associated to newly enabled transitions only have to meet lower and upper bounds on their firing times. We can now show that for Waiting nets, the set of successors of a state class is finite and can be effectively computed despite waiting transitions and non-determinism.

Consider the waiting net of Fig. 1-b. This net starts in a configuration $C_0 = (M_0.N_0, v_0)$ with $M_0(p_0) = M_0(p_1) = M_0(p_3) = 1$ $M_0(p) = 0$ for every other place, and $N_0(p_2) = 0$. From this configuration, one can let an arbitrary amount of time $\delta \in \mathbb{R}^{\geq 0}$ elapse. If $0 \leq \delta < 3$, then the value of clock x_1 is still smaller than the upper bound $\beta(t_1) = 3$. Then, if t_0 fires from $C_0' = (M_0.N_0, v_0 + \delta)$, the net reaches a new configuration $C_1 = (M_1.N_1, v_1)$ where $M_1(p_1) = M_1(p_3) = 1$, $M_1(p) = 0$ for every other place, and $N_0(p_2) = 1$. We have $v_1(x_0) = 0, v_1(x_1) = v_1(x_2) = \delta$. One can still wait before firing t_1 in configuration, i.e., t_1 is not urgent and can fire immediately of within a duration $3 - \delta$. Now, if $3 \leq \delta < 5$, then $v_1(x_1) = 3, v_1(x_2) < 5$ so transition t_1 is urgent and must fire, and transition t_2 still has to wait before firing. Hence, choosing $3 \leq \delta < 5$ forces to fire t_1 immediately after t_0. Conversely, if $\delta \geq 5$ then after firing t_0, the net is in configuration $C_2 = (M_1.N_1, v_2)$ where $v_2(x_1) = 3$ and $v_2(x_2) \in [5, 6]$, forcing t_1 or t_2 to fire immediately without elapsing time. This example shows that the time elapsed in a configuration has to be considered when computing successors of a state class. We have to consider whether the upper bound of a waiting transition has been reached or not, and hence to differentiate several cases when firing a single transition t. Fortunately, these cases are finite, and depend only on upper bounds attached to waiting transitions by domain D.

Definition 10 (Upper Bounds Ordering). *Let $M.N$ be a marking, D be a firing domain with constraints of the form $a_i \leq \theta_i \leq b_i$. Let $B_{M.N,D} = \{b_i \mid t_i \in \mathsf{enabled}(M)\}$. We can order bounds in $B_{M.N,D}$, and define bnd_i as the ith bound in $B_{M.N,D}$. We also define $bnd_0 = 0$ and $bnd_{|B_{M.N,D}|+1} = \infty$.*

Consider a transition t_f firable from $C = (M.N, D)$. This means that there is a way to choose a delay θ_f that does not violate urgency of all other transitions. We use $B_{M.N,D}$ to partition the set of possible values for delay θ_f in a finite set of intervals, and find which transitions reach their upper bound when θ_f belongs to an interval. Recall that $\theta_f \leq \theta_j$ for every fully enabled transition t_j. This means that when considering that t_f fires after a delay θ_f such that $bnd_i \leq \theta_f \leq bnd_{i+1}$, as D also gives a constraint of the form $a_f \leq \theta_f \leq b_f$, considering an interval such that bnd_i is greater than $\min\{b_j \in B_{M.N,D} \mid t_j \in \mathsf{FullyEnabled}(M.N)\}$ or smaller than a_f leads to inconsistency of constraint $D_{|full} \cup \bigwedge_{t_j \in \mathsf{FullEnabled}(M.N)} \theta_f \leq \theta_j \wedge bnd_i \leq \theta_f \leq bnd_{i+1}$. We denote by $B^{t_f}_{M.N,D}$ the set of bounds $B_{M.N,D}$ pruned out from these inconsistent bound values. Now, choosing a particular interval $[bnd_i, bnd_{i+1}]$ for the possible values in θ_f indicates for which waiting transitions $t_1, \ldots t_k$ the clocks $x_{t_1}, \ldots x_{t_k}$ measuring time elapsed since enabling has reached upper bounds $\beta(t_1), \ldots \beta(t_k)$. The values of these clocks become irrelevant, and hence the corresponding θ_i's have to be eliminated from the domains.

Definition 11 (Time progress (to the next bound)). *Let $M.N$ be a marking, D be a firing domain, and $b = \min B_{M.N,D}$ be the smallest upper bound for enabled transitions. The domain reached after progressing time to bound b is the domain D' obtained by:*

- *replacing every variable θ_i by expression $\theta'_i - b$*
- *eliminating every θ'_k whose upper bound is b,*
- *computing the normal form for the result and renaming all θ'_i to θ_i*

Progressing time to the next upper bound allows to remove variables related to waiting transitions whose clocks have reached their upper bounds from a firing domain. We call these transitions *timed-out transitions.* For a transition $t_k \in$ *waiting*$(M.N)$ if $v(x_{t_k}) = \beta(t_k)$, variable θ_k, that represents the time needed to reach the upper bound of the interval is not meaningful any more: either t_k gets disabled in the future, or is fired with $\theta_k = 0$. So the only information to remember is that t_k will be urgent as soon as it becomes fully enabled.

Definition 12 (Successors). *A* successor *of a class $C = (M.N, D)$ after firing of a transition t_f is a class $C' = (M'.N', D')$ such that $M'.N'$ is the marking obtained after firing t_f from $M.N$, and D' is a firing domain reached after firing t_f in some interval $[b_r, b_{r+1}]$ with b_r, b_{r+1} consecutive in $B^{t_f}_{M.N,D}$.*

Given C and a firable transition t_f, we can compute the set $\mathsf{Post}(C, t_f)$ of successors of C, i.e. $\mathsf{Post}(C, t_f) := \{(M'.N', \mathsf{next}_r(D, t_f)) \mid b_r \in B^{t_f}_{M.N,D} \cup \{0\}\}$. The next marking is the same for every successor and is $M'.N' = M.N - {}^{\bullet}t_f + t_f^{\bullet}$. We then compute $\mathsf{next}_r(D, t_f)$ as follows:

1) Time progress: We successively progress time from D to bounds $b_1 < b_2 < \cdots < b_r$ to eliminate variables of all enabled transitions reaching their upper bounds, up to bound r. We call D^r the domain obtained this way. Every transition t_k in *Enabled*$(M.N)$ that has no variable θ_k in D^r is hence a waiting transition whose upper bound has been reached.

2) Firing condition: We add to D^r the following constraints: we add the inequality $(b_r \leq \theta_f \leq b_{r+1})$, and for every transition $t_j \in \mathsf{FullyEnabled}(M) \setminus \{t_f\}$, we add to D^r the inequality $\theta_f \leq \theta_j$. This means that no other transition was urgent when t_f has been fired. Let D^u be the new firing domain obtained this way. If any fully enabled transition t_j has to fire before t_f, then we have a constraint of the form $a_j \leq \theta_j \leq b_j$ with $b_j < a_f$, and D^u is not satisfiable. As we know that t_f is firable, this cannot be the case, and D^u has a solution, but yet, we have to include in the computation of the next firing domains reached after firing of t_f the constraints on θ_f due to urgency of other transitions.

3) Substitution of variables: As t_f fires after elapsing θ_f time units, the time to fire of other transitions whose clocks did not yet exceed their upper bounds decreases by the same amount of time. Variables of timed-out transitions have already been eliminated in D^u. So for every $t_j \neq t_f$ that has an associated constraint $a_j \leq \theta_j \leq b_j$ we do a variable substitution reflecting the fact that the new time to fire θ'_j decreases w.r.t the former time to fire θ_j. We set $\theta_j := \theta_f + \theta'_j$. When this is done, we obtain a domain D'^{u,b_r} over a set of variables $\theta'_{i_1}, \dots \theta'_{i_k}$, reflecting constraints on the possible remaining times to upper bounds of all enabled transitions that did not timeout yet.

4) Variable Elimination: As t_f fired at time θ_f, it introduced new relationships between remaining firing times of other transitions, i.e. other $\theta'_i \neq \theta_f$, that must be preserved in the next state class. However, as t_f is fired, in the next class, it is either newly enabled, or not enabled. We hence need to remove θ_f from inequalities, while preserving an equivalent set of constraints. This is achieved by elimination of variable θ_f from D'^{u,b_r}, for instance with the well known Fourier-Motzkin technique (see [17] for details). We proceed similarly with variable θ'_i for every transition t_i that is enabled in marking $M.N$ but not in $M.N - {}^\bullet(t_f)$. After elimination, we obtain a domain D'^{E,b_r} over remaining variables.

5) Addition of new constraints: The last step to compute the next state classes is to introduce fresh constraints for firing times of newly enabled transitions. For every $t_i \in\uparrow \mathsf{enabled}(M.N, t_f)$ we add to D'^{E,b_r} the constraint $\alpha(t_i) \leq \theta'_i \leq \beta(t_i)$. For every timed-out transition t_k that becomes fully enabled, we add to D'^{E,b_r} the constraint $\theta_k = 0$. Timed-out transitions that become fully enabled are hence urgent in the next class. After adding all constraints associated to newly enabled transitions, we obtain a domain, in which we can rename every θ'_i to θ_i to get a domain D'^{F,b_r}. Notice that this domain needs not be minimal, so we do a last normalization step (see Definition 8) to obtain a final canonical domain $\mathsf{next}_r(D, t_f) = D'^{F,b_r\,*}$.

More than one transition can fire from $(M.N, D)$, and for a given firable transition t_f, $Post(D, t_f)$ contains one domain per bound in $B^{t_f}_{M.N,D}$. It is hence clear that a state class can have more than one successor, with different markings and domains. Now, if a waiting net has no control place, transitions are either enabled or fully enabled in every configuration. Step 1 of successor construction leaves the starting domain D unchanged, and consequently the state class built is exactly the standard construction for TPNs (see [7,20]). Let $\mathsf{Post}(C)$ be the set of successors of a class C. Then $|\mathsf{Post}(C)| \leq |\mathsf{enabled}(M.N)|^2$. Computing successors can be repeated from each class in $Post(C)$. For a given net $\mathcal{W}$, and a given marking $M_0.N_0$, we denote by $\mathcal{C}(W)$ the set of classes that can be built inductively. This set need not be finite, but we show next that this comes from markings, and that the set of domains appearing in state classes is finite.

Definition 13. *(State Class Graph) The* State Class Graph *of a waiting net $\mathcal{W}$ is a graph $SCG(\mathcal{W}) = (\mathcal{C}(W), C_0, \longrightarrow)$ where $C_0 = (M_0.N_0, D_0)$, and $C \xrightarrow{t} C'$ iff $C' \in Post(C, t)$.*

Let $\rho = (M_0.N_0, v_0) \xrightarrow{d_1} (M_0.N_0, v_0 \oplus d_1) \xrightarrow{t_1} (M_1.N_1, v_1) \dots (M_k.N_k, v_k)$ be a run of $\mathcal{W}$ and $\pi = (M'_0.N'_0, D_0).(M'_1.N'_1, D_1) \dots (M'_k.N'_k, D_k)$ be a path in $SCG(\mathcal{W})$. We will say that ρ and π *coincide* iff $\forall i \in 1..k, M_i.N_i = M'_i.N'_i$, and for every step $(M_i.N_i, v_i) \xrightarrow{d_i} (M_i.N_i, v_i \oplus d_i) \xrightarrow{t_i} (M_{i+1}.N_{i+1}, v_{i+1})$, there exists an interval $[b_r, b_{r+1}]$ such that $d_i \in [b_r, b_{r+1}]$ and $D_{i+1} = \mathsf{next}_r(D_i, t_i)$.

Proposition 1 (Completeness). *For every run $\rho = (M_0.N_0, v_0) \dots (M_k.N_k, v_k)$ of $\mathcal{W}$ there exists a path π of $SCG(\mathcal{W})$ such that ρ and π coincide.*

Proof (sketch). By induction on the length of runs. For the base case, we can easily prove that any transition firing from the initial configuration after some delay d gives a possible solution for D_0 and a successor class, as D_0 does not contain constraints of the form $\theta_i - \theta_j \leq c_{ij}$. The induction step is similar, and slightly more involved, because domains contain constraints involving pairs of variables. However, we can show (Lemma 2 in [17]) that along run ρ for every pair of steps composed of a time elapsing of duration d_i followed by the firing of a transition t_f, we have $d_i \in [a_{i,f}, b_{i,f}]$, where $a_{i,f}$ is the lower and $b_{i,f}$ the upper bound on variable θ_f at step i of the run. Hence, for every run of $\mathcal{W}$ there is a path that visits the same markings and maintains consistent constraints. □

Proposition 2 (Soundness). *Let π be a path of $SCG(\mathcal{W})$. Then there exists a run ρ of $\mathcal{W}$ such that ρ and π coincide.*

Proposition 1 shows that every marking reached by a run of a waiting net appears in its state class graph. The proof of Proposition 2 uses a similar induction on runs length, and shows that we do not introduce new markings. These propositions show that the state class graph is a sound and complete abstraction, even for unbounded nets. We can show a stronger property, which is that the set of domains appearing in a state class graph is finite.

Proposition 3. *The set of firing domains in $SCG(\mathcal{W})$ is finite.*

Proof (sketch). Domains are of the form $\{a_i \leq \theta_i \leq b_i\}_{t_i \subseteq T} \cup \{\theta_i - \theta_j \leq c_{i,j}\}_{t_i, t_j \subseteq T}$. We can easily adapt proofs of [7] (lemma 3 page 9) to show that every domain generated during the construction of the SCG has inequalities of the form $a_i \leq \theta_i \leq b_i$ and $\theta_i - \theta_j \leq c_{ij}$, where $0 \leq a_i \leq \alpha(t_i)$, $0 \leq b_i \leq \beta(t_i)$ and $-\alpha(t_i) \leq c_{ij} \leq \beta(t_i)$. This does not yet prove that the set of domains is finite. We define domains that are *bounded and linear*, i.e. upper and lower bounded by some constants, and where constants appearing in inequalities are linear combinations of a finite set of constant values. Domain D_0 is bounded and linear, and a series of technical lemmas (given in [17]) show that variable elimination, reduction to a canonical form, etc. preserve bounds and linearity (a similar result was shown in [7] for domains of TPNs). The set of bounded linear domains between fixed bounds is finite, so the set of domains of a waiting net is finite. □

This property of waiting nets is essential, as waiting nets allow to stop clocks. Bounded Petri nets with stopwatches do not have a finite state class representation, because clock differences in domains can take any value. WPNs do not have this kind of problem because clocks are stopped at a predetermined instant (when they reach the upper bound of an interval).

Corollary 1. *If $\mathcal{W}$ is a bounded waiting net then $SCG(\mathcal{W})$ is finite.*

Proof. States of $SCG(\mathcal{W})$ are of the form $(M.N, D)$ where $M.N$ is a marking and D a domain for time to fire of enabled transitions. By definition of boundedness, there is a finite number of markings appearing in $SCG(\mathcal{W})$. By Proposition 3, the set of domains appearing in $SCG(\mathcal{W})$ is finite, so $SCG(\mathcal{W})$ is finite. □

More precisely, if a net is k_P-bounded, there are at most k_P^P possible markings, and the number of possible domains is bounded by $(2 \cdot K_{\mathcal{W}} + 1)^{|T+1|^2}$, where $K_{\mathcal{W}} = \max_{i,j} \lfloor \frac{\beta_i}{\alpha_j} \rfloor$ is an upper bound on the number of linear combinations of bounds appearing in domains. Hence the size of $SCG(\mathcal{W})$ is in $O(k_P^P \cdot (2 \cdot K_{\mathcal{W}} + 1)^{|T+1|^2})$. A direct consequence of Proposition 1, Proposition 2, and Corollary 1 is that many properties of bounded waiting nets are decidable.

Corollary 2 (Reachability and Coverability). *The reachability and coverability problems for bounded waiting nets are decidable and PSPACE-complete.*

Proof. For membership, given a target marking $M_t.N_t$ it suffices to explore non-deterministically runs starting from $(M_0.N_0, D_0)$ of length at most $|SCG(\mathcal{W})|$ to find marking $M_t.N_t$, or to find a marking that covers $M_t.N_t$. Such reachability questions are known to be in NLOGSPACE w.r.t. the size of the explored graph, whence the NPSPACE=PSPACE complexity. For hardness, we already know that reachability for 1-safe Petri nets is PSPACE-Complete [12], and a (bounded) Petri net is a (bounded) waiting net without control places and with $[0, \infty)$ constraints. Similarly, given 1-safe Petri net and a place p, deciding if a marking with $M(p) = 1$ (which is a coverability question) is reachable is PSPACE-complete [15]. This question can be recast as a coverability question for waiting nets, thus establishing the hardness of coverability. □

5 Expressiveness

A natural question is the expressiveness of waiting nets w.r.t other models with time. There are several ways to compare expressiveness of timed models: One can build on relations between models such as isomorphism of their underlying timed transition systems, timed similarity, or bisimilarity. In the rest of this section, we compare models w.r.t. the timed languages they generate. For two particular types of model $\mathcal{M}_1$ and $\mathcal{M}_2$, we will write $\mathcal{M}_1 \leq_{\mathcal{L}} \mathcal{M}_2$ when, for every model $X_1 \in \mathcal{M}_1$, there exists a model X_2 in $\mathcal{M}_2$ such that $\mathcal{L}(X_1) = \mathcal{L}(X_2)$. Similarly, we will write $\mathcal{M}_1 <_{\mathcal{L}} \mathcal{M}_2$ if $\mathcal{M}_1 \leq_{\mathcal{L}} \mathcal{M}_2$ and there exists a model $X_2 \in \mathcal{M}_2$ such that for every model $X_1 \in \mathcal{M}_1$, $\mathcal{L}(X_2) \neq \mathcal{L}(X_1)$. Lastly, we will says that $\mathcal{M}_1$ and $\mathcal{M}_2$ are equally expressive and write $\mathcal{M}_1 =_{\mathcal{L}} \mathcal{M}_2$ if $\mathcal{M}_1 \leq_{\mathcal{L}} \mathcal{M}_2$ and $\mathcal{M}_1 \leq_{\mathcal{L}} \mathcal{M}_2$. In the rest of this section, we compare bounded and unbounded waiting nets with injective/non-injective labelling, with or without silent transitions labelled by ϵ to timed automata, TPNs, Stopwatch automata, and TPNs with stopwatches.

We first have obvious results. It is worth nothing that every model with non-injective labeling is more expressive than its injective counterpart. Similarly, every unbounded model is strictly more expressive than its bounded subclass. Waiting nets can express any behavior specified with TPNs. Indeed, a WTPN without control place is a TPN. One can also remark that (unbounded) TPNs, and hence WTPNs are not regular. It is also well known that the timed language of a bounded TPN can be encoded by a time bisimilar timed automaton [11,20]. We show next that one can extend the results of [20], i.e. reuse the state class construction of Sect. 4 to build a finite timed automaton $\mathcal{A}_{\mathcal{W}}$ that recognizes the same language as a waiting net $\mathcal{W}$. As shown by Proposition 1 and Proposition 2, the state class graph $SCG(\mathcal{W})$ is sound and complete. State class graphs abstract away the exact values of clocks and only remember constraints on remaining time to fire. If we label moves by the name of the transition used to move from a state class to the next one, we obtain an automaton that recognizes the untimed language of $\mathcal{W}$. Further, one can decorate a state class graph with clocks and invariants to recover the timing information lost during abstraction.

Definition 14 (Extended State Class). *An* extended state class *is a tuple $C_{ex} = (M.N, D, \chi, trans, XP)$, where $M.N$ is a marking, D a domain, χ is a set of real-valued clocks, $trans \in (2^T)^\chi$ maps clocks to sets of transitions and $XP \subseteq T$ is a set of transitions which upper bound have already been reached.*

Extended state classes were already proposed in [20] as a building step for state class timed automata recognizing languages of bounded TPNs. Here, we add information on transitions that have been enabled for a duration that is at least their upper bound. This is needed to enforce urgency when such transitions become firable. In extended state classes, every clock $x \in \chi$ represents the time since enabling of several transitions in $trans(x)$, that were enabled at the same instant. So, for a given transition t, the clock representing the valuation $v(x_t)$ is $trans^{-1}(t_i)$. Let $\mathbb{C}^{ex}$denote the set of all state classes. We can now define the state class timed automaton $SCTA(\mathcal{W})$ by adding guards and resets to the transitions of the state class graph, and invariants to state classes.

Definition 15 (State Class Timed Automaton). *The state class timed automaton of $\mathcal{W}$ is a tuple $SCTA(\mathcal{W}) = (L, l_0, X, \Sigma, Inv, E, F)$ where:*

- *$L \subseteq \mathbb{C}^{ex}$ is a set of extended state classes. $l_0 = (M_0.N_0, D_0, \{x_0\}, trans_0, XP_0)$, where $trans_0(x_0) = \mathsf{Enabled}(M_0.N_0)$ and $XP_0 = \emptyset$.*
- *$\Sigma = \lambda(T)$, and $X = \bigcup_{(M.N,D,\chi,trans)} \chi \subseteq \{x_1, \dots x_{|T|}\}$ is a set of clocks*
- *E is a set of transitions of the form $(C_{ex}, \lambda(t), g, R, C'_{ex})$. In each transition, $C_{ex} = (M.N, D, \chi, trans, XP)$ and $C'_{ex} = (M'.N', D', \chi', trans', XP')$ are two extended state classes such that $(M'.N', D') \longrightarrow (M.N, D)$ is a move of the STG with $D' = next_r(D, t)$.*
  We can compute the set of transitions disabled by the firing of t from $M.N$, denoted $\mathsf{Disabled}(M.N, t)$ and from there, compute a new set of clocks χ'. We have $\chi' = \chi \setminus \{x \in \chi \mid trans(x) \subseteq \mathsf{Disabled}(M.N, t)\}$ if firing t does not enable new transitions. If new transitions are enabled, we have $\chi' = \chi \setminus \{x \in \chi \mid trans(x) \subseteq \mathsf{Disabled}(M.N, t)\} \cup \{x_i\}$, where i is the smallest index for a clock in X that is not used. Similarly, we can set

$trans'(x_k) = \begin{cases} trans(x_k) \setminus \mathsf{Disabled}(M.N,t) \text{ } if\ trans(x_k) \not\subseteq \mathsf{Disabled}(M.N,t) \\ \uparrow \mathsf{enabled}(M.N,t) \text{ } if\ x_k = x_i \\ Undefined\ otherwise \end{cases}$

$XP' = XP \cap \mathsf{Enabled}(M - {}^\bullet(t)) \setminus \mathsf{FullyEnabled}(M'.N')$
$\cup \{t_k \in \mathsf{Enabled}(M'.N') \mid \theta_k \notin D'\}$

The guard g *is set to* $\alpha(t) \leq trans^{-1}(t)$. *Let* $\mathsf{Urgent}(C_{ex}, t, C'_{ex}) = XP \cap \mathsf{Enabled}(M - {}^\bullet(t)) \cap \mathsf{FullyEnabled}(M'.N')$. *The set of clocks reset is* $R = \{x_i\}$ *if some clock is newly enabled, and* $R = \emptyset$ *otherwise. For the invariant, we have two cases. If* $\mathsf{Urgent}(C_{ex}, t, C'_{ex}) = \emptyset$ *i.e. if there is no transition of* XP *that becomes fully enabled (and hence urgent) after firing* t*, the invariant* Inv' *is set to* $\bigwedge_{\substack{x_j \in trans^{-1}(\mathsf{FullyEnabled}(M'.N')),\\ t_k \in trans(x_j) \cap \mathsf{FullyEnabled}(M'.N')}} x_j \leq \beta(t_k)$. *Conversely, if* $\mathsf{Urgent}(C_{ex}, t, C'_{ex}) \neq \emptyset$ *the invariant is set to* $\bigwedge_{t_k \in \mathsf{Urgent}(C_{ex},t,C'_{ex})} trans^{-1}(t_k) \leq 0$

Proposition 4. *Let* $\mathcal{W}$ *be a waiting net. Then* $\mathcal{L}(SCTA(\mathcal{W})) = \mathcal{L}(\mathcal{W})$.

Proof (sketch). Obviously, every sequence of transitions in $\mathcal{L}(SCTA(\mathcal{W}))$ is a sequence of transitions of the STG, and hence there exists a timed word that corresponds to this sequence of transitions. Furthermore, in this sequence, every urgent transition is fired in priority before elapsing time, and the delay between enabling and firing of a transition t lays between the upper and lower bound of the time interval $[\alpha_t, \beta_t]$ if some time elapses in a state before the firing of t, and at least β_t time units if t fires immediately after reaching some state in the sequence (it is an urgent transition, so the upper bound of its interval has been reached, possibly some time before full enabling). Hence, every timed word of $SCTA(\mathcal{W})$ is also a timed word of $\mathcal{W}$. We can reuse the technique of Proposition 1 and prove by induction on the length of runs of $\mathcal{W}$ that for every run of $\mathcal{W}$, there exists a run of $SCTA(\mathcal{W})$ with the same sequence of delays and transitions. □

We are now ready to compare expressiveness of waiting nets and their variants w.r.t other types of time Petri nets, and with timed automata. For a given class $\mathcal{N}$ of net, we will denote by $B-\mathcal{N}$ the bounded subclass of $\mathcal{N}$, add the subscript ϵ if transitions with ϵ labels are allowed in the model, and a superscript $\overline{inj}$ if the labeling of transitions is non-injective. For instance $B - WTPN_\epsilon^{\overline{inj}}$ denotes the class of bounded waiting nets with non-injective labeling and ϵ transitions. It is well known that adding ϵ moves to automata increases the expressive power of the model [13]. Similarly, allowing non-injective labeling of transitions increases the expressive power of nets. Lastly, adding stopwatches to timed automata or bounded time Petri nets make them Turing powerful [10].

Theorem 1. $BWTPN <_{\mathcal{L}} TA(\leq, \geq)$.

Proof. From Proposition 4, we can translate every bounded waiting net $\mathcal{W}$ to a finite timed automaton $SCTA(\mathcal{W})$. Notice that $SCTA(\mathcal{W})$ uses only constraints of the form $x_i \geq a$ in guards and of the form $x_i \leq b$ in invariants. Thus,

$BWTPN \subseteq TA(\leq, \geq)$. This inclusion is strict. Consider the timed automaton $\mathcal{A}_1$ of Fig. 3. Action a can occur between date 2 and 3 and b between date 4 and 5.The timed language of $\mathcal{A}_1$ cannot be recognized by a BWTPN with only two transitions t_a and t_b, because t_a must be firable and then must fire between dates 2 (to satisfy the guard) and 3 (to satisfy the invariant in s_1). However, in TPNs and WTPNs, transitions that become urgent do not let time elapse, and cannot be disabled without making a discrete move. As t_b is the only other possible move, but is not yet allowed, no WTPN with injective labeling can encode the same behavior as $\mathcal{A}_1$.

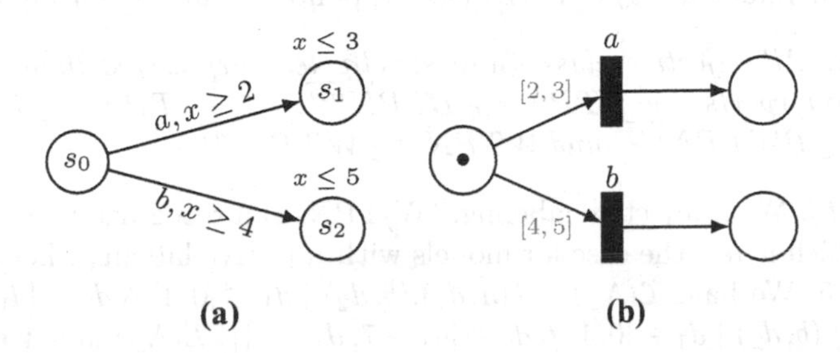

Fig. 3. a) A timed automaton $\mathcal{A}_1$ b) an equivalent time<u>d</u> Petri net

Remark 2. It was proved in [6] that timed automata (with $\epsilon-$transitions) have the same expressive power as bounded TPNs with $\epsilon-$transitions. These epsilon transitions can be used to "steal tokens" of a waiting transition, and prevent it from firing after a delay. This cannot be done with waiting nets without ϵ. Hence, bounded TPN with $\epsilon-$transitions are strictly more expressive than waiting nets, and than waiting net with non-injective labeling.

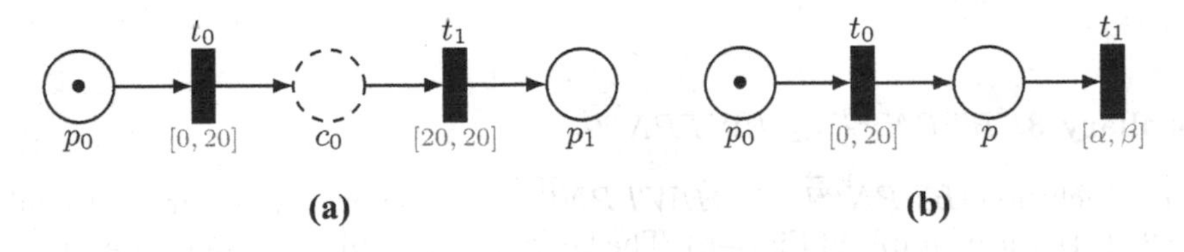

Fig. 4. a) A waiting net $\mathcal{W}$ b) a part of TPN needed to encode $\mathcal{L}(\mathcal{W})$.

Remark 3. Another easy result is that timed Petri nets and waiting nets are incomparable. Indeed, timed Petri nets cannot encode urgency of TPNs, and as a consequence some (W)TPNs have no timed Petri net counterpart, even in the bounded case. Similarly, one can design a timed Petri net in which a transition is firable only in a bounded time interval and is then disabled when time elapses. We have seen in Fig. 3-a) that $\mathcal{L}(\mathcal{A}_1)$ cannot be recognized by a waiting net. However, it is easily recognized by the timed Petri net of Fig. 3-b).

Theorem 2. *$TPN <_{\mathcal{L}} WTPN$ and $BTPN <_{\mathcal{L}} BWTPN$.*

Proof (sketch). TPNs are WTPNs without control places so $TPN \leq_{\mathcal{L}} WTPN$ and $BTPN \leq_{\mathcal{L}} BWTPN$. We can show that inclusions are strict with the net $\mathcal{W}$ of Fig. 4, that recognizes language $\mathcal{L}(W) = \{(t_0, d_0)(t_1, 20) \mid 0 \leq d_0 \leq 20\}$. Assuming that a TPN recognizes this language, it must contain the subnet of Fig. 4-b), for some values α, β. However, there is no assignment for α, β allowing to consider all values for d_0 in $\mathcal{L}(W)$ (see Appendix G in [17] for details). □

Theorem 3. *All injective classes are strictly less expressive than their non-injective counterparts, i.e. $BTPN <_{\mathcal{L}} BTPN^{\overline{inj}}$, $TPN <_{\mathcal{L}} TPN^{\overline{inj}}$, $BWTPN <_{\mathcal{L}} BWTPN^{\overline{inj}}$, and $WTPN <_{\mathcal{L}} WTPN^{\overline{inj}}$.*

Proof (sketch). With injective labeling, (W)TPNs can recognize unions of timed language, which is not the case for models with injective labeling. Let $\mathcal{N}_2$ be the TPN of Fig. 5. We have $\mathcal{L}(\mathcal{N}_2) = \{(a, d_1).(b, d_2) \mid d_1 \in [0, 1] \wedge d_2 \in [d_1 + 4, d_1 + 5]\} \cup \{(a, d_1).(b, d_2) \mid d_1 \in [0, 1] \wedge d_2 \in [d_1 + 7, d_1 + 8]\}$. $\mathcal{L}(\mathcal{N}_2)$ is not recognized by any (waiting) net with injective labeling. □

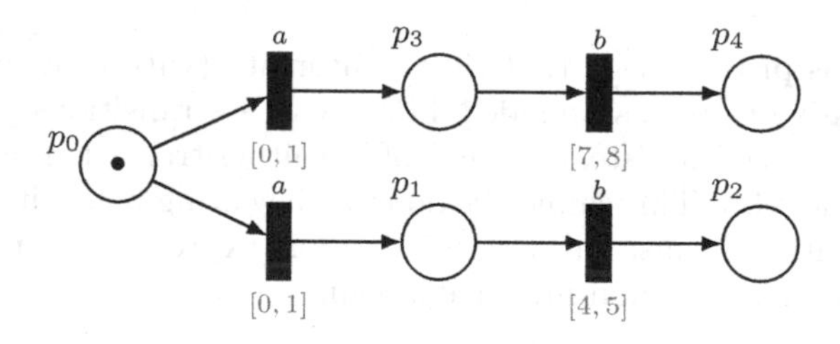

Fig. 5. A TPN $\mathcal{N}_2$ with non-injective labeling.

Corollary 3. *$BTPN^{\overline{inj}} <_{\mathcal{L}} BWTPN^{\overline{inj}}$*

Proof. Inclusion $BTPN^{\overline{inj}} \leq_{\mathcal{L}} BWTPN^{\overline{inj}}$ is straightforward from Definition 2. Take the example of Fig. 4-a). The language recognized cannot be encoded with a non-injective TPN, for the reasons detailed in the proof of Theorem 2. □

To conclude on the effects of non-injective labeling, we can easily notice that $BWTPN^{\overline{inj}} <_{\mathcal{L}} TA(\leq, \geq)$ because the automaton construction of Definition 15 still works (one labels transitions of the automaton with labels attached to transitions and keep the same construction). The last point to consider is whether allowing silent transitions increases the expressive power of the model. It was shown in [13] that timed automata with epsilon transitions are strictly more expressive than without epsilon. We hence have $TA(\leq, \geq) <_{\mathcal{L}} TA_\epsilon(\leq, \geq)$. We can also show that differences between WTPNs, TPN, and automata disappear when silent transitions are allowed.

Theorem 4. $TA_\epsilon(\leq, \geq) =_{\mathcal{L}} BTPN_\epsilon =_{\mathcal{L}} BWTPN_\epsilon$

Proof. The equality $TA_\epsilon(\leq, \geq) = BTPN_\epsilon$ was already proved in [6]. Given $BWTPN_\epsilon$, one can apply the construction of Definition 15 to obtain a state class timed automaton (with ϵ transitions) recognizing the same language. □

Figure 6 shows the relations among different classes of nets and automata, including TPNs and automata with stopwatches. An arrow $\mathcal{M}_1 \longrightarrow \mathcal{M}_2$ means that $\mathcal{M}_1$ is strictly less expressive than $\mathcal{M}_2$, and this relation is transitively closed. All extensions with clocks and stopwatches allow the considered model to simulate runs of Turing Machines. Actually, it has been shown that these models can encode two-counters machines (and then Turing machines). Obviously, all stopwatch models can simulate one another. Hence, these models are equally expressive in terms of timed languages as soon as they allow ϵ transitions. The red dashed line in Fig. 6 is the frontier for Turing powerful models, and hence also for decidability of reachability or coverability.

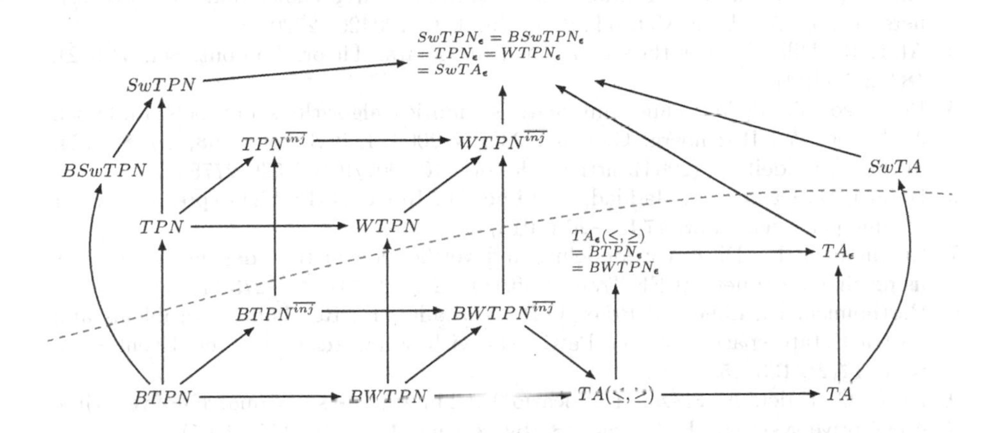

Fig. 6. Relation among net and automata classes, and frontier of decidability.

6 Conclusion

We have proposed waiting nets, a new variant of time Petri nets, that measure time elapsed since enabling of a transition while waiting for additional control allowing its firing. This class obviously subsumes Time Petri nets. More interestingly, expressiveness of bounded waiting nets lays between that of bounded

TPNs and timed automata. Waiting nets allow for a finite abstraction of the firing domains of transitions. A consequence is that one can compute a finite state class diagram for bounded WTPNs, and decide reachability and coverability.

As future work, we will investigate properties of classes of WTPN outside the bounded cases. In particular, we should investigate if being free-choice allows for the decidability of more properties in unbounded WTPNs [3]. A second interesting topic is control. Waiting nets are tailored to be guided by a timed controller, filling control places in due time to allow transitions firing. A challenge is to study in which conditions one can synthesize a controller to guide a waiting net in order to meet a given objective.

References

1. Abdulla, P.A., Nylén, A.: Timed petri nets and BQOs. In: Colom, J.-M., Koutny, M. (eds.) ICATPN 2001. LNCS, vol. 2075, pp. 53–70. Springer, Heidelberg (2001). https://doi.org/10.1007/3-540-45740-2_5
2. Akshay, S., Genest, B., Hélouët, L.: Decidable classes of unbounded petri nets with time and urgency. In: Kordon, F., Moldt, D. (eds.) PETRI NETS 2016. LNCS, vol. 9698, pp. 301–322. Springer, Cham (2016). https://doi.org/10.1007/978-3-319-39086-4_18
3. Akshay, S., Hélouët, L., Phawade, R.: Combining free choice and time in Petri nets. J. Log. Algebraic Methods Program. **110**, 100426 (2020)
4. Alur, R., Dill, D.L.: A theory of timed automata. Theor. Comput. Sci. **126**(2), 183–235 (1994)
5. Bengtsson, J., Yi, W.: Timed automata: semantics, algorithms and tools. In: Desel, J., Reisig, W., Rozenberg, G. (eds.) ACPN 2003. LNCS, vol. 3098, pp. 87–124. Springer, Heidelberg (2004). https://doi.org/10.1007/978-3-540-27755-2_3
6. Bérard, B., Cassez, F., Haddad, S., Lime, D., Roux, O.H.: The expressive power of time petri nets. TCS **474**, 1–20 (2013)
7. Berthomieu, B., Diaz, M.: Modeling and verification of time dependent systems using time petri nets. IEEE Trans. Software Eng. **17**(3), 259–273 (1991)
8. Berthomieu, B., Lime, D., Roux, O.H., Vernadat, F.: Reachability problems and abstract state spaces for time Petri nets with stopwatches. Discret. Event Dyn. Syst. **17**(2), 133–158 (2007)
9. Bucci, G., Fedeli, A., Sassoli, L., Vicario, E.: Timed state space analysis of real-time preemptive systems. IEEE Trans. Software Eng. **30**(2), 97–111 (2004)
10. Cassez, F., Larsen, K.: The impressive power of stopwatches. In: Palamidessi, C. (ed.) CONCUR 2000. LNCS, vol. 1877, pp. 138–152. Springer, Heidelberg (2000). https://doi.org/10.1007/3-540-44618-4_12
11. Cassez, F., Roux, O.H.: Structural translation from time petri nets to timed automata. J. Syst. Softw. **79**(10), 1456–1468 (2006)
12. Cheng, A., Esparza, J., Palsberg, J.: Complexity results for 1-safe nets. Theor. Comput. Sci. **147**(1&2), 117–136 (1995)
13. Diekert, V., Gastin, P., Petit, A.: Removing ϵ-transitions in timed automata. In: Reischuk, R., Morvan, M. (eds.) STACS 1997. LNCS, vol. 1200, pp. 583–594. Springer, Heidelberg (1997). https://doi.org/10.1007/BFb0023491
14. Dill, D.L.: Timing assumptions and verification of finite-state concurrent systems. In: Sifakis, J. (ed.) CAV 1989. LNCS, vol. 407, pp. 197–212. Springer, Heidelberg (1990). https://doi.org/10.1007/3-540-52148-8_17

15. Esparza, J.: Decidability and complexity of petri net problems — an introduction. In: Reisig, W., Rozenberg, G. (eds.) ACPN 1996. LNCS, vol. 1491, pp. 374–428. Springer, Heidelberg (1998). https://doi.org/10.1007/3-540-65306-6_20
16. Escrig, D.F., Ruiz, V.V., Alonso, O.M.: Decidability of properties of timed-arc petri nets. In: Nielsen, M., Simpson, D. (eds.) ICATPN 2000. LNCS, vol. 1825, pp. 187–206. Springer, Heidelberg (2000). https://doi.org/10.1007/3-540-44988-4_12
17. Hélouët, L., Agrawal, P.: Waiting nets (extended version). Technical report, INRIA, CMI & ENS Paris Saclay (2022). https://hal.inria.fr/hal-03613598
18. Jacobsen, L., Jacobsen, M., Møller, M.H., Srba, J.: Verification of timed-arc petri nets. In: Černá, I., et al. (eds.) SOFSEM 2011. LNCS, vol. 6543, pp. 46–72. Springer, Heidelberg (2011). https://doi.org/10.1007/978-3-642-18381-2_4
19. Jones, N.D., Landweber, L.H., Lien, Y.E.: Complexity of some problems in petri nets. Theor. Comput. Sci. **4**(3), 277–299 (1977)
20. Lime, D., Roux, O.H.: Model checking of time petri nets using the state class timed automaton. Discret. Event Dyn. Syst. **16**(2), 179–205 (2006)
21. Mayr, E.W.: An algorithm for the general Petri net reachability problem. In: Proceedings of the 13th Annual ACM Symposium on Theory of Computing, Milwaukee, Wisconsin, USA, 11–13 May 1981, pp. 238–246. ACM (1981)
22. Merlin, P.M.: A study of the recoverability of computing systems. Ph.D. thesis, University of California, Irvine, CA, USA (1974)
23. Parrot, R., Briday, M., Roux, O.H.: Timed petri nets with reset for pipelined synchronous circuit design. In: Buchs, D., Carmona, J. (eds.) PETRI NETS 2021. LNCS, vol. 12734, pp. 55–75. Springer, Cham (2021). https://doi.org/10.1007/978-3-030-76983-3_4
24. Popova-Zeugmann, L.: On time petri nets. J. Inf. Process. Cybern. **27**(4), 227–244 (1991)
25. Rackoff, C.: The covering and boundedness problems for vector addition systems. Theor. Comput. Sci. **6**, 223–231 (1978)
26. Reynier, P.-A., Sangnier, A.: Weak time petri nets strike back! In: Bravetti, M., Zavattaro, G. (eds.) CONCUR 2009. LNCS, vol. 5710, pp. 557–571. Springer, Heidelberg (2009). https://doi.org/10.1007/978-3-642-04081-8_37
27. Ruiz, V.V., Gomez, F.C., de Frutos-Escrig, D.: On non-decidability of reachability for timed-arc petri nets. In: PNPM, p. 188. IEEE Computer Society (1999)

Computing Transience Bounds of Emergency Call Centers: A Hierarchical Timed Petri Net Approach

Xavier Allamigeon, Marin Boyet, and Stéphane Gaubert(✉)

INRIA and CMAP, École Polytechnique, IP Paris, CNRS, Palaiseau, France
{Xavier.Allamigeon,Marin.Boyet,Stephane.Gaubert}@inria.fr

Abstract. A fundamental issue in the analysis of emergency call centers is to estimate the time needed to return to a congestion-free regime after an unusual event with a massive arrival of calls. Call centers can generally be represented by timed Petri nets with a hierarchical structure, in which several layers describe the successive steps of treatments of calls. We study a continuous approximation of the Petri net dynamics (with infinitesimal tokens). Then, we show that a counter function, measuring the deviation to the stationary regime, coincides with the value function of a semi-Markov decision problem. We establish a finite time convergence result, exploiting the hierarchical structure of the Petri net. We obtain an explicit bound for the transience time, as a function of the initial marking and sojourn times. This is based on methods from the theory of stochastic shortest paths and non-linear Perron–Frobenius theory. We illustrate the bound on a case study of a medical emergency call center.

Keywords: Timed Petri Nets · Continuous Petri Nets · Stationary Regimes · Transience bound · Emergency Call Centers · Semi-Markov Decision Processes · Stochastic Shortest Path

1 Introduction

Context. The handling of emergency calls in dedicated call centers features various concurrency and synchronization patterns, breaking down in a stepwise process with tasks involving multiple agents, to be performed with as little delay as possible. A fundamental question is to fix the staffing so that calls be swiftly handled, taking into account the customary demand, as well as scenarios allowing sudden bulk of calls, originating for instance from exceptional events. In particular, one needs to compute the time needed to absorb such a bulk of calls, depending on the center characteristics and on the number of agents of various types. The treatment of incoming calls may be delayed during the transient phase, and so, the duration of this phase appears to be a critical performance measure.

Timed Petri nets have been used in [ABG15, ABG21] to model emergency call centers. Along the lines of [CGQ95, ABG21], timed Petri nets can be modelled by

L. Bernardinello and L. Petrucci (Eds.): PETRI NETS 2022, LNCS 13288, pp. 90–112, 2022.
https://doi.org/10.1007/978-3-031-06653-5_5

counter variables that give the number of firings of the transitions as a function of time. The discrete dynamics is generally hard to analyze, and so, a continuous approximation, with infinitesimal firings and stationnary routings, has been studied there. Then, the evolution of the counter variables is determined by a dynamic programming equation of semi-Markov type. In this way, first order performance measures like the long run throughput, were computed analytically as a function of resources [ABG21]. However, finer key performance evaluation issues, like the understanding of the transient behavior—in particular the estimation of the "catch-up time", i.e., the time needed to return to a stationary regime—have not been addressed so far in this setting.

Contribution. We consider a class of continuous timed Petri nets with a single input. In a reference scenario, this input is an affine function of time; this represents a regular arrival of calls in our application. Then, we study the behavior of the dynamical system under a deviation from this scenario, induced for instance by a bulk of arrivals.

Our main result shows that when the system is sufficiently staffed, meaning that the intrinsic throughput of the Petri net exceeds the input flow, the Petri net trajectory ultimately catches up a stationary regime driven by the input, see Theorem 4. This result is obtained by showing that the deviation to the stationary regime coincides with the value function of a stochastic shortest path problem (SSP), see Theorem 2. In this way, we are reduced to quantifying the convergence time of SSP problems. In Theorem 6, we characterize the SSP configurations for which the convergence occurs in finite time. This characterization involves the existence of a partial order over states, such that all optimal policies make moves that decrease the order. Then, we consider a hierarchical class of timed Petri nets, for which this partial order is known *a priori*—in the application to call centers, this corresponds to the natural ordering of tasks in the chain of treatment. In Theorem 7, we obtain an explicit upper bound on the catch-up times (or transience times) to recover from a perturbation. We illustrate our results on the emergency call center application throughout the paper. Section 3 provides the needed background and tools on semi-Markov decision processes; a novelty here, of a somewhat technical nature, is to handle the case of instantaneous transitions, which arises in our applications.

Related Work. The question of convergence of the earliest behavior of timed event graphs to a periodic or stationary regine has received much attention in the discrete event systems literature, see [BCOQ92] (especially Th. 3.109), and also [HOvdW05, Ch. 8 and 9], with an application to the Dutch railway network. For timed event graphs, the duration of the transient behavior, also sometimes called "coupling time", has been extensively studied, in particular by techniques of max-plus spectral theory, see [BG01, SS12, MNS14, MNS21]. The same problem has arisen in the setting of deterministic dynamic programming [HA99], and in the analysis of distributed algorithms [CBFN13],

Here, we extend the bounds given in the above references, passing from timed event graphs to continuous Petri nets with stationary routings, or, equivalently, passing from deterministic (semi-)Markov decision processes to stochastic semi-Markov decision processes. By comparison with the "deterministic" case, our proofs require new tools, coming from non-linear Perron-Frobenius theory and from the theory of stochastic shortest paths. In particular, Theorem 4 on the asymptotic convergence time builds on the work of [BT91] on stochastic shortest path problems, and extends some of these results to the semi-Markov case. Theorem 6 exploits techniques of non-linear Perron–Frobenius theory [AGN11, AGQS19] in order to characterize the property of convergence in finite time. To our knowledge, no characterization of the finite time convergence was known, even in the setting of Markov decisions processes.

Estimating the speed of convergence to the stationary regime for Markov decision processes is indeed a difficult and classical issue. General asymptotic convergence results, like the ones of [SF78b, SF78a, SF79], show that a convergence does occur with a ultimately geometric rate. However, they lead to bounds and speed estimates that are nonconstructive. An explicit bound of the time needed to enter in the geometric convergence regime was given in [Bon07], for shortest path stochastic configurations, supposing that all costs are positive. In contrast, we consider here the property of *finite time* convergence, leading to different bounds.

We rely on a correspondence between continuous timed Petri nets and semi-Markov decisions processes, developed in [CGQ98, CGQ95, ABG21]. This allows one to obtain asymptotic theorems and develop computational methods to solve performance evaluation issues by means of Markov decision techniques. Related analytical results were obtained in [GG04] with a different approach. Alternative approaches to the question of "absorption time" may rely on stochastic models, or network calculus. We leave the treatment of such aspects for further work, noting that the continuous Petri net behaviors studied arise as scaling limits of discrete deterministic or stochastic models. Hence, capturing probabilistic or network calculus aspects is expected to add one layer of difficulty – see e.g. [BR19] for a probabilistic treatment of emergency call centers.

Finally, we refer the reader to [Yus82, Fei94, Ros70] for background on semi-Markov decision processes, a.k.a, renewal programs [DF68]. See also [HG11] for a recent reference.

The proofs of our main results, together with additional explanatory or illustrative materials, can be found in [ABG22]. The essential tools and ideas, however, are in the body of the paper.

2 Petri Net Model of a Medical Emergency Call Center

A specific motivation in computing transience bounds for timed Petri nets originates from a real-life case study of the medical emergency call centers of the Paris area (SAMU 75, 92, 93 and 94), that is discussed in details in [ABG21].

We consider a medical emergency call center with three types of agents: *medical regulation assistants* (MRAs) who pick up calls and orient patients through the system, *emergency physicians* who handle the calls deemed to be the most serious by the MRAs, after which it can be decided to send an ambulance or a mobile intensive care unit, and *general practicioners*. Once a MRA has detected that a patient should talk with some emergency physician, the former stays on line with the patient and waits for one of the latter to be available; at this point a very short conversation happens between the MRA and the doctor, in which the MRA summarizes the case. Calls not passed to emergency physicians by the MRAs are either transfered to general practitioners if they concern less serious medical matters, and in this case the patient is put on hold in a virtual waiting room, or these calls are hang up if it was not a health-related distress call. In the last two situations, the MRA is able to pick up immediately a new incoming call. We focus on the next fundamental tasks of the emergency chain of treatment.

TASK 0: an emergency inbound call arrives;
TASK 1: an inbound call is picked up by a medical regulation assistant, who will decide if the call should be passed on to an emergency physician or not;
TASK 2: the instruction by the MRA of a call not requiring to talk with the emergency physician is completed;
TASK 3: the instruction by the MRA of a call requiring to talk with the emergency physician is completed **and** communication with the latter is initiated;
TASK 4: the short briefing between the MRA and the emergency physician is over, phone consultation between the patient and the physician starts;
TASK 5: the consultation of the patient with the physician ends.

Depending on the need to talk with an emergency physician or not, calls will follow TASKS 0, 1 and 2 or TASKS 0, 1, 3, 4 and 5. For $t \geq 0$, we denote by $z_i(t)$ the number of TASKS i completed up to time t, starting with no completed tasks at the instant $t = 0$.

We assume that calls arrive at a rate λ, and that a fraction π of them will require a discussion with an emergency physician. There are a total of N_A MRAs and N_P emergency doctors. The conversation between the patient and the MRA takes a time t_1, the synchronization step between an MRA and a physician takes a time t_2, and the consultation between the physician and the patient consumes a time t_3 (all these durations are assumed constant for sake of simplicity).

In [ABG21], we modeled the previous organization by a continuous timed Petri net [CGQ95], shown on Fig. 1. So $z_i(t)$ represents the number of firings of the transition labeled z_i, up to time t. We use the approximation in [ABG21], allowing "infinitesimal firings", so that $z_i(t)$ is a real nonnegative number, rather than an integer. In this setting, the functions z_i are governed by the following dynamics:

$$\begin{cases} z_1(t) = z_0(t) \wedge \left(N_A + z_2(t) + z_4(t)\right) \\ z_2(t) = (1-\pi) z_1(t-t_1) \\ z_3(t) = \pi z_1(t-t_1) \wedge \left(N_P + z_5(t)\right) \\ z_4(t) = z_3(t-t_2) \\ z_5(t) = z_4(t-t_3) \end{cases} \tag{EMS}$$

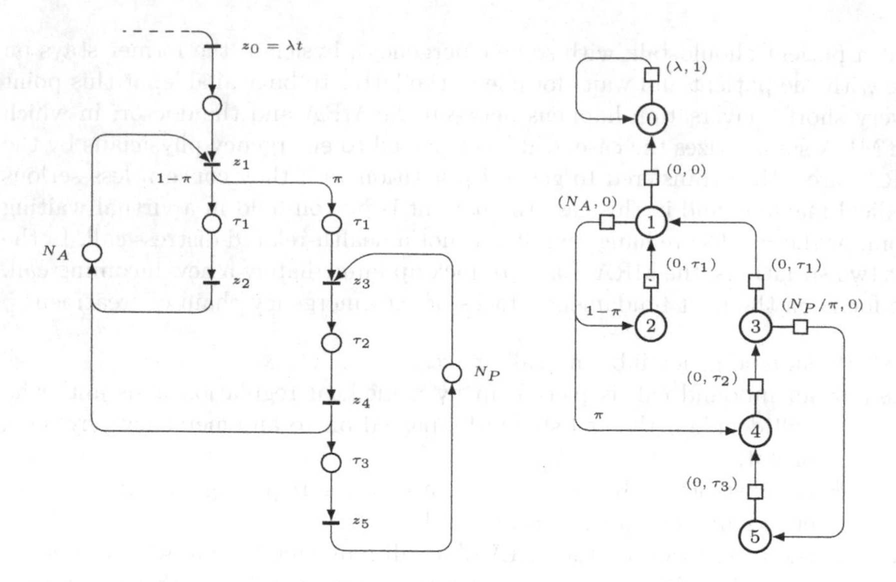

Fig. 1. Petri net representing an emergency call center [ABG21] (left). The corresponding undiscounted SMDP (right), see Sect. 3 for more information: states (resp. actions) are depicted by circles (resp. squares). A pair of the form (cost, sojourn time) is attached to each action. Transition probabilities from actions to states are given along the arcs if not equal to one.

where $\wedge$ stands for the minimum operation. A minimum of several terms arises if a task requires the synchronization of multiple resources (e.g., an inbound call and the availability of a MRA, or availability of both a MRA and a doctor). Delayed terms (of the form $z_i(t-\tau)$) are induced by the completion time of tasks. Weights (e.g., π and $1-\pi$) occur when calls are split in several categories.

The analysis methods developed in this paper also apply to more complex examples of emergency call centers, leading to larger Petri nets. Such examples can be found in [Boy22]. An extension of the formalism, incorporating non-constant holding times, is also analyzed there.

3 The Equivalence Between Semi-Markov Decision Processes and Continuous Time Petri Nets

We next recall the correspondence between the dynamics of continuous timed Petri nets with stationary routings and semi-Markov decision processes, developed in [ABG21].

3.1 The SMDP Model

In *Semi-Markov Decision Processes* (SMDPs) [Put14, Yus82], sometimes referred to as Markov renewal programs, the decision epochs are allowed to occur at any real-valued time. Indeed, between two successive moves, a *sojourn time* attached to states and actions must elapse.

Here, the finite set of states is denoted by S, and for all $i \in S$ the finite set of playable actions from state i is denoted by A_i. We denote $A := \biguplus_{i \in S} A_i$ the disjoint union of the $(A_i)_{i \in S}$. As a result of playing action a from state i, the player immediately incurs a deterministic cost c^a, is held in the state i for a nonnegative and deterministic (so possibly null) time t^a, and finally goes to state $j \in S$ with probability p_j^a. Actions are pulled as soon as a state is reached, and we assume that $\sum_{j \in S} p_j^a = 1$ for all $i \in S$. We depict in Fig. 1 an SMDP with six states and eight actions. We will see that this is precisely the SMDP corresponding to our running example of a medical emergency call center, in the sense that the counter equations of the Petri nets are equivalent to the dynamic programming equations of the SMDP [ABG21].

Recall that a *strategy* [Put14] is a mapping that associate choices of actions to play based on *histories*, i.e. possible realizations of already visited states, pulled actions, incurred costs and sojourn times etc. We denote by $\mathbb{F}$ the set of strategies. A strategy f in $\mathbb{F}$ and an initial state $i \in S$ induce a probability measure $\mathbb{P}_i^f$ on the set of histories of the game. The strategy f and the initial state i also give rise to a random process $(\widehat{i}_k, \widehat{a}_k)_{k \in \mathbb{N}}$ (with $\widehat{i}_0 = i$) of visited states and chosen actions. Denoting by $h = (i_k, a_k)_{k \in \mathbb{N}}$ a general trajectory realized by this process, we denote by $\widehat{c}_k$ (resp. $\widehat{t}_k$) the random variable such that $\widehat{c}_k(h) = c^{a_k}$ (resp. $\widehat{t}_k(h) = t^{a_k}$) for all k in $\mathbb{N}$.

The *value function* $v^* : S \times \mathbb{R} \to \mathbb{R}$ of the game in finite horizon is then defined as follows, so that for i in S and t in $\mathbb{R}$, $v^*(i,t)$ denotes the minimum (over all strategies) expected cost incurred by the player up to time t by starting in state i at the instant 0:

$$v^*(i,t) := \inf_{f \in \mathbb{F}} \mathbb{E}_i^f \left(\sum_{k=0}^{\widehat{N_t}} \widehat{c}_k \right) \tag{1}$$

where $\mathbb{E}_i^f$ denotes the expectation operator relatively to $\mathbb{P}_i^f$ and $\widehat{N_t}$ is the random variable with values in $\mathbb{N}$ such that $\widehat{N_t}(h) = \sup \left\{ n \in \mathbb{N} \mid \sum_{k=0}^{n-1} \widehat{t}_k(h) \leq t \right\}$. indeed observe that the cost $\widehat{c}_n$ is incurred at time $\sum_{k=0}^{n-1} \widehat{t}_k(h)$. We shall mainly focus on cases where $t \geq 0$, since the above definition implies $v^*(i,t) = 0$ for all i in S and $t < 0$.

Allowing zero sojourn times and yet immediate incurred costs raises the question of the well-posedness of the value function in (1). To prevent what we call *Zeno behaviors*, i.e., the accumulation of an infinite cost over a finite time period, a restriction is in order. We characterize *non-Zeno* SMDPs using the standard notion of *policies*: a policy σ is a map from S to A such that $\sigma(i)$ lies in A_i for every state i of S (some authors refer to this object as a *decision rule*). We denote by $\mathfrak{S} := \prod_{i \in S} A_i$ the finite set of all the policies. If σ is a policy, P^σ denotes the $|S| \times |S|$ matrix with entries $(p_j^{\sigma(i)})_{i,j \in S}$, while c^σ (resp. t^σ) is the vector with entries $(c^{\sigma(i)})_{i \in S}$ (resp. $(t^{\sigma(i)})_{i \in S}$).

Assumption A. *For all policies σ in $\mathfrak{S}$, for all finite classes $F \subset S$ of the Markov chain induced by σ, there is at least one state i in F with $t^{\sigma(i)} > 0$, i.e. with positive sojourn time attached to the chosen action $\sigma(i)$ from i.*

This assumption, milder than imposing an almost surely positive sojourn time for all actions of the game, is the same as the one that Schweitzer and Federgruen considered in their study [SF78b] of an average-cost criterion for SMDPs. Under this assumption, one can check that the value function v^* is well defined. In the rest of the paper, we shall denote by $t_{\max}$ the maximum positive sojourn time t^a attached to an action a in A.

In accordance with the dynamic programming principle, the value function satsifies the following recursive optimality equation:

$$\forall t \geq 0, \ \ \forall i \in S, \quad v(i,t) = \min_{a \in A_i} \left\{ c^a + \sum_{j \in S} p_j^a \, v(j, t - t^a) \right\}. \qquad \text{(DP)}$$

The *correspondence theorem*, established in [ABG21, Th. 6.1], shows that the counter functions of priority-free continuous timed Petri nets with preselection routings are always of this form. When $t^a \equiv 1$, corresponding to holding times equal to 1 in all places, we recover the standard dynamic programming equation of Markov decision processes. However, allowing *real* soujourn times is natural in the Petri-net applications, leading to a semi-Markov framework.

When $t^a \equiv 1$, an initial condition $v(-1)$ fully determines the value function $v(t)$ for all $t \in \mathbb{N}$. In the case of SMDPs, the knowledge of the value function over the whole interval $[t - t_{\max}, t)$ is needed to compute $v(t)$, where $t_{\max}$ is the maximum positive sojourn time t^a attached to an action $a \in A$, and we get an infinite dimensional dynamics. This motivates the introduction of the space $\mathcal{V}$ of bounded $|S|$-dimensional vector-valued functions over $[-t_{\max}, 0)$ in order to characterize the initial conditions of (DP). The following proposition shows that, under such an initial condition, there is a unique trajectory satisfying the dynamic programming equations (DP). The main difficulty is to handle null sojourn times, then terms of $v(\cdot, t)$ can appear in both sides of (DP). We recall that a function of a real variable is *càdlàg* if it is right continuous and admits a limit at the left of every point.

Proposition 1. *Let $v^0 \in \mathcal{V}$. Under Assumption A, there is a unique function v defined on $[-t_{\max}, \infty)$ satisfying the dynamic programming equation* (DP) *and for all $t \in [-t_{\max}, 0)$, $v(t) = v^0(t)$. In addition, if v^0 is càdlàg (resp. piecewise-constant, piecewise-affine), then v is càdlàg (resp. piecewise-constant, piecewise-affine).*

Building in Proposition 1, we establish that v^* satisfies (DP).

Theorem 1. *The value function v^* defined by* (1) *is the unique function null on $[-t_{\max}, 0)$ that verifies the dynamic programming equations* (DP).

The last theorem and Proposition 1 show in particular that v^* is piecewise-constant.

Example. In the case of the SMDP depicted in Fig. 1, corresponding to the medical emergency call center, the dynamic programming equations (DP) specialize to:

$$v^*(0,t) = v^*(0,t-t_0) + \lambda t_0 \tag{2a}$$
$$v^*(1,t) = v^*(0,t) \wedge \big(N_A + (1-\pi)v^*(2,t) + \pi v^*(4,t)\big) \tag{2b}$$
$$v^*(2,t) = v^*(1,t-t_1) \tag{2c}$$
$$v^*(3,t) = v^*(1,t-t_1) \wedge \big(N_P/\pi + v^*(5,t)\big) \tag{2d}$$
$$v^*(4,t) = v^*(3,t-t_2) \tag{2e}$$
$$v^*(5,t) = v^*(4,t-t_3) \tag{2f}$$

We denote $(e_0 := 1, e_1 := 1, e_2 := 1-\pi, e_3 := \pi, e_4 := \pi, e_5 := \pi)$. As shown in [ABG21], under steady call arrivals with rate λ so that $z_0(t) = \lambda t$, it can be seen that $z_i(t)/e_i = v^*(i,t)$, since initial conditions (nullity before $t = 0$) and dynamics coincide. This is a consequence of the main result of [ABG21] which states that priority free timed Petri nets with a stoichiometric invariant are in correspondence with SMDPs and share common dynamics equations (actually the correspondence $v^*(0,t) = \lambda t$ holds for $t_0 \to 0$ but further computations shall not require to take this limit, see Sect. 4.1).

3.2 Properties of the Dynamics Governing the Value Function

To portray the effect of applying equations (DP) to a function of $\mathcal{V}$, it shall come in handy to introduce the following evolution operator.

Definition 1. *For* $t \geq 0$, *we define the* evolution operator $\mathscr{S}_t$ *as the self-map of the set of functions* $\mathcal{V}$, *propagating an initial condition* v^0 *by* t *time units, i.e.,*

$$\mathscr{S}_t : v^0 \longmapsto \begin{cases} [-t_{\max}, 0) \to \mathbb{R}^{|S|} \\ \qquad s \mapsto v(t+s), \end{cases} \tag{3}$$

where v *is the function uniquely determined by the initial condition* v^0 *and the equations* (DP), *in accordance with Proposition 1.*

The operator $\mathscr{S}_t$ for SMDPs with $t \geq 0$ plays the same role as the t-fold iterate of the Bellman operator for MDPs (with in that case $t \in \mathbb{N}$). It follows from Theorem 1 and (3) that for all $t \geq 0$ and sufficiently small ε, we have $v^*(t) = \mathscr{S}_{t+\varepsilon}[\tilde{\mathbf{0}}](-\varepsilon)$, where $\tilde{\mathbf{0}}$ denotes the null function of $\mathcal{V}$. It readily follows that the family $(\mathscr{S}_t)_{t\geq 0}$ constitutes a one-parameter semi-group, i.e., $\mathscr{S}_0$ is the identity map of $\mathcal{V}$, and for all $t, t' \geq 0$, we have $\mathscr{S}_t \circ \mathscr{S}_{t'} = \mathscr{S}_{t+t'}$.

We list below some important properties of the operators $(\mathscr{S}_t)_{t\geq 0}$. The space $\mathcal{V}$ is equipped with the usual pointwise partial ordering $\leq$, and with the sup-norm $\|v^0\|_\infty := \sup_{s\in[-t_{\max},0)} \|v^0(s)\|_\infty$. We denote by $\tilde{\mathbf{1}}$ the constant function of $\mathcal{V}$ with all coordinates equal to 1.

Proposition 2. *Let $t \geq 0$. The operator $\mathscr{S}_t$ is (i) additively homogeneous: $\forall \alpha \in \mathbb{R}$, $\mathscr{S}_t[v^0 + \alpha\tilde{1}] = \mathscr{S}_t[v^0] + \alpha\tilde{1}$; (ii) monotone (or order-preserving): $v^0 \leq v'^0 \implies \mathscr{S}_t[v^0] \leq \mathscr{S}_t[v'^0]$; (iii) nonexpansive: $\|\mathscr{S}_t[v'^0] - \mathscr{S}_t[v^0]\|_\infty \leq \|v'^0 - v^0\|_\infty$; (iv) continuous in both the uniform and pointwise convergence topologies.*

The following proposition states the existence of an affine stationary regime, i.e., a particular initial condition function of $\mathcal{V}$ for which evolution under (DP) amounts to a translation in time.

Proposition 3 ([ABG21]). *Suppose that Assumption A holds. Then,*

(i) there exists two vectors $\chi(\mathscr{S})$ and h in $\mathbb{R}^{|S|}$ such that the affine function v^{aff} of $\mathcal{V}$ defined by $v^{\mathrm{aff}} : s \mapsto \chi(\mathscr{S})s + h$ satisfies $\mathscr{S}_t[v^{\mathrm{aff}}](s) = v^{\mathrm{aff}}(t+s)$;
(ii) for all v^0 in $\mathcal{V}$, we have for all s in $[-t_{\max}, 0)$, $\mathscr{S}_t[v^0](s) \underset{t\to\infty}{=} \chi(\mathscr{S})t + O(1)$. In particular, we have $v^*(t) \underset{t\to\infty}{=} \chi(\mathscr{S})t + O(1)$.

As it appears in item (ii) above, the *growth rate* $\lim_{t\to\infty} \mathscr{S}_t[v^0]/t$ is independent of the choice of v^0 in $\mathcal{V}$ and relates only to the topology and the parameters of the SMDP (or its evolution semi-group $\mathscr{S}$), hence the notation $\chi(\mathscr{S})$. This vector is none other than the solution of the minimal time-average cost problem associated with our SMDP, and the following proposition provides an explicit formula to compute it in terms of the policies and the associated Markov chains, see also [SF78a]:

Proposition 4. *Suppose that Assumption A holds. Then, the vector $\chi(\mathscr{S})$ featured in Proposition 3 is unique and we have for all i in S*

$$\chi(\mathscr{S})_i = \min_{\sigma \in \mathfrak{S}} \sum_{F \in \mathcal{F}(\sigma)} \phi_i^F \frac{\langle \mu_F^\sigma, c^\sigma \rangle}{\langle \mu_F^\sigma, t^\sigma \rangle}, \tag{4}$$

where for a policy σ in $\mathfrak{S}$, $\mathcal{F}(\sigma)$ denotes the set of final classes of the Markov chain induced by σ, and for a final class F in $\mathcal{F}(\sigma)$, μ_F^σ (resp. ϕ_i^F) denotes the invariant nonnegative measure of class F (resp. the probability to reach F starting from i under σ).

Observe that the conditions of the non-Zeno Assumption A are precisely those which make $\chi(\mathscr{S})$ well-defined in Proposition 4.

Remark 1. It shall turn convenient in Sect. 5.2 to consider an SMDP induced by a subset of actions. This may also require to restrict ourselves to an adequate subset of states. To that purpose, we say that a pair $(S', A' := \biguplus_{i\in S'} A'_i)$ with $S' \subset S$ and $\emptyset \neq A'_i \subset A_i$ for all i in S is a *consistent subset of states and actions* if for all $a \in A'$, we have $\sum_{j \in S'} p_j^a = 1$.

If (S', A') is such a pair, the equations (DP) restricted to actions of A' and to states of S' induce an evolution semi-group that we may denote by $\left(\mathscr{S}_t|_{S',A'}\right)_{t\geq 0}$ on the set of "initial condition" functions with suitable dimension, as set out in Proposition 1 and Definition 1. Any restricted semi-group of this form also enjoys the properties listed in Proposition 2.

4 Deviation of the Value from the Congestion-Free Regime

4.1 Reducing to a Stochastic Shortest Path Problem

We deal with systems that have to meet a known inbound demand. In the framework of SMDP, such an input is conveniently modeled by distinguishing a sink state denoted by 0, for which the value function $t \mapsto v^*(0,t)$ is prescribed for all $t \geq 0$. We will assume that $v^*(0,t) = \lambda t$ for all $t \geq 0$, where $\lambda > 0$ is constant rate of tasks to perform per unit of time. This input profile may be realized by our SMDP framework by first supposing that a single action a_0 is playable from state 0 such that $p_0^{a_0} = 1$, $t^{a_0} > 0$ and $c^{a_0} = \lambda t^{a_0}$, and also that the dynamics is initialized with a function of $\mathcal{V}_\lambda := \{v \in \mathcal{V} \mid v(0,s) = \lambda s \text{ for all } s \in [-t_{\max}, 0)\}$. In what follows, we shall refer to this setting as *λ-sink SMDPs.*

By an abuse of language and notation, we shall still refer to the *value function* and denote by v^* the solution of the dynamics of a λ-sink SMDP uniquely determined by the initial condition v^0 in $\mathcal{V}$ such that $v^0|_{S\setminus\{0\}} = \tilde{\mathbf{0}}$. Indeed, this initial condition is not null on state 0 on $[-t_{\max}, 0)$, as required by Theorem 1, but the function it generates coincides on $[0,\infty)$ with the value function obtained by prescribing $v^*(0,t) = 0$ for $t \in [-t_{\max}, 0)$ and $v^*(0,t) = \lambda t$ for $t \geq 0$, provided that the actions giving access to state 0 are equipped with null sojourn time, which involves no loss of generality up to considering intermediary states.

Proposition 3 and Proposition 4 show that the growth rate of the solutions of (DP) is given by the one of an *average-cost optimal* policy; the final classes of this policy indicate which parts of the system, by staying within them, enable the player to minimize his/her costs in the long run. In our emergency call center application, these final classes correspond to the slowest part of the treatment chain and therefore indicate which resources are bottleneck *via* the computation of the growth rate. Observe that in a λ-sink SMDP, the singleton $\{0\}$ is a final class of all policies of $\mathfrak{S}$ and that 0 must always have a growth-rate of λ. We want to focus on "congestion-free regimes" of these SMDPs, i.e., for which the evolution under (DP) is driven by the input, resulting in $\chi(\mathscr{S}) = \lambda\mathbf{1}$ (where $\mathbf{1}$ denotes the vector with all components set to 1). According to Proposition 4 and the previous remark, it is relevant to introduce the minimum possible growth-rate in the SMDP except λ, denoted by $\underline{\chi}$ and defined as

$$\underline{\chi} := \min_{\sigma \in \mathfrak{S}} \min_{\substack{F \in \mathcal{F}(\sigma) \\ F \neq \{0\}}} \left\{ \frac{\langle \mu_F^\sigma, c^\sigma \rangle}{\langle \mu_F^\sigma, t^\sigma \rangle} \right\} . \tag{5}$$

It is therefore seen from Proposition 4 that $\chi(\mathscr{S}) = \lambda\mathbf{1}$ can occur only if $\underline{\chi} \geq \lambda$, i.e., the least growth-rate of the cost is achieved by accessing the sink state 0; and imposing $\underline{\chi} > \lambda$ ensures that the less costly final class of all policies is always $\{0\}$. Conversely, we note that the value function of a state i can be driven by the input only if this state has access to 0, captured by the term $\phi_i^{\{0\}}$ in (4) (recall that for two states i and j of the SMDP, we say that i *has access to* j iff there are states $i_0, i_1, \dots, i_k, i_{k+1}$ of S with $i_0 = i$ and $i_{k+1} = j$, and actions $a_0, a_1, \dots, a_k$ of $A_{i_0} \times A_{i_1} \times \cdots \times A_{i_k}$ such that $p^{a_\ell}_{i_{\ell+1}} > 0$ for all $0 \leq \ell \leq k$). This motivates the statement of the next assumption to characterize the congestion-free regimes of λ-sink SMDPs.

Assumption B. *Consider a* λ*-sink SMDP.*

(1) All the states of S *have access to* 0,
(2) $\underline{\chi} > \lambda$.

To investigate if a better result than the mere $v^*(t) \underset{t\to\infty}{=} \lambda t\mathbf{1} + O(1)$ can be obtained under Assumption B, we focus on the deviation $\Delta v^*(t) := v^*(t) - \lambda t\mathbf{1}$. We remark that our setting bears much resemblance with the subclass of problems associated with MDPs known as *Stochastic Shortest Path* (SSP) problems. An *SSP configuration* refers to a MDP in which there is a distinguished sink state denoted by 0 such that any playable action from state 0 has null cost and forces to stay in 0 (for all $a \in A_0$, $c^a = 0$ and $p^a_0 = 1$). It is therefore seen that as soon as one reaches state 0, the accumulated cost no longer evolves and the game virtually stops; in this case it is licit to study the limit of the value function in (1) when t tends to ∞. We point out that the notion of SSP configuration carries over SMDPs. However, we are not aware of any study of the stochastic shortest path problem in the semi-Markov setting.

We now show that studying the deviation Δv^* reduces to a SSP problem.

Theorem 2. *Let* $\mathcal{P}$ *be a* λ*-sink SMDP. The SMDP* $\mathcal{P}'$ *obtained from* $\mathcal{P}$ *by changing the costs to the* reduced costs $c^a - \lambda t^a$ *for all* $a \in A$ *is in SSP configuration. Moreover,* Δv^* *satisfies the dynamic programming equations of* $\mathcal{P}'$*, i.e., for all* $t \geq 0$ *and* $i \in S$:

$$\Delta v^*(i,t) = \min_{a \in A_i} \left\{ \left(c^a - \lambda t^a\right) + \sum_{j \in S} p^a_j \, \Delta v^*(j, t - t^a) \right\} . \tag{6}$$

The key elements that make the SSP configuration arise under the cost reduction transformation featured in Theorem 2 are the fact that 0 remains an absorbing sink state but in addition becomes cost-free.

The mere SSP configuration in terms of topology is not sufficient to obtain convergence of Δv^*. Instead, to ensure that the SSP problem is well-posed and that ultimate reachability of state 0 is guaranteed, the notion of *proper* policy is often introduced.

Definition 2. *A policy* σ *in* $\mathfrak{S}$ *is* proper *if for all* i *in* S, $\lim_{n\to\infty} \left[\left(P^\sigma\right)^n\right]_{i0} = 1$. *A non-proper policy is called* improper.

The next proposition links our setting of congestion-free regime for λ-sink SMDPs to the most standard assumptions made in the SSP literature, expressed in terms of proper and improper policies.

Proposition 5. *Let $\mathcal{P}$ be a λ-sink SMDP. The two assumptions of B are equivalent to the following two conditions on the SMDP $\mathcal{P}'$ in SSP configuration constructed in Theorem 2:*

(1') There exists a proper policy.
(2') For every improper policy, the expectation of the accumulated reduced cost incurred up to time t converges to $+\infty$ as $t \to \infty$, for at least one initial state.

This entails that the SMDP with reduced costs satisfies precisely the condition [BT91, Assumption 1].

Theorem 3 (Corollary of [BT91]**).** *Suppose that conditions (1') and (2') of Proposition 5 hold. Then, the equations*

$$\forall i \in S \backslash \{0\}, \quad u(i) = \min_{a \in A_i} \left\{ (c^a - \lambda t^a) + \sum_{j \in S} p_j^a u(j) \right\}, \qquad u(0) = 0 \quad (7)$$

admit a unique solution in $\mathbb{R}^{|S|}$.

In what follows, if Assumption B is verified, we denote by u^* the unique solution of Eq. (7). Bertsekas and Tsitsiklis show in [BT91] that the total cost of the SSP problem specified to the case of MDPs (when all the delays $(t^a)_{a \in A}$ are equal to 1) coincides with the solution of (7), and that it arises as the limit of iterates of the associated Bellman operator applied to any starting vector. We establish in the next theorem that these results carry over the semi-Markov framework, by considering the evolution semigroup $(\mathscr{S}_t^{\Delta})_{t \geq 0}$ associated with the reduced dynamic programming Eq. (6). The latter naturally acts on the set of initial conditions $\mathcal{V}_0 := \{v \in \mathcal{V} \mid v_0(s) = 0 \text{ for all } s \in [-t_{\max}, 0)\}$.

Theorem 4. *Suppose that Assumptions A and B hold. Then, for all v^0 in $\mathcal{V}_0$, for all s in $[-t_{\max}, 0)$, we have $\lim_{t \to \infty} \mathscr{S}_t^{\Delta}[v^0](s) = u^*$.*

In other words, Theorem 4 states that the minimum ultimate total expected cost of an SMDP in SSP configuration (such as the decision process $\mathcal{P}'$ featured in Theorem 2) is the same as in a corresponding MDP with unit transition times. Indeed, we study the infinite-horizon limit of a total reduced cost, which is a time indifferent quantity. In particular, the optimality equality characterizing the limit cost u^* has precisely the same form in the MDP and in the SMDP case, the only change being that the delays $(t^a)_{a \in A}$ can take non unit values in the SMDP case.

We obtain a direct corollary of Theorem 4 for the study of the value function, which indeed improves the result of Proposition 3 in the congestion-free regime. We insist on the importance of the first result for applications: whatever the initial condition, the trajectory of a non-Zeno λ-sink SMDP in congestion-free regime always ultimately catches up the input $t \mapsto \lambda t$, up to a constant delay u^*.

Corollary 1. *For a λ-sink SMDP, if Assumptions A and B hold, we have for all v^0 in $\mathcal{V}_\lambda$ and s in $[-t_{\max}, 0)$:* $\mathscr{S}_t[v^0](s) \underset{t\to\infty}{=} \lambda(t+s)\mathbf{1} + u^* + o(1)$. *In particular,* $v^*(t) \underset{t\to\infty}{=} \lambda t\mathbf{1} + u^* + o(1)$.

Example. We come back to our running example of a medical emergency call center, and we first acknowledge that the SMDP depicted in Fig. 1 is indeed a λ-sink SMDP. It is easy to verify in this figure that all the states have access to 0 and thus Assumption B-(1) is satisfied. The computation of the scalar $\underline{\chi}$, using closed forms of the probabilistic invariants associated with the final classes of the four policies of the SMDP, provides $\underline{\chi} = \lambda_A \wedge \lambda_P$, where

$$\lambda_A := \frac{N_A}{t_1 + \pi t_2} \qquad \text{and} \qquad \lambda_P := \frac{N_P}{\pi(t_2 + t_3)}.$$

These two quantities are fundamental in the analysis of the SMDP behavior and closely tied to real organization. The first one, λ_A, can be interpreted by the calls handling speed of MRAs, indeed there are N_A of them and a fraction $1-\pi$ (resp. π) of their work is to perform tasks 1 and 2 (resp. 1, 2, 3 and 4) which consumes a time t_1 (resp. $t_1 + t_2$), hence accounting for an average cycle time of $t_1 + \pi t_2$. The quantity λ_P can similarly be interpreted as the handling speed of the N_P emergency physicians, having a cycle time of $t_2 + t_3$ (to achieve tasks 3, 4 and 5) for a fraction π of all the calls. Depending on the choice of N_A and N_P (but also of the other parameters), there may or may not be enough agents (whether MRAs or doctors) to perform without delay all the tasks arriving with flow λ.

The medical emergency call center that we model by means of this SMDP thus behaves in its congestion-free regime if and only if Assumption B-(2) is met, i.e., if $\lambda_A > \lambda$ and $\lambda_P > \lambda$. In this case, the call center is correctly staffed and there are enough agents to ultimately pick up all the calls with no delay, i.e., all the functions $t \mapsto z_i(t)$ in (EMS) admit their maximum throughput λe_i. The policy that selects $(\sigma(1), \sigma(3)) = (a_1^-, a_3^-)$ is the one that achieves minimal throughput (only the calls arrival itself is bottleneck). It is associated with the unique final class $\{0\}$ and an affine stationary regime $t \mapsto \lambda t\mathbf{1} + u^*$, where $u^* = -\lambda(0, 0, t_1, t_1, t_1 + t_2, t_1 + t_2 + t_3)$. As shown in Corollary 1, thanks to the sufficient staffing, this congestion-free regime where all calls are handled with no delays is always ultimately reached.

If $\lambda_A < \lambda$ or $\lambda_P < \lambda$, this means that there are either not enough MRAs or not enough emergency physicians to perform all the required tasks, and due to the synchronization step (Task 3), the whole chain of treatment is slowed down. Limit cases ($\lambda_A = \lambda$ or $\lambda_P = \lambda$) would require a more detailed analysis.

4.2 Transience Time Needed to Catch-Up the Input

In the congestion-free regime, the function $w : t \mapsto \lambda t\mathbf{1} + u^*$ of $\mathcal{V}_\lambda$ featured in Corollary 1 gives rise to an affine stationary regime in the sense that for all

$t \geq 0$ and s in $[-t_{\max}, 0)$, we have $\mathscr{S}_t[w](s) = w(t+s)$ with no error term, like introduced in Proposition 3-(i); any such regime actually differs from w by a multiple of $\tilde{\mathbf{1}}$.

We leverage on the previous framework to study the effect of a perturbation on a steady-flow input in a λ-sink SMDP. For sake of simplicity, suppose that the system is initialized in its stationary affine congestion-free regime w (this entails little loss of generality since this behavior is always ultimately reached according to Corollary 1). We suppose that at $t = \underline{t}$, the input incurs a step of cost, so that $v(0,t) = \lambda t + M\tilde{H}(t - \underline{t})$ for all $t \geq 0$, where $M \geq 0$ and $\tilde{H}$ denotes the Heaviside step function. In our emergency call center application, this can be used to simulate the sudden arrival at time $\underline{t}$ of M new calls to handle on top of the usual demand with rate λ. The next result formalizes the fact that these requirements on $v(0,\cdot)$ actually define a unique trajectory of the λ-sink SMDP.

Proposition 6. *Under Assumption A, there is a unique function* $v\colon [-t_{\max}, \infty) \to \mathbb{R}^{|S|}$ *such that*

- *for all* t *in* $[-t_{\max}, \infty)$, $v(0,t) = \lambda t + M\tilde{H}(t - \underline{t})$,
- *for all* t *in* $[-t_{\max}, 0)$, $v\big|_{S\setminus\{0\}} = \tilde{\mathbf{0}}$,
- v *satisfies Eq.* (DP) *for states in* $S\setminus\{0\}$.

Moreover, if Assumption B holds, we have $v(t) \underset{t\to\infty}{=} \lambda t\mathbf{1} + u^* + M\mathbf{1} + o(1)$.

Building on the last statement of Proposition 6, which guarantees that in spite of the perturbation, the trajectory of the system still catches up the input (including the step of magnitude M), we want to study the transience time before this final regime is reached. To this purpose, we define for the state-by-state catch-up times (or transience times) $(\theta_i)_{i\in S}$ by:

$$\forall i \in S, \quad \theta_i := \inf\left\{t \geq \underline{t} \;\middle|\; v(i,t) = \lambda t + u^*(i) + M\right\}.$$

In Sect. 5, we shall study cases where the catching-up is exact and thus occurs in finite time; but we may as well have defined the $(\theta_i)_{i\in S}$ by the time needed for $v(i,t)$ to approach $\lambda t + u^*(i) + M$ up to a chosen precision ε.

The next theorem shows that the problem of characterizing the $(\theta_i)_{i\in S}$ reduces to the study of catch-up times in an SMDP in SSP configuration starting from a particular initial condition.

Theorem 5. *Suppose that Assumptions A and B hold. Let* v' *be the function uniquely determined by the initial condition* $s \mapsto u^* - M\mathbf{1}_{S\setminus\{0\}}$ *in* $\mathcal{V}_0$ *and the dynamics of the reduced-costs SMDP in SSP configuration featured in Theorem 2. Then, for all* i *in* S, *we have* $\theta_i = \underline{t} + \theta'_i$, *where* $\theta'_i := \inf\left\{t \geq 0 \;\middle|\; v'(i,t) = u^*(i)\right\}$.

5 Finite Time Convergence of the Semi-Markov SSP-value

In this section, we tackle the purified version of our problem brought to light in Theorem 5, letting aside λ-sink SMDP to solely focus on SMDPs in SSP configurations. In particular, the evolution semigroups considered hereafter such as $(\mathscr{S}_t)_{t\geq 0}$ are associated with dynamics of type (6) with null costs on state 0. Furthermore, observing that the definition of $\underline{\chi}$ in (5) for λ-sink SMDPs carries over to SSP configurations (that are essentially 0-sink SMDPs). In consequence, we rephrase the two conditions of Assumption B in the special case of SMDPs in SSP configuration.

Assumption C. *Consider a SMDP in SSP configuration.*

(1) All the states of S *have access to* 0*,*
(2) $\underline{\chi} > 0$.

The following lemma ensures that Assumption C corresponds to Assumption B in the sense of the reduction stated in Theorem 2.

Lemma 1. *Let* $\mathcal{P}$ *be a* λ*-sink SMDP, and* $\mathcal{P}'$ *be the SMDP in SSP configuration built in Theorem 2. Then* $\mathcal{P}$ *satisfies Assumption B if and only if* $\mathcal{P}'$ *satisfies Assumption C.*

We point out that this result follows from the application of Proposition 5 to $\mathcal{P}$ and $\mathcal{P}'$, and the fact that $\mathcal{P}'$ is precisely the reduced cost SMDP associated to itself.

The purpose of this section is to provide quantitative and constructive upper bounds on the $(\theta_i')_{i\in S}$ introduced in Theorem 5. This, in turn, yields upper bounds on the catch-up times of the original λ-sink SMDP, thanks to Theorem 5.

5.1 The Different Convergence Phases

Recall as stated in Theorem 4 that a function determined by an initial condition v^0 in $\mathcal{V}_0$ and satisfying the dynamic programming equations of a SMDP in SSP configuration converges towards u^* as considered horizons approach infinity. We first provide a qualitative insight on the speed of convergence associated with this result.

For all i in $S\backslash\{0\}$, we denote by A_i^* the nonempty subset of A_i composed of *optimal actions*, i.e. those achieving minimality in (7). Similarly, we introduce $\mathfrak{S}^* := A_0 \times \prod_{i\in S\backslash\{0\}} A_i^*$ the set of *optimal policies*, included in $\mathfrak{S}$. In accordance with Remark 1, optimal actions induce an evolution semi-group $\mathscr{S}^*$, which corresponds to applying only optimal policies. The following proposition shows that there is always an instant t^* such that: 1. before t^*, either optimal or non-optimal actions can be picked, and thus the evolution of the dynamics amounts to applying the semigroup $\mathscr{S}$; 2. after t^*, only optimal actions are chosen, so that the finer dynamics associated with the semigroup $\mathscr{S}^*$ are actually used. Moreover, the second phase induces a geometric speed of convergence controlled by the spectral radii of the probability matrices of optimal policies. The geometric convergence result is deduced from [AGQS19, Th. 1 and Th .2], using tools from non-linear Perron–Frobenius theory [AGN11]. In particular, we refer to [AGQS19] for background on weighted sup-norms.

Proposition 7. *Suppose Assumptions A and C hold. Let $v^0 \in \mathcal{V}_0$ and denote by v the solution of* (DP) *it determines. Then, there exists $t^* \geq 0$ such that for all $t \geq t^*$ and all s in $[-t_{\max}, 0)$, we have $v(t+s) = \mathscr{S}^*_{t-t^*}\left[\mathscr{S}_{t^*}[v^0]\right](s)$. Moreover, if we define $\nu := \max_{\sigma \in \mathfrak{S}^*} \rho\big(P^\sigma|_{(S\setminus\{0\})\times(S\setminus\{0\})}\big) < 1$, where $\rho(\cdot)$ denotes the spectral radius of a matrix, then for all $\varepsilon > 0$ small enough, there is a weighted sup norm in which $\mathscr{S}^*_{t_{\max}}$ is a contraction of rate $\nu + \varepsilon$.*

We point out that this proposition is reminiscent of the different phases of convergence established by Schweitzer and Federgruen in their work [SF79] of the deviation $v(t) - \chi(\mathscr{S})t$ in the case of MDPs when this quantity admits a limit.

We are interested in cases where convergence occurs in finite time, i.e., *without geometric residual*, which is desirable in our call center application. The next theorem shows that such a finite time convergence cannot be expected unless all probability matrices associated with optimal proper policies and restricted to states of $S\setminus\{0\}$ are nilpotent, hence making the rate of geometric convergence ν featured in Proposition 7 null. The latter is also equivalent to some restrictions on the SMDP topology, that we may interpret as requiring a form of *hierarchy* within the set of states that is compatible with the moves made by the optimal policies; these should always "descend" in the hierarchy until finally reaching 0, the minimal element.

Theorem 6. *Suppose Assumptions A and C hold. Then, the following are equivalent:*

(1) for all v^0 in $\mathcal{V}_0$ and associated solution v of (DP), *there exists t^* in $\mathbb{R}$ such that for all $t \geq t^*$, $v(t) = u^*$*
(2) for all proper optimal policies σ in $\mathfrak{S}^*$, we have $\rho\big(P^\sigma|_{(S\setminus\{0\})\times(S\setminus\{0\})}\big) = 0$,
(3) there exists a partial ordering $(\leq)$ on S such that for all σ in $\mathfrak{S}^*$ and for all i in $S\setminus\{0\}$, $\mathrm{supp}(\sigma(i)) \subset \{j \in S\,,\, j < i\}$.

5.2 Hierarchical SSP Configurations

The set $\mathfrak{S}^*$ of optimal policies that control the ultimate rate of convergence according to the previous theorem is in general not known, and depends on the costs and sojourn times attached to the actions of the SMDP. It is desirable to identify conditions of a more topological nature under which finite time convergence occurs regardless of optimal character of some policies. Building on the statement of Theorem 6-(iii), we choose to enforce the existence of a partial ordering $(\leq)$ on S such that applying some actions necessarily make the order strictly decrease. This is formalized in the next assumption.

Assumption D. *There is a partial ordering $(\leq)$ on S, such that for all state $i \in S$, there is a partition of the set of playable actions from i in the form $A_i = A_i^- \uplus A_i^+$, with the condition that if $a \in A_i^-$, then $(p_j^a > 0 \implies j < i)$ and if $a \in A_i^+$, then $(p_j^a > 0 \implies i \leq j)$, in which $j < i$ means that $j \leq i$ and $j \neq i$. In addition, $A_i^- \neq \emptyset$ for all $i \in S\setminus\{0\}$.*

In other words, Assumption D requires that actions either strictly "descend" or weakly "ascend" relatively to the states hierarchy, the first case being always possible. In what follows, we shall refer to $a \in A_i^-$ (resp. $a \in A_i^+$) as a "descending action" (resp. an "ascending action"). Remark that imposing in Assumption D that $A_i^- \neq \emptyset$ for all $i \in S\backslash\{0\}$ implies the condition (1') of Proposition 5, since any policy which makes use of only descending actions is proper. This implication turns into an equivalence (i.e., any proper policy must make use of only descending actions) if the next condition is also met.

Assumption E. *For all $\sigma \in \mathfrak{S}$, if there exists $i \in S$ such that $\sigma(i) \in A_i^+$, then σ is improper.*

Under Assumptions D and E (stronger than the condition (iii) of Theorem 6), we obtain that $A_i^* \subset A_i^-$ for all $i \in S$ and according to Theorem 6, convergence towards u^* arises in finite time.

Example. For the medical emergency call center depicted on Fig. 1, these two hierarchical assumptions are satisfied, as highlighted on the first picture of Fig. 2. The states hierarchy is the one naturally given by the (partial) order in which the different tasks are performed in the call center.

Our goal is to leverage on the hierarchical structure brought by Assumptions D and E to bound the catch-up times $(\theta_i')_{i \in S}$ defined in Theorem 5. To this purpose, we remark that the state 0 is the bottom element of the order $(\leq)$, and that the states that lie in low layers of the hierarchy shall have "quicker" access to state 0 than states located higher in the hierarchy (since the latter may need to pass through the former). From this perspective, if i is a state of S, it is natural to study the catch-up time θ_i' after the states located lower than i in the hierarchy already caught up the input. It amounts to determining the $(\theta_i')_{i \in S}$ by following an inductive scheme given by the partial ordering $(\leq)$.

This reasoning would result in straightforward bounds in the absence of ascending actions, i.e., which make the player move in states located higher in the hierarchy and thus slow down the catching-up of the input at the lowermost state 0. We know under Assumption C and E that these actions are ultimately non optimal because they are necessarily associated with improper policies and it is less costly to end up playing a proper policy. However for short decision horizons, some of these ascending actions may actually be optimal.

We tackle this difficulty by considering the collection of state-accessibility graphs of our SMDP in which particular states cannot use descending actions. More precisely, given $i \in S\backslash\{0\}$ such that $A_i^+ \neq \emptyset$, we introduce the graph $\mathcal{G}_S^{(i)}$ with nodes set S and with edge set $\{(k,\ell) \in S^2 \mid \exists a \in A_k \backslash A_k^- \text{ s.t. } p_\ell^a > 0\}$. We denote by $\mathcal{C}(i)$ the strongly connected component of state i in the graph $\mathcal{G}_S^{(i)}$. The Fig. 2 represents the two such subgraphs of our running example SMDP introduced in Fig. 1. It can be seen on this example that for i in S such that $A_i^+ \neq \emptyset$, not selecting actions in A_i^- amounts to staying within states of $\mathcal{C}(i)$ with no way of accessing 0. The next lemma formalizes this fact.

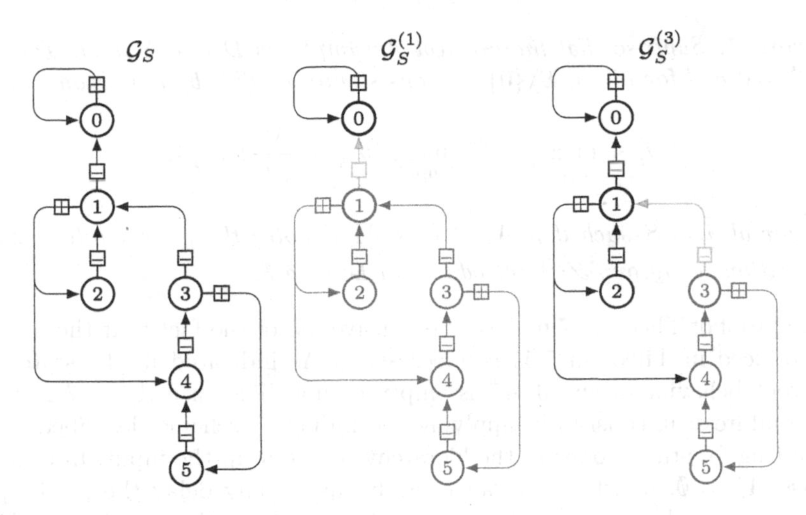

Fig. 2. Left: a hierarchical Stochastic Shortest Path configuration (satisfying Assumptions D and E) with six states. We have $0 \leq 1 \leq 2$ and $0 \leq 1 \leq 3 \leq 4 \leq 5$, while other pairs need not to be comparable. Middle (resp. right): the remaining sub-graph when ruling out actions of A_1^- (resp. A_3^-), with strongly connected component $\mathcal{C}(1)$ (resp. $\mathcal{C}(3)$) outlined in blue. (Color figure online)

Lemma 2. *Suppose that Assumptions C, D and E hold. For all $i \in S\backslash\{0\}$ such that $A_i^+ \neq \emptyset$, the following properties hold:*

(i) 0 is not accessible from i in $\mathcal{G}_S^{(i)}$,
(ii) every state accessible from i in $\mathcal{G}_S^{(i)}$ also has access to i via descending actions.

Let $i \in S\backslash\{0\}$ such that $A_i^+ \neq \emptyset$. To prove that it cannot be more interesting in the long run to stay in the class $\mathcal{C}(i)$ than playing descending actions to reach state 0, we focus on the particular evolution associated with (DP) in $\mathcal{C}(i)$. Observe that the pair $(\mathcal{C}(i), A_{\mathcal{C}(i)})$, with $A_{\mathcal{C}(i)} := A_i^+ \uplus \bigcup_{j \in \mathcal{C}(i), j \neq i} A_j$, is a consistent subset of states and actions in the sense of Remark 1. We thus introduce the associated evolution semigroup $\big(\mathscr{S}_t\big|_{\mathcal{C}(i),\, A_{\mathcal{C}(i)}}\big)_{t \geq 0}$, denoted for short by $(\widetilde{\mathscr{S}_t^{(i)}})_{t \geq 0}$.

Lemma 3. *Suppose Assumptions D and E hold. Let i in S such that $A_i^+ \neq \emptyset$. Then, the growth rate of $\widetilde{\mathscr{S}^{(i)}}$ is uniform on $\mathcal{C}(i)$, i.e., of the form $\chi^{(i)}\mathbf{1}$, where $\chi^{(i)} \geq \underline{\chi}$.*

We are now ready to formulate the bounds on the catch-up times $(\theta_i')_{i \in S}$. Recall that these are defined in Theorem 5 under Assumption B for λ-sink SMDPs, or equivalently under Assumption C for SMDPs in SSP configuration. In particular we have $\underline{\chi} > 0$.

Theorem 7. *Suppose that hierarchical Assumptions D and E hold. Then, we have $\theta'_0 = 0$ and for all $i \in S\backslash\{0\}$, θ'_i consistently verifies by induction:*

$$\theta'_i \leq \max_{a \in A_i^-} \{t^a + \max_{j \in \mathrm{supp}(a)} \theta'_j\} + \frac{M}{\chi^{(i)}} \mathbf{1}_{A_i^+ \neq \emptyset},$$

where for all i in S such that $A_i^+ \neq \emptyset$, $\chi^{(i)}\mathbf{1}$ denotes the growth rate vector of the evolution semigroup $\widetilde{\mathscr{S}}^{(i)}$ introduced in Lemma 3.

The proof of Theorem 7 makes an extensive use of the fact that the function v' introduced in Theorem 5 is nondecreasing. As indicated in the statement, a different behavior occurs if A_i^+ is empty or not. The case $A_i^+ = \emptyset$ is rather easy to address: it consists in applying the inductive scheme described earlier and waiting for states lower in the hierarchy to catch up the input. In contrast, the case $A_i^+ \neq \emptyset$, in which an "adverse" behavior may delay the catch-up by using actions of A_i^+, is harder to analyze. The key ingredient of the proof is to bound below the evolution of v' on the class $\mathcal{C}(i)$ by an affine stationary regime of the evolution semigroup $\widetilde{\mathscr{S}}^{(i)}$. Since the latter has positive growth rate, and v' eventually reaches a constant value on states located lower in the hierarchy, actions of A_i^+ cannot be selected for a too long duration.

We set $\theta^* := \max_{i \in S} \theta'_i$ the maximum of the catch-up times, so that after time θ^*, all states have caught up the input.

Example. We suppose that in addition to the usual calls arrival rate λ, our medical emergency call center undergoes the sudden arrival of M extra calls (for instance, corresponding to an event with many casualties) at time $\underline{t} = 0$. As proved in Corollary 1, if the call center is well-staffed, this peak of calls will ultimately be absorbed and the system shall return to a "cruise regime", where the policy $(\sigma(1), \sigma(3)) = (a_1^-, a_3^-)$ is applied (recall that the latter is the only proper policy of the system). As already observed and commented in Fig. 2, our SMDP modeling the call center satisfies hierarchical Assumptions D and E. As a result, the previous peak is overcome in a finite time θ^* and we can apply Theorem 7 to bound this duration. We obtain

$$\theta^* = \frac{M}{(\lambda_A \wedge \lambda_P) - \lambda} + t_1 + \frac{M}{\lambda_P - \lambda} + t_2 + t_3. \tag{8}$$

In this transience bound, we identify three terms coming from the mere communication delays between states (the times t_1, t_2 and t_3), and two terms originating from the states that could play ascending actions. The term $M/((\lambda_A \wedge \lambda_P) - \lambda)$ corresponds to a maximum time needed for state 1 to choose action a_1^-. We check that it is proportional to the amount of extra calls to pick up and is governed by the minimum residual handling speed of agents after they performed all the usual tasks, i.e. the throughput on class $\mathcal{C}(1)$ of semigroup $(\widetilde{\mathscr{S}}_t^{(1)})_{t \geq 0}$, given by either $\lambda_A - \lambda$ or $\lambda_P - \lambda$ depending on which policy realizes the minimal average-cost between $(\sigma(1), \sigma(3)) = (a_1^+, a_3^-)$ and $(\sigma(1), \sigma(3)) = (a_1^+, a_3^+)$. Similarly, the term $M/(\lambda_P - \lambda)$ bounds the time needed for state 3 to choose action a_3^- after $\theta_1' + t_1$; the denominator corresponds to the throughput on class $\mathcal{C}(3) = \{3, 4, 5\}$ of semigroup $(\widetilde{\mathscr{S}}_t^{(3)})_{t \geq 0}$ (applying policy $\sigma(1), \sigma(3)) = (a_1^+, a_3^+))$.

Building on Theorem 7, we may give a coarser and simple bound on the global catch-up time θ^* that emphasizes the tree structure of hierarchical SMDPs. We denote by d the maximal length of a descending path in S relatively to the ordering $(\leq)$, and by d^+ the maximal number of states with non-empty set of ascending playable actions along a descending path.

Corollary 2. *Under conditions of Theorem 7, convergence towards u^* occurs in a time θ^* such that $\theta^* \leq d \times t_{\max} + d^+ \times (M/\underline{\chi})$.*

This upper bound on the total time of convergence is governed by the ratio $M/\underline{\chi}$, i.e., it is proportional to the magnitude of the bulk that affected the input, and inversely proportional to the growth-rate of the inner part of the system (no taking into account the input). Rephrasing this result in terms of the λ-sink SMDPs, such as done in the example above, we get that $\underline{\chi} = \lambda' - \lambda$, where λ' is an intrinsic throughput of the system and λ is the input throughput. In the emergency call center example, λ' represents the maximal admissible input rate of calls. E.g., in (8), $\lambda' = \lambda_A \wedge \lambda_P$ is the maximal input flow that does not exceed the capacities of treatment of the MRA and physicians. Hence, the difference $\lambda' - \lambda$ has an intuitive interpretation as a *security margin*, which increases with the staffing.

Although the results of Theorem 7 and Corollary 2 were brought to address the question raised in Theorem 5 of the transience time to catch-up a specific Heaviside-type perturbation in λ-sink SMDPs, the derived bounds bear some generality. Indeed, any perturbation taking the form of an extra positive cost on the input—or equivalently an extra negative cost on states of $S \backslash \{0\}$—possibly not instantaneous but spread over a time window of length smaller than $t_{\max}$ can be bounded by such a template. This addresses the realistic case where the inner part of the system suffers a delay with respect to the input due to the perturbation. Besides, note that our techniques also allow us to bound the transience time of going back to an input-driven regime starting "ahead of it". This would correspond to a "negative bulk" of arrivals (e.g., a reduced rate of arrivals, or an absence of arrivals, over a short time period), i.e., the opposite situation to the one considered in our motivating application. As an examination of our proofs shows, this case is easier to handle because it is always non-optimal to pick ascending actions, thus resulting in bounds without the $M/\underline{\chi}$ terms.

Concluding Remarks

We provided an explicit upper bound for the time needed for an emergency call center to absorb a bulk of calls, relying on a continuous time Petri net model. This is based on an analysis of the conditions for convergence in finite time for semi-Markov decision processes. We showed that, under a "hierarchical" assumption on the topology of Petri nets, satisfied by emergency call centers, there is a finite absorption time, bounded by an expression whose essential term is of the form $d^+ M/(\lambda' - \lambda)$, where λ is the input rate, λ' is an intrinsic throughput of the system (with an explicit monotone dependence in the staffing), assuming that $\lambda' > \lambda$, M is the bulk size, and d^+ is a constant depending on the topology of the system but not on the staffing. Whereas the order $M/(\lambda' - \lambda)$ of our bound is optimal, we believe there is still room for improvement for the multiplicative constant d^+ that we obtained. We plan to address this issue in further work.

Acknowledgements. The second author was partially supported by the PhD fellowship program of DGA. We also acknowledge the support of IHEMI, and of the IROE-PRMO program of Fondation Mathématique Jacques Hadamard, sponsored by EDF. The modelling of the medical emergency call center described in Sect. 2 was carried out as part of a collaboration with the SAMU of AP-HP. We thank especially, Pr. P. Carli, Dr. E. Chanzy, Dr. E. Lecarpentier, Dr. Ch. Leroy, Dr. Th. Loeb, Dr. J.-S. Marx, Dr. N. Poirot and Dr. C. Telion for making this work possible, for their support and for insightful comments. We also thank all the other personals of the SAMU, in particular Dr. J. Boutet, J.-M. Gourgues, I. Lhomme, F. Linval and Th. Pérennou. This work also strongly benefited from the experience acquired, since 2014, on the analysis of the new platform "PFAU" (answering to the emergency numbers 17, 18 and 112), developed by PP. We thank especially Lcl S. Raclot and R. Reboul.

References

[ABG15] Allamigeon, X., Bœuf, V., Gaubert, S.: Performance evaluation of an emergency call center: tropical polynomial systems applied to timed Petri nets. In: Sankaranarayanan, S., Vicario, E. (eds.) FORMATS 2015. LNCS, vol. 9268, pp. 10–26. Springer, Cham (2015). https://doi.org/10.1007/978-3-319-22975-1_2

[ABG21] Allamigeon, X., Boyet, M., Gaubert, S.: Piecewise affine dynamical models of timed Petri nets-application to emergency call centers. Fundam. Inform. **183**(3–4), 169–201 (2021). Extended version of an article published in the International Conference on Applications and Theory of Petri Nets and Concurrency, 2020, Springer LNCS

[ABG22] Allamigeon, X., Boyet, M., Gaubert, S.: Computing transience bounds of emergency call centers: a hierarchical timed Petri net approach. Extended version of the present article. arXiv:2202.02752 (2022)

[AGN11] Akian, M., Gaubert, S., Nussbaum, R.: A Collatz-Wielandt characterization of the spectral radius of order-preserving homogeneous maps on cones. arXiv:1112.5968 (2011)

[AGQS19] Akian, M., Gaubert, S., Qu, Z., Saadi, O.: Solving ergodic Markov decision processes and perfect information zero-sum stochastic games by variance reduced deflated value iteration. In: 2019 IEEE 58th Conference on Decision and Control (CDC), pp. 5963–5970. IEEE (2019)

[BCOQ92] Baccelli, F., Cohen, G., Olsder, G.-J., Quadrat, J.-P.: Synchronization and Linearity. Wiley, Hoboken (1992)

[BG01] Bouillard, A., Gaujal, B.: Coupling time of a (max, plus) matrix. In: Proceedings of the Workshop on Max-Plus Algebras, a satellite event of the first IFAC Symposium on System, Structure and Control (Praha, 2001). Elsevier (2001)

[Bon07] Bonet, B.: On the speed of convergence of value iteration on stochastic shortest-path problems. Math. Oper. Res. **32**(2), 365–373 (2007)

[Boy22] Boyet, M.:. Piecewise affine dynamical systems applied to the performance evaluation of emergency call centers. Ph.D. manuscript, Ecole polytechnique (2022)

[BR19] Boeuf, V., Robert, Ph.: A stochastic analysis of a network with two levels of service. Queueing Syst. **92**(3-4), 30 (2019)

[BT91] Bertsekas, D.P., Tsitsiklis, J.N.: An analysis of stochastic shortest path problems. Math. Oper. Res. **16**(3), 580–595 (1991)

[CBFN13] Charron-Bost, B., Függer, M., Nowak, T.: Transience bounds for distributed algorithms. In: Braberman, V., Fribourg, L. (eds.) FORMATS 2013. LNCS, vol. 8053, pp. 77–90. Springer, Heidelberg (2013). https://doi.org/10.1007/978-3-642-40229-6_6

[CGQ95] Cohen, G., Gaubert, S., Quadrat, J.P.: Asymptotic throughput of continuous timed Petri nets. In: Proceedings of the 34th Conference on Decision and Control, New Orleans, December 1995

[CGQ98] Cohen, G., Gaubert, S., Quadrat, J.P.: Algebraic system analysis of timed Petri nets. In: Gunawardena, J. (ed.) Idempotency, Publications of the Isaac Newton Institute, pp. 145–170. Cambridge University Press, Cambridge (1998)

[DF68] Denardo, E.V., Fox, B.L.: Multichain Markov renewal programs. SIAM J. Appl. Math. **16**, 468–487 (1968)

[Fei94] Feinberg, E.A.: Constrained semi-Markov decision processes with average rewards. Z. Oper. Res. **39**(3), 257–288 (1994)

[GG04] Gaujal, B., Giua, A.: Optimal stationary behavior for a class of timed continuous Petri nets. Automatica **40**(9), 1505–1516 (2004)

[HA99] Hartmann, M., Arguelles, C.: Transience bounds for long walks. Math. Oper. Res. **24**(2), 414–439 (1999)

[HG11] Huang, Y., Guo, X.: Finite horizon semi-Markov decision processes with application to maintenance systems. Eur. J. Oper. Res. **212**(1), 131–140 (2011)

[HOvdW05] Heidergott, B., Olsder, G.-J., van der Woude, J.: Max-Plus at Work. Princeton University Press, Princeton (2005)

[MNS14] Merlet, G., Nowak, T., Sergeev, S.: Weak CSR expansions and transience bounds in max-plus algebra. Linear Algebra Appl. **461**, 163–199 (2014)

[MNS21] Merlet, G., Nowak, Th., Sergeev, S.: On the tightness of bounds for transients of weak CSR expansions and periodicity transients of critical rows and columns of tropical matrix powers. Linear Multilinear Algebra 1–31 (2021)

[Put14] Puterman, M.L.: Markov Decision Processes: Discrete Stochastic Dynamic Programming. Wiley, Hoboken (2014)

[Ros70] Ross, S.M.: Average cost semi-Markov decision processes. J. Appl. Probab. **7**(3), 649–656 (1970)

[SF78a] Schweitzer, P.J., Federgruen, A.: The functional equations of undiscounted Markov renewal programming. Math. Oper. Res. **3**(4), 308–321 (1978)

[SF78b] Schweitzer, P.J., Federgruen, A.: The asymptotic behavior of undiscounted value iteration in Markov decision problems. Math. Oper. Res. **2**, 360–381 (1978)

[SF79] Schweitzer, P.J., Federgruen, A.: Geometric convergence of value-iteration in multichain Markov decision problems. Adv. Appl. Prob. **11**, 188–217 (1979)

[SS12] Sergeev, S., Schneider, H.: CSR expansions of matrix powers in max algebra. Trans. AMS **364**(11), 5969–5994 (2012)

[Yus82] Yushkevich, A.A.: On semi-Markov controlled models with an average reward criterion. Theory Probab. Appl. **26**(4), 796–803 (1982)

Tools

Kong: A Tool to Squash Concurrent Places

Nicolas Amat(✉) and Louis Chauvet

LAAS-CNRS, Université de Toulouse, CNRS, INSA, Toulouse, France
namat@laas.fr

Abstract. Kong, the Koncurrent places Grinder, is a tool designed to compute the concurrency relation of a Petri net by taking advantage of structural reductions. The specificity of Kong is to rely on a state space abstraction, called polyhedral abstraction in previous works, that involves a combination of structural reductions and linear arithmetic constraints between the marking of places.

Keywords: Petri nets · Abstraction techniques · Reachability problems

1 Introduction

Kong, the *Koncurrent places Grinder*, is a recent formal verification tool for Petri nets that can take advantage of structural reductions to accelerate the verification of reachability properties. We made our code freely available under the GPLv3 license and all the software, scripts and data used in this paper are available on GitHub.

In a nutshell, Kong can compute a reduced Petri net, (N', m'), from an initial one, (N, m), and prove properties about the initial net by exploring only the state space of the reduced one. A difference with previous works on structural reductions [4,15], is that our approach is not tailored to a particular class of properties—such as safety or the absence of deadlocks—but could be applied to more general problems. In this paper, we focus on a particular problem supported by Kong, called the *concurrent places problem*.

The correctness of our tool relies on two main theoretical notions. First, a new state space abstraction method, that we called *polyhedral abstraction* in [1], which involves a combination of structural reductions and linear arithmetic constraints between the marking of places. Second, a new data structure, called *Token Flow Graph* (TFG) in [2], that can be used to compute properties based on a polyhedral abstraction. We give a short overview of these two notions in this paper. Nonetheless, our main objective here is to describe the features implemented in our tool.

The basic operation involved in our approach is to compute reductions of the form $(N, m) \rhd_E (N', m')$ where: N is an initial Petri net (that we want to

L. Bernardinello and L. Petrucci (Eds.): PETRI NETS 2022, LNCS 13288, pp. 115–126, 2022.
https://doi.org/10.1007/978-3-031-06653-5_6

analyse); N' is a residual net (hopefully simpler than N); and E is a system of linear equations. The goal is to preserve enough information in E so that we can rebuild the reachable markings of N knowing only those of N'. We say in this case that N and N' are E-equivalent. While there are many examples of the benefits of structural reductions when model-checking Petri nets, the use of an equation system (E) for tracing back the effect of reductions is new.

In our approach, the computation of structural reductions is delegated to a separate tool. We mention two possibilities in this paper. First the tool REDUCE, which is a new addition to the Tina model-checking toolbox since version 3.7 (https://projects.laas.fr/tina). We also describe, with more details, a new open-source framework called SHRINK. This is a highly customizable tool, and also a library, that we hope can be reused and improved in other contexts.

A TFG is a graph-like data structure that can be built from an E-equivalence statement, $(N, m) \rhd_E (N', m')$, and that embodies the structure of the equations occurring in E. KONG can build a TFG from sequences of reductions computed using SHRINK or REDUCE, and use it to symbolically explore the state space of the initial net.

We describe two applications of TFGs. The main application [2] is to compute the *concurrency relation* of a Petri net; what is also known as the concurrent places problem [8]. The goal is to enumerate all pairs of places that can be marked together in some reachable marking. This problem has practical applications, for instance because of its use for decomposing a safe Petri net into the product of concurrent processes [8,9]. It also provides an interesting example of safety property that nicely extends the notion of *dead places*; meaning places that can never be marked. To illustrate the versatility of our approach, we also describe a new feature, implemented in KONG, for checking the reachability of a given marking. Our method exploits the TFG to compute a reduced, projected marking that needs to be found in the reduced net. This is a direct illustration of the philosophy behind KONG, which is solving complex problems by first reducing a Petri net; then solving the problem in a "lower dimension"; before finally transposing this solution to the initial net.

Outline. The rest of the paper is organized as follows. In Sect. 2, we detail how to install and use KONG. Section 3 describes the architecture of KONG and SHRINK. We illustrate the workings of KONG on a concrete example, in Sect. 4. Finally, we validate our tool by discussing the results of experiments performed with nets used in the 2021 edition of the Model Checking Contest (MCC).

2 Commands, Basic Usage and Installation

KONG is an open-source tool made freely available on GitHub (https://github.com/nicolasAmat/Kong). The code repository also provides all the material to reproduce the experiments described in Sect. 5.

Dependencies. KONG is written in Python and requires a version 3.5 or higher. It also requires the `graphviz` Python library in order to output a graphical description of Token Flow Graphs (optional). Scripts and models included in the repository are used for benchmarking and for continuous testing. KONG is intended to be as understandable as possible; the code is heavily documented and we provide many tracing and debugging options that can help understand its inner workings.

We support two different tools to compute polyhedral abstractions, REDUCE and SHRINK, that both use the same input and output formats. REDUCE is a tool developed inside the Tina toolbox [14], since version 3.7. It is currently used by the TINA.TEDD and SMPT model-checkers, that both compete in the Model Checking Contest (MCC) [3,13], albeit on different examinations. SHRINK is an open-source alternative, on which we focus in Sect. 3. KONG runs REDUCE if the executable is in the current `PATH` environment variable, but automatically switches to SHRINK otherwise. It is still possible to enforce the use of SHRINK by using the `--shrink` option. It is also possible to directly provide a precomputed result of structural reductions with the option `--reduced-net`.

Concurrent and Dead Places. KONG is a CLI tool organized around subcommands to expose its different features. The tool provides several options that are described in the documentation using `--help`. We give a brief description of some of them in the following sections.

The main subcommands of KONG are `conc` and `dead` for, respectively, computing the concurrent relation and the list of dead places in a net. When computing a concurrency matrix, KONG relies on an external tool to compute the concurrency matrix of the reduced net. This is currently done using CÆSAR.BDD, part of the CADP toolbox [7,12], which is the state-of-the-art tool for the concurrent places problem [7,12].

KONG takes as inputs ordinary, safe Petri nets defined using either the Petri Net Markup Language (PNML) [11], or the Nest-Unit Petri Net (NUPN) format [9]. (The file format is automatically detected from the file extension.) The use of a NUPN decomposition, which provides information about the concurrent structure of the net, can bring a significant performance improvement. The tool was designed to be fully compatible with Petri net instances used in the MCC. For instance, we can make use of NUPN information added to a PNML model using its tool-specific extension mechanism.

KONG can be executed as a Python script or converted into a standalone executable using `cx_Freeze`. Each subcommand only requires the path to the input Petri net (with a `.pnml` or `.nupn` extension). Hence a typical call to KONG is of the form '`./kong.py conc model.pnml`'. We also provide two main options to limit the exploration performed by CÆSAR.BDD: `--bdd-timeout` to set a time limit and `--bdd-iterations` to limit the number of iterations. Debugging options are described in Sect. 4.

The concurrency relation of a Petri net, denoted C, is encoded as a symmetric matrix of dimension $|P|$, where $|P|$ is the number of places in the net. We also use the name *concurrency matrix*. We use the notation $C[p, q] = 1$ when places p, q can be marked together in a reachable state, and 0 otherwise. In some cases, we may need to work with "partial relations"; for example when we impose a time limit. We say that the concurrency matrix is *incomplete* in this case and use the value '·' (a dot) for pairs of places where the relation is undecided.

Our output format for the concurrency matrix is taken from CÆSAR.BDD. We can output our results using a compressed format, based on a run-length encoding (RLE) of the rows of C. For the sake of readability, it is possible to disable this encoding using option `--no-rle`. It is also possible to print the place ordering with option `--place-names`.

A call to '`kong.py conc`' delegates the computation of the concurrent relation on the reduced net to the tool CÆSAR.BDD. It can also take as input a precomputed concurrency matrix of the reduced net, using option `--reduced-matrix`. Likewise, the `dead` subcommand provides option `--reduced-vector` if we have a precomputed list of dead places for the reduced net.

Marking Reachability. The `reach` subcommand provides a procedures to check if a given marking is reachable. Like previously, this command relies on an external tool to check if a marking is reachable in the reduced net. To this end, we use SIFT, which is an explicit-state model-checker for Petri nets from the Tina toolbox, that can check reachability properties on the fly.

The tool takes as input a Petri net—not necessarily safe, ordinary or bounded—described either in the PNML or the NET format. (NET is the specification format of the Tina toolbox). The target marking is defined using a simple textual format, as a space-separated list of place identifiers with their multiplicities, of the form `p*k`, where `p` is a place and `k` is a positive integer. By default, places that are not listed contain no tokens. The path to the file describing the target marking is given using option `--marking`.

3 Architecture of Kong

Our tool is basically composed of three modules: `kong.py` the front-end program in charge of parsing command-line options; `pt.py` a Petri net parser; and `tfg.py` the data structure and computational module based on Token Flow Graphs. We illustrate the architecture of KONG in Fig. 1, where we describe the different steps involved during a typical computation. The first step is to reduce the input Petri net, say (N, m), using the SHRINK tool. SHRINK outputs a reduced net (N', m') and a system of linear equations E. We display in Fig. 2 a sequence of structural reductions, with their equations, computed using SHRINK. By construction, the result of this first stage is guaranteed to be a polyhedral abstraction.

Then we build a Token Flow Graph, $[\![E]\!]$, from the set of linear equations in E. The TFG is a Directed Acyclic Graph (DAG), capturing the specific structure of the equations in E, that allows us to reason about the reachable markings by playing a token game on this graph.

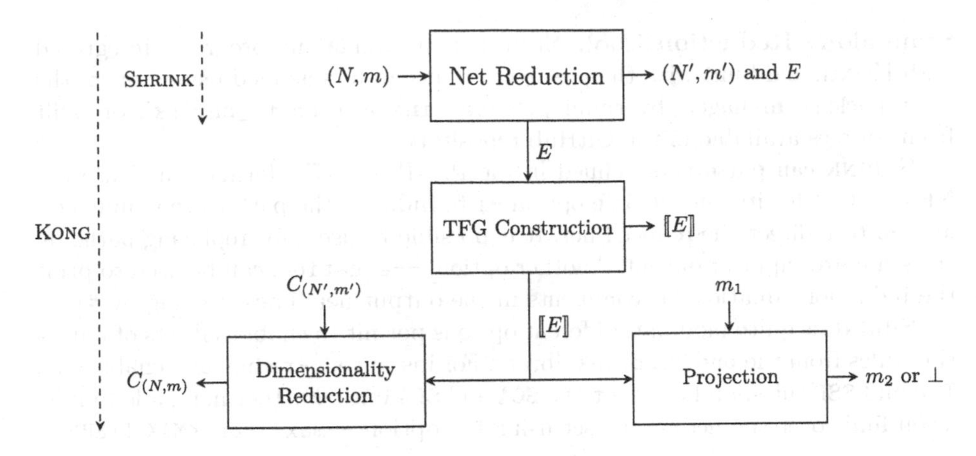

Fig. 1. KONG's architecture.

At this stage, we must distinguish two possible cases. First, the net could be fully reduced, meaning the resulting net is "empty"; it has no remaining places. In this case, the set of markings of (N, m) is exactly the solutions of the linear system E. Hence the TFG is enough to compute the concurrency matrix using an algorithm that we call *dimensionality reduction*, or to decide if a given marking is reachable. Otherwise, we have a non-trivial reduced net, in which case we need to compute the concurrency matrix of (N', m') or to check the reachability of the *projection* of our marking of interest.

The first module of our pipeline relies on two Rust libraries, based on a common crate called `pnets`, that defines functions for parsing, manipulating and reducing Petri nets. This code is freely available on GitHub (https://github.com/Fomys/pnets), under the MIT license.

Petri Nets Library. The `pnets` library is the core for parsing and manipulating Petri nets. It supports both standard and timed Petri nets. Internally, Petri nets are stored using adjacency lists, ensuring a low memory footprint and fast iterations over connected places and transitions. The toolbox includes two sublibraries for parsing nets: `pnets_pnml` and `pnets_tina`, respectively for the PNML and NET formats.

Structural Reduction Library. The `pnets_shrink` library implements reductions rules described in [5,6]. It implements a large subset of the reductions included in REDUCE, such as (definitions refers to the ones in [6]): `T` - Redundant transitions (def. 1), `P` - Redundant places (def. 2), `SCA` - Simple chain agglomeration (def. 5), `SLA` - Simple loop agglomeration (def. 6), and `SSP` - Source-sink pair (def. 10).

Standalone Reduction Tool. SHRINK is a standalone program, integrated with KONG, and built with the `pnets` crate. It can be installed using `cargo`, the Rust package manager, by running '`cargo install pnets_shrink`', or built from sources available in the GitHub repository.

SHRINK can parse nets defined in the PNML or NET formats, and use the NET format for its output. Use option `-i` to indicate the path to the input net, and `-o` to redirect the reduced net. It is possible to use `-` for replacing paths by the standard input or output. Another option, `--equation`, can be used to print the reduction equations as comments in the output net (lines starting by `#`).

SHRINK is quite modular, different options permit to enable subsets of reduction rules from the `pnets_shrink` library. For instance, `--redundant` enables the `T`, `P` and `SSP` rules, and `compact` the `SCA` and `SLA` ones. Furthermore, a loop iteration limit over the net can be set using the option `--max-iter <MAX_ITER>`.

4 Concrete Example

The simplest way to illustrate the usage of KONG is to look at a concrete example. This is also a good opportunity to show the debugging options provided by our tool. Assume (N, m) is the net in top left position in Fig 2.

Net Reduction. Structural reduction is performed iteratively, until no new reductions are possible. We display, Fig. 2, a sequence of four reductions that leads to the result computed with SHRINK; the marked net at the bottom-right. Each row is an example of reduction, and its associated equation. First, it is always safe to remove a *redundant place*, e.g. a place with the same pre and post conditions than another one. This is the case with places p_4, p_5. Redundant places can sometimes be found by looking at the structure of the net, but we can use more elaborate methods to find redundant places by solving an integer linear programming problem [16]. After the removal of p_5, we obtain the equation $p_4 = p_5$, and we are left with the residual net at the left part of row 2. In this case, we can use an agglomeration rule, which states that we can fuse places inside a "deterministic sequence" of transitions. For instance to simplify places p_1 and p_2 into a new place, a_{12}. Similar situations, where we can aggregate several places together, can be found by searching patterns in the net. After this step, we find a new opportunity to reduce a redundant place, based on the structural invariant $a_{12} = p_3 + p_4$. We conclude by agglomerating places p_3 and p_4 into a new place, a_{13}.

At the end of these reductions, we obtain the reduced net, (N', m'), with only 3 places instead of 6. We also obtain a system of four linear equations $E \triangleq (p_5 = p_4), (a_{12} = p_1 + p_2), (a_{12} = p_3 + p_4), (a_{13} = p_3 + p_4)$.

KONG provides an option, `--save-reduced-net`, to save the reduced net into a specific file. Additionally, we can print the reduction equations with the option `--show-equations`.

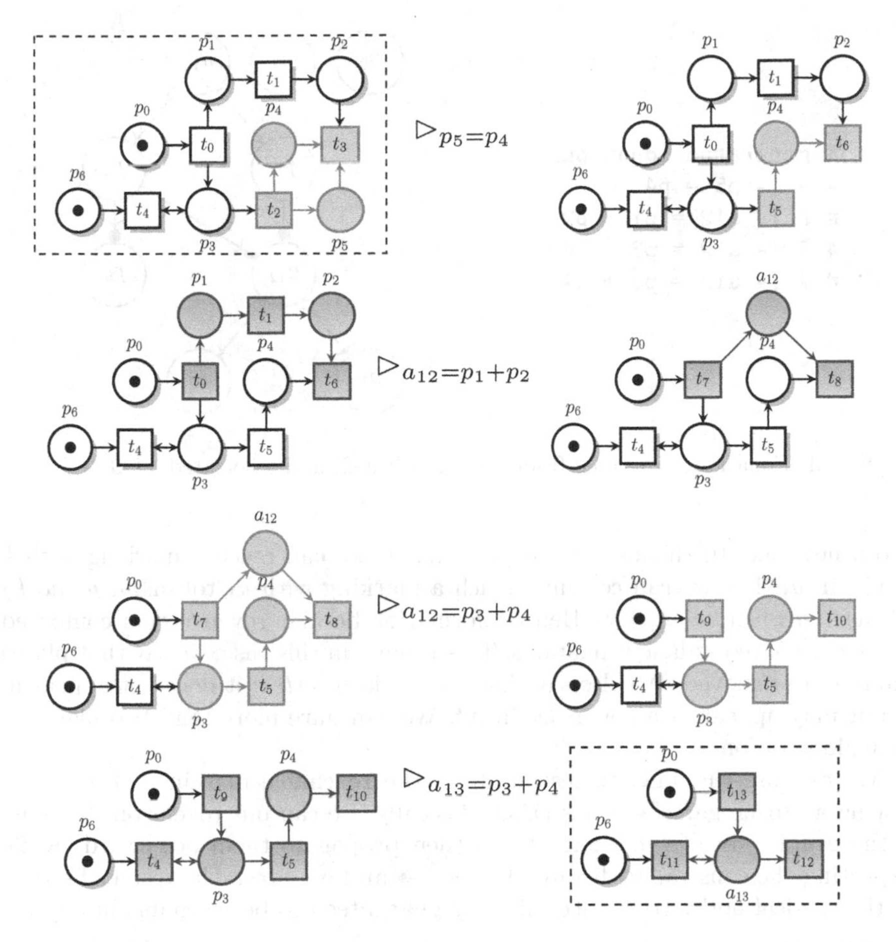

Fig. 2. Example of sequence of four reductions leading from the net N to N'.

TFG Construction. KONG can build the TFG associated with the linear system E; see Fig. 3. It is possible to output a graphical version of the TFG using option `--draw-graph`. The TFG is a DAG where the vertices are the places of the input and reduced net, in addition to the free variables from E. The set of roots (nodes with no predecessor) is exactly the set of places of the reduced net N'. Arcs in the TFG are used to depict the relation induced by equations in E.

A TFG includes two different kinds of arcs. Arcs for *redundancy equations*, $q \rightarrow\!\!\bullet\, p$, to represent equations of the form $p = q$ (or $p = q + r + \dots$), corresponding to redundant places. In this case, we say that place p is *removed* by arc $q \rightarrow\!\!\bullet\, p$, because the marking of q may influence the marking of p, but not necessarily the other way round.

The second kind of arcs, $a \circ\!\!\rightarrow p$, is for *agglomeration equations*. It represents equations of the form $a = p + q$, generated when we agglomerate several places

```
# generated equations
# R |- p5 = p4
# A |- a12 = p1 + p2
# R |- a12 = p3 + p4
# A |- a13 = p3 + p4
```

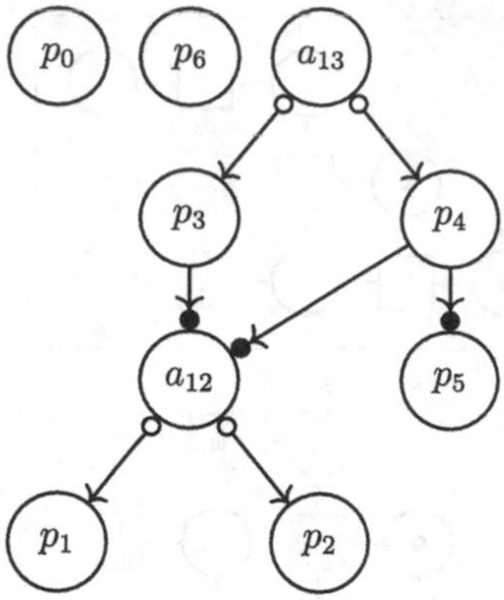

Fig. 3. Equations generated from net N, in Fig. 2, and associated TFG $[\![E]\!]$.

into a new one. In this case, we expect that if we can reach a marking with k tokens in a, then we can certainly reach a marking with k_1 tokens in p and k_2 tokens in q when $k = k_1 + k_2$. Hence information flows in reverse order compared to the case of redundancy equations. This is why, in this case, we say that places p and q are removed. We also say that node a is *inserted*; it does not appear in N but may appear as a new place in N'. We can have more than two places in an agglomeration.

We can use the TFG to reason about the reachable markings of a net by playing a "token game" on this DAG. Basically, we can put tokens on the roots of the graph (given a marking of N') then propagate them downwards while respecting the constraints dictated by the $\rightarrow\!\!\bullet$ and $\circ\!\!\rightarrow$ arcs. The result observed on the $\circ\!\!\rightarrow$-leaf nodes (the places of N) is guaranteed to be reachable in (N, m).

Concurrent Places Algorithm. With subcommand `conc`, the final stage is to compute the concurrency matrix of the input net, $C_{(N,m)}$, from the one of the reduced net, $C_{(N',m')}$. Currently, Kong uses CÆSAR.BDD to compute $C_{(N',m')}$. But we could adapt Kong to use any other tool that can compute the concurrency relation, such as [17]. It is possible to output this matrix with option `--show-reduced-matrix` (resp. `--show-reduced-vector` if we use subcommand `dead`).

We can give an intuition for our *Dimensionality Reduction* algorithm using our example. For instance, we have that place a_{13}, in the reduced net N' of Fig. 2, is non-dead (because we can fire t_9). As a consequence, all the successors nodes of a_{13} in the TFG (that are also places in N) must also be non-dead, meaning $C[p_i, p_i] = 1$ for all i in 1..5. Also, we can deduce that p_4 is concurrent to p_5 (meaning $C[p_4, p_5] = 1$), because of the redundancy $p_5 = p_4$, and p_1, p_2 are concurrent to p_3, p_4, p_5. A detailed description of our algorithm can be found in [2].

Marking Reachability Decision. With subcommand `reach`, the final step is to project the marking of interest into a new marking defined on the reduced net, and to check its reachability in the reduced net (N', m').

We illustrate this procedure by taking two concrete examples on the marked net N given in Fig. 2 (first row, left). Assume we want to check if marking $m_1 \triangleq (p_0 = 0, p_1 = 1, p_2 = 1, p_3 = 1, p_4 = 1, p_5 = 1, p_6 = 0)$ is reachable in (N, m). This marking can be mapped to a unique marking of N', namely $m_2 \triangleq (p_0 = 0, p_6 = 0, a_{13} = 2)$. (Use option `--show-projected-marking` to output this marking.) Deciding if marking m_1 is reachable in (N, m) is equivalent to deciding if $m_2 \triangleq (p_0 = 0, a_2 = 2, p_6 = 0)$ is reachable in (N', m') (which it is not). Observe that m_1 would be reachable if the initial marking m was $(p_0 = 2, p_6 = 1)$ and the other places empty.

The "marking projection" algorithm can also directly return with a contradiction ($\perp$), meaning that the target marking cannot be reached. Assume we want to check the reachability of a marking m'_1 such that $m'_1(p_4) = 2$ and $m'_1(p_1) = m'_1(p_2) = 0$. It is not possible to project this marking into N' while respecting the constraint given in the TFG. In this case, we directly obtain that m'_1 is not reachable in (N, m).

5 Performance

We used the database of models provided by the Model Checking Contest [3,13] to study the performances of Kong. For the computation of concurrent matrices, among the 562 safe and ordinary instances used in the MCC'2021, we kept only the ones with reduction opportunities; which amount to 424 nets in total. And we selected 426 instances (among 1 411) to evaluate the marking reachability procedure, for which we generated 5 queries that are markings found using a "random walk" on the state space of the net. We used Reduce to compute net reductions, we computed the concurrency matrices on the reduced net with Cæsar.bdd, version 3.6, part of CADP version 2022-b "Kista", published in February 2022 and used Sift to check the reachability of the projected marking.

To understand the impact of reductions on the computation time, we compare Cæsar.bdd and Sift alone, on the initial net, and Kong + Reduce + Cæsar.bdd or Sift on the reduced net. We display our results in the charts of Fig. 4, which gives the number of feasible instances, for each tool, when we change the timeout value. (To reproduce the experiments follow the instructions from the README file in the `benchmark/` directory of the repository).

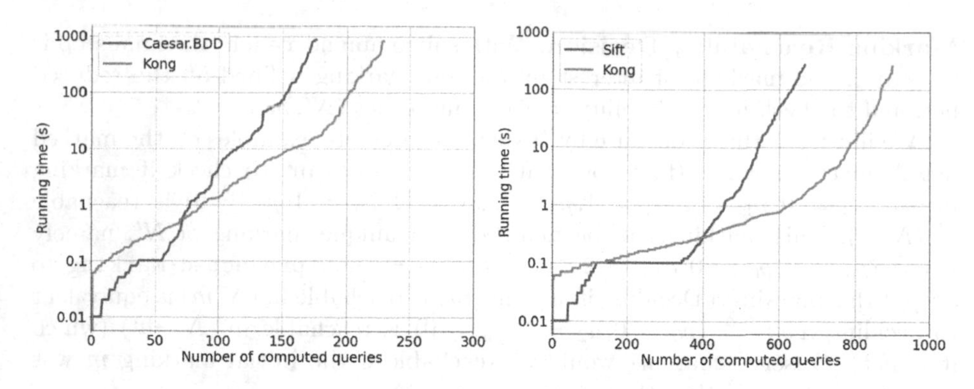

Fig. 4. Minimal timeout to compute a given number of instances: (left) concurrency matrices, (right) reachable markings.

We observe that net reductions have a clear impact on the speed-up and that we can compute more instances with reductions than without: 229 matrices for KONG against 176 for CÆSAR.BDD alone, with a timeout of 15 min. The same observations holds for the reachability procedure: 901 queries solved for KONG against 667 for SIFT alone, with a timeout of 5 min. Furthermore concerning the tool REDUCE, we obtained on safe instances a mean reduction ratio—that is the quotient between how many places can be removed and the number of places in the initial net—of 40% (median of 26%), computed in an average time of 0.7 s (median of 0.2 s).

6 Future Work

Both KONG and SHRINK are destined to evolve. For instance, we want to experiment with more challenging problems using KONG and the TFG data-structure. We are particularly interested in answering reachability queries expressed using a boolean combination of constraints over place markings. Another interesting problem would be to support the verification of Generalized Mutual Exclusion Constraints, like in [10], that requires checking invariants involving a weighted sums over the marking of places, of the form $\sum_{p \in P} w_p.m(p) \leqslant k$, with $w_1, \ldots, w_n, k$ constants in $\mathbb{Z}$.

We also want to explore new reduction rules using our polyhedral abstraction framework SHRINK. We already developed new reduction rules for specific models from the MCC, such as `Election2020` and `ViralEpidemic`, and plan to look at more specific examples of reduction rules.

To conclude, we are convinced that there is still a lot of work to be done to compute polyhedral abstractions, and to apply them on useful and complex problems.

Acknowledgements. We would like to thank Bernard Berthomieu and Silvano Dal Zilio for their help on the development of our reduction library, and Pierre Bouvier for his remarks that helped improve the quality of KONG.

References

1. Amat, N., Berthomieu, B., Dal Zilio, S.: On the combination of polyhedral abstraction and SMT-based model checking for Petri nets. In: Buchs, D., Carmona, J. (eds.) PETRI NETS 2021. LNCS, vol. 12734, pp. 164–185. Springer, Cham (2021). https://doi.org/10.1007/978-3-030-76983-3_9
2. Amat, N., Dal Zilio, S., Le Botlan, D.: Accelerating the computation of dead and concurrent places using reductions. In: Laarman, A., Sokolova, A. (eds.) SPIN 2021. LNCS, vol. 12864, pp. 45–62. Springer, Cham (2021). https://doi.org/10.1007/978-3-030-84629-9_3
3. Amparore, E., et al.: Presentation of the 9th edition of the model checking contest. In: Beyer, D., Huisman, M., Kordon, F., Steffen, B. (eds.) TACAS 2019. LNCS, vol. 11429, pp. 50–68. Springer, Cham (2019). https://doi.org/10.1007/978-3-030-17502-3_4
4. Berthelot, G.: Transformations and decompositions of nets. In: Brauer, W., Reisig, W., Rozenberg, G. (eds.) ACPN 1986. LNCS, vol. 254, pp. 359–376. Springer, Heidelberg (1987). https://doi.org/10.1007/978-3-540-47919-2_13
5. Berthomieu, B., Le Botlan, D., Dal Zilio, S.: Petri net reductions for counting markings. In: Gallardo, M.M., Merino, P. (eds.) SPIN 2018. LNCS, vol. 10869, pp. 65–84. Springer, Cham (2018). https://doi.org/10.1007/978-3-319-94111-0_4
6. Berthomieu, B., Le Botlan, D., Dal Zilio, S.: Counting Petri net markings from reduction equations. Int. J. Softw. Tools Technol. Transf. **22**(2), 163–181 (2019). https://doi.org/10.1007/s10009-019-00519-1
7. Bouvier, P., Garavel, H.: Efficient algorithms for three reachability problems in safe Petri nets. In: Buchs, D., Carmona, J. (eds.) PETRI NETS 2021. LNCS, vol. 12734, pp. 339–359. Springer, Cham (2021). https://doi.org/10.1007/978-3-030-76983-3_17
8. Bouvier, P., Garavel, H., Ponce-de-León, H.: Automatic decomposition of Petri nets into automata networks – a synthetic account. In: Janicki, R., Sidorova, N., Chatain, T. (eds.) PETRI NETS 2020. LNCS, vol. 12152, pp. 3–23. Springer, Cham (2020). https://doi.org/10.1007/978-3-030-51831-8_1
9. Garavel, H.: Nested-unit Petri nets. J. Log. Algebraic Methods Program. **104**, 60–85 (2019). https://doi.org/10.1016/j.jlamp.2018.11.005
10. Giua, A., DiCesare, F., Silva, M.: Generalized mutual exclusion contraints on nets with uncontrollable transitions. In: IEEE International Conference on Systems, Man, and Cybernetics. IEEE (1992). https://doi.org/10.1109/ICSMC.1992.271666
11. Hillah, L.M., Kordon, F., Petrucci, L., Trèves, N.: PNML framework: an extendable reference implementation of the Petri net markup language. In: Lilius, J., Penczek, W. (eds.) PETRI NETS 2010. LNCS, vol. 6128, pp. 318–327. Springer, Heidelberg (2010). https://doi.org/10.1007/978-3-642-13675-7_20
12. INRIA: CADP (2020). https://cadp.inria.fr/
13. Kordon, F., et al.: Complete results for the 2020 edition of the model checking contest (2021). http://mcc.lip6.fr/2021/results.php
14. LAAS-CNRS: Tina Toolbox (2020). http://projects.laas.fr/tina

15. Murata, T., Koh, J.: Reduction and expansion of live and safe marked graphs. IEEE Trans. Circ. Syst. **27**(1), 68–71 (1980). https://doi.org/10.1109/TCS.1980.1084711
16. Silva, M., Terue, E., Colom, J.M.: Linear algebraic and linear programming techniques for the analysis of place/transition net systems. In: Reisig, W., Rozenberg, G. (eds.) ACPN 1996. LNCS, vol. 1491, pp. 309–373. Springer, Heidelberg (1998). https://doi.org/10.1007/3-540-65306-6_19
17. Wiśniewski, R., Wiśniewska, M., Jarnut, M.: C-exact hypergraphs in concurrency and sequentiality analyses of cyber-physical systems specified by safe Petri nets. IEEE Access **7**, 13510–13522 (2019). https://doi.org/10.1109/ACCESS.2019.2893284

IOPT-Tools – From Executable Models to Automatic Code Generation for Embedded Controllers Development

Fernando Pereira[1,4], Filipe Moutinho[2,4], Anikó Costa[2,4], João-Paulo Barros[3,4], Rogério Campos-Rebelo[3,4], and Luis Gomes[2,4(✉)]

[1] Instituto Superior de Engenharia de Lisboa, Lisbon, Portugal
fjp@deea.isel.ipl.pt

[2] NOVA School of Science and Technology, Caparica, Portugal
{fcm,akc,lugo}@fct.unl.pt

[3] Polytechnic Institute of Beja, Beja, Portugal

[4] Centre of Technology and Systems-UNINOVA, Caparica, Portugal
{jpb,rcr}@uninova.pt

Abstract. This paper describes the IOPT-Tools cloud-based tool-chain, offering a complete set of tools supporting design automation for embedded controller's development, benefiting from adopting a model-driven development attitude. The tools are freely available online at http://gres.uninova.pt/IOPT-Tools/. The tool-chain relies on IOPT nets (Input-Output Place-Transition nets) to describe the controller behavior, allowing an explicit representation of constraints on input and output signals and events, which is necessary for the development of controllers. The IOPT-Tools tool-chain includes tools for interactive graphical IOPT nets models editing, simulation and test (token-player, timing diagram, remote debugging), as well as a state-space generator, state-space visualization, and a query system for properties verification. The tool-chain also supports the automatic generation of execution code to be directly deployed in the controllers' implementation platforms, such as FPGA boards, as well as Arduino, Raspberry, and other Linux-based boards. Most notably, it is possible to obtain C code and VHDL code to be directly deployed into the referred boards without writing/changing a line in the generated code. The tool-chain uses the PNML format for storing the models and can import PNML models generated by other frameworks, automatically generating, if necessary, an associated graphical representation. After, it is possible to add input and output signals and events to create controller models. IOPT-Tools also supports net operations, namely net addition, allowing composition of sub-models, and net splitting, which in conjunction with the use of clock domains and dedicated communication channels support the development of distributed controllers.

Keywords: Embedded controllers · Design Automation · Code Generation · Petri nets

This work was partially financed by Portuguese Agency FCT - Fundação para a Ciência e Tecnologia, in the framework of project UIDB/00066/2020.

L. Bernardinello and L. Petrucci (Eds.): PETRI NETS 2022, LNCS 13288, pp. 127–138, 2022.
https://doi.org/10.1007/978-3-031-06653-5_7

1 Introduction

Traditionally, Petri nets are sometimes classified as *autonomous* or *non-autonomous*. Autonomous nets are used when their semantics are independent of the external environment, namely when the transition enabling and firing is not constrained by any element coming from the environment. On the other hand, the Petri net classes usually named as *non-autonomous* have semantics taking into consideration elements from the external environment, being augmented with dependencies on time, inputs sensors, and output actuators.

For several engineering areas, particularly when one wants to model the behavior of a controller for discrete-event systems using Petri nets, it is of paramount importance to explicitly model the relation between the controller (and associated Petri net model) and the environment (that is under control). This relation includes dependencies on input and output signals and events and the need to change the model execution semantics accordingly.

To that end, several classes of Petri nets have been proposed during the last decades, such as interpreted and synchronized nets [1–3], as well as several other classes having factory automation applications in mind [4–6].

However, according to the Petri Nets Tools Database [7], very few tools provide adequate support for the modeling of external dependencies, especially when automatic code generation is considered, for common microcontrollers or programmable devices hardware for implementation.

The tools framework presented in this paper (IOPT-Tools), freely available at [8], is intended to mitigate the referred gap.

The structure of the paper is as follows. Section 2 presents the motivation and objectives. After, Sect. 3 provides background information on IOPT nets, main characteristics and execution semantics. Section 4 briefly presents the IOPT-Tools architecture and associated development flow, while the main functionalities of the IOPT-Tools framework are presented in Sect. 5. Finally, Sect. 6 concludes and identifies some future works.

2 Motivation and Objectives

The IOPT-Tools framework [9,10] relies on a non-autonomous class of Petri nets, tailored for the specification, analysis, and synthesis of controllers: the IOPT nets - Input-Output Place-Transition nets [11–13]. These nets allow the automatic generation of executable code for discrete event-driven controllers, most notably C for software frameworks, and VHDL for hardware implementations.

In this paper, several new functionalities of the tool-chain are presented (when compared with [9,10]), particularly new features allowing simulation, remote debugging and testing, operations on nets (namely net addition and net decomposition), and new support for interoperability with external tools' frameworks.

In this sense, the main goal of the IOPT-Tools framework is to provide support for all phases of the development of controllers for discrete-event systems.

The development starts with the behavioral description through an IOPT net model and ends with the automatic generation of execution code ready to be directly deployable into implementation platforms. Also, the IOPT net model can be simulated and validated using integrated tools or relying on external tools interoperability.

3 Preliminaries

IOPT nets extend Place-Transition nets, adding dependencies on several non-autonomous elements of the environment under control. In particular, input and output signals and events are available as annotations to the graph, and the execution semantics was changed accordingly. The syntax and semantics of IOPT nets can be found elsewhere [11,13].

Transition firing is also constrained by input events occurrence and conditioned by the evaluation of guard functions, which depend on input signals values. Upon a transition firing, output events associated with the transition can be generated and output signals updated. Output signals activation can also be associated with places with non-empty markings.

Test (or read) arcs are available as a major convenience to allow compact modeling for specific situations. Priorities can be associated with transitions, supporting strategies to automatically solve conflicts and allow deterministic execution, even at the expense of an unfair/unbalanced decision.

Ensuring a deterministic execution is very important for embedded controllers in many application areas, namely industrial systems and safety-critical systems. In order to ensure it, a step-based execution semantics is adopted relying upon the following aspects:

- Maximal step execution, as all transitions enabled and ready will fire in the same execution step;
- Cycle-accurate execution, delaying the acquisition of input events to the beginning of the next execution step, considering that the execution of a step by the controller will take some time (non-instantaneous);
- Single server semantics, transitions will fire only once per execution step, even if enabled multiple times.

Input events can be generated due to changes in input signals (comparing the value of the signal in two consecutive steps) or as autonomous events, independent of any signal and helpful for simulations, inter-subsystem communication, or both. Similarly, output events generated by transition firing can be associated with output signals (forcing their updating) or as autonomous events.

Finally, support for modeling of Globally Asynchronous Locally Synchronous (GALS) systems and distributed execution of the model is achieved introducing decomposition of the model into sub-models (using the net splitting operation [14]) and the concept of time-domain [15], which is associated with the nodes of a subnet (ultimately, each subnet will be associated with a specific implementation platform). The different subnets can be interconnected using a new type of arcs, named *channel arcs* and a new type of node having place semantics and representing *communication channels*. Five types of *communication channels* are available; due to space limitations, further information can be obtained in [13,15].

4 IOPT-Tools Architecture and Availability

As already mentioned, IOPT-Tools [9,10] is a cloud-based platform, freely available at [8]. It relies on IOPT nets and includes a set of tools for developing controllers of discrete event-driven systems. This set of tools, partly presented in Fig. 1, includes an interactive graphical Petri net editor, simulator, remote debugger and tester, property verification tools based on reachability graph generation, and automatic code generation tools. The automatically generated C code or VHDL hardware descriptions are amenable to be directly deployed (without writing additional lines of code) into most common controllers' implementation platforms, such as Arduino series, Raspberry Pi series, Intel Edison, Red-Pitaya, Coral-dev-board, Zed-boards, and other Linux-based platforms and FPGA boards. It is important to note that other implementation platforms are also supported for software implementations if a C compiler is available. It is also possible to generate other types of execution code associated with the model (but not explicitly analyzed in this paper), such as Instruction List (IL) code amenable to be deployed into a specific line of industrial Programmable Logic Controllers (PLC) from Siemens, as well as Simulink System Block amenable to be integrated with MatLab/Simulink projects.

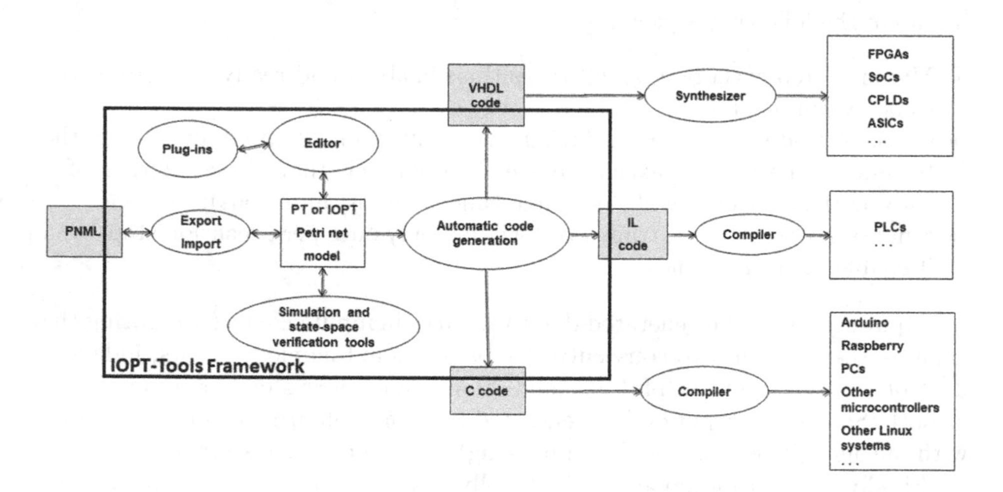

Fig. 1. IOPT-Tools: Block diagram of tools.

All tools share a common web-based graphical user interface. No software installation is required, taking advantage of standard W3C technologies, such as

SVG (Scalable Vector Graphics), AJAX (Asynchronous Javascript and XML), and XSLT (eXtensible Stylesheet Language Transformations).

Three types of user accounts, all having access to all functionalities of the framework, can be considered for usage of the IOPT-Tools framework:

- Default user account «guest» (password «guest») is a common area, but without privacy, as all models can be used/changed/deleted by anyone;
- Default user account «models» (password «models») is also a common area with similar privileges as «guest» account, but containing a set of read-only pre-prepared reference models providing to the novice a brief introduction to IOPT-Tools;
- Personal user accounts are free and only require an email address (recommended usage).

5 IOPT-Tools Functionality

After logging in at [8], the list of tools in the framework are made available and associated with a set of buttons presented in the interface, as shown in Fig. 2.

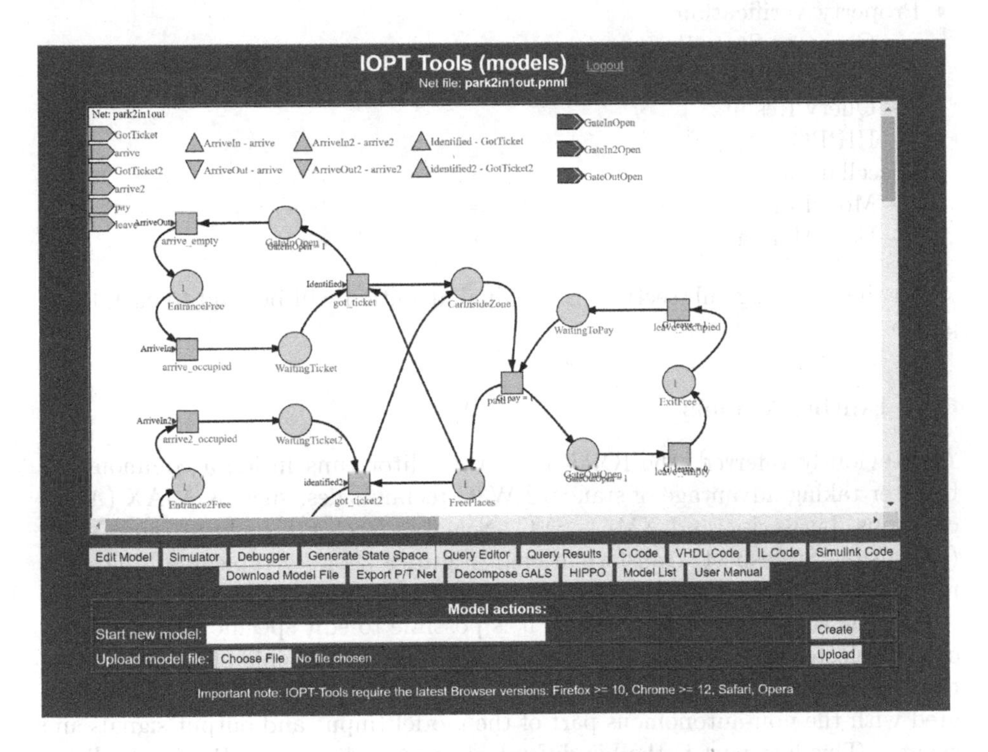

Fig. 2. IOPT-Tools: User interface main page.

Overall functionalities can be grouped as follows (some of the listed functionalities are available inside one of the tools):

- Editing Models
  - Edit model
  - Net addition
  - Net split
  - Decompose GALS
- Automatic generation of execution code
  - C Code
  - VHDL Code
  - IL Code
  - Simulink Code
- Simulators
  - Simulator (Token player and timing Diagram)
  - (Remote) Debugger
- Create, Export and Import models
  - Start new model
  - Upload Model file
  - Download Model File
  - Export P/T Net
- Property Verification
  - Generate State Space
  - Query Editor
  - Query Results
  - HIPPO
- Miscellaneous
  - Model List
  - User Manual

In the following sub-sections, individual attention will be paid to each listed group.

5.1 Editing Models

As previously referred, the IOPT Petri net editor runs inside a common web browser taking advantage of standard W3C technologies, such as AJAX (Asynchronous Javascript and XML), SVG (Scalable Vector Graphics), and XSLT (eXtensible Stylesheet Transformations). Figure 3 shows a typical editing environment, where the model is edited in the central window.

On the right side of the window, it is possible to edit specific characteristics of the selected element, the one that is highlighted. Possible elements include those associated with the graph (places, transitions, and arcs) and those associated with the non-autonomous part of the model (input and output signals and events). This last part is the big difference compared to other Petri net editing tools. In the following sub-sections, some of these attributes will be referred to whenever necessary.

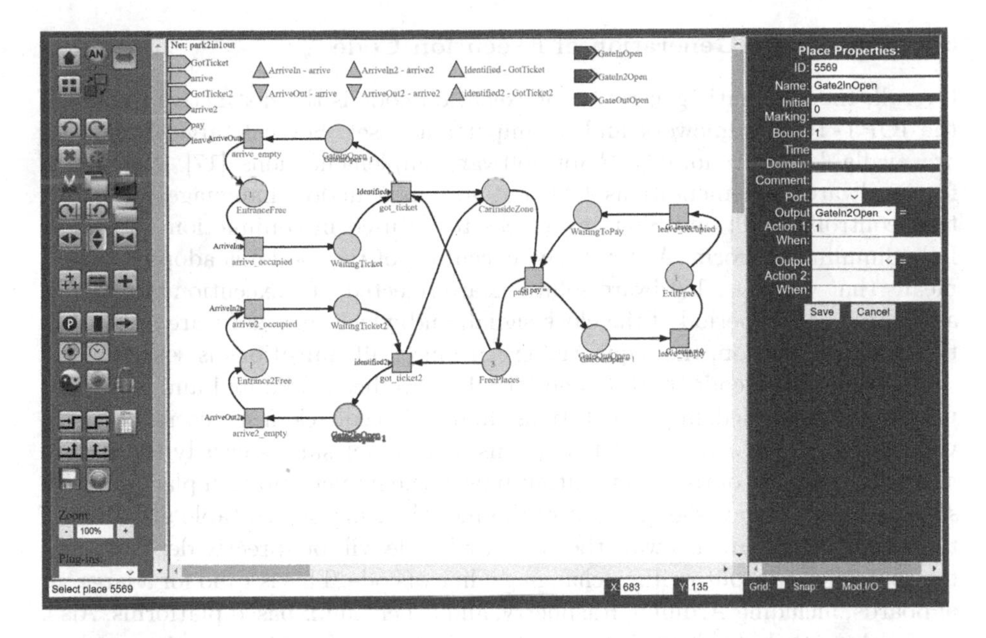

Fig. 3. IOPT net editor page.

A coloring scheme, which received very favorable comments from users, combined with a distinctive graphical representation for showing the different elements of the model was adopted, allowing a clear perception of the type of input/output signal/event.

On the left side of the window, several buttons are ready to support specific editing tasks. These include selecting elements, undo, redo, cut, copy, paste, rotation, mirror, generation of complementary places, semaphores, and other graphical conveniences, and editing of places, transitions, arcs, and input and output signals and events (as already anticipated).

Of particular interest, there is also direct support for net inclusion (disjoint addition), allowing the reading of other models into the editing area, as well as the addition of several nets through the definition of fusion sets (a form of place and transition fusion proposed elsewhere [16]). This is a significant convenience for the user, as it supports the re-utilization of models within new models.

On the other hand, support for decomposition of a model into several sub-models is possible using the net splitting plug-in, considering a set of rules proposed elsewhere [14]. This decomposition technique could be used as the starting point to produce a set of sub-models. This paves the way to support the distributed execution of the model, where each sub-model can be associated with a different time domain and interconnected with other sub-models through a new type of node with the semantic of a place (the communication channel), as proposed in [15]. When coming to the implementation level, this set of sub-models will be deployed as a Globally Asynchronous Locally Synchronous system into a network of controllers.

5.2 Automatic Generation of Execution Code

Overall, the automatic generation of execution code is the distinctive feature of the IOPT-Tools framework and an important asset. Several target languages are available, most notably C for software implementations [17] and VHDL for hardware implementations [18], but also Instruction Language for industrial controllers [19] and Simulink blocks to be used in conjunction with MatLab/Simulink platform. A step-based execution of the model is adopted, which means that whenever hardware solutions are selected, the execution of a step is associated with a period of the clock signal, and whenever a software implementation is the solution, the sequential execution of all transitions is selected.

The generated code, both C and VHDL, have been validated and compared with manually-coded implementations, and the feedback from users has been very positive. One aspect that biased this opinion for sure is clearly the facility to associate the ID of the input/output pins of the implementation platform with specific input/output signal/event of the model (using one editable attribute of the signal/event). In this way, the generated code will be directly deployed into the controller, without writing/changing a line of code. This is valid for a large set of boards, including Arduino, Raspberry, and other Linux-based platforms. Also, some other attributes have a strong impact in terms of implementation, such as the *Bound* attribute associated with places, which will be considered when the execution code is generated to determine the amount of resources necessary for the implementation (particularly relevant whenever hardware implementation is of interest).

5.3 Simulation and Testing

The IOPT-Tools framework also supports the simulation of the model. Two forms are available. The first one is just the usual interactive token-player, where benefiting from an interface with some similarities with Fig. 3, the user can change the values of the input signals/events and be exposed to the animation of the model. As in the editing tasks, the user will take advantage of the coloring of places and transitions to get a direct perception of the place's marking or the condition of a transition if enabled or/and ready. The second way to perform the simulation is by providing the timing diagram associated with the evolution of input signals and events. The simulation can then be executed automatically or interactively step by step. The tool generates the timing diagram of all inputs and outputs and all place marking and transition firings. This provides the traditional view in electronic circuits simulation/testing. It is important to refer that it is possible switching from one view (token-player) to the other (timing diagram), or vice-versa.

The three steps described– editing the model, simulating the model, and generating executable implementation code– support a sound path for rapid prototyping of controllers for discrete event systems. This has been extensively used in several courses at NOVA University Lisbon, Portugal [20].

The IOPT-Tools framework also supports *remote debugging*: the monitoring and remote operation of controllers running code generated by the framework. The user interface of the remote debugger is similar to the interactive token-player simulator, but the animation of the model is achieved through the real-time acquisition of the values of signals and events provided by the controller at the plant level. The communication with the remote controller is achieved through a minimalist http server (also generated automatically by IOPT-Tools framework). The user interface also allows overwriting of values acquired from the plant. More specifically, it is possible to perform a remote step-by-step testing procedure if some malfunction is detected at the controller end.

5.4 Create, Export, and Import Models

The capability of interchange models between different tool frameworks is the main idea behind the proposal of PNML standard (Petri net markup language) [21]. IOPT-Tools fully support this approach. In this sense, it is possible to import PNML files of place-transition models (using *Upload model file* option). The functionality has been used to ensure interoperability with external tools. It is also a way to allow users of other tool frameworks to benefit from the unique characteristics of IOPT-Tools, namely the automatic code generation for implementation (after augmenting their models with the non-autonomous elements). Important to refer that whenever the model to import does not have a graphical representation of the nodes and arcs of the graph, the open source graph visualization software Graphviz [22] is used to produce this information.

Additionally, it is possible to obtain the PNML file associated with the IOPT net model (using the *Download Model File* button), as well as the PNML file associated with the underlying P/T net model (using the *Export P/T Net* button). The latter is obtained by removing all references and dependencies on signals and events.

5.5 Property Verification

The model-checking subsystem, allowing verification of properties of the model, is composed by a state-space generator, complemented by a query system able to produce answers to specific questions [23,24]. When the state-space graph is small, it can be directly presented in the browser. Yet, in most cases, state-space graphs are too large to be visualized; in those cases, it is possible to download the state-space graph (as an XML file), allowing its inspection by external tools. The query editor allows the construction of mathematical expressions that will be evaluated while the spate-space is being constructed. The editor of the mathematical expressions (similar to the ones used to fill in several attributes of places and transitions, such as transitions guards and expressions associated with output signals) is straightforward when the type of questions is simple. However, it falls short when more complex questions are of interest. This is an aspect where currently IOPT-Tools users rely more on external tools (using the export feature presented before) than on the model-checking subsystem. One of these links with external tools is available through a direct connection with the tool HIPPO [25], where some answers on invariants can be obtained.

Overall, the model-checking subsystem has an important role in three situations:

- When generating the state-space graph and computing the maximum number of tokens for all places; this information will be back-annotated into the model, in the *Bound* attribute, which will be used by the code generator tools (as referred in Sect. 5.2);
- As the physical systems running the code generated automatically may suffer from malfunctions or cause hazardous situations, the model-checking subsystem may be used to automate the detection of problematic situations and verify if undesired states are reachable;
- As an important support for teaching activities.

5.6 Miscellaneous

At any moment, it is possible to return to the *Model List* and select another model. There is also a *User Manual*, very important for newcomers.

6 Conclusions and Future Works

The framework was built together with the definition of the IOPT nets class to allow the generation of executable code from Petri net models targeting controllers. Hence, the primary motivation to use this set of tools remains the same: a formal and precise graphical language to simulate and generate code for discrete-event systems controllers. Additionally, several cooperating controllers executing in distinct time domains can also be modeled, thus allowing simulation and code generation for GALS systems.

As all tool developers know very well, the job is never completed if one wants to keep the tool live and updated, giving adequate answers to new challenges. This is also the case for IOPT-Tools, where one primary source of user feedback is the continuous usage of IOPT-Tools for teaching purposes in engineering courses at several universities, most notably at NOVA University Lisbon, Portugal.

Current tool support for verification and property analysis has been extensively used for teaching purposes in engineering courses. Yet, it is still far from an industrial-strength tool regarding robustness and efficiency. Hence, as future work, we intend to prioritize the following developments:

- Robust support for interoperability with industrial-strength external verification tools, e.g., the Spin model-checker;
- Improve the graphical user interface usability and design;
- Better support for modularization, including hierarchical composition/decomposition of models.

References

1. Silva, M.: Las Redes de Petri: en la Automática y la Informática. Editorial AC, Madrid (1985)
2. David, R., Alla, H.: Petri Nets & Grafcet; Tools for Modelling Discrete Event Systems. Prentice Hall International (UK) Ltd. (1992)
3. David, R., Alla, H.: Discrete, Continuous, and Hybrid Petri Nets, 2nd edn. Springer, Heidelberg (2010). https://doi.org/10.1007/978-3-642-10669-9
4. Frey, G., Minas, M.: Editing, visualizing, and implementing signal interpreted Petri nets. In: Proceedings of the AWPN 2000, Koblenz, pp. 57–62, October 2000
5. Hanisch, H.-M., Lüder, A.: A signal extension for Petri nets and its use in controller design. Fund. Inform. **41**(4), 415–431 (2000)
6. Venkatesh, K., Zhou, M., Caudill, R.J.: Comparing ladder logic diagrams and petri nets for sequence controller design through a discrete manufacturing system. IEEE Trans. Industr. Electron. **41**(6), 611–619 (1994)
7. Petri nets tool database. http://www.informatik.uni-hamburg.de/TGI/PetriNets/tools/db.html. Accessed 26 Jan 2022
8. IOPT-Tools website. http://gres.uninova.pt/IOPT-Tools/. Accessed 26 Jan 2022
9. Gomes, L., Moutinho, F., Pereira, F.: IOPT-tools - a web based tool framework for embedded systems controller development using Petri nets. In: 2013 23rd International Conference on Field Programmable Logic and Applications, September 2013
10. Pereira, F., Moutinho, F., Gomes, L.: IOPT-tools - towards cloud design automation of digital controllers with Petri nets. In: ICMC 2014- International Conference on Mechatronics and Control (2014)
11. Gomes, L., Barros, J.-P., Costa, A., Nunes, R.: The input-output place-transition Petri net class and associated tools. In: 2007 5th IEEE International Conference on Industrial Informatics, vol. 1, pp. 509–514, June 2007
12. Gomes, L., Moutinho, F., Pereira, F., Ribeiro, J., Costa, A., Barros, J.-P.: Extending input-output place-transition petri nets for distributed controller systems development. In: ICMC 2014 - International Conference on Mechatronics and Control, 3–5 July 2014, Jinzhou, China, pp. 1099–1104 (2014)
13. Gomes, L., Barros, J.P.: Refining IOPT Petri nets class for embedded system controller modeling. In: IECON 2018–44th Annual Conference of the IEEE Industrial Electronics Society, pp. 4720–4725 (2018)
14. Costa, A., Gomes, L.: Petri net partitioning using net splitting operation. In: 2009 7th IEEE International Conference on Industrial Informatics, June 2009
15. Moutinho, F., Gomes, L.: Distributed Embedded Controller Development with Petri Nets: Application to Globally-Asynchronous Locally-Synchronous Systems, 1st edn. Springer, Heidelberg (2015). https://doi.org/10.1007/978-3-319-20822-0
16. Barros, J.-P., Gomes, L.: Net model composition and modification by net operations: a pragmatic approach. In: INDIN'2004 - 2nd IEEE International Conference on Industrial Informatics, 24–26 June 2004, Berlin, Germany (2004). https://doi.org/10.1109/INDIN.2004.1417350
17. Gomes, L., Rebelo, R., Barros, J.-P., Costa, A., Pais, R.: From Petri net models to C implementation of digital controllers. In: ISIE 2010 - IEEE International Symposium on Industrial Electronics, Bari, Italy, 4–7 July 2010
18. Pereira, F., Gomes, L.: Automatic synthesis of VHDL hardware components from IOPT Petri net models. In: IECON 2013 - The 39th Annual Conference of the IEEE Industrial Electronics Society, 10–13 November 2013, Vienna, Austria (2013)

19. Feio, R., Rosas, J., Gomes, L.: Translating IOPT Petri net models into PLC ladder diagrams. In: ICIT 2017 - 18th International Conference on Industrial Technology, 22–25 March 2017, Toronto, Canada (2017)
20. Gomes, L., Costa, A.: Cloud based development framework using IOPT Petri nets for embedded systems teaching. In: ISIE 2014 - 2014 IEEE International Symposium on Industrial Electronics, 1–4 June 2014, Istanbul, Turkey, pp. 2202–2206 (2014). https://doi.org/10.1109/ISIE.2014.6864959
21. Billington, J., et al.: The Petri net markup language: concepts, technology, and tools. In: van der Aalst, W.M.P., Best, E. (eds.) ICATPN 2003. LNCS, vol. 2679, pp. 483–505. Springer, Heidelberg (2003). https://doi.org/10.1007/3-540-44919-1_31
22. Ellson, J., Gansner, E., Koutsofios, L., North, S.C., Woodhull, G.: Graphviz—open source graph drawing tools. In: Mutzel, P., Jünger, M., Leipert, S. (eds.) GD 2001. LNCS, vol. 2265, pp. 483–484. Springer, Heidelberg (2002). https://doi.org/10.1007/3-540-45848-4_57
23. Pereira, F., Moutinho, F., Gomes, L., Campos-Rebelo, R.: IOPT Petri net state space generation algorithm with maximal-step execution semantics. In: INDIN 2011 - 9th IEEE International Conference on Industrial Informatics, 26–29 July 2011, Caparica, Lisbon, Portugal, pp. 789–795 (2011). ISBN 978-1-4577-0434-5. https://doi.org/10.1109/INDIN.2011.6034958
24. Pereira, F., Moutinho, F., Gomes, L., Ribeiro, J., Campos-Rebelo, R.: An IOPT-net state-space generator tool. In: INDIN 2011 - 9th IEEE International Conference on Industrial Informatics, 26–29 July 2011, Caparica, Lisbon, Portugal, pp. 383–389 (2011). ISBN 978-1-4577-0434-5. https://doi.org/10.1109/INDIN.2011.6034907
25. HIPPO website. http://hippo.iee.uz.zgora.pl/index.php. Accessed 26 Jan 2022

OCπ: Object-Centric Process Insights

Jan Niklas Adams[1(✉)] and Wil M. P. van der Aalst[1,2]

[1] Process and Data Science, RWTH Aachen University, Aachen, Germany
{niklas.adams,wvdaalst}@pads.rwth-aachen.de
[2] Fraunhofer Institute for Applied Information Technology, Sankt Augustin, Germany

Abstract. Process mining uses event sequences recorded in information systems to discover and analyze the process models that generated them. Traditional process mining techniques make two assumptions that often do not find correspondence in real-life event data: First, each event sequence is assumed to be of the same type, i.e., all sequences describe an instantiation of the same process. Second, events are assumed to exclusively belong to one sequence, i.e., not being shared between different sequences. In reality, these assumptions often do not hold. Events may be shared between multiple event sequences identified by objects, and these objects may be of different types describing different subprocesses. Assuming "unshared" events and homogeneously typed objects leads to misleading insights and neglects the opportunity of discovering insights about the interplay between different objects and object types. Object-centric process mining is the term for techniques addressing this more general problem setting of deriving process insights for event data with multiple objects. In this paper, we introduce the tool OCπ. OCπ aims to make the process behind object-centric event data transparent to the user. It does so in two ways: First, we show frequent process executions, defined and visualized as a set of event sequences of different types that share events. The frequency is determined with respect to the activity attribute, i.e., these are object-centric variants. Second, we allow the user to filter infrequent executions and activities, discovering a mainstream process model in the form of an object-centric Petri net. Our tool is freely available for download (http://ocpi.ai/).

Keywords: Process Mining · Object-Centric Petri Net · Process Discovery · Object-Centric Variants

1 Introduction

Process mining is an umbrella term describing techniques to derive data-driven insights into processes. The data come in an event log, describing the event sequences of many process executions and their associated data. Typically, three different process mining fields are considered: process discovery, conformance checking, and process enhancement [1]. Techniques from process discovery aim to

L. Bernardinello and L. Petrucci (Eds.): PETRI NETS 2022, LNCS 13288, pp. 139–150, 2022.
https://doi.org/10.1007/978-3-031-06653-5_8

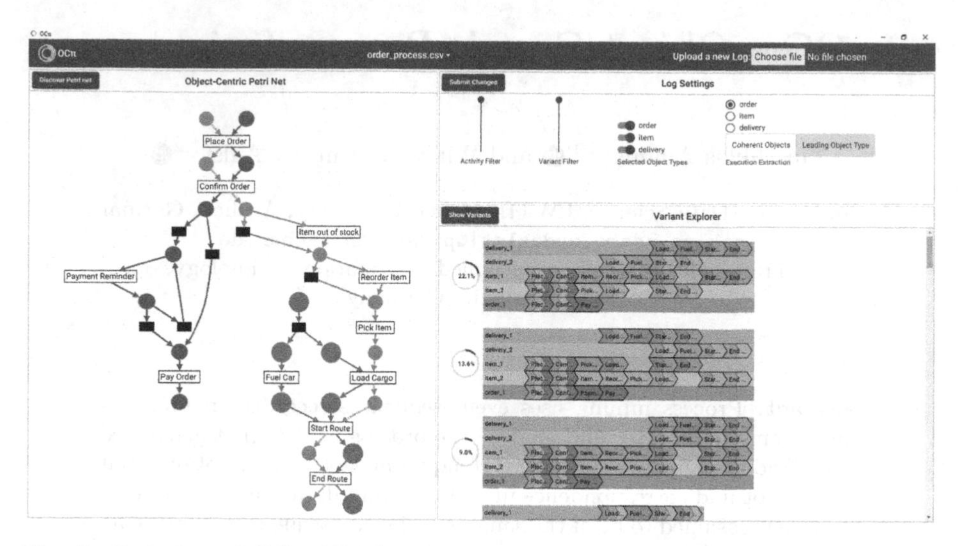

Fig. 1. Overview of OCπ: The log management is integrated into the toolbar, the user can further interact with the tool by submitting the desired settings through the log settings component. Petri net and object-centric variants can be explored through scrolling, zooming and panning.

construct a model from the event log, e.g., a Petri net or a BPMN model. Such a model aims to describe the event sequences contained in the event log with only one comprehensive model. Conformance checking deals with quantifying and describing how well a model corresponds to an event log. Process enhancement aims to deliver data-driven process improvements.

Process mining techniques make some assumption about the nature of event logs and the event sequences event logs contain. Most techniques make the following two assumptions: **(1)** An event log contains process executions related to individual objects, often called cases. Therefore, every event sequence describes an execution of a single case. Each case is of the same case notion. **(2)** Event sequences for cases are independent of each other, i.e., two cases do not share events.

In reality, event logs often violate these assumptions. There is often no clear case notion in an event log. Events can be associated with multiple cases [2]. Imagine an ordering process: An order of some items is placed in a system. These items are later delivered. Events can refer to a case notion of an order, an item, or a delivery. Furthermore, some events are shared between different cases of different case notions, e.g., an event that describes the placing of an order of two items.

In traditional process mining, we cannot derive insights from such event data as a whole. One case notion would need to be chosen, and events referring to multiple objects of that notion would need to be duplicated, called *flattening* [2]. This procedure removes essential information about interactions between different case notions and objects. These problems are the motivation for *object-*

centric process mining [3]. By dropping the two mentioned assumptions and adapting process mining techniques, object-centric process mining aims to deliver algorithms that are able to exploit event logs with multiple case notions and shared events fully.

Different methods for dealing with multiple case notion processes exist. Some, like artifacts [7,8] and proclets [11], deal with the problem mostly from a modeling perspective. Object-centric process mining [2] takes the object-centric event data [12] as a starting point to discover process models and insights. So far, the discovery of process models in the form of object-centric Petri nets [3] has been introduced. A discussion of sound object-centric workflow nets has recently been published [14]. Furthermore, basic conformance checking techniques for object-centric Petri nets and event logs [4], and performance analysis measures cite[16] have been introduced. On the tool side, tools to extract object-centric Petri nets [3] and object-centric directly-follows graphs [5] as well es storing and querying multiple case notion event data in the form of graph databases [10] have been introduced. Furthermore, object-centric Petri nets have been used to model digital twins of organizations [15].

However, some key ingredients of traditional process mining are, so far, not available to users. Here, we focus on variant visualization and discovering process models for frequent variants. Process executions can be equivalent if they describe the same execution sequences of event activities. The equivalence classes they form are commonly known as *variants* [9]. Each variant has a frequency determined by the number of process executions in this variant. Filtering and exploring frequent variants provides the user with insights into the mainstream behavior of the underlying process and yields a mainstream model.

Therefore, with OCπ (cf. Fig. 1), we provide a tool that augments object-centric process discovery in the following two ways:

1. We allow the user to filter the least frequent variants of process executions from the retrieved event log to discover an object-centric Petri net that shows the mainstream behavior of the event log.
2. We provide a variant explorer that allows the user to retrieve and explore the variants of process executions and their frequencies.

The remainder of this paper is structured as follows. In Sect. 2, we introduce some basic concepts on which this tool is built, i.e., object-centric event logs, extraction of executions, variants, and object-centric Petri nets. We explain the algorithmic foundations and some concepts relevant for the understanding and usage of the tool in Sect. 3. We provide an extensive overview of the functionalities, the implementation and installation requirements in Sect. 4. We conclude this tool paper in Sect. 5.

2 Object-Centric Process Mining

In this section, we introduce some of the basic concepts on which this tool is built. The tool takes input in the form of an object-centric event log, extracts

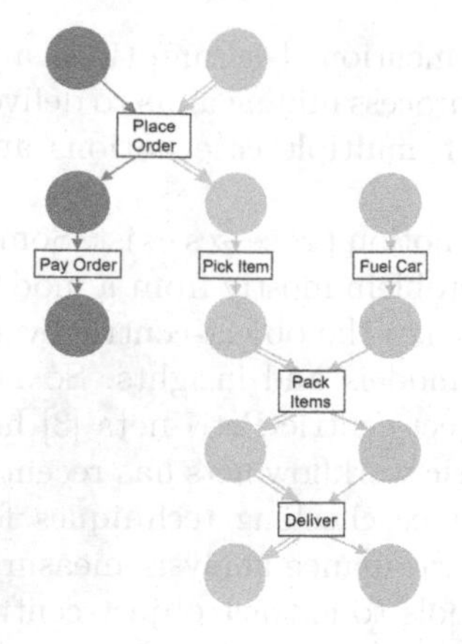

Fig. 2. Example of an object-centric Petri net. Places are colored according to the object type they belong to; variable arcs (double lined) can consume a variable amount of tokens.

Table 1. Example of an object-centric event log. Each event can be associated to multiple objects of different object types. In this log, order, item and delivery are the object types.

Event	Activity	Order	Item	Delivery
e_1	Place Order	o1	i1,i2	
e_2	Pick item		i2	
e_3	Pick item		i1	
e_4	Fuel car			d1
e_5	Pay order	o1		
e_6	Pack items		i1,i2	d1
e_7	Deliver		i1,i2	d1

process executions from it, determines frequent equivalence classes, i.e., variants, and discovers and displays an object-centric Petri net to the user. Therefore, we give a short formal introduction of these concepts in this section.

An object-centric event log can be seen as an extension to traditional event logs used in process mining [1] that records multiple case notions (object types) for each event and allows referencing to multiple cases (objects) of each object type of an event.

Definition 1 (Object-Centric Event Log). *Let U_E be the universe of event identifiers, U_{OT} be the universe of object types, U_O be the universe of objects and U_A be the universe of activities. $\mathcal{P}(X)$ denotes the power set of a set X. $\pi_{ot} : U_O \to U_{OT}$ maps an object to its object type. An object-centric event log is a tuple $L = (E, OT, O, A, \pi_o, \pi_a, \prec)$ consisting of event identifiers $E \subseteq U_E$, object types $OT \subseteq U_{OT}$, objects $O \subseteq U_O$, activities $A \subseteq U_A$, a mapping function from events to objects $\pi_o : E \to \mathcal{P}(O)$ and a mapping function from events to activities $\pi_a : E \to A$. The event identifiers are subject to a total order $\prec$.*

An example of an object-centric event log in table format is given in Table 1. Each event has a unique identifier e_i and an activity[1]. Furthermore, each event has reference to a set of objects. Each object is associated with one object type of *order, item* or *delivery.* In traditional process mining, each process execution is associated with exactly one object, i.e., each process execution is one sequence. The notion of a process execution can be generalized for object-centric event

[1] We omit the timestamp and additional attributes as they are not relevant for the capabilities described in this paper.

data, involving the sequences multiple objects sharing events. An extraction technique retrieves a set of process executions from an event log.

Definition 2 (Process Execution). *Let* $L = (E, OT, O, A, \pi_o, \pi_a, \prec)$ *be an object-centric event log.* $P_L = \{(E', O') \mid E' \subseteq E \wedge O' \subseteq O \wedge e \in E' \Leftrightarrow \pi_o(e) \cap O' \neq \emptyset \wedge (O', \{(o, o') \in O' \times O' \mid \exists_{e \in E'} o, o' \in \pi_o(e)\})$ *is a connected graph*$\}$ *is the set of process executions of an event log. An extraction technique* $f_{extract} : L \rightarrow \mathcal{P}(P_L)$ *extracts a subset of all process executions.*

For the example of the event log excerpt in Table 1, a single process executions could be all seven events and the four objects. Another possible process execution would be a subset of these objects and their events, e.g., only the order o1 and the two items i1 and i2. Different methods are available to retrieve subsets of all process executions. The process executions in the extracted subset should have some similar characteristics to be comparable. We discuss two different extraction techniques in Subsect. 3.2.

Variants in process mining summarize multiple cases (or process executions) with the same control-flow behavior. This is translated to the object-centric setting by determining equivalency of process executions concerning the event activity attribute and grouping equivalent executions in one class, i.e., variant.

Definition 3 (Equivalent Process Executions). *Let* $P = \{p_1, p_2, \ldots p_n\}$ *be a set of process executions. An oracle* $f_{equiv} : P \rightarrow \{1, \ldots, m\}$ *maps executions to* $m \in \mathbb{N}$ *classes of equivalent executions considering the event's activity. Each class is one variant* $V_i = \{p_j \in P \mid f_{equiv}(p_j) = i\}$ *for* $i \in \{1, \ldots, m\}$.

We can discover an object-centric Petri net from an object-centric event log [3]. Object-centric Petri nets borrow from colored Petri nets [13] to be able to model different object types and how they interact.

Definition 4 (Object-Centric Petri Net). *Let* $N = (P, T, F, l)$ *be a Petri net with places* P*, transitions* T*, a flow relation* $F \subseteq T \times P \cup P \times T$ *with* $T \cap P = \emptyset$ *and a labelling function* $l : T \nrightarrow U_A$ *and let* $OT \subseteq U_{OT}$ *be a set of object types. An object-centric Petri net* $OCPN = (N, pt, f_{var})$ *is a tuple of a Petri net* N*, a mapping function from places to object types* $pt : P \rightarrow OT$ *and* $f_{var} \subseteq F$ *describing a subset of arcs which are variable arcs, i.e., they can consume and produce more than one token.*

An example of an object-centric Petri net is given in Fig. 2. This Petri net describes the process used to generate the event log of Table 1. Each place has a color corresponding to one object type. Each arc can either be a standard arc or a variable arc. When playing the token game in such a Petri net, a binding execution of a transition would consume tokens associated with objects in the input places. These could be multiple tokens in the case of a variable arc. The consumed tokens are then produced in the output places of the corresponding object types of the input places.

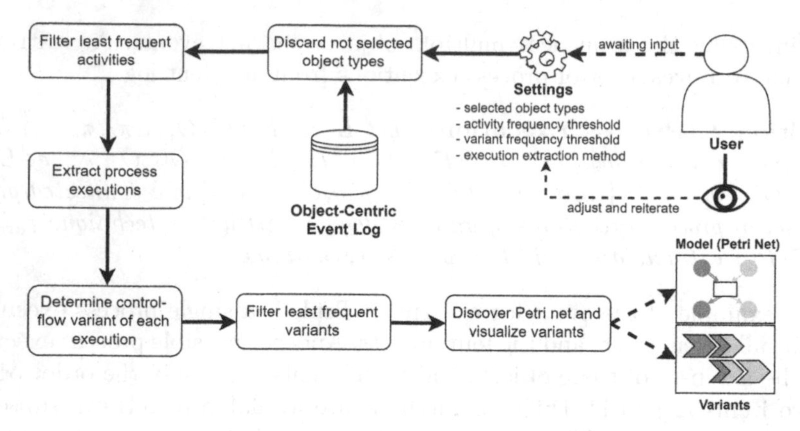

Fig. 3. Overview over the user interaction and the algorithmic steps in OCπ.

3 Algorithmic Concept

The technique to discover an object-centric Petri net from an object-centric event log is described in [3]. We focus on preprocessing the underlying event log for process discovery according to some user input. The general signature of the processing we apply is depicted below.

$$L \xrightarrow{f_{extract}} \{p_1, \ldots, p_n\} \xrightarrow{f_{equiv}} \{V_1, \ldots, V_m\} \rightarrow \text{Petri net \& Variants}$$

The detailed algorithmic concept of our tool is depicted in Fig. 3. The user provides an object-centric event log and makes a choice about several settings: The selected object types, the activity threshold, the variant threshold, and the technique to extract process executions. Based on this, our tool applies different filtering and algorithmic steps. Process executions are extracted from the object-centric event log, and their equivalence classes, i.e., variants, are calculated. Subsequently, the user can explore the object-centric Petri net and the variants, adjust the input settings, and reiterate until the result is sufficient for the user. In the following sections, we provide deeper insights into the different algorithmic steps conducted by our tool that are of importance for the understanding of the user on how to interpret the results.

3.1 Filtering

The filtering possibilities included in our tool, by activity frequency and variant frequency, both follow the same method: The user selects a threshold between 0 and 1.0. Subsequently, the minimum number of behavioral observations, i.e., either activities or variants, is collected such that the cumulative frequency of these observations exceeds this threshold. The events that are not associated with these observations are filtered out. For the example of activity filtering, the relative frequencies of each activity in the event log are calculated. After the user sets a threshold, the most frequent activities are greedily added to a set of activities that should be kept in the event log until the threshold is met. All events with activities not in this set will be discarded.

3.2 Process Execution Extraction

The process execution extraction technique determines how process executions are retrieved from the object-centric event log and is, therefore, important for the variants retrieved as well as the results of the variant filtering. We provide two execution extraction techniques: *coherent objects* and *leading object type*.

For a brief explanation of these techniques, we use a concept of direct relations between objects: if two objects share an event, they are directly related. Two objects can be transitively related if a chain of direct relations leads from one object to the other. The length of the chain is the level of transitivity. Coherent objects take all objects directly and transitively related into one process execution. The underlying assumption is that they are all dependent on each other by sharing events. However, this might lead to process executions that are too extensive for some logs and some users. Imagine the ordering process from Table 1. If multiple items of multiple orders end up in the same delivery, all of these orders would be one process execution. While they, indeed, all depend on each other, this might be too extensive for the user who may only be interested in the execution of an order and the associated objects or a delivery and the associated objects. Because of this, we include leading object type as a technique for execution extraction. It constructs executions by taking each object of the leading object type and recursively adding directly related objects until objects of the same type have already been added on a lower level of transitivity. These objects are not added anymore, and their directly related objects are not further traversed. The events of the selected objects are one process execution. This execution extraction cuts off some dependencies to have more understandable and comprehensive process executions. Some events may end up in multiple process executions. Our implementation determines the equivalence of process executions by testing for automorphism between the different execution graphs given by the process executions.

3.3 Variant Visualization

Each execution is associated with one variant, describing the sequence of activities for each involved object. We visualize these variants by giving each object one lane. Each object gets a color, dependent on the object type's base color, which is slightly altered for each object. We draw a chevron for each event. The activity is depicted inside the chevron. If an event is shared between objects, the corresponding chevron is drawn on each lane and colored with all colors of all involved objects. Generally, the chevrons are of the same width. Only if a chevron is placed between two shared chevrons, the width is adjusted to fit the gap. Except for the shared chevrons, the chevron ordering and horizontal

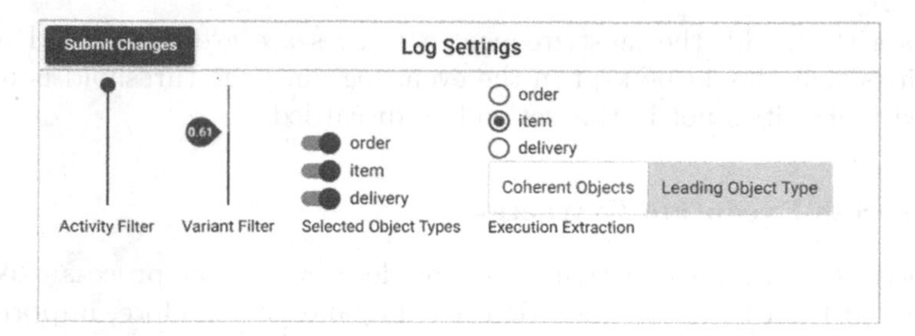

Fig. 4. The log settings component, the central point of interaction with the users.

positioning gives information about the time-ordering within one object, not between objects.

4 Usage and Functionalities

In this section, we introduce the core functionalities of our tool. A complete view of the tool is depicted in Fig. 1. Our tool is separated into four components: two for user interaction and two for exploring the output. The user interaction components are the log management (integrated into the toolbar on top) and the log settings. The output components are the object-centric Petri net explorer and the variant explorer. We are going to introduce each of the components in the following sections.

4.1 Event Log Management

OCπ offers extensive event log management. The file formats *jsonocel* and *jsonxml* introduced in the OCEL standard [12] as well as a *csv* import are supported. The CSV file should contain an "event_activity" and "event_timestamp" column. As a supportive element for uploading CSV, we implemented functionality to choose potential object types to prevent unwanted columns from becoming object types. The file's encoding should comply with UTF-8 encoding. The separator will automatically be detected. Each log can also be deleted at an arbitrary point in time. Each event log will be uniquely identified by its name. Two event logs with the same filename cannot be uploaded.

4.2 Log Settings

The log setting component is the main point of input from the user and is depicted in Fig. 4. We provide two filtering thresholds: One to filter out infrequent activities and one to filter out infrequent variants. Furthermore, we allow the user to discard some object types for their scope of analysis, effectively removing these object types and their objects from the event log. We provide two different techniques to extract process executions. The first one is called *coherent objects*,

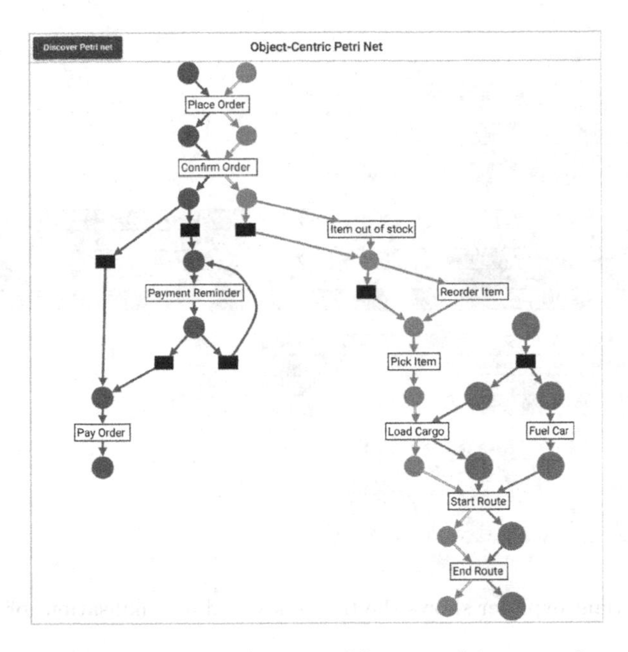

Fig. 5. The object-centric Petri net discovered from an object-centric event log according to the settings provided by the user.

the second technique is called *leading object type*. Subsection 3.2 provides a detailed explanation of the execution extraction. The settings can be submitted to the back-end by pressing the *submit changes* button.

4.3 Process Model Explorer

Based on the event log that is processed according to the setting provided by the user, our tool discovers an object-centric Petri net and displays it to the user as a process model. The component is depicted in Fig. 5. The visualization can interactively be explored, navigation by zooming and dragging/panning is supported. These functionalities help make large Petri nets with many object types and transitions accessible to users. Every object type has one globally assigned color, which is also used by the variant explorer component described later. The Petri net can be discovered by pressing the button. If the settings were changed and the Petri net is not consistent with the current submitted settings, this button changes its color to red to indicate a necessary update.

4.4 Variant Explorer

The variant explorer displays the variants of the process executions extracted from the event log based on the provided settings. It is depicted in Fig. 6. The colors of the object types are consistent with the colors in the object-centric

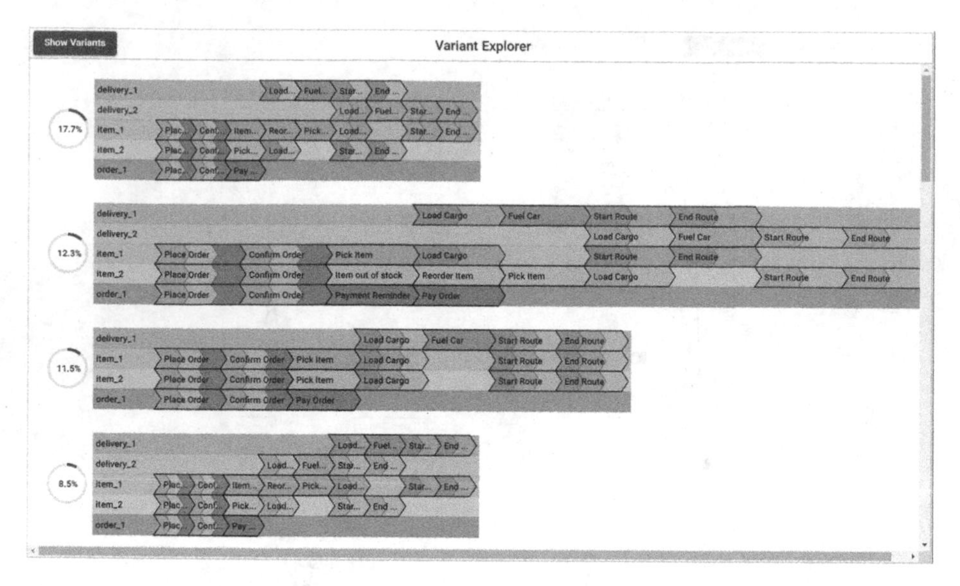

Fig. 6. The variant explorer shows the frequency and a visualisation for each variant.

Petri net component. Each object of an object type is colored in a different shade of the object types' base color. Each object gets a lane to describe its event sequence, lanes of the same object types are grouped. The frequency of the variant (with respect to the event log after the filtering settings are applied) is depicted on the left-hand side of the variant. With a click on the variant, the variant is unfolded such that the full activity labels are visible. The explorer allows scrolling in both directions, vertically and horizontally.

4.5 Implementation

We implemented this tool on a technology stack of Python, Django, Angular, and D3.js and GraphViz[2] for visualization. The core algorithmic functionality is taken from the OCPA[3] and the PM4Py library [6]. The tool can be run on Windows by downloading it from http://ocpi.ai/ and running the executable named *OCpi.exe*.

5 Conclusion

In this paper, we introduced the tool OCπ. This tool enables users to load object-centric event data and explore novel insights: The process execution variants contained in the object-centric event data and their frequencies. Furthermore, we enable the user to filter out infrequent variants, infrequent activities, and

[2] GraphViz needs to be installed. See: https://graphviz.org/download/.
[3] https://github.com/gyunamister/ocpa.

unwanted object types to discover an object-centric Petri net according to the chosen settings. This allows a user to interactively explore an object-centric process model and its most frequent variants.

Acknowledgements. We thank the Alexander von Humboldt (AvH) Stiftung for supporting our research.

References

1. van der Aalst, W.M.P.: Process Mining: Data Science in Action. Springer, Heidelberg (2016). https://doi.org/10.1007/978-3-662-49851-4
2. van der Aalst, W.M.P.: Object-centric process mining: dealing with divergence and convergence in event data. In: Ölveczky, P.C., Salaün, G. (eds.) SEFM 2019. LNCS, vol. 11724, pp. 3–25. Springer, Cham (2019). https://doi.org/10.1007/978-3-030-30446-1_1
3. van der Aalst, W.M.P., Berti, A.: Discovering object-centric Petri nets. Fundam. Inform. **175**(1–4), 1–40 (2020). https://doi.org/10.3233/FI-2020-1946
4. Adams, J.N., van der Aalst, W.M.P.: Precision and fitness in object-centric process mining. In: 3rd International Conference on Process Mining, ICPM 2021, Eindhoven, Netherlands, 31 October–4 November 2021, pp. 128–135. IEEE (2021). https://doi.org/10.1109/ICPM53251.2021.9576886
5. Berti, A., van der Aalst, W.M.P.: Extracting multiple viewpoint models from relational databases. In: Ceravolo, P., van Keulen, M., Gómez-López, M.T. (eds.) SIMPDA 2018-2019. LNBIP, vol. 379, pp. 24–51. Springer, Cham (2020). https://doi.org/10.1007/978-3-030-46633-6_2
6. Berti, A., van Zelst, S.J., van der Aalst, W.M.P.: Process mining for python (PM4Py): bridging the gap between process- and data science. CoRR abs/1905.06169 (2019), http://arxiv.org/abs/1905.06169
7. Calvanese, D., Montali, M., Estañol, M., Teniente, E.: Verifiable UML artifact-centric business process models. In: Li, J., Wang, X.S., Garofalakis, M.N., Soboroff, I., Suel, T., Wang, M. (eds.) Proceedings of the 23rd ACM International Conference on Conference on Information and Knowledge Management, CIKM 2014, Shanghai, China, 3–7 November 2014, pp. 1289–1298. ACM (2014). https://doi.org/10.1145/2661829.2662050
8. Cohn, D., Hull, R.: Business artifacts: a data-centric approach to modeling business operations and processes. IEEE Data Eng. Bull. **32**(3), 3–9 (2009)
9. Dumas, M., Rosa, M.L., Mendling, J., Reijers, H.A.: Fundamentals of Business Process Management, 2nd edn. Springer, Heidelberg (2018). https://doi.org/10.1007/978-3-662-56509-4
10. Esser, S., Fahland, D.: Multi-dimensional event data in graph databases. J. Data Semant. **10**(1–2), 109–141 (2021). https://doi.org/10.1007/s13740-021-00122-1
11. Fahland, D.: Describing behavior of processes with many-to-many interactions. In: Donatelli, S., Haar, S. (eds.) PETRI NETS 2019. LNCS, vol. 11522, pp. 3–24. Springer, Cham (2019). https://doi.org/10.1007/978-3-030-21571-2_1
12. Ghahfarokhi, A.F., Park, G., Berti, A., van der Aalst, W.M.P.: OCEL: a standard for object-centric event logs. In: Bellatreche, L., et al. (eds.) ADBIS 2021. CCIS, vol. 1450, pp. 169–175. Springer, Cham (2021). https://doi.org/10.1007/978-3-030-85082-1_16

13. Jensen, K., Kristensen, L.M., Wells, L.: Coloured Petri nets and CPN tools for modelling and validation of concurrent systems. Int. J. Softw. Tools Technol. Transf. **9**(3–4), 213–254 (2007). https://doi.org/10.1007/s10009-007-0038-x
14. Lomazova, I.A., Mitsyuk, A.A., Rivkin, A.: Soundness in object-centric workflow Petri nets. CoRR abs/2112.14994 (2021). https://arxiv.org/abs/2112.14994
15. Park, G., van der Aalst, W.M.P.: Realizing a digital twin of an organization using action-oriented process mining. In: 3rd International Conference on Process Mining, ICPM 2021, Eindhoven, Netherlands, 31 October–4 November 2021, pp. 104–111. IEEE (2021). https://doi.org/10.1109/ICPM53251.2021.9576846
16. Park, G., Adams, J.N., van der Aalst, W.M.P.: OPerA: object-centric performance analysis. CoRR. **abs/2204.10662** (2022). https://doi.org/10.48550/arXiv.2204.10662

Applications

From Graphs to the Science Computer of a Space Telescope

The Power of Petri Nets in Systems Engineering

Rafal Graczyk[1(✉)], Waldemar Bujwan[2], Marcin Darmetko[2], Marcin Dziezyc[3], Damien Galano[4], Konrad Grochowski[3], Michal Kurowski[3], Grzegorz Juchnikowski[2], Marek Morawski[2], Michal Mosdorf[3], Piotr Orleanski[2], Cedric Thizy[5], and Marcus Völp[1]

[1] Interdisciplinary Centre for Security, Reliability and Trust, University of Luxembourg, Esch-sur-Alzette, Luxembourg
rafal.graczyk@uni.lu

[2] Centrum Badań Kosmicznych Polskiej Akademii Nauk, Warsaw, Poland
piotr.orleanski@cbk.waw.pl

[3] N7 Space, Warsaw, Poland
mmosdorf@n7space.com

[4] European Space Agency, Noordwijk, The Netherlands
damien.galano@esa.int

[5] Center Spatial de Liege, Liege, Belgium
cedric.thizy@uliege.be

Abstract. Space system engineering has to follow a rigorous design process to manage performance/risk trade-offs at each development stage and possibly across several functional and organizational domains. The process is further complicated by the co-development of multiple solutions, each contributing differently to the goal and with different trade-offs. Moreover, the design process is iterative, involving both changing requirements and specifications along the different ways that lead to the set goal of the mission. The above requires rigorous modeling that, in addition, must be easily extendible and maintainable across organizational units. On the example of the PROBA-3 science computer (instrument control unit, CCB DPU), we show how Petri Nets can serve as such a simple-to-maintain, holistic model, combining finite-state characterizations with dynamic system behavior caused by hardware-software interactions, to express the component-state dependent end-to-end performance characteristics of the system. The paper elaborates on how the proposed Petri-Net-modeling scheme allows for system architecture optimization that result in safely reduced technical margins and in turn substantial savings in components costs. We show that performance metrics, obtained from simulation, correlate well with the real performance characteristics of the flight model of PROBA-3's science computer.

The Proba-3 ASPIICS project is developed under the auspices of the ESA's General Support Technology Programme (GSTP) and the ESA's Prodex Programme thanks to the sponsorships of seven member states: Belgium, Poland, Romania, Italy, Ireland, Greece, and the Czech Republic. This work is also supported by the Fond Nationale de Recherche, Luxembourg, through grant CS20/IS/14689454 - HERA.

L. Bernardinello and L. Petrucci (Eds.): PETRI NETS 2022, LNCS 13288, pp. 153–174, 2022.
https://doi.org/10.1007/978-3-031-06653-5_9

Keywords: Petri Net · Systems Engineering · Performance Modeling · On-Board Computer · Scientific Payload

1 Introduction

While space missions may have a clearly specified goal, there are many ways of achieving it. Not all solution pathways are equal: some entail higher implementation costs, others are limited in the performance they achieve, introduce unacceptable risks, or utilize spacecraft resources too extensively. Designing spacecrafts and their subsystems is a gradual, iterative, incremental process and has to happen concurrently across functional and organizational boundaries. Gradual advances and small increments lead the project from initial requirements and specification up to construction, test and deployment. At all times, subsequent steps in the development process have to be evaluated from the perspective of risk mitigation, reducing unknowns and removing technical obstacles.

Classically, space projects in the ecosystem of the European Space Agency (ESA) progress through the following seven phases:

Phase 0 - *Early Conceptualization*
Phase A - *Mission Definition*
Phase B - *Preliminary Technical Design*
Phase C - *Critical Technical Design*
Phase D - *Flight Equipment Manufacturing, Assembly, Integration and Test*
Phase E - *Deployment in Flight*
Phase F - *Disposal*

Subsequent phases thereby build upon design decisions, trade-offs and experiments from earlier phases and the high formalization of the process has the goal of reducing programmatic risk. Examples of the latter include overruns on costs or of the schedule, which are often caused by wrong requirements or assumptions, that were discovered too late for quick and easy fixes. This calls for the deployment and active use of all possible means for simulating, testing, and verifying assumptions, solutions and decisions, as early as possible and sensible.

Space projects often experience a clash between opposing forces: on the one hand, there is a demand for large technical margins on designed subsystems to ensure with a very large likelihood that evolving requirements will be met; on the other hand, there is the need to optimize the design, reduce the typically very large costs incurred by an extensive use of high reliability components and resources (mass, energy, processing power, memory size) utlization. This dilemma drives the strong need to analyze, evaluate and experiment with the designed system to judge technical decisions and to ensure that, as the project matures, risks are indeed reduced while striving for optimal utilization of available resources.

In this paper, we would like to share our experience of using Petri Nets in practice for modeling the processing performance of the Instrument Control Unit for Coronagraph, a sun observing telescope, which has been the primary payload of ESA's Proba-3 mission. Application of Petri Nets has been essential in the early stages of the project (Phases B and C) where architectural decisions and optimization attempts had been confronted with the need to prove the system would be capable of meeting required performance.

Aside from our experience report, this paper contributes:

- the design of Coronagraph's instrument control unit architecture,
- an initial Petri Net model and simulation exploring the system's design space,
- an advanced Petri Net model correlated with flight equipment,
- a concept for the hardware-software co-design of embedded systems beyond the space domain, and
- the tangible system optimization results we achieved applying the above methods.

Parts of this work (Sects. 3.1, 3.2 and 4) have been previously published as a PhD thesis [14].

2 Background

Modeling and simulation of space systems, be it avionics, command and data handling systems, or mission payloads, is crucial for their correct high-level design, for their technical specification and, finally, for an initial verification of their key requirements. Early modeling and simulation helps making tested functionalities traceable for future user needs and for ensuring a common understanding on what is actually being built and why. Modeling and simulation of the system design allows for idea feasibility checks, architecture trade-offs, initiates early prototyping, brief requirements verification, and de-risking any critical aspects of the system [4,6,7,15,19,26]. Two modeling approaches are frequently used for spacecrafts, which we describe in the following.

2.1 Analytical and Mathematical Models

Every system operation can be modeled by means of affecting the energy, mass or information flow (in sense of enabling or disabling the flow, accumulation or transformation of resource). Modeling along those lines starts with capturing the basic building blocks of the system, defining transfer functions and their transient response, which leads to a description of components in form of a mix of algebraic and differential equations. From these equations, one further derives system balancing equations to describe the system dynamics in simplified form, by linking storage, flow or transformation of mass, energy or information. Balance equations are a valid method for modeling fluid systems, attitude determination and control, and some electrical systems (e.g., power systems) are conveniently modeled by means of flow and storage. In this approach, the ultimate goal is to describe the state space of a system by formulating state equations, which define the internal state of the system, and output equations, which define the system response as a function of its current system state and received inputs.

2.2 Dynamic and Functional Models

While flow or variation of physical attributes are conveniently expressed in terms of differential equations, the modeling of computation or data flow needs to take into account the stateful and event-driven nature of the data processing and control systems. State charts or graphs and finite-state machines (FSMs) are mathematical models used to represent such computation in classic logic devices or, when extended like in UML [16], SDL [2] or AADL [9], to represent software execution paths and software component relations in an embedded system [18]. FSMs can be in only one state at a time. External events induce state changes. This makes FSMs a convenient way of representing a single activity over time and showing the dependency of modeled systems on transition triggering conditions. The operation of one or more FSMs over time, and their interaction can be extended into flow networks. A flow network is a particular example of a directed graph where each edge has its maximum capacity and has some temporary flow value. Flow values cannot exceed the maximum capacity of an edge. Flow has to follow the preservation rules, meaning that effective network-node inflow must be equal to outflow (with the exception of source and sink nodes). This simple methodology allows for brief analysis of the dynamic behavior of systems. Flow networks are especially useful for modeling system aspects related to transportation like electric current, liquid or heat flow or data transfer. Flow networks are useful not only for the analysis of system evolutions over time, but also for finding the maximum flow capability of the whole network. Quite often, computing and control systems have to be analyzed as an evolution over time (in continuous domain), while system state transitions occur at discrete events, when associated conditions trigger desired reactions. Such events exhibit a competition against other triggers and each one of them, typically, has its own stochastic mechanisms that govern determining a new system state. For each state transition, new events may be scheduled and previously scheduled events may be canceled. Petri Nets provide a versatile analysis framework for this kind of modeling, especially if the modeled system exhibits randomness, state-transitions, concurrency and scheduling [13,22,23,29]. Some works on Petri Nets applied to space systems engineering can be found, including software modeling [21], satellite constellation modeling [8] and instrument simulation [20].

2.3 Hybrid Models

Hybrid models, combining discrete and dynamic aspects, such as PDL [5,10] and hybrid-systems state machines [25] have been investigated in the research community, but are not yet widely deployed in space system design. We therefore leave the investigation of such models as future work.

2.4 State-of-the-Art Modeling Approaches

A representative example of a state-of-the-art modeling approach for embedded cyber-physical systems, used in ESA, is TASTE [1,3] and its toolchain. TASTE focuses on the software aspect of system operations and supports the creation of systems using formal methods and automated code generation. The system under evaluation shall be defined in AADL, but a large number of other tools and languages are supported as well, including SDL, VHDL, ASN.1, SCADE, and Simulink. Cheddar and MAST are used for model verification. Verified models can be converted into Ada, C, or C++ and deployed on the target hardware. The TASTE framework targets the development of safety and mission-critical communication, control and data processing systems and associated real-time applications. The goal for using TASTE is to facilitate the understanding of a specification or design to get an early executable representation of the system and allow independent testing and verification at different levels of abstractions [24].

3 The Proba-3 Mission, ASPIICS Coronagraph and Instrument Control Unit

Proba-3 includes, among others, the ASPIICS (Association of Spacecraft for Polarimetric and Imaging Investigation of the Corona of the Sun) primary payload. The mission is devoted to demonstrating in-orbit the precision formation flying of two satellites. The first one produces a nearly perfect eclipse allowing the second one—the PROBA-3 Coronagraph ASPIICS—to observe the sun corona closer to the rim than ever before. To achieve that, both satellites need to keep, their distance and alignment, precisely and accurately, during the observations. The coronagraph will cover the range of radial distances between ~1.1 and 3 solar radii, thus providing continuous observational conditions very close to those during a total solar eclipse, but without the effects of Earth's atmosphere. ASPIICS will provide novel solar observations capabilities to achieve two major science objectives in the area of solar physics [11,27]:

1. understanding the physical processes that govern the quiescent solar corona; and
2. understanding the physical processes that lead to coronal mass ejections (CMEs) and that thus determine space weather.

The Coronagraph Control Box (CCB) is the electronic controller of the ASPIICS Coronagraph Instrument (CI). The CCB consists of a compact housing that contains:

- a Data Processing Unit (DPU), i.e., an embedded payload computer module that is capable of processing and buffering data and of executing management and control algorithms. The DPU is responsible for interacting with the Proba-3 Coronagraph Satellite on-board computer, called Advanced Data and Power Management System (ADPMS);

- a Power Conditioning Unit (PCU), i.e., a power supply module that provides all the voltages required by the CI, along with voltage/current measurement capabilities for telemetry data generation and protection circuits. The PCU is switched on at the moment when the ADPMS provides power, and supplies all units instantaneously;
- an Ancillary Electronics Unit (AEU) contains switches for ASPIICS power on and off control, as well as advanced actuation control for Filter Wheel Assembly (FWA) and Front Door Assembly (FDA) stepper motors, the Coronagraph Optical Box (COB) heater system and the ADCs to gather telemetry data. AEU functionality is controlled by the DPU;

The Data Processing Unit (DPU) of the Coronagraph Instrument is responsible for all control and scientific processing algorithms. It is built on two main components. One, is a processor (CPU, GR712RC), executing the control software (Boot Software and, target, Application Software). The other is a configurable logic device (FPGA, RTAX2000S), that acts as the processor's co-processor, implementing all features that are not available in the processor. The processor interfaces to Flash (for boot-image and application software storage) and to SDRAM as operations memory and as storage for scientific data. The FPGA is equipped with an external SRAM acting as a cache for scientific data packet formation. The processor's main functions are:

- scientific data acquisition, processing and transfer to On-Board Computer
- flight software and operations schedule execution
- motor and temperature control, power management
- fault detection, isolation and recovery

Main FPGA functions are:

- extending the CPU features through Space Wire
- providing missing communication interfaces (Packet Wire)
- Shadow Position Sensor data acquisition and preprocessing

See Fig. 1 for an overview over the DPU functional blocks and their relation to the overall system architecture.

3.1 Classic Approach to Performance Modeling

The system architecture, as described at the beginning of Sect. 3, shall be evaluated during its development against the system requirements. System characteristics estimation is needed to, first, understand the feasibility of the proposed solution, and, second, to identify the system's technical margins, which are needed to understand the project's technical risks.

The classical approach for modeling complex embedded system like the CCB DPU, would be to create a simple pipeline model. Such a model assumes tiles (chunks of images taken by Coronagraph) are transferred from one buffering place to another, immediately, with maximum rate. Typically, such a model will

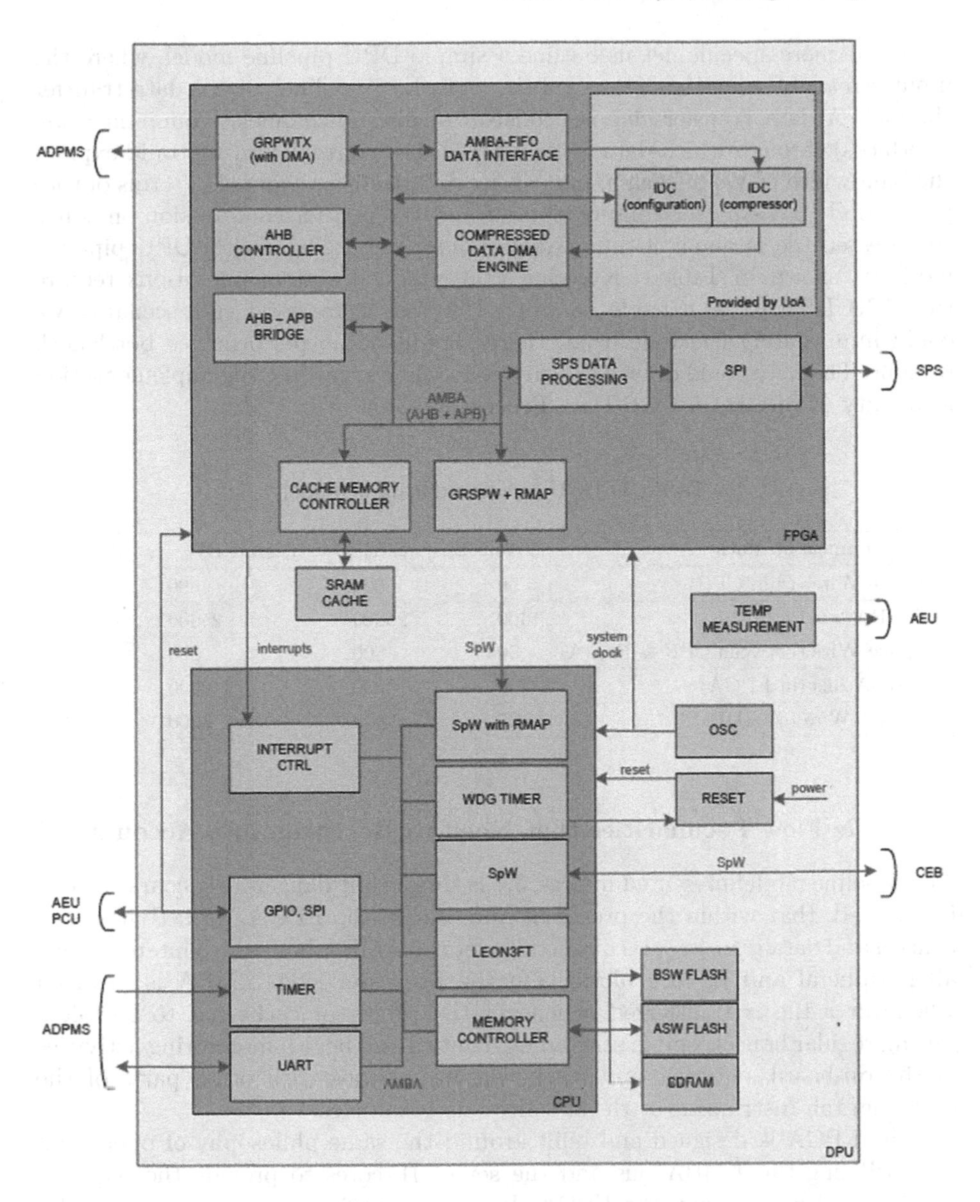

Fig. 1. Data Processing Unit (DPU) architecture

not include any functional dependencies or blocking operations and is therefore limited to giving theoretical maximum performance estimates of a given hardware configuration of the system. It can help detect of performance bottlenecks, which are typically located at the communication interface with the lowest bandwidth.

To be more specific, let us outline a simple DPU pipeline model, where the input stream of scientific data is fed through the "pipeline" over a data transfer channel. A data transfer channel consists of all sub-channels (communication interfaces) through which data is transferred. For convenience, we could express the bandwidth (BW) of each of part of the data transfer channel in terms of tiles per second (BW_{Coff} for compression off and BW_{Con} for compression on, since compressed tile is about 3 times smaller than raw). The simple DPU pipeline model is shown in Table 1. Knowing that planned system operations require the CCB DPU to be capable handling 192 tiles/s, in worst case scenario, we could immediately conclude, that, there should be no performance bottleneck present. Then, we could draw a positive conclusion regarding the implementation feasibility of PROBA-3 CCB Data Processing Unit.

Table 1. DPU pipeline dataflow model

Communication Path	BW [Mbps]	BW_{Coff} [tiles/s]	BW_{Con} [tiles/s]
Space Wire (from CEB)	50	800	800
AMBA bus (in CPU)	1600	25600	25600
Space Wire (between CPU & FPGA)	50	800	800
AMBA bus (in FPGA)	200	3200	9600
Packet Wire (to ADPMS)	25	400	3200

3.2 Tile-Flow Peculiarities that Need to Be Taken into Account

The pipeline model presented in Sect. 3.1 is simple but dangerously coarse. It can be noticed, that within the processor and within the FPGA, all activity—i.e., control and data transfer—is conducted over the AMBA bus, which interconnects all peripheral and IP-core blocks. On the processor side, AMBA is involved whenever a tile is transferred, whenever the processor cache has to be filled, during regular housekeeping, scheduling, control and health monitoring activities of the on-board software, and whenever interactions with other parts of the Coronagraph Instrument or the satellite bus take place.

The FPGA is designed and built around the same philosophy of operation, also utilizing the AMBA bus and the set of IP-cores to provide the required functionalities. However, the FPGA does not contain any processor itself. Its operation therefore crucially depends on remote configuration by the main processor (CPU). Therefore, although some bus transactions inside the FPGA (like DMA transfers) happen automatically, they have to be configured beforehand by the CPU. In consequence, the CPU remains in full control of scientific data flows, both in the GR712RC processor and in the RTAX 2000 FPGA.

The FPGA further reports back to the CPU any detected events related to the communication and Shadow Position Sensor (SPS) operation. Reporting is via 3 interrupt lines originating from the Packet Wire block denoting successful transmission of a tile to the spacecraft mass memory, from the IDC data compression engine, denoting end-of-tile processing, and from the SPS Readout Engine block. This generates additional asynchronous events, on top of those received by the CPU directly (e.g., in relation to the communication with the satellite's On-Board Computer and the ADPMS).

Let us take a brief glimpse how the AMBA bus works to understand a major issue we have to address. AMBA (also called here AHB) is a multiplexed bus (see Fig. 2). A multiplexer, connecting the master (i.e. source of data) to a slave (i.e. data sink), blocks any connection from happening until the bus is freed again. Only one type of operation can be executed at a time and by only one master at a time. That is either a tile can be transferred from Space Wire to SDRAM, or from SDRAM to the FPGA for compression, provided the FPGA is configured or interacts with the CPU. The same holds for the FPGA side AMBA bus: either the IDC compression engine (or the CPU) feeds tiles directly into the SRAM cache or tiles are sent through Packet Wire to the On-Board Computer. On a given AMBA bus (of the CPU or FPGA), each AMBA operation blocks all other operations from being executed. Introducing multiple separate buses allows increasing the degree of parallelism, but the dependence on the CPU to manage data flows continues to significantly affect system performance. This aspect is ignored in classic modeling, causing it to be potentially flawed.

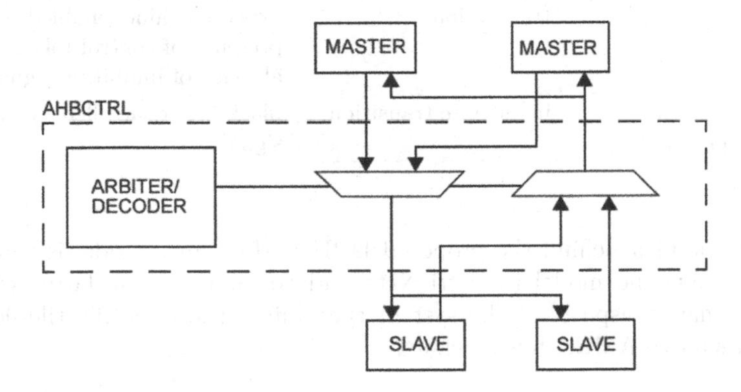

Fig. 2. AMBA AHB controller functional block diagram

4 Processing Performance Modeling

In order to build a Petri Net model capable of correctly characterizing the CCB DPU processing performance, we need to map DPU functional components and other aspects of the system to Petri Net (PN) primitives.

We used a Extended Deterministic and Stochastic Petri Net (eDSPN) to model the CCB DPU. Table 2 summarizes the PN primitives and the physical reality they represent. All the models presented in the paper were built and simulated with the TimeNet software [12,17,28].

Table 2. Physical system element to Petri Net primitive mapping.

Physical system	Petri Net	Comment
tile, control signal	token	tokens are all the same, token role is interpreted depending on topology of places and transactions
memory buffer, cache, processing block, process mode change	place	place function is subject to abstract interpretation
buffer or communication interface capacity	inhibitor arc	arc ended with a circle, connecting place and inhibited transaction, activated when number of tokens in starting place exceeds defined threshold
communication interface	deterministic transaction	black rectangle, firing rate inversely proportional to tile throughput, enabled in presence of control token, absence of inhibiting signals
communication bus	exponential transaction	white rectangle, firing rate expected value, enabled in presence of control token, absence of inhibiting signals
software process preemption	immediate transition	black bar, controlled by guard signals

The basic idea behind the proposed DPU performance model is to convert the classic pipeline model to Petri Nets and to augment the latter with the missing dynamic aspects of the system that affect the scientific tile-flow and that characterize AMBA bus blocking.

4.1 Initial Processing Performance Model

The proposed CCB DPU Petri Net model is shown on Fig. 3. It characterizes the CCB DPU configuration where compression is enabled. In the model, the places P_CEB and P_ADPMS represent the scientific data stream source and the sink, respectively. To obtain a more flexible simulation, we introduced a parameter N (typically set to 10000) for the number of tokens (or tiles) that are to be transferred. Simulation ends once N tokens arrive at P_ADPMS. The time to transfer all tokens thereby serves as a performance indicator of the system throughput, as shown in Eq. 1.

$$System\_Throughput = \frac{N}{time\_to\_transfer\_N\_tokens} \tag{1}$$

Tokens need to traverse along the whole DPU science data path to be stored in P_ADPMS. Some places along this path represent internal buffers of the communication interfaces (P_SPW[0..2]), two places represent the key buffers of the DPU system - the SDRAM memory attached to CPU (P_SDRAM) and SRAM memory attached to FPGA (P_SRAM). The IDC compression engine, assumed to process 1 tile at a time, is labeled as P_IDC. The maximum buffer size (in number of tiles it can hold) is enforced by the use of transition feedback, an inhibitor arc.

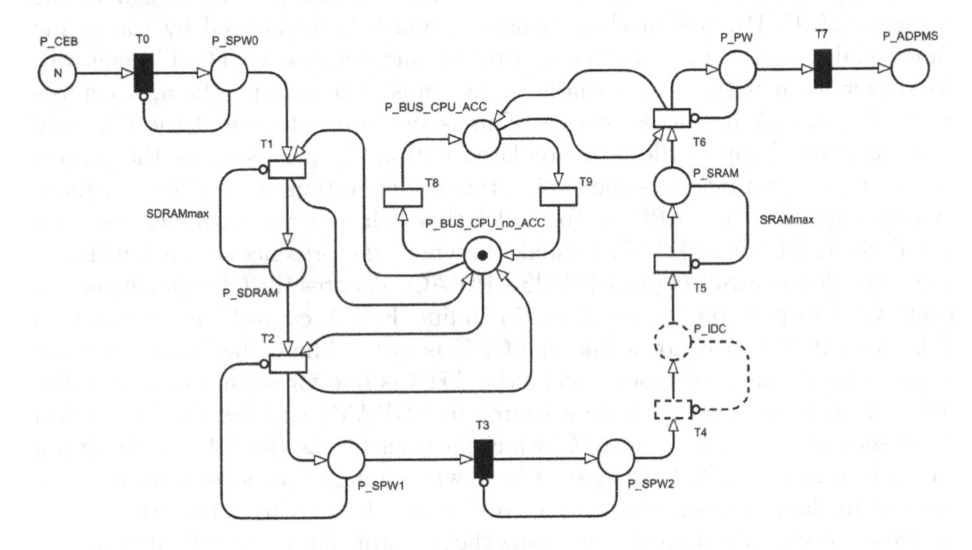

Fig. 3. Data Processing Unit (DPU) performance model. Compression engine, disabled in raw mode, marked with dashed line.

Petri Net places are connected using transitions, representing the throughput of the interfaces and buses that are involved in transferring tiles between the respective buffers and processing nodes in physical system. The transitions T0 and T3 are Space Wire interfaces from the CEB and between the CPU and the FPGA. They operate at 50 Mbps. Transition T7 is a Packet Wire interface to ADPMS and operates at 25 Mbps. Transition T1 and T2 represent the AHB transfers to and from the SRAM buffer. Similarly, T4, T5, T6 represent AHB transitions in the FPGA. The AHB bus in the processor operates at 1600 Mbps and the AHB in the FPGA at 200 Mbps. Equation 2 denotes the firing rate (i.e., expected time to fire).

$$Time\_to\_Fire = \frac{Tile\_Size}{Throughput} \tag{2}$$

There are three modes of scientific data flow processing: compression, bit-stuffing and raw. Compression and bit-stuffing make use of the IDC engine to perform their activities. This part of the system's Petri Net model can remain the same. Only the transition firing rate needs to change in order to reflect changes in the tile throughput in the FPGA. This is because tile sizes are reduced after compression, utilizing only a fraction of the bandwidth. For raw mode the location P_IDC and transmission T4 are removed, T5 then connects directly to P_SPW2, and firing rates are modified to reflect the transition of full size tiles.

In order to model the dynamic behavior of the system, at the stage of development, where perhaps only hardware prototypes are available, but no software, simplifying assumptions have to be made. Plain throughput, as shown in the pipeline model, is not sufficient to validate architecture design in the presence of AMBA bus blocking behavior, which is introduced by competing functional chains (science vs SPS control vs maintenance vs TC/TM control). Moreover, even within the scientific data transfer functional chain, a competition for AMBA bus access occurs. This is because only one of the following two situations happen: tiles are circulated within the processor; or the processor configures and monitors the semi-automatic operation of the Direct Memory Access engines in the FPGA. To model this behavior we introduce the subnet P_BUS_CPU{∅,_no}_ACC, T8 and T9, where the presence of a token (interpreted as flow control) in place P_BUS_CPU_ACC denotes the CPU having access over AHB to perform its activities (including FPGA control) and a token in P_BUS_CPU{∅,_no} denotes that the CPU is either idle or performs activities using only it's cache memories, while the AHB is free for science data transfer. The T1 transfer (from CEB data source to SDRAM) will fire only if a token is present at P_BUS_CPU_no_ACC, which in turn blocks the T2 transfer (from SDRAM buffer to FPGA via Space Wire), which mimics the same limitations as present in the processor. Firing of T1 or T2, besides moving tokens that represent tiles of scientific data moving along the scientific datapath, will also put the flow control token back to P_BUS_CPU_no_ACC. A similar concept of operation is implemented for enabling T6, with small but important difference, that this transition can fire only when the CPU takes control over the GR712RC AMBA bus and the FPGA Space Wire link with RMAP enabled to configure the Packet Wire DMA engine.

4.2 Simulation of the Initial Processing Performance Model

Since the initial model is designed to validate the architectural considerations before actual hardware is built and software deployed on it, there is not much gain in simulating performance to evaluate overall absolute values. Obtained measures would be questioned during technical reviews as being too affected by model assumptions. Much more appealing at this early stage is leveraging the model for exploring numerous design variants and configurations, to see which of them are promising to fulfill system and, eventually, mission requirements.

To evaluate the CCB DPU architecture, one needs to show the system performance goal can be met and under what conditions. Therefore, on the one hand, simulation needs to show that the most demanding (data-intensive) observation scenario can be served within the expected payload operation window (here 25–30% of the orbital period) and that the influence of AMBA bus blocking on the scientific data flow is tolerable. The first can be obtained by simulating the time required to transfer and process the selected amount of tokens from CEB to ADPMS, extrapolating the result to estimate the time required for transferring all tiles generated during the CME-Watch scenario (~2.9M tiles). The second measure can be obtained by defining *AMBA_blocking_ratio* (Eq. 3) as the fraction of time during which the token is in place P_BUS_CPU_ACC in contrast to being in either in P_BUS_CPU_ACC or P_BUS_CPU_no_ACC.

$$AMBA\_blocking\_ratio = \frac{T9}{(T8 + T9)} * 100\% \tag{3}$$

Figure 4 shows DPU's performance obtained from simulating the initial model outlined above, configured with (a) and without (b) compression. The blue-shaded area at the bottom of the figures denotes the time limit available for DPU operations: 21600 s. s. Scientific data streaming to ADPMS is considered successful when it finishes before this time limit. The plots show the time to stream data to ADPMS that was generated in the worst-case CME-watch scenario for different CPU AMBA bus blocking ratios (i.e., for different fractions of total time that the AMBA bus is not used for transferring scientific data tiles, refer to Eq. 3). Simulations are performed for few cases, showing fours orders of magnitude of AMBA blocking ratio granularity. The more the AMBA bus is blocked for other activities than transferring tiles from the CEB and further into the DPU and ADPMS (horizontal axes), the more time it takes to complete the whole data dump to the on-board computer. This actually confirms initial suspicions, since when tile-reception is blocked at the GR712RC, then tile inflow to the DPU is limited as well and no advantage can be drawn from the fast internal DPU interfaces.

Another insight into CCB DPU operations can be obtained in relation to the impact of the CPU AMBA bus mode switching time granularity. The larger this granularity, the more time is spent in a given mode before switching and the lower will be DPU's capability to stream the data. Keeping mode switching granularity high and ensuring buffers are emptied quickly, prevents tile pile-ups and appears to be a key to achieving the required performance and drove further hardware and software codesign.

5 Implementation

Validating the architecture with the help of Petri Nets, as described in the previous section, confirms the initial assumptions and indicate that it will be feasible to implement the CCB DPU. They provided insight into the system's internal operation and allowed to better understand the involved hardware-software interactions, allowing the project to advance into subsequent phases, where the focus is on building and deploying the actual equipment.

Subsequent CCB DPU models are developed in an iterative approach, where each following model tests the assumptions, solutions and design decisions that have been made with the previous model and that are most critical. The outcome

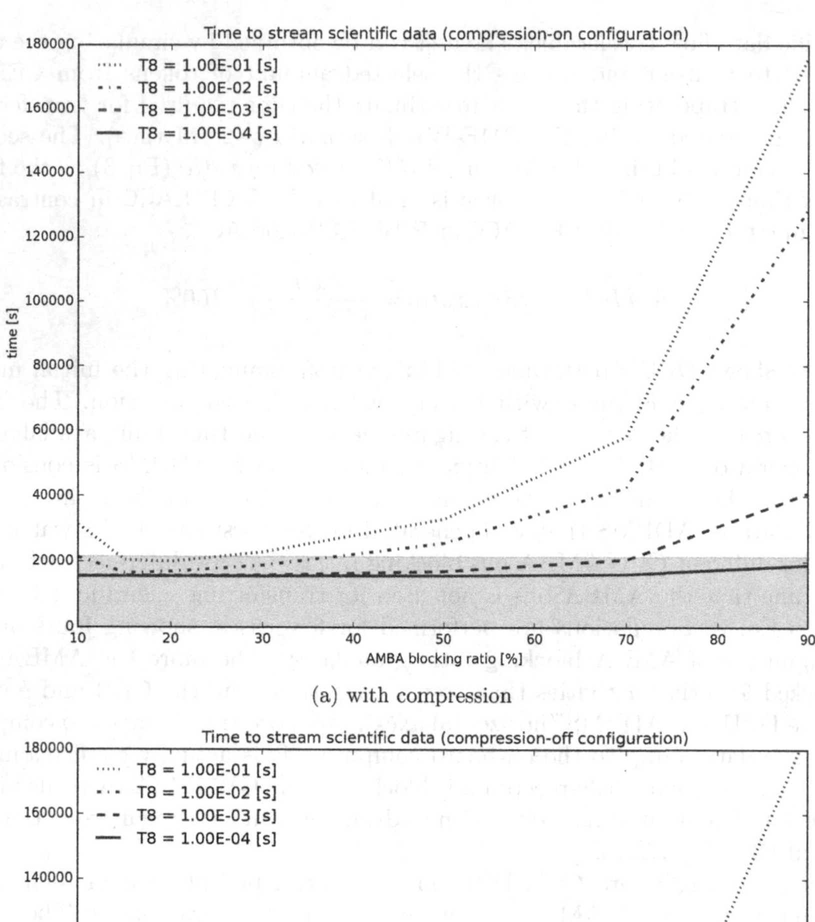

(a) with compression

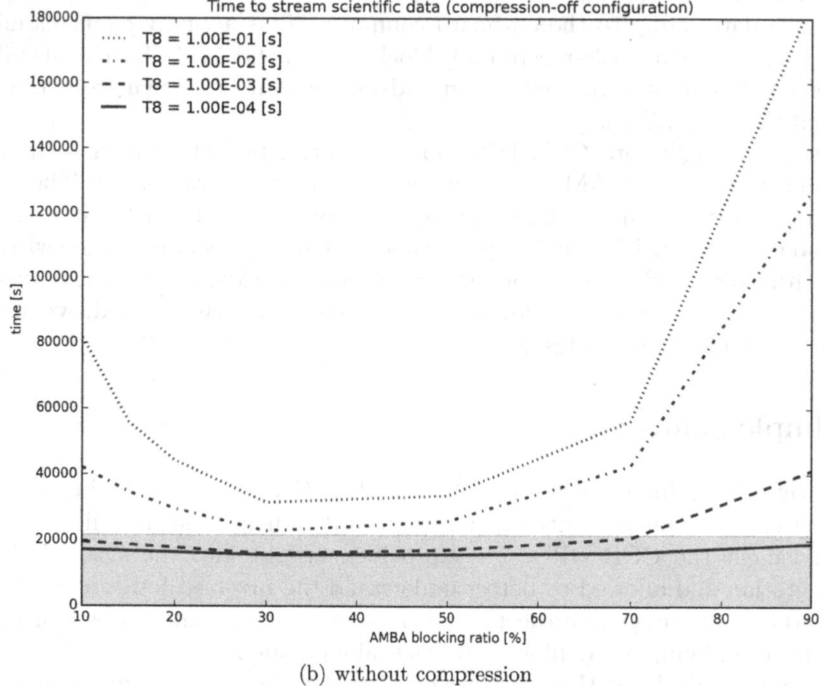

(b) without compression

Fig. 4. Data Processing Unit (DPU) performance simulation

of analyzing such a refinement may require discarding certain design decisions or they may remain uncertain in the sense that still multiple options may be viable. The best viable state in the final phase is the desired system. A summary of the models built within the CCB DPU project is provided in Table 3. For clarity, only the models that are most relevant for the performance evaluation are shown.

Table 3. CCB DPU models

Model	Phase	Applicability	Purpose
Software Mock-Up	B & C	Prototyping	Used for verifying schedulability of software components
Development Model	B & C	Prototyping	Used for testing the hardware and software, verification of initial assumptions, getting acquainted with technology stack
Engineering Model	C & D	Advanced Prototyping	Built be as close to Flight Model as possible, used for testing critical functionality and interfacing with the rest of the system. Intermediate software revisions deployed.
Proto-Flight Model	D & E	To conduct the mission	Full flight configuration, materials, quality and processes, tested to acceptance levels. Flight software revision deployed

The DPU Development Model allowed for early prototyping and was built for reducing the risks associated with technology and components selection. It further served to create first mock-ups of the critical hardware and software parts and for testing initial assumptions about operations, including first performance measurements. For example, at this stage, measurements already indicated a tile throughput of 252 tiles/s on average, for which we had to implement only part of the final software functionality.

Following the above, we conducted (as part of our Verification and Validation campaign) a performance verification of the DPU processing capabilities on a physical implementation of the models (the Engineering Model and Proto-Flight Model, shown in Fig. 5). The campaign attempted to answer whether the requirements defining the system have been met, but also, what were the actual limits of the hardware and of the software scientific data processing power. The data flow test involved verifying that the rate of transfer of compressed science data from the CCB to the ADPMS memory module met the requirements set out in the system requirements specification. The scenario measured the average number of tiles processed by the CCB and sent to the ADPMS each second based on the events logged by the software for each tile. The numbers obtained from validation test logs, collected from runs performed on the engineering model

Fig. 5. Flight model of Coronograph Control Box (CCB) Data Processing Unit (DPU)

of the CCB, connected to the electrical ground support equipment, oscillated around 205 tiles per second. It therefore met the requirement of at least 192 tiles per second, as defined in the requirements specification.

However the DPU engineering model, at time of test, did not operate under the final version of flight software. Similar tests conducted with the Proto-Flight Model in its final software configuration revealed a scientific tile throughput of 233 tiles per second (all values for compression mode enabled), still well within the specification requirements.

6 Correlating the Petri Net Performance Model with the Physical Model

It is worth noticing, that the performance model and actual verification measurements were organized around different methodologies. The Petri Net performance model was targeting simulation of what is happening inside the DPU, while it is fed with a worst case expected data stream to figure out what are required buffer sizes and what control-performance bottlenecks might be encountered in the system. Modeling was performed for the purpose of reducing uncertainty (and associated programmatic risks) and for optimizing the technical margins of the system. The verification measurements, however, were targeting the discovery of actual performance limits in the system. Therefore it was fed with the maximum data stream allowed by the input interface bandwidth, measuring the resulting processing performance.

To correlate these performance results, both have to follow the same experimentation philosophy. However, since at the time of writing, the CCB underwent the integration process with the satellite platform, it was easier to adjust the Petri Net model rather than repeating the performance measurements with the payload, although the latter will have to happen anyway as part of the validation campaign of the integrated satellite.

A quick look at the Petri Net models in Fig. 3 allows noticing that to replicate the performance experiments, which had been performed on the implemented equipment, it suffices to modify transaction T1, defining the data stream input rate, from 192 tiles/s to 800 tiles/s which is about the maximum that can be handled by the SpaceWire interface at a configured bit rate of 50 Mbps.

The other aspect that has to be adjusted in our models is the T8/T9 transactions to resemble the behavior of the CCB Application Software. In the initial modeling of the CCB DPU, the expected value of the T8 and T9 transactions firing time was of main interest, but not their absolute values, which at that time would have to be guessed. The analysis had therefore focused on the relation between T8 and T9 and on the order of magnitude of the firing time (granularity of translation activation) as described in Sect. 4.2. Now, after the Flight Software (ASW) is available and deployed in target architecture, it is possible to estimate these values more accurately.

T8 indicates for how long the CPU manages tile-inflow to the system, receiving in the first part of the data pipeline the tiles from imager electronics (CEB) and within the CPU itself (storing in SDRAM, passing to FPGA, configuring FPGA DMAs and IDC operation). T8 is defined by execution times of ASW components summed up in Table 4.

T9 defines how long the CPU manages tile-outflow of the system in the second part of the data pipeline, pushing out the tiles stored in the FPGA SRAM to On-Board Computer(ADPMS) through PacketWire. T9 is defined by execution time of ASW components (summed up as well in Table 4).

Table 4. Relevant ASW components average execution times, modeled by respective transactions.

Transaction	Component	Average execution time [ms]
T8	*tile_related_rmap_transaction*	0.012
	tile_spw_transmission	1.250
	fpga_idc_compression	0.660
	Total:	1.922
T9	*fpga_pw_transmission*	2.73
	Total:	2.73

As for T8's and T9's expected firing times, the key architecture quality factor derived from initial modeling, AMBA blocking factor (Eq. 3) is 59% and it has a very fine granularity (order of milliseconds) which are the exact circumstances that have been discovered as offering sufficient CCB DPU performance margins and flexibility to meet the system requirements (refer to Sect. 4.2).

However, setting realistic values of T8 and T9 is necessary, but not sufficient, to ensure adequate similarity between Proto-Flight Model and Petri Nets. At this stage, we also needed to accommodate the fact that the Application Flight Software is much more complex than it was considered (and feasible for taking into account) at initial modeling time. Application Flight Software is responsible for real-time SPS sensor measurements and their delivery further down the Guidance, Navigation and Control loop. This is 2 Hz period process and some associated processing load. Among other loads there is a telecommand decoding and telemetry packet generation, as well as, all the housekeeping and control activities and, last but not least, fault detection, isolation and recovery functions (internal monitoring).

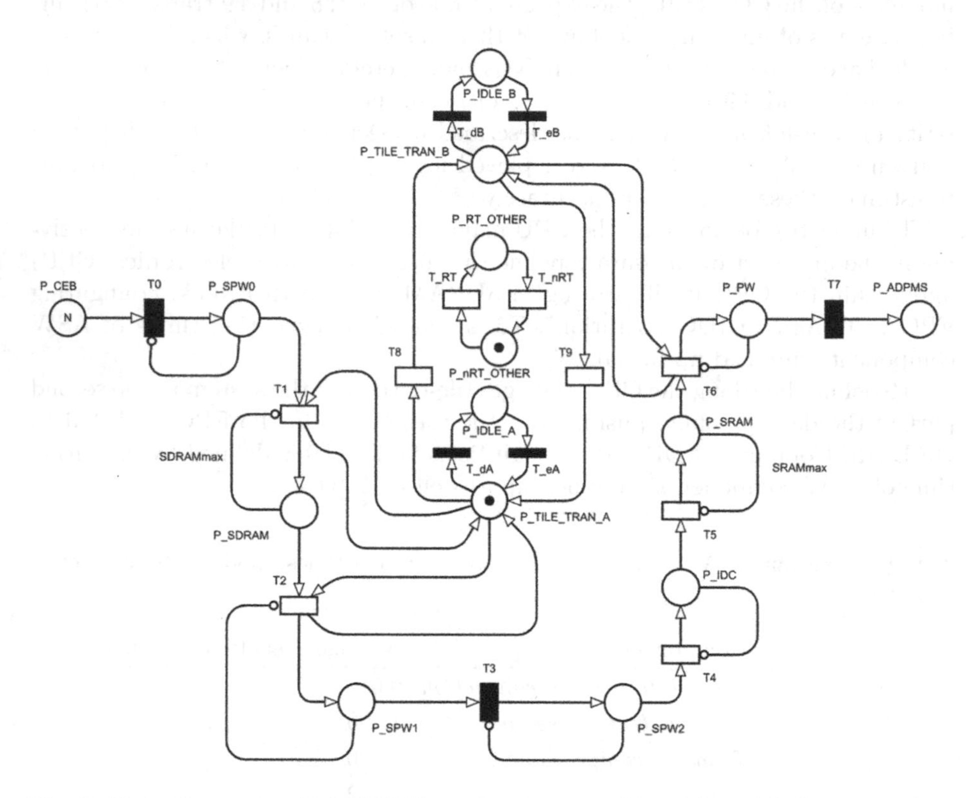

Fig. 6. Data Processing Unit (DPU) performance model, with enabled compression, correlated with Proto-Flight Model

All in all, those real-time (deterministic scheduling) and asynchronous (interrupt based) loads block, when executed, the AMBA bus for tile transfer and this effect can be modeled by controlling the presence of tile transmission enabling token rotation in P_TILE_TRAN_A & T8 & P_TILE_TRAN_B & T9 path. This is done by adding the P_IDLE_A and P_IDLE_B places which siphons in (by immediate transitions T_dA and T_dB) or siphons out (by immediate transitions T_eA and T_eB) the token so none of the transition responsible for modeling actual tile transfer can fire. Disabling the tile transmission in the system, effectively is performed by activating T_dA or T_dB. Both these transition are controlled by an enabling function, asserted when the newly introduced place P_RT_OTHERS contains a token. By symmetry, the tile transmission within the system recovers when T_eA or T_eB are activated, by asserting the enabling function when place P_nRT_OTHER contains the token. It shall now be obvious that a token in the place P_RT_OTHER models the times when the CPU executes the scheduled and asynchronous loads, while a token in P_nRT_OTHER models the situation when the CPU performs best-effort operations and tile-transfers. The transition T_RT ensures that periodic and other loads are executed in simulation every 500 ms and the T_nRT sets the expected load execution time to about 200 ms, which estimate is based on execution loads of major components of ASW. The model is presented in Fig. 6. The enabling connections from P_RT_OTHERS to T_dA and T_dB, as well as from P_nRT_OTHER to T_eA and T_eB are omitted for maintaining clarity of figure.

Simulation of the updated models in TimeNet software yields that, in order to transfer 10000 tiles the DPU would need 41 and 59 s, respectively, for models with compression enabled and disabled. This corresponds to about 244 and 169 tiles/s peak performance capability of DPU for compression-on and -off mode. For the mode of operation involving the bit-stuffing (each pixel is coded with 16 bits but contains only 14 bits of information) time to transfer while simulated tile batch is about 51 s, which corresponds to 196 tiles/s on average.

Simulation results correlate well with experimental results, as presented in Table 5, in the whole spectrum of DPU modes of operation. Following models are taken into account: Development Model (DM), Engineering Model (EM), Proto-Flight Model (PFM) and Petri Net model (P/N-model).

Table 5. The processing performance (in tiles per second) of CCB DPU models.

Mode of operation	DM	EM	PFM	P/N-model
raw tiles	—	—	180	169
bit-stuffed tiles	—	—	202	196
compressed tiles	252	205	233	244

7 Gained Insights and Conclusions

Petri Nets are not new to industry. However, they are not used very widely. They can serve addressing the increasing importance, in logistics, network and computer architectures design and analysis and workflow systems.

The work presented in this paper documents the way Petri Nets were involved in systems engineering of complex control and data processing systems. This method of modeling has been proven as a useful tool for obtaining insights in architecture trade-offs, but more importantly, the model could be easily updated to track and reflect the changes in the system, as development gains maturity and up to the stage of quite accurate correlation of simulation results with the measurements of processing performance of DPU flight hardware and software. Our main achievements of the CCB DPU modeling process using Petri Nets can be summed up in following way:

- DPU Petri Net performance model revealed that DPU has to be treated as streaming, not buffering, device, as in worst-case operation scenario of "CME-Watch" there will be no time to send all scientific data tiles after the end of observations;
- to the last point, DPU amount of on-board SDRAM memory has been reduced significantly, which in terms of Flight Units delivery is equivalent of buying 9 SDRAM memory modules less (savings of roughly 50 thousand Euro in components costs);
- DPU performance model analysis provided insights in how the Application Software controlling the DPU has to manage the scientific data flow and circulation of tiles from CEB, to compression engine and to ADPMS Mass Memory Modules in order to meet the processing performance demands;
- the Petri Net model validity has been proven by, first, building the flight equipment meeting performance requirements, second, accurately correlating the updated Petri Net model with the measurements performed on the DPU Flight Model.

All in all, we found that time and effort spent on developing modeling tools, such as Petri Nets, are likely to pay back in the future. Modeling, if it allows early concept prototyping in order to fail and pivot or to consolidate and move forward, without any doubt, is an invaluable support, that is directly traceable to savings in time and money spent on the project. The described creation of Proba-3 Coronagraph Control Box and its Digital Processing Unit in particular, provides solid evidence for this claim.

The presented work contributes a small step towards providing new systems engineering tools for use in the aerospace or space industry. It is a perfect moment for such discussions, as space business undergoes deep changes, starting from the New Space revolution up to evolution of Model-based Systems Engineering into a Digital Twin Spacecraft concept. Petri Nets, thanks to their high level of abstraction and versatility, tackle a large class of system engineering issues, especially in the early design phases.

References

1. TASTE. https://taste.tools/. Accessed 26 Jan 2022
2. Blommestijn, R., Fuchs, J.: Specification and description language (SDL), July 1999. z.100
3. Blommestijn, R., Fuchs, J.: Technical Dossier on System Modelling and Simulation Tools, July 2012. iss 2, rev 2A

4. Bluff, R.: Avionic system modelling. In: International Conference on Simulation 1998 (Conf. Publ. No. 457), pp. 11–18, September 1998. https://doi.org/10.1049/cp:19980610
5. Bollig, B., Fortin, M., Gastin, P.: Communicating finite-state machines, first-order logic, and star-free propositional dynamic logic. J. Comput. Syst. Sci. **115**, 22–53 (2021). https://doi.org/10.1016/j.jcss.2020.06.006
6. Eickhoff, J.: Simulating Spacecraft Systems. Springer, Heidelberg (2009)
7. Eickhoff, J.: Onboard Computers, Onboard Software and Satellite Operations: An Introduction, 2012 edn. Springer, New York (2011)
8. Ereau, J.F., Saleman, M.: Modeling and simulation of a satellite constellation based on Petri nets, pp. 66–72, January 1996. https://doi.org/10.1109/RAMS.1996.500644
9. Feiler, P.: Model-Based Engineering with AADL: An Introduction to the SAE Architecture Analysis & Design Language, 1st edn. Addison-Wesley Professional (2012)
10. Fischer, M.J., Ladner, R.E.: Propositional dynamic logic of regular programs. J. Comput. Syst. Sci. **18**(2), 194–211 (1979). https://doi.org/10.1016/0022-0000(79)90046-1
11. Galano, D., et al.: Development of ASPIICS: a coronagraph based on Proba-3 formation flying mission. In: Lystrup, M., MacEwen, H.A., Fazio, G.G., Batalha, N., Siegler, N., Tong, E.C. (eds.) Space Telescopes and Instrumentation 2018: Optical, Infrared, and Millimeter Wave, vol. 10698, pp. 906–918. International Society for Optics and Photonics, SPIE (2018). https://doi.org/10.1117/12.2312493
12. German, R., Kelling, C., Zimmermann, A., Hommel, G.: TimeNET-a toolkit for evaluating non-Markovian stochastic Petri nets. In: Proceedings of the Sixth International Workshop on Petri Nets and Performance Models, pp. 210–211, October 1995. https://doi.org/10.1109/PNPM.1995.524333
13. Girault, C., Valk, R.: Petri Nets for Systems Engineering: A Guide to Modeling, Verification, and Applications. Springer, Heidelberg (2003)
14. Graczyk, R.: Reliability and performance modeling of configurable electronic systems for unmanned spacecraft. Editorial Series on Accelerator Science, Warsaw University of Technology Publishing House, Warsaw (2016)
15. Hall, A.D.: A Methodology for Systems Engineering. Van Nostrand (1962)
16. Holt, J.: UML for systems engineering (2004)
17. Kelling, C., German, R., Zimmermann, A., Hommel, G.: TimeNET: evaluation tool for non-Markovian stochastic Petri nets. In: Proceedings of IEEE International Computer Performance and Dependability Symposium, September 1996. https://doi.org/10.1109/IPDS.1996.540206
18. Kordon, F., Canals, A., Dohet, A.: Embedded systems analysis and modeling with SysML, UML and AADL. Electonics Engineering Series, ISTE, London (2013)
19. Kossiakoff, A., Sweet, W.N.: Systems Engineering Principles and Practice, 1st edn. Wiley-Interscience, New York (2002)
20. Lloret, J.C., Roux, J.L., Algayres, B., Chamontin, M.: Modelling and evaluation of a satellite system using EVAL, a Petri Net based industrial tool. In: Jensen, K. (ed.) ICATPN 1992. LNCS, vol. 616, pp. 379–383. Springer, Heidelberg (1992). https://doi.org/10.1007/3-540-55676-1_23
21. Malott, L., Palangpour, P.: Small spacecraft software modeling: a Petri net-based approach. In: AIAA/USU Conference on Small Satellites, August 2013
22. Molloy: Performance analysis using stochastic petri nets. IEEE Trans. Comput. **C-31**(9), 913–917 (1982). https://doi.org/10.1109/TC.1982.1676110

23. Murata, T.: Petri nets: properties, analysis and applications. Proc. IEEE **77**(4), 541–580 (1989). https://doi.org/10.1109/5.24143
24. Perrotin, M., et al.: TASTE in action. In: 8th European Congress on Embedded Real Time Software and Systems (ERTS 2016). TOULOUSE, France, January 2016. https://hal.archives-ouvertes.fr/hal-01289678
25. Platzer, A.: Logical Analysis of Hybrid Systems Proving Theorems for Complex Dynamics. Springer, Heidelberg (2010)
26. Rechtin, E.: Systems Architecting: Creating and Building Complex Systems. Prentice Hall (1991)
27. Renotte, E., et al.: Recent achievements on ASPIICS, an externally occulted coronagraph for PROBA-3. In: MacEwen, H.A., Fazio, G.G., Lystrup, M., Batalha, N., Siegler, N., Tong, E.C. (eds.) Space Telescopes and Instrumentation 2016: Optical, Infrared, and Millimeter Wave, vol. 9904, pp. 1112–1126. International Society for Optics and Photonics, SPIE (2016). https://doi.org/10.1117/12.2232695
28. Zimmermann, A.: Modeling and evaluation of stochastic Petri nets with TimeNET 4.1. In: 2012 6th International Conference on Performance Evaluation Methodologies and Tools (VALUETOOLS), pp. 54–63, October 2012
29. Zurawski, R., Zhou, M.: Petri nets and industrial applications: a tutorial. IEEE Trans. Ind. Electron. **41**(6), 567–583 (1994). https://doi.org/10.1109/41.334574

Petri Nets Semantics of Reaction Rules (RR)

A Language for Ecosystems Modelling

Franck Pommereau[1(✉)], Colin Thomas[1,2], and Cédric Gaucherel[2]

[1] IBISC, Univ. Évry, Univ. Paris-Saclay, 91020 Évry-Courcouronne, France
franck.pommereau@univ-evry.fr

[2] AMAP-INRA, CIRAD, CNRS, IRD, Univ. Montpellier, 34398 Montpellier, France

Abstract. The EDEN framework provides formal modelling and analysis tools to study ecosystems. At the heart of the framework is the *reaction rules* (RR) modelling language, that is equipped with an operational semantics and can be translated into Petri nets with equivalent semantics. In this paper, we formally define the RR language and its semantics, detailing the initial definition from [8] and extending it with a notion of *constraints* that allows to model mandatory events. Then, we consider in turn two classes of Petri nets: *priority Petri nets* (PPN), which are safe place/transition Petri nets equipped with transitions priorities, and *extended Petri nets* (EPN) which are PPN further extended with read arcs, inhibitor arcs, and reset arcs. For each of these classes, we define the translation of an RR system into a Petri net and prove that the state-space generated with the RR operational semantics is equivalent to the marking graph of the Petri net resulting from the translation. We use a very strong notion of equivalence by considering *labelled transition systems* (LTS) isomophism with states and labels matching.

1 Introduction

The framework EDEN has been developed and used for ecological studies for more than five years [4–8,12]. It provides tools and methods to formally model ecosystems, and analyse them through an interactive method that lets the users explore their models dynamics and draw understanding progressively. Properties of interest include searching the root causes (events, conditions, or states) leading to trajectories of interest, structural stabilities, or collapses. In addition to the published works, EDEN has been used by more than a dozen Master interns in ecology, who modelled and analysed varied ecosystems.

At the heart of EDEN is the *reaction rules* modelling language (RR). An RR system consists of (1) a set of *Boolean variables* representing the functional presence (on) or absence (off) of an entity in an ecosystem, and (2) a set of *actions* representing the possible events that lead to observable changes in the ecosystem (*i.e.* assign new values to variables). A species is functionally present if its presence enables observable effects on the ecosystem, otherwise it is functionally

L. Bernardinello and L. Petrucci (Eds.): PETRI NETS 2022, LNCS 13288, pp. 175–194, 2022.
https://doi.org/10.1007/978-3-031-06653-5_10

```
inhabitants:                               constraints:
    Rp+: reproductives                       Fg- >> Te-                      # C1
    Wk-: workers                           rules:
    Sd-: soldiers                            Rp+ >> Ec+                      # R1
    Te-: termitomyces (fungi)                Rp+, Ec+ >> Wk+                 # R2
structures:                                  Wk+ >> Wd+, Te+, Fg+, Ec+       # R3
    Ec-: egg chambers                        Wk+, Wd+ >> Sd+, Rp+            # R4
    Fg-: fungal gardens                      Wk+, Te+ >> Wd-                 # R5
resources:                                   Wd- >> Wk-, Te-                 # R6
    Wd-: wood                                Wk- >> Fg-, Sd-                 # R7
competitors:                                 Wk-, Rp- >> Ec-                 # R8
    Ac*: ant competitors                     Ac+, Sd- >> Wk-, Rp-            # R9
```

Fig. 1. A toy model of a termite colony, adapted from [8]. Variables are defined in the left column, dispatched into user-chosen sections (that play the role of comments). Each variable is given a name, an initial value (+ for on, - for off, * for both values), and a description. For instance, variable Rp is initially on and models the reproductives (queen and king). Constraints and rules, collectively referred to as actions, are defined in the right column, each has a left-hand side that corresponds to its guard and a right-hand side that corresponds to its effect. We have named the actions for reference using a comment "# ...". For instance, rule R9 states that if Ac is on and Sd is off, then rule R9 may be executed, yielding a state in which Wk and Rp are set to off.

absent. Actions are divided into *constraints* and *rules*, the only difference being that the former have the priority over the latter (*i.e.* no rule can be executed if a constraint can). Constraints are useful in particular to model cascading events or transient states. For instance, if a pond dries its inhabitants will rapidly die. Such a situation may be modelled by a constraint, and the state where the pond is dry but its inhabitants are still present has to be transient. RR models can be seen as an analogue for ecology to Boolean networks for systems biology [21], but with important differences that will be underlined later on. An example of an RR system is given in Fig. 1. Note that an RR system may have several initial states, for instance this one has two (one with Ac+ and another with Ac-).

The contribution of this paper is graphically summarised in Fig. 2. In Sect. 3, we give a formal definition of RR systems, detailing the initial definition from [8] and enriching it with the notion of constraints as well as with the possibility to have more than one initial state. We then define the semantics of RR systems in terms of *labelled transitions systems* (LTS). At the end of Sect. 3, we define two transformations on an RR system, *normalisation* and *elementarisation*, and we prove that they generate RR systems whose semantics are equivalent to that of the original RR system. In Sect. 4, we define the translation of *elementary* RR systems into *priority Petri nets* (PPN: regular Petri nets extended with transitions priorities), and we prove that the nets resulting from this translation yield *marking graphs* (MG) that are equivalent to the semantics of the translated elementary RR system. This corresponds to the right-most column of the diagram depicted in Fig. 2. Finally, in Sect. 5, we define the translation of *normal* RR systems into *extended Petri nets* (EPN: PPN further extended with read-, inhibitor-

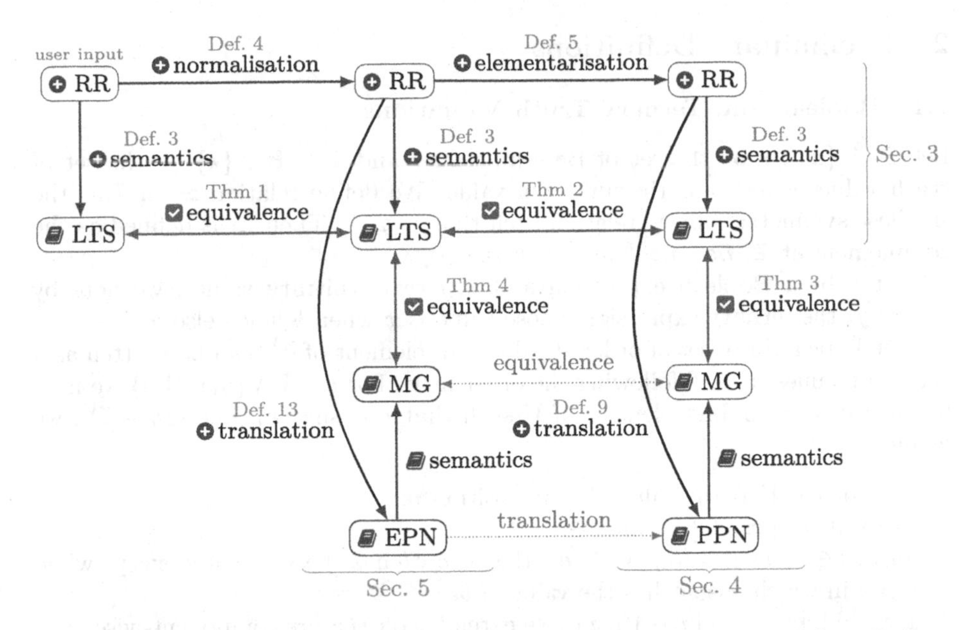

Fig. 2. Visual summary of the paper, where ⊕ marks the elements we define, 📘 those that exist already in the literature, and ☑ those we prove. RR stands for *reaction rules*, LTS for *labelled transition systems*, MG for *marking graph*, EPN for *extended Petri nets*, and PPN for *priority Petri nets*.

and reset-arcs), and we prove that the nets resulting from this translation yield marking graphs that are equivalent to the semantics of the translated normal RR system. This corresponds to the middle column of the diagram depicted in Fig. 2.

From the proved equivalences, we also have that the marking graph of an EPN translated from a normal RR system is equivalent to the marking graph of the PPN translated from the elementarisation of the normal RR system (upper-most gray dotted edge in the diagram from Fig. 2). Finally, note that there exists a translation from EPN to PPN that is well known in the Petri net community[1] and corresponds exactly to what we use in the current paper: safe places are translated into pairs of complementary places, which allows to implement reset and inhibitor arcs as regular arcs, while read arcs are implemented with side-loops.

To start with, Sect. 2, provides the basic definitions upon which the rest is defined; in particular the definition of *labelled transitions systems* LTS and their equivalence through isomorphism with states and labels matching.

[1] We did not find its formal definition and proof of correctness in the literature.

2 Preliminary Definitions

2.1 Boolean and Ternary Truth Valuations

Let $\mathbb{B} \stackrel{\text{df}}{=} \{\bot, \top\}$ be the set of Boolean values and $\mathbb{T} \stackrel{\text{df}}{=} \mathbb{B} \uplus \{\star\}$ be the set of truth values where $\star$ is the unknown value. We define relation $\not\simeq$ on $\mathbb{T}$ as the smallest symmetric binary relation such that $\top \not\simeq \bot$. Then, $\simeq$ is defined as the complement of $\not\simeq$, *i.e.*, $a \simeq b$ iff $\neg(a \not\simeq b)$.

Let b be a Boolean expression and x, y two arbitrary values, we note by $\langle b \,?\, x : y\rangle$ the ternary expression whose value is x when b holds else y.

Let V be a finite set of ordered values, an element of $\mathbb{T}^V$ can be written as a vector of values from $\mathbb{T}$ following the order in V. For $v \in V \setminus \{\max(V)\}$ we note by $\text{succ}(v)$ the smallest element of V such that $v < \text{succ}(v)$. Let $x, y \in \mathbb{T}^V$, we define:

- $x[v]$ for $v \in V$ is the value of x at position v;
- $x \simeq y$ iff $x[v] \simeq y[v]$ for all $v \in V$;
- for $a, b \in \mathbb{T}$, $a \lhd b \stackrel{\text{df}}{=} \langle b \neq \star \,?\, b : a\rangle$, that is $a \lhd b$ has the value of a except when $b \neq \star$ in which case it has the value of b.
- $x \lhd y \stackrel{\text{df}}{=} [x[v] \lhd y[v] \mid v \in V]$, *i.e.* we extend $\lhd$ on vectors component-wise;
- for $a \in \mathbb{T}$ and $v \in V$, we note by $a^{[v]}$ the element of $\mathbb{T}^V$ such that: $a^{[v]}[v] = a$ and $a^{[v]}[v'] = \star$ for all $v' \in V \setminus \{v\}$, i.e., $a^{[v]} \stackrel{\text{df}}{=} [\langle u = v \,?\, a : \star\rangle \mid u \in V]$.

x may be considered as the set of indexes where it valuates to $\top$, *i.e.*, x may be viewed as set $\{v \in V \mid x[v] = \top\}$. Operation $x \lhd y$ will be used to compute a new state by applying on a state x the effect y of an executed action. This is why it is defined as x except for some values that are updated as in y. Thus, $\star$ in y at some position v means that the executed action has no effect on variable v.

For example, take $x \stackrel{\text{df}}{=} [\star, \bot, \top]$, $y \stackrel{\text{df}}{=} [\bot, \top, \star]$, and $z \stackrel{\text{df}}{=} [\top, \star, \top]$ with $V \stackrel{\text{df}}{=} [u, v, w]$ we have:

- $x[u] = \star$, $x[v] = \bot$, and $x[w] = \top$;
- $x \simeq z$, but $x \not\simeq y$ because $x[v] = \bot \not\simeq y[v] = \top$;
- $x \lhd y = [\bot, \top, \top]$ that is y except on variable w where $y[w] = \star$ and thus we use the value of $x[w]$;
- similarly, we have $y \lhd z = [\top, \top, \top]$;
- $\top^{[u]} = [\top, \star, \star]$ and $\bot^{[v]} = [\star, \bot, \star]$.

2.2 Multisets

A multiset m over a domain D is a function $m : D \to \mathbb{N}$ (natural numbers), where, for $d \in D$, $m(d)$ is the number of occurrences of d in the multiset m. We note by D^* the set of all multisets over D. The empty multiset is noted by $\mathbb{0}$ and is the constant function $\mathbb{0} : D \to \{0\}$. Similarly we define $\mathbb{1} : D \to \{1\}$ the unit multiset. A set X may be used as multiset $\mathbb{1}$ over X. A multiset m over D may be naturally extended to any domain $D' \supset D$ by defining $m(d) \stackrel{\text{df}}{=} 0$ for all $d \in D' \setminus D$, which explains why we generally do not need to be precise about multisets domains. If m_1 and m_2 are two multisets over D, we define:

- $m_1 \leq m_2$ iff $m_1(d) \leq m_2(d)$ for all $d \in D$;
- $m_1 + m_2$ is the multiset over D defined by $(m_1 + m_2)(d) \stackrel{\text{df}}{=} m_1(d) + m_2(d)$ for all $d \in D$;
- $m_1 - m_2$ is the multiset over D defined by $(m_1 - m_2)(d) \stackrel{\text{df}}{=} \max(0, m_1(d) - m_2(d))$ for all $d \in D$;
- m_1/m_2 is the multiset over D defined by $(m_1/m_2)(d) \stackrel{\text{df}}{=} \langle m_2(d) = 0?0 : m_1(d)\rangle$ for all $d \in D$. This operation nullifies m_1 where m_2 is zero. For $D' \subseteq D$ we may use m_1/D' by treating D' as a multiset as explained above.
- for $d \in D$, we note by $d \in m_1$ the fact that $m_1(d) > 0$.

2.3 Labelled Transition Systems

A labelled transition system (LTS) is a tuple $(S, I, A, \rightarrow)$ such that:

- S is the set of states;
- $I \subseteq S$ is the set of initial states;
- A is the set of labels;
- $\rightarrow \subseteq S \times A \times S$ if the set of transitions.

We note by $s \xrightarrow{a} s'$ the fact that $(s, a, s') \in \rightarrow$.

It should be stressed that we use a definition where several initial states are allowed, which will be the case for all our formalisms.

Let $L \stackrel{\text{df}}{=} (S, I, A, \rightarrow)$ and $L' \stackrel{\text{df}}{=} (S', I', A', \rightarrow')$ be two LTS, they are called equivalent through (g, h) iff:

- g is a bijection from S to S';
- h is a function from A' to A;
- $I' = g(I)$;
- for all $x, y \in S$, $x \xrightarrow{a} y$ iff $g(x) \xrightarrow{a'} g(y)$ with $h(a') = a$.

Thus g is an isomophism between the two LTS, and h defines a matching on the labels of the transitions. Assuming that L is an "original" LTS to which we compare a "transformed" LTS L', h maps every action label in L' to the original label it was obtained from. We note by id the identity function that may be later used as g. This definition results in a very strong notion of equivalence that requires matching the states as well as the transitions labels.

2.4 Regular Petri Nets

A regular Petri net (RPN) is a tuple (P, T, W) where:

- P is the finite set of places, depicted as circle-shaped nodes;
- T is the finite set of transitions, depicted as rectangle- or square-shaped nodes;
- $W \in ((P \times T) \cup (T \times P))^*$ is the weight of arcs, arcs with non-zero weights are depicted as directed edges, labelled by the weight when it is greater that 1.

A marking m of a RPN is a multiset over P, $m(p)$ is called the marking of place p and is the number of tokens held by p. Tokens are depicted as black bullets $\bullet$ inside p. Given $t \in T$, we define ${}^\bullet t \stackrel{\text{df}}{=} \{p \mapsto W(p,t) \mid (p,t) \in W\} \in P^*$ that is the preset of t, and $t^\bullet \stackrel{\text{df}}{=} \{p \mapsto W(t,p) \mid (t,p) \in W\} \in P^*$ that is the postset of t. Both are multisets of places in which the multiplicity of each place p is the weight of the arc from/to t.

A transition $t \in T$ is enabled at a marking m iff ${}^\bullet t \leq m$. In such a case, t may fire, leading to the marking $m' \stackrel{\text{df}}{=} m - {}^\bullet t + t^\bullet$, which is noted by $m \xrightarrow{t} m'$. The state graph of a RPN (P, T, W) with respect to a set of initial markings $M \subseteq P^*$ is a the smallest LTS $(S, M, T, \rightarrow)$ such that $M \subseteq S$ and, if $m \xrightarrow{t} m'$ in the RPN, then $m' \in S$ and $(m, t, m') \in \rightarrow$ in the LTS as well. This LTS is generally referred to as the marking graph, or the reachability graph, but we call it the state graph to streamline the comparison between our formalisms. Similarly, we have generalised the definition to allow a set of initial markings instead of just one as it is usually the case.

A RPN is safe (or 1-safe) with respect to a set M of initial markings iff for all marking m of its state graph we have $m \leq \mathbb{1}$. In the following, all our nets will be safe so that most multisets in the definitions will be without repetitions (and thus equivalent to sets). But since we will have to prove that our nets are safe, we must state the definitions in the more general context of non-safe Petri nets.

3 Reaction Rules (RR)

3.1 Definition and Syntax

An RR system consists of a set of *Boolean variables* together with *actions* that can change the variables when their values meet the action preconditions. Actions are separated into *constraints* and *rules*, the former having a higher priority.

Definition 1 (RR systems). *An* RR *system is a tuple* $(\mathcal{V}, \mathcal{I}, \mathcal{C}, \mathcal{R})$ *where:*

- $\mathcal{V}$ *is a finite set of ordered variables;*
- $\mathcal{I} \subseteq \mathbb{B}^{\mathcal{V}}$ *is the set of initial states;*
- $\mathcal{C}$ *is a finite set of constraints;*
- $\mathcal{R}$ *is a finite set of rules, disjoint from* $\mathcal{C}$*;*
- $\mathcal{A} \stackrel{\text{df}}{=} \mathcal{C} \uplus \mathcal{R}$ *is the set of actions, and each action is a pair* $(\ell, r) \in \mathbb{T}^{\mathcal{V}} \times \mathbb{T}^{\mathcal{V}}$*.*

RR systems were originally defined in [8] using a concrete syntax that is easy to edit in simple text files. Formalisation was done using much heavier notations that we simplified here thanks to ternary truth values. Figure 3 shows the grammar of the concrete syntax.

In the concrete syntax, using `Ac*`, like in Fig. 1, means that we have two initial states, one with `Ac+` and another with `Ac-`. So that initial states are indeed in $\mathbb{B}^{\mathcal{V}}$. However, the abstract syntax allows more varied initial states than it is possible to define using the concrete syntax. For instance, for two variables, it is possible to have $\mathcal{I} = \{[\top, \bot], [\bot, \top]\}$ which cannot be obtained using the concrete syntax. This is not a problem since we base all the following on the abstract syntax, but rather a practical limitation when using the concrete syntax.

```
rr ::= { vardecl } [ constraints ] [ rules ]
vardecl ::= word ":" "\n" { varinit ":" line "\n" }
varinit ::= word ( "*" | "+" | "-" )
word ::= /[A-Z][A-Z0-9_]*/i
line ::= /[^\n]*/
constraints ::= "constraints:\n" { action }
rules ::= "rules:\n" { action }
action ::= varstate { "," varstate } ">>" varstate { "," varstate } "\n"
varstate ::= word ( "+" | "-" )
```

Fig. 3. Concrete syntax for RR systems in BNF notation. The left-hand side (resp. right-hand side) of an actions correspond to the ℓ (resp. r) part in the definition. A variable that do not appear in one side of an action is assumed to be $\star$ so that the ℓ and r parts of the actions are fully defined using this syntax.

3.2 Operational Semantics

The execution of an RR system is straightforward: starting from a set of initial states, we can reach new states by applying constraints first, then rules from states where no constraint can be applied. Note that we explicitly forbid self-loops, *i.e.* action applications that would not change the state. This contrasts, in particular, with Boolean networks where self-loops are not only allowed but desired and searched for as they usually correspond to stable states of the system of interest (*e.g.* a phenotype of a cell is often modelled as such a stable state). In our setting, such states will be deadlocks.

Definition 2 (RR firing rule). *Let $(\mathcal{V},\mathcal{I},\mathcal{C},\mathcal{R})$ be an* RR *system, with $\mathcal{A} \stackrel{\text{df}}{=} \mathcal{C} \uplus \mathcal{R}$ its set of actions. Let $a \stackrel{\text{df}}{=} (\ell, r) \in \mathcal{A}$ be an action and $s \in \mathbb{B}^{\mathcal{V}}$ be a state. Then:*

1. *Action a is enabled at s iff $s \simeq \ell$.*
2. *If a is a constraint, it can be fired from s yielding a new state $s' \stackrel{\text{df}}{=} s \lhd r$ iff it is enabled at s, which is noted by $s \xrightarrow{a} s'$.*
3. *If a is a rule, it can be fired from s yielding a new state $s' \stackrel{\text{df}}{=} s \lhd r$ iff it is enabled at s and no constraint in $\mathcal{C}$ is enabled at s, which is noted by $s \xrightarrow{a} s'$.*

The semantics of an RR system is expressed as expected in terms of a LTS obtained by firing actions from the initial states until saturation.

Definition 3 (RR state graph). *Let $(\mathcal{V},\mathcal{I},\mathcal{C},\mathcal{R})$ be an* RR *system, with $\mathcal{A} \stackrel{\text{df}}{=} \mathcal{C} \uplus \mathcal{R}$ its set of actions. Its state graph is the smallest* LTS *$(S, \mathcal{I}, \mathcal{A}, \rightarrow)$ such that $\mathcal{I} \subseteq S$ and, if $s \xrightarrow{a} s'$ in the* RR *system with $s \neq s'$, then $s' \in S$ and $(s, a, s') \in \rightarrow$.*

Note that we forbid self-loops in the definition of an RR state graph instead of in the definition of actions enabling. Both approaches would be correct but the one we have chosen will simplify the comparison with Petri nets state graphs. From now on, we always consider that we are within a LTS and thus we forbid firing actions when this would create a self-loop.

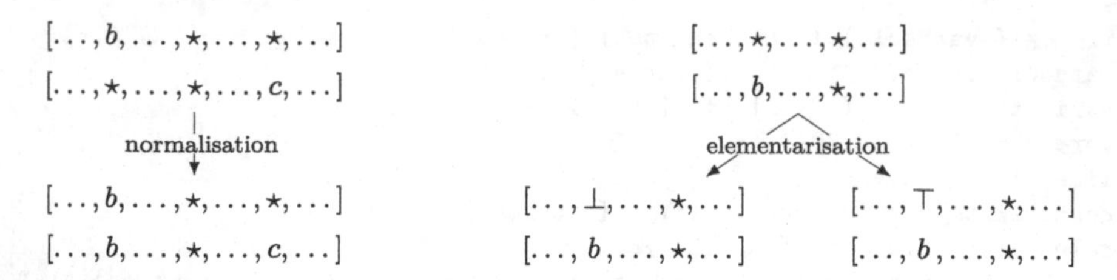

Fig. 4. Illustration of normalisation (left) and elementarisation (right), where $b, c \in \mathbb{B}$. Each action (ℓ, r) is depicted with ℓ drawn above r.

3.3 Normal and Elementary RR Systems

As defined above, the actions of an RR system have implicit elements. For instance, taking `A`, `B`, `C` as the variables, when one writes `A+, B- >>C+` in the textual syntax, this corresponds to an action $([\top, \bot, \star], [\star, \star, \top])$ in the definition. Firing this action can be done only from state $[\top, \bot, \bot]$ (otherwise it would be a self-loop), yielding new state $[\top, \bot, \top]$, which results in $A = \top$ and $B = \bot$ while this is not explicit in the action.

An equivalent writing of the same action would be `A+, B- >>A+, B-, C+`, which explicitly specifies which values `A` and `B` get upon firing. This latter version of the action is called *normal*, *i.e.* all of its left-hand side variables appear in the right-hand side, and we show below that any action can be rewritten this way without changing the semantics.

An even more explicit writing of this action would be `A+, B-, C- >>A+, B-, C+`, which clearly states that `C` has to be $\bot$ to fire the action. Such an action, with exactly the same variables on both sides, is called *elementary*. We show below that any normal rule can be rewritten as a set of equivalent elementary rules without changing the semantics. The elementarisation of an action is a one-to-many transformation since there may exist different states from which a normal action can be fired. Consider for instance `A+ >>A+, B+, C+`. It is normal and can be fired whenever $\mathtt{A} = \top$ and $(\mathtt{B}, \mathtt{C}) \neq (\top, \top)$, which corresponds to the three distinct states $[\top, \bot, \bot]$, $[\top, \top, \bot]$, and $[\top, \bot, \top]$ (with the fourth possibility $[\top, \top, \top]$ yielding a self-loop).

Definition 4 (normalisation). *Let* $R \stackrel{\text{df}}{=} (\mathcal{V}, \mathcal{I}, \mathcal{C}, \mathcal{R})$ *be an* RR *system, with* $\mathcal{A} \stackrel{\text{df}}{=} \mathcal{C} \uplus \mathcal{R}$ *its set of actions. R is called normal iff for all* $a \stackrel{\text{df}}{=} (\ell, r) \in \mathcal{A}$ *and for all* $v \in \mathcal{V}$ *we have:* $r[v] = \star \implies \ell[v] = \star$.

The normalisation of R, noted by $\mathsf{norm}(R)$, *is obtained by replacing all its actions* (ℓ, r) *with* $\mathsf{norm}(\ell, r) \stackrel{\text{df}}{=} (\ell, [\langle r[v] = \star\,?\,\ell[v] : r[v]\rangle \mid v \in \mathcal{V}])$.

How $\mathsf{norm}(\ell, r)$ works is illustrated on the left of Fig. 4: for each $v \in \mathcal{V}$, if $r[v] = \star$ then it is replaced by $\ell[v]$ (which may be $\star$ also).

Theorem 1. *Let* $R \stackrel{\text{df}}{=} (\mathcal{V}, \mathcal{I}, \mathcal{C}, \mathcal{R})$ *be an* RR *system. Then, R and* $\mathsf{norm}(R)$ *generate equivalent state graphs.*

Proof. We first note that systems R and $R' \stackrel{\text{df}}{=} \mathsf{norm}(R)$ are defined on the same variables so that every state of R is also a valid state of R' and vice-versa. Moreover, they have the same initial states by definition. Let s and s' be two states of R or R', and let $a \stackrel{\text{df}}{=} (\ell, r)$ be an action of R and $a' \stackrel{\text{df}}{=} \mathsf{norm}(\ell, r) \stackrel{\text{df}}{=} (\ell, r')$. We will prove that $s \xrightarrow{a} s'$ in R iff $s \xrightarrow{a'} s'$ in R'. As a consequence, starting from the same initial states the LTSs of R and R' are equivalent through (id, h) with $h \stackrel{\text{df}}{=} \{\mathsf{norm}(a) \mapsto a \mid a \in \mathcal{A}\}$.

($\Rightarrow$) assume $s \xrightarrow{a} s'$. By Definition 2 we have $s \simeq \ell$ and $s' = s \lhd r$. Since the left-hand side of a' is ℓ, we also have a' enabled by s. So we need to prove that $s' = s \lhd r = s \lhd r'$, or equivalently, that for all variable v we have $s[v] \lhd r[v] = s[v] \lhd r'[v]$. From Definition 4, there are three cases:

- if $\ell[v] = \star = r[v]$, we also have $r'[v] = \ell[v]$ and thus $r'[v] = \star = r[v]$, hence the result;
- if $r[v] \neq \star$ then $r'[v] = r[v]$ hence the result;
- if $\ell[v] \neq \star = r[v]$, then $r'[v] = \ell[v]$, and since a is enabled we have $\ell[v] = s[v]$, hence the result.

($\Leftarrow$) assume $s \xrightarrow{a'} s'$. The proof is essentially the same, by exchanging a (resp. r) with a' (resp. r'). □

Definition 5 (elementarisation). *Let $R \stackrel{\text{df}}{=} (\mathcal{V}, \mathcal{I}, \mathcal{C}, \mathcal{R})$ be an* RR *system, with $\mathcal{A} \stackrel{\text{df}}{=} \mathcal{C} \uplus \mathcal{R}$ its set of actions. R is called elementary iff for all $a \stackrel{\text{df}}{=} (\ell, r) \in \mathcal{A}$ and for all $v \in \mathcal{V}$ we have: $\ell[v] = \star \iff r[v] = \star$. Consequently, an elementary* RR *system is also normal.*

Assuming that R is normal, its elementarisation, noted by $\mathsf{elem}(R)$ is the RR *system $(\mathcal{V}, \mathcal{I}, \mathcal{C}', \mathcal{R}')$ where*

$$\mathcal{C}' \stackrel{\text{df}}{=} \bigcup_{a \in \mathcal{C}} \mathsf{elem}(a) \quad \textit{and} \quad \mathcal{R}' \stackrel{\text{df}}{=} \bigcup_{a \in \mathcal{R}} \mathsf{elem}(a)\,,$$

with $a \stackrel{\text{df}}{=} (\ell, r)$, $\mathsf{elem}(a) \stackrel{\text{df}}{=} \mathsf{elem}(\ell, r, \min(\mathcal{V}))$ and $\mathsf{elem}(\ell, r, v)$ defined as:

- $\{(\ell \lhd \bot^{[v]}, r), (\ell \lhd \top^{[v]}, r)\}$ *if $v = \max(\mathcal{V})$ and $\ell[v] = \star \neq r[v]$;*
- $\{(\ell, r)\}$ *if $v = \max(\mathcal{V})$ and $\neg(\ell[v] = \star \neq r[v])$;*
- $\mathsf{elem}(\ell \lhd \bot^{[v]}, r, \mathrm{succ}(v)) \cup \mathsf{elem}(\ell \lhd \top^{[v]}, r, \mathrm{succ}(v))\}$ *if $v < \max(\mathcal{V})$ and $\ell[v] = \star \neq r[v]$;*
- $\mathsf{elem}(\ell, r, \mathrm{succ}(v))$ *if $v < \max(\mathcal{V})$ and $\neg(\ell[v] = \star \neq r[v])$.*

How $\mathsf{elem}(\ell, r)$ works is illustrated on the right of Fig. 4: for each $v \in \mathcal{V}$ in turn, every $\ell[v] = \star$ such that $r[v] \neq \star$ is replaced by either $\bot$ or $\top$, yielding two new actions.

Theorem 2. *Let $R \stackrel{\text{df}}{=} (\mathcal{V}, \mathcal{I}, \mathcal{C}, \mathcal{R})$ be a normal* RR *system. Then, R and $\mathsf{elem}(R)$ generate equivalent state graphs.*

Proof. As with normalisation, R and R' may have the same states and they do have the same initial states. Let s and s' be two states of R or R'. Let $a \stackrel{\text{df}}{=} (\ell, r)$

be an action of R. We will prove that $s \xrightarrow{a} s'$ in R iff $s \xrightarrow{a'} s'$ in R' for some $a' \in \mathsf{elem}(\ell, r)$. As a consequence, starting from the same initial states the LTSs of R and R' are equivalent through (id, h) with $h \stackrel{\text{df}}{=} \{a' \mapsto a \mid a \in \mathcal{A} \wedge a' \in \mathsf{elem}(a)\}$. The right-hand sides of a and all $a' \in \mathsf{elem}(a)$ are the same, so we just need to prove that a is enabled at s iff some $a' \in \mathsf{elem}(a)$ is enabled at s.

($\Rightarrow$) assume $s \xrightarrow{a} s'$. Take $\ell' \stackrel{\text{df}}{=} [\langle \ell[v] = \star \neq r[v]\ ?\ s[v] : \ell[v] \rangle \mid v \in \mathcal{V}]$, we have $(\ell', r) \in \mathsf{elem}(a)$ and it is enabled at s because ℓ and ℓ' only differ on positions v where $\ell[v] = \star$ and at these positions we have $\ell'[v] = s[v]$.

($\Leftarrow$) take $a' \stackrel{\text{df}}{=} (\ell', r) \in \mathsf{elem}(a)$ and assume $s \xrightarrow{a'} s'$. As previously, ℓ and ℓ' only differ on positions v where $\ell[v] = \star$, which does not restrict enabling. □

4 Priority Petri Nets

Regular Petri nets may be extended with transitions priorities. In our setting, we just need two levels of priorities, so we distinguish a set of urgent transitions whose firing is always preferred above that of non-urgent transitions (hence the used of letter U below). The former will be used to implement constraints while the latter will be used to implement rules.

Definition 6 (PPN). *A priority Petri net (PPN) is a tuple (P, T, W, U) where (P, T, W) is a RPN, called the underlying RPN, and $U \subseteq T$ is the set of urgent transitions.*

Definition 7 (PPN firing rule). *Let (P, T, W, U) be a PPN and m a marking of it. A transition $t \in U$ is enabled at m iff it is enabled at m in the underlying RPN. A transition $t \in T \setminus U$ is enabled at m iff it is enabled at m in the underlying RPN and no $u \in U$ is enabled at m. If t is enabled at m then we may have $m \xrightarrow{t} m - {}^\bullet t + t^\bullet$ just like in the underlying RPN.*

Definition 8 (PPN state graph). *Let (P, T, W, U) be a PPN and M the set of its initial markings. Its state graph is the smallest LTS $(S, M, T, \rightarrow)$ such that $M \subseteq S$ and if $m \xrightarrow{t} m'$ in the PPN with $m \neq m'$, then $m' \in S$ and $(m, t, m') \in \rightarrow$ in the LTS as well. The PPN is safe with respect to M iff for all $m \in S$ we have $m \leq \mathbb{1}$.*

Note that, as for RR systems, we restrict the LTS semantics of PPN to avoid self-loops (hence the condition $m \neq m'$).

Translation from an elementary RR system to a PPN is made by creating a pair of complementary places $p_v^\top$ and $p_v^\perp$ for each variable v and then, each action a gives rise to a transition t_a linked to these places as depicted in Fig. 5. The figure also depicts the translation to PPN of four elementary rules, each being depicted separately from the others for the sake of readability.

Another way to avoid self-loops would be to remove from this translation all transitions that do not change the marking (i.e., transitions t such that ${}^\bullet t = t^\bullet$). This is probably the most practical solution but, as discusses later, it will not work for EPN so we prefer to use the same approach for both classes of Petri nets. Moreover, doing so, we guarantee that every action in an elementary RR system has a corresponding transition in the PPN translation.

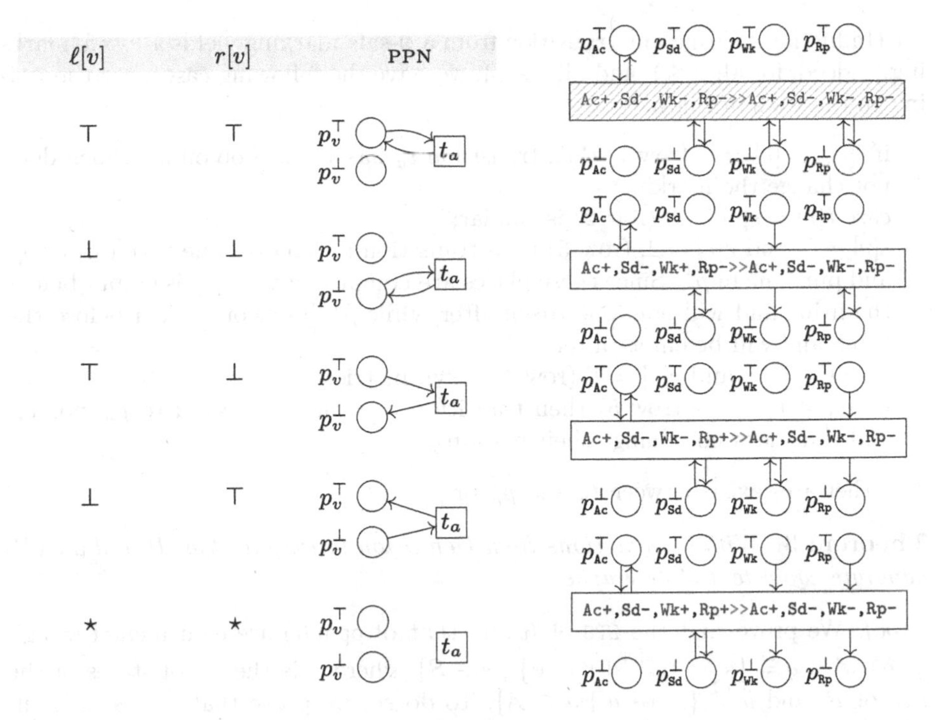

Fig. 5. (Left) Depiction of all the possible relations between a variable v and an action $a \stackrel{\mathrm{df}}{=} (\ell, r)$ of an elementary RR system, and how this is translated to PPN. **(Right)** The four elementary rules resulting from rule `R9` (`Ac+, Sd- >>Wk-, Rp-`) in the termite model of Fig. 1 translated to PPN (presented separately to improve readability). The top-most transition is never fired because it does not change the marking.

Definition 9 (elementary RR systems to PPN). *Let $R \stackrel{\mathrm{df}}{=} (\mathcal{V}, \mathcal{I}, \mathcal{C}, \mathcal{R})$ be an elementary* RR *system, with $\mathcal{A} \stackrel{\mathrm{df}}{=} \mathcal{C} \uplus \mathcal{R}$ its set of actions. R can be translated to a* PPN $\mathsf{ppn}(R) \stackrel{\mathrm{df}}{=} (P, T, W, U)$ *and a set of initial markings M as follows:*

- $P \stackrel{\mathrm{df}}{=} \{p_v^\top, p_v^\bot \mid v \in \mathcal{V}\}$;
- $T \stackrel{\mathrm{df}}{=} \{t_a \mid a \in \mathcal{A}\}$;
- $W \stackrel{\mathrm{df}}{=} \{(p_v^{\ell[v]}, t_a), (t_a, p_v^{r[v]}) \mid a \stackrel{\mathrm{df}}{=} (\ell, r) \in \mathcal{A} \wedge v \in V \wedge \neg(\ell[v] = \star = r[v])\}$;
- $U \stackrel{\mathrm{df}}{=} \{t_a \mid a \in \mathcal{C}\}$;
- $M \stackrel{\mathrm{df}}{=} \{\{p_v^{s[v]} \mid v \in \mathcal{V}\} \mid s \in \mathcal{I}\}$.

Proposition 1. *With the notations from Definition 9, we have that $\mathsf{ppn}(R)$ is a safe* PPN *with respect to M.*

Proof. We prove by induction that all the reachable markings are safe and that the pairs of places $p_v^\top$ and $p_v^\bot$ are complementary places (i.e., they together hold exactly one token). For brevity below, we call v-safe such a marking.

(Basis.) Every marking in M is v-safe because for all $s \in \mathcal{I}$ and all $v \in \mathcal{V}$, depending on the value of $s[v]$, we put exactly one token in either $p_v^\top$ or $p_v^\bot$.

(Induction.) Firing any transition from a v-safe marking yields a v-safe marking. Indeed, for all $v \in \mathcal{V}$ and all $a \in \mathcal{A}$, we have the following cases, corresponding to the rows of Fig. 5:

- if $\ell[v] = r[v] = \top$ (row 1) then transition t_a has a side loop on $p_v^\top$ which does not change the marking;
- case $\ell[v] = r[v] = \bot$ (row 2) is similar;
- $\ell[v] = \top$ and $r[v] = \bot$ (row 3) then transition t_a removes one token from $p_v^\top$ and puts one in $p_v^\bot$. Since these places are complementary, $p_v^\bot$ is empty before the firing and will hold one token after while $p_v^\top$ holds one token before the firing and will be empty after;
- case $\ell[v] = \bot$ and $r[v] = \top$ (row 4) is symmetric;
- if $\ell[v] = r[v] = \star$ (row 5) then transition t_a is not connected to $p_v^\top$ nor $p_v^\bot$ and thus does not change their markings.

No other arcs exist between t_a and $p_v^\top$ or $p_v^\bot$. □

Theorem 3. *With the notations from Definition 9, we have that R and $\mathsf{ppn}(R)$ generate equivalent state graphs.*

Proof. We prove that the LTS of R and that of $\mathsf{ppn}(R)$ are equivalent through (g, h) with $g \stackrel{\text{df}}{=} \{s \mapsto \{p_v^{s[v]} \mid v \in \mathcal{V}\} \mid s \in S\}$ where S is the set of states of the LTS of R, and $h \stackrel{\text{df}}{=} \{t_a \mapsto a \mid a \in \mathcal{A}\}$. To do so, we prove that $s \xrightarrow{a} s'$ in R iff $g(s) \xrightarrow{t_a} g(s')$ in $\mathsf{ppn}(R)$ with $s \neq s'$ two states of R.

($\Rightarrow$) assume $s \xrightarrow{a} s'$, with $a \stackrel{\text{df}}{=} (\ell, r)$. Since a is enabled, we have $\ell[v] = s[v]$ or $\ell[v] = \star = r[v]$ for all $v \in \mathcal{V}$, and thus ${}^\bullet t_a = \{p_v^{s[v]} \mid v \in \mathcal{V} \wedge \neg(\ell[v] = \star = r[v])\}$ by definition of W. Moreover, we have $g(s) = \{p_v^{s[v]} \mid v \in \mathcal{V}\}$ by definition. Thus ${}^\bullet t_a \leq g(s)$ and t_a is enabled. Take m' the marking such that $g(s) \xrightarrow{t_a} m' \stackrel{\text{df}}{=} g(s) - {}^\bullet t + t^\bullet$, it remains to prove that $m' = g(s')$. From the definitions we have

$$
\begin{aligned}
m' = \{p_v^{s[v]} \mid v \in \mathcal{V}\} & \qquad (1)\\
- \{p_v^{\ell[v]} \mid v \in \mathcal{V} \wedge \neg(\ell[v] = \star = r[v])\} & \qquad (2)\\
+ \{p_v^{r[v]} \mid v \in \mathcal{V} \wedge \neg(\ell[v] = \star = r[v])\} & \qquad (3)
\end{aligned}
$$

and we can consider each $v \in \mathcal{V}$ separately. There are three cases, corresponding to the rows in Fig. 5:

- if $\ell[v] = r[v] \neq \star$ (rows 1–2) we also have $s[v] = \ell[v]$ because s enables a, thus one token exists in $p_v^{s[v]}$ at (1), it is removed at (2) and another is added at (3) in $p_v^{\ell[v]}$ so it is marked in m'. Moreover we have $s'[v] = s[v]$ by definition of RR firing, so $g(s')$ has one token in $p_v^{s[v]}$ and none in its complementary place hence the result;
- if $\ell[v] \neq r[v]$ (rows 3–4) then because R is elementary none of them is $\star$. Assume $\ell[v] = \top$ and $r[v] = \bot$ (the other case is symmetric). The token in $p_v^\top$ from (1) is removed at (2), and one token is added to $p_v^\bot$ at (3). Moreover, by definition of RR firing, we have $s'[v] = r[v] = \bot$ so that $g(s')$ has one token in $p_v^\bot$ and none in $p_v^\top$ hence the result;

- if $\ell[v] = \star = r[v]$ (row 5) then place $p_v^{s[v]}$ is left untouched with one token inside because it is not connected to t_a. Moreover we have $s'[v] = s[v]$ by definition of RR firing, so $g(s')$ has one token in $p_v^{s[v]}$ and none in its complementary place hence the result.

($\Leftarrow$) assume $g(s) \xrightarrow{t_a} g(s')$. Since t_a is enabled, we have ${}^\bullet t_a = \{p_v^{\ell[v]} \mid v \in \mathcal{V} \wedge \neg(\ell[v] = \star = r[v])\} \leq g(s) = \{p_v^{s[v]} \mid v \in \mathcal{V}\}$. So, for each v such that $\neg(\ell[v] = \star = r[v])$ we have $\mathbb{0} < \{p_v^{\ell[v]}\} \leq \{p_v^{s[v]}\} \leq \mathbb{1}$ and thus $\ell[v] = s[v]$. Moreover, each v such that $\ell[v] = \star = r[v]$ has no influence on the enabling of a. So a is enabled at s. It remains to show that $s \xrightarrow{a} s'$. Taking $m' = g(s')$, Eq. (1–3) above still holds and we consider each v separately. There are three cases, corresponding to the rows in Fig. 5:

- if $\ell[v] = s[v] = r[v]$ (rows 1–2), one token is removed and another is added from $p_v^{s[v]}$ while $p_v^{\neg s[v]}$ remains empty, thus $s[v] = s'[v]$ which is what firing a with $\ell[v] = s[v] = r[v]$ yields;
- if $\ell[v] = s[v] \neq r[v]$ (rows 3–4), the token in $p_v^{s[v]}$ from (1) is removed at (2) and one is added to $p_v^{\neg s[v]}$ at (3). Because the net is safe, $p_v^{\neg s[v]}$ is empty in $g(s)$ and holds exactly one token in $g(s')$. Firing a from s with $\ell[v] \neq r[v]$ changes the value of v thus $s'[v] = \neg s[v]$, which corresponds to the marking;
- if $\ell[v] = r[v] = \star$ (row 5), token in $p_v^{s[v]}$ from (1) is not removed at (2) and no token is added in $p_v^{s[v]}$ nor $p_v^{\neg s[v]}$ at (3). $g(s')$ has one token in $p_v^{s[v]}$ and $s[v] = s'[v]$, which is what firing a with $\ell[v] = r[v] = \star$ yields. □

5 Extended Petri Nets

Priority Petri nets may be further extended with:

- read arcs (depicted as bare edges) that allow to test for the presence of tokens without consuming them (letter Z below is for "Zero tokens consumed");
- inhibitor arcs (depicted with a white dot at the transition side) that allow to test for the absence of tokens in a place (letter H below is for "inHibitor");
- reset arcs (depicted with a black diamond at the transition side) that allow to consume all the tokens from a place, if any (letter F below is for "Flush").

Considering this class rather that PPN allows a translation from RR systems without resorting to elementarisation, only normalisation is needed. On the good side, one action written by the modeller is being translated to one EPN transition and normalisation can be kept invisible. On the bad side, we need to cope with a more complex class of Petri nets for which fewer tools may be available.

Definition 10 (EPN). *An extended Petri net (*EPN*) is a tuple* (S, T, W, U, Z, H, F) *where* (S, T, W, U) *is a* PPN, *called the underlying* PPN, *and:*

- $Z \in (P \times T)^*$ *defines the (weighted) read arcs and for* $t \in T$ *we define* ${}^*t \stackrel{\text{df}}{=} \{p \mapsto Z(p,t) \mid (p,t) \in Z\}$ *the places from which* t *reads tokens;*

- $H \in (P \times T)^*$ *defines the (weighted) inhibitor arcs and for* $t \in T$ *we define* $^\circ t \stackrel{\text{df}}{=} \{p \mapsto H(p,t) \mid (p,t) \in H\}$ *the multiset of places from which* t *checks the absence of too much tokens;*
- $F \subseteq (P \times T)$ *defines the reset arcs and for* $t \in T$ *we define* $^\blacklozenge t \stackrel{\text{df}}{=} \{p \in P \mid (p,t) \in F\}$ *the set of places whose marking is reset by* t.

Definition 11 (EPN firing rule). *Let* (S,T,W,U,Z,H,F) *be an* EPN *and* m *a marking of it. A transition* $t \in T$ *is enabled at* m *iff it is also enabled in the underlying* PPN, *and we have* $^*t \leq m$ *and* $m/^\circ t < {}^\circ t$. *If* t *is enabled at* m *then we may have* $m \xrightarrow{t} m'$ *with* $m' \stackrel{\text{df}}{=} m/(P \setminus {}^\blacklozenge t) - {}^\bullet t + t^\bullet$.

The intuition behind this firing rule is as follows:

- t must be enabled in the underlying PPN, that is: there are enough tokens to be consumed by the regular arcs, and priorities are respected;
- $^*t \leq m$ checks that there are enough tokens to be tested by the read arcs. Read arcs are weighted so for instance $^*t(p) = 2$ means that two tokens will be tested in p, thus the inequality;
- $m/^\circ t < {}^\circ t$ checks that there are not too much tokens with respect to the inhibitor arcs. We consider $m/^\circ t$ instead of m because a weight zero on a inhibitor arc corresponds to the absence of such an arc. So, this condition can be read as "for every place p such that there is an inhibitor arc between t and p with weight $w > 0$, p must be marked by less than w tokens";
- $m' \stackrel{\text{df}}{=} m/(P \setminus {}^\blacklozenge t) - {}^\bullet t + t^\bullet$ is similar to $m' \stackrel{\text{df}}{=} m - {}^\bullet t + t^\bullet$ in PPN but instead of computing m' from m, we compute it from m restricted to the places that are not connected to t through a reset arc. In other words, $m/(P \setminus {}^\blacklozenge t)$ is m in which we emptied all the places connected to t through a reset arc.

Definition 12 (EPN state graph). *Let* (S,T,W,U,Z,H,F) *be an* EPN *and* M *the set of its initial markings. Its state graph is the smallest* LTS $(S,M,T,\rightarrow)$ *such that* $M \subseteq S$ *and if* $m \xrightarrow{t} m'$ *in the* EPN *with* $m \neq m'$, *then* $m \in S$ *and* $(m,t,m') \in \rightarrow$ *in the* LTS *as well. The* EPN *is safe with respect to* M *iff for all* $m \in S$ *we have* $m \leq \mathbb{1}$.

As with PPN, we have restricted the semantics to avoid self-loops in an EPN state graph.

Translation from a normal RR system to an EPN is made by creating one place p_v for each variable v and one transitions t_a for each action a that is connected to each p_v as depicted in Fig. 6. An example of such a translation is depicted on the right of the figure.

Definition 13 (normal RR systems to EPN). *Let* $R \stackrel{\text{df}}{=} (\mathcal{V},\mathcal{I},\mathcal{C},\mathcal{R})$ *be a normal* RR *system, with* $\mathcal{A} \stackrel{\text{df}}{=} \mathcal{C} \uplus \mathcal{R}$ *its set of actions.* R *can be translated to a* EPN $\mathsf{epn}(R) \stackrel{\text{df}}{=} (P,T,W,U,Z,H,F)$ *and a set of initial markings* M *as follows, with reference to Fig. 6 displayed at the end of lines:*

- $P \stackrel{\text{df}}{=} \{p_v \mid v \in \mathcal{V}\}$;
- $T \stackrel{\text{df}}{=} \{t_a \mid a \in \mathcal{A}\}$;

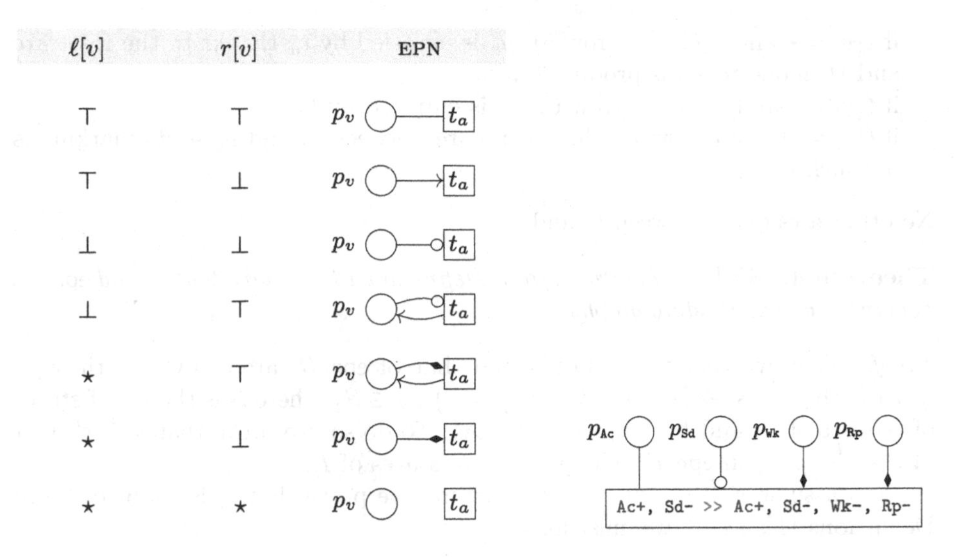

Fig. 6. (Left) Depiction of all the possible relations between a variable v and an action $a \stackrel{\text{df}}{=} (\ell, r)$ of a normal RR system, and how this is translated to EPN. **(Right)** The normal rule resulting from rule **R9** in the termite model of Fig. 1 translated to EPN.

- $W \stackrel{\text{df}}{=} \{(p_v, t_a) \mid v \in \mathcal{V} \wedge a \stackrel{\text{df}}{=} (\ell, r) \in \mathcal{A} \wedge \ell[v] = \top \wedge r[v] = \bot\}$ *(row 2)*
  $+ \{(t_a, p_v) \mid v \in \mathcal{V} \wedge a \stackrel{\text{df}}{=} (\ell, r) \in \mathcal{A} \wedge \ell[v] \neq \top \wedge r[v] = \top\}$; *(rows 4–5)*
  $U \stackrel{\text{df}}{=} \{t_a \mid a \in \mathcal{C}\}$;
- $Z \stackrel{\text{df}}{=} \{(p_v, t_a) \mid v \in \mathcal{V} \wedge a \stackrel{\text{df}}{=} (\ell, r) \in \mathcal{A} \wedge \ell[v] = r[v] = \top\}$; *(row 1)*
- $H \stackrel{\text{df}}{=} \{(p_v, t_a) \mid v \in \mathcal{V} \wedge a \stackrel{\text{df}}{=} (\ell, r) \in \mathcal{A} \wedge \ell[v] = \bot\}$; *(rows 3–4)*
- $F \stackrel{\text{df}}{=} \{(p_v, t_a) \mid v \in \mathcal{V} \wedge a \stackrel{\text{df}}{=} (\ell, r) \in \mathcal{A} \wedge \ell[v] = \star \wedge r[v] \neq \star\}$; *(rows 5–6)*
- $M \stackrel{\text{df}}{=} \{\{p_v \mid v \in \mathcal{V} \wedge s[v] = \top\} \mid s \in \mathcal{I}\}$.

Proposition 2. *With the notations from Definition 13, we have that* $\mathsf{epn}(R)$ *is a safe* EPN *with respect to* M.

Proof. We prove by induction that all reachable markings are safe.

(Basis.) Every marking in M is safe because for all $s \in \mathcal{I}$ and all $v \in \mathcal{V}$, depending on the value of $s[v]$, we put at most one token in p_v.

(Induction.) Firing any transition from a safe marking yields a safe marking. Indeed, for all $v \in \mathcal{V}$ and all $a \in \mathcal{A}$, we have the following cases, corresponding to the rows of Fig. 6:

- if $\ell[v] = r[v] = \top$ (row 1), only a read arc exists between t_a and p_v, which does not change its marking;
- if $\ell[v] = \top$ and $r[v] = \bot$ (row 2), a token is consumed by t_a from p_v which keeps it safe;
- if $\ell[v] = r[v] = \bot$ (row 3), only an inhibitor arc exists between t_a and p_v, which does not change its marking;
- if $\ell[v] = \bot$ and $r[v] = \top$ (row 4), a token is produced by t_a into p_v but only if it is empty thanks to the inhibitor arc between t_a and p_v;

- if $\ell[v] = \star$ and $r[v] = \top$ (row 5), p_v is emptied by t_a thanks to the reset arc and then one token is produced in p_v;
- if $\ell[v] = \star$ and $r[v] = \bot$ (row 6), p_v is emptied by t_a;
- if $\ell[v] = r[v] = \star$ (row 7), there is no arc between t_a and p_v so its marking is untouched.

No other arcs exist between t_a and p_v. □

Theorem 4. *With the notations from Definition 13, we have that* R *and* $\mathsf{epn}(R)$ *generate equivalent state graphs.*

Proof. We prove that the LTS of R and that of $\mathsf{epn}(R)$ are equivalent through (g, h) with $g \stackrel{\text{df}}{=} \{s \mapsto \{p_v \mid v \in \mathcal{V} \wedge s[v] = \top\} \mid s \in S\}$ where S is the set of states of the LTS of R, and $h \stackrel{\text{df}}{=} \{t_a \mapsto a \mid a \in \mathcal{A}\}$. To do so, we prove that $s \xrightarrow{a} s'$ in R iff $g(s) \xrightarrow{t_a} g(s')$ in $\mathsf{epn}(R)$ with $s \neq s'$ two states of R.

($\Rightarrow$) assume $s \xrightarrow{a} s'$, with $a \stackrel{\text{df}}{=} (\ell, r)$. First we prove that t_a is enabled. From Definitions 11 and 13, we must have:

$$^{\bullet}t \stackrel{\text{df}}{=} \{p_v \mid v \in \mathcal{V} \wedge \ell[v] = \top \wedge r[v] = \bot\} \leq g(s) \stackrel{\text{df}}{=} \{p_v \mid v \in \mathcal{V} \wedge s[v] = \top\} \quad (4)$$
$$^{*}t \stackrel{\text{df}}{=} \{p_v \in \mathcal{V} \mid \ell[v] = r[v] = \top\} \leq g(s) \quad (5)$$
$$^{\circ}t \stackrel{\text{df}}{=} \{p_v \in \mathcal{V} \mid \ell[v] = \bot\} > g(s)/^{\circ}t \quad (6)$$

(4) and (5) hold because when a is enabled we must have $s[v] = \top$ for every v such that $\ell[v] = \top$. (6) holds because when a is enabled we must have $s[v] = \bot$ and thus $g(s)(p_v) = 0$ for every v such that $\ell[v] = \bot$. Take m' the marking such that $g(s) \xrightarrow{t_a} m' \stackrel{\text{df}}{=} g(s)/(P \setminus {}^{\blacklozenge}t) - {}^{\bullet}t + t^{\bullet}$. It remains to prove that $m' = g(s')$. From the definitions we have:

$$\begin{aligned} m' = \{p_v \mid v \in \mathcal{V} \wedge s[v] = \top \wedge (\ell[v] \neq \star \vee r[v] = \star)\} \quad & (7) \\ - \{p_v \mid v \in \mathcal{V} \wedge \ell[v] = \top \wedge r[v] = \bot\} \quad & (8) \\ + \{p_v \mid v \in \mathcal{V} \wedge \ell[v] \neq \top \wedge r[v] = \top\} \quad & (9) \end{aligned}$$

where $\ell[v] \neq \star \vee r[v] = \star$ corresponds to $p_v \notin {}^{\blacklozenge}t$ by definition of F in the translation. Then we can consider each $v \in \mathcal{V}$ separately and there are seven cases, corresponding to the rows in Fig. 6:

- if $\ell[v] = r[v] = \top$ (row 1), then we have $s[v] = \top$ because a is enabled and $s'[v] = \top$ by definition of RR firing. On the Petri net side, the token in p_v from m is kept at (7) and not removed at (8), and no token is added at (9), so the marking is not changed, hence the result;
- if $\ell[v] = \top$ and $r[v] = \bot$ (row 2), then we have $s[v] = \top$ and $s'[v] = \bot$. Moreover, the token in p_v from m is kept at (7), it is removed at (8), and no token is added at (9), hence the result;
- if $\ell[v] = r[v] = \bot$ (row 3), then we have $s[v] = s'[v] = \bot$. Moreover, there is no token in p_v from m thus none can be kept at (7), and none is added at (9), hence the result;

- if $\ell[v] = \bot$ and $r[v] = \top$ (row 4), then we have $s[v] = \bot$ and $s'[v] = \top$. Moreover, there is no token in p_v from m thus none is kept at (7), and one is added at (9), hence the result;
- if $\ell[v] = \star$ and $r[v] = \top$ (row 5), then we do not know what $s[v]$ is but we have $s'[v] = \top$. Moreover, any token in p_v from m is not kept at (7), and one is added at (9), hence the result;
- if $\ell[v] = \star$ and $r[v] = \bot$ (row 6), then we do not know what $s[v]$ is but we have $s'[v] = \bot$. Moreover, any token in p_v from m is not kept at (7), and none is added at (9), hence the result;
- if $\ell[v] = r[v] = \star$ (row 7), then $s[v] = s'[v]$. Moreover, any token in p_v from m is kept at (7), not removed at (8), and no other token is added at (9), hence the result.

($\Leftarrow$) assume $g(s) \xrightarrow{t_a} g(s')$, and thus relations (4–6). Consider each $v \in \mathcal{V}$ separately, we have four cases to prove that a is enabled at s:

- if $\ell[v] = r[v] = \top$ then from (5) we have $0 < {}^{*}t(p_v) \leq g(s)(p_v)$ thus $s[v] = \top$;
- if $\ell[v] = \top$ and $r[v] = \bot$ then from (4) we have $0 < {}^{\bullet}t(p_v) \leq g(s)(p_v)$ and thus $s[v] = \top$;
- if $\ell[v] = \bot$ then from (6) we have $1 = {}^{\circ}t(v) > g(s)(p_v)$ thus $s[v] = \bot$;
- if $\ell[v] = \star$ then v has no influence on the enabling of a.

It remains to show that $s \xrightarrow{a} s'$, taking $m' = g(s')$, Eq. (7–9) still holds and we consider each v separately. There are five cases corresponding to the rows in Fig. 6:

- if $\ell[v] = r[v]$ (rows 1, 3, and 7), then a possible token in p_v is kept at (7) and not removed at (8) while none is added at (9), and from the definition of RR firing we have $s'[v] = s[v]$, hence the result;
- if $\ell[v] = \top$ and $r[v] = \bot$ (row 2), then one token is consumed from p_v at (8) and none is added at (9) so that $m'(p_v) = 0$. From the definition of RR firing we have $s'[v] = \bot$, hence the result;
- if $\ell[v] = \bot$ and $r[v] = \top$ (row 4), then $m(p_v) = 0$ and one token is added at (9), so that $m'(p_v) = 1$. Moreover, we have $s'[v] = \top$, hence the result;
- if $\ell[v] = \star$ and $r[v] = \top$ (row 5), then no token is copied from m into p_v at (7) and one is added at (9), so that $m'(p_v) = 1$. Moreover, we have $s'[v] = \top$, hence the result;
- if $\ell[v] = \star$ and $r[v] = \bot$ (row 6), then any token in p_v is skipped at (7) and none is added at (9) so that $m'(p_v) = 0$. Moreover, we have $s'[v] = \bot$, hence the result. □

6 Conclusion

We have presented a modelling language for ecosystems called *reaction rules* (RR) that has been developed and used for more than five years. This is a simple rule-based language in which ecological entities are modelled as Boolean variables, and the potential events in the ecosystem are modelled as if-then rules. This language is equipped with an operational semantics expressed in terms of *labelled transitions systems* (LTS). Then, we have proposed two alternative denotational semantics: (1) a translation to Petri nets extended with transitions priorities (PPN) that is obtained through an *elementarisation* of the translated system's rules, and (2) a translation to PPN further extended with read, inhibitor, and reset arcs (EPN) that is obtained trough a *normalisation* of the translated system's rules. The main result of this paper is to prove that all these semantics are strongly equivalent, which is expressed in terms of the isomorphism of the corresponding LTS, with states and labels matching. We have defined in proofs constructive mappings that can be used in practice to translate one kind of LTS into another. The overall contribution is summarised in Fig. 2 page 3.

The main interest of having several consistent semantics is the ability to chose one or another depending on the situation. For example, the operational semantics can be presented in intuitive terms directly on the RR concrete syntax, and thus it is suitable to be explained to ecologists. However, no implementation exists for it so it cannot be used to compute state-spaces. On the other hand, the Petri net semantics allows to use one of the numerous tools readily available for Petri nets. For instance, in [5–8], we have used TINA [1] to compute explicit state-spaces from the PPN semantics of RR since TINA supports transitions priorities. We also have used the EPN semantics in [4,12] through a translation of extended Petri nets into GAL systems [19, Sec. 5] in order to compute symbolic state-spaces using libDDD and ITS-tools [18,19]. Another use of the EPN semantics could be through the SNAKES [15] library for interactive simulation.

6.1 Related Works

The design of the RR modelling language has been made by computer scientists working together with ecologists, with the goal to provide a language that is both as simple as possible and also sufficiently descriptive for actual use by ecologists. Actually, modelling ecosystems by discrete systems with *if-then* rules was proposed in the early 90's in [16,17], but using multi-valued variables. These works have then evolved towards cellular automata and, to the best of our knowledge, remained focused on simulation-based analysis.

RR being based on Boolean variables, it may appear similar to Boolean networks that are widely used in systems biology [14,21]. However, both languages have several important differences. First of all, they greatly differ in the modelling philosophy: Boolean networks are centred onto how each variable is influenced by the others and thus present the system as an interaction network; RR is centred onto the potential events in the system and thus presents the system as a rule-based model. Then, RR allows to express non-determinism at the level of rules while, in Boolean networks, it arises only in the semantics from the update mode of the variables [3]. This is a crucial feature to model ecosystems that often exhibit such non-deterministic behaviours where the same causes (as far as they can be observed) may lead to distinct consequences. Finally, from the LTS perspective, RR is strictly more expressive than Boolean networks. It has been proved that it can generate any LTS based on Boolean variables while Boolean networks cannot generate LTS in which a state has successors with incompatible

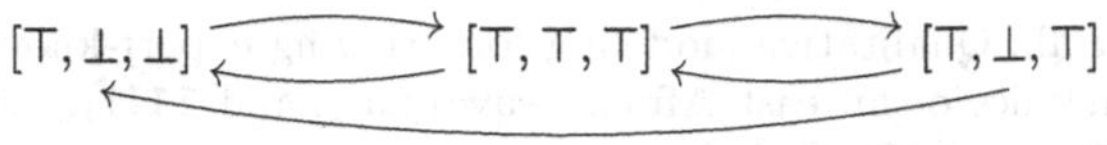

$$[\top,\bot,\bot] \rightleftarrows [\top,\top,\top] \rightleftarrows [\top,\bot,\top]$$

Fig. 7. A LTS on three Boolean variables that can be modelled by an RR system but not by a Boolean network. (This is left to the reader.) This LTS has been taken in the literature in ecology and corresponds to actual observations [9].

updates of some variables [20], as for example that depicted in Fig. 7. From this it appears that Boolean networks offer a trade-off between modelling complexity and generality, by allowing modellers to focus on the evolution of each variable. RR on the other hand, is focused on the events and may lead to more detailed, and thus more complex, models.

6.2 Future Works

Several extensions of the RR language will be considered in the future. In particular, we have preliminary results based on an extension with explicit spatial information, which allows to model ecosystems taking into account their "geography". Another extension that is demanded by some users is the ability to have multi-valued variables, for example to represent ecosystems where some species play different roles depending on several thresholds of their population. Finally, we are working on a compact semantics that would remove from the state-space the constraints and the states from which they are executed. This is motivated by the fact that constraints are usually introduced to skip states that are only transient and should be discarded when studying the long term dynamics. While such semantics is quite easy to obtain in explicit state-spaces, it is more tricky for symbolic state-spaces. Moreover, the properties preserved or not by this transformation are still to be precisely characterised.

Another trend of research addresses more particularly the EPN semantics of RR systems. A PhD is in progress about obtaining unfoldings *à la* McMillan [13] for such Petri nets, with the aim to be able to apply in ecology the techniques developed in [2,10,11] for a Petri nets semantics of Boolean networks.

References

1. Berthomieu, B., Ribet, P.O., Vernadat, F.: The tool TINA-construction of abstract state spaces for Petri nets and time Petri nets. Int. J. Prod. Res. **42**(14), 2741–2756 (2004)
2. Chatain, T., Haar, S., Kolčák, J., Paulevé, L., Thakkar, A.: Concurrency in Boolean networks. Nat. Comput. **19**(1), 91–109 (2019). https://doi.org/10.1007/s11047-019-09748-4
3. Chatain, T., Haar, S., Paulevé, L.: Boolean networks: beyond generalized asynchronicity. In: Baetens, J.M., Kutrib, M. (eds.) AUTOMATA 2018. LNCS, vol. 10875, pp. 29–42. Springer, Cham (2018). https://doi.org/10.1007/978-3-319-92675-9_3

4. Cosme, M., et al.: Qualitative modeling for bridging expert-knowledge and social-ecological dynamics of an east African savanna. Land **11**(1), 42 (2022). https://doi.org/10.3390/land11010042
5. Di Giusto, C., Gaucherel, C., Klaudel, H., Pommereau, F.: Analysis of discrete models for ecosystem ecology. In: Roque, A., et al. (eds.) BIOSTEC 2019. CCIS, vol. 1211, pp. 242–264. Springer, Cham (2020). https://doi.org/10.1007/978-3-030-46970-2_12
6. Gaucherel, C., Carpentier, C., Geijzendorffer, I., Noûs, C., Pommereau, F.: Discrete-event models for conservation assessment of integrated ecosystems. Ecol. Inform. **61** (2021). https://doi.org/10.1016/j.ecoinf.2020.101205
7. Gaucherel, C., Pommereau, F., Hély, C.: Understanding ecosystem complexity via application of a process-based state space rather than a potential surface. Complexity **2020** (2020). https://doi.org/10.1155/2020/7163920
8. Gaucherel, C., Pommereau, F.: Using discrete systems to exhaustively characterize the dynamics of an integrated ecosystem. Methods Ecol. Evol. **10**(9), 1615–1627 (2019)
9. Liao, C.: Complexity in the open grazing system: rangeland ecology, pastoral mobility and ethnobotanical knowledge in Borana, Ethiopia. Ph.D. thesis, Cornell University (2016)
10. Mandon, H., Su, C., Haar, S., Pang, J., Paulevé, L.: Sequential reprogramming of Boolean networks made practical. In: Bortolussi, L., Sanguinetti, G. (eds.) CMSB 2019. LNCS, vol. 11773, pp. 3–19. Springer, Cham (2019). https://doi.org/10.1007/978-3-030-31304-3_1
11. Mandon, H., Su, C., Pang, J., Paul, S., Haar, S., Paulevé, L.: Algorithms for the sequential reprogramming of Boolean networks. IEEE/ACM Trans. Comput. Biol. Bioinform. **16**(5), 1610–1619 (2019)
12. Mao, Z., Centanni, J., Pommereau, F., Stokes, A., Gaucherel, C.: Maintaining biodiversity promotes the multifunctionality of social-ecological systems: holistic modelling of a mountain system. Ecosystem Serv. **47** (2021). https://doi.org/10.1016/j.ecoser.2020.101220
13. McMillan, K.L., Probst, D.K.: A technique of state space search based on unfolding. Formal Methods Syst. Des. **6**(1), 45–65 (1995)
14. Naldi, A., et al.: Cooperative development of logical modelling standards and tools with CoLoMoTo. Bioinformatics **31**(7) (2015). https://doi.org/10.1093/bioinformatics/btv013
15. Pommereau, F.: SNAKES: a flexible high-level petri nets library (tool paper). In: Devillers, R., Valmari, A. (eds.) PETRI NETS 2015. LNCS, vol. 9115, pp. 254–265. Springer, Cham (2015). https://doi.org/10.1007/978-3-319-19488-2_13
16. Rykiel, E.J.: Artificial intelligence and expert systems in ecology and natural resource management. Ecol. Model. **46**(1) (1989). https://doi.org/10.1016/0304-3800(89)90066-5
17. Starfield, A.M.: Qualitative, rule-based modeling. BioScience **40**(8) (1990). https://doi.org/10.2307/1311300
18. Thierry-Mieg, Y.: Homepage of ITS-tools. http://lip6.github.io/ITSTools-web
19. Thierry-Mieg, Y.: From symbolic verification to domain specific languages. Habilitation thesis, UPMC (2016)
20. Thomas, C.: Model checking applied to discrete models of ecosystems. Master's thesis, ENS Paris-Saclay (2019)
21. Thomas, R.: Boolean formalization of genetic control circuits. J. Theor. Biol. **42**(3) (1973). https://doi.org/10.1016/0022-5193(73)90247-6

Modelling the Next Generation ALICE Grid Middleware Using Coloured Petri Nets

Maxim Storetvedt[1(✉)], Latchezar Betev[2], Nikola Hardi[2], Håvard Helstrup[1], Kristin Fanebust Hetland[1], and Bjarte Kileng[1]

[1] Faculty of Engineering and Science, Western Norway University of Applied Sciences, Bergen, Norway
msto@hvl.no

[2] CERN, Geneva, Switzerland

Abstract. JAliEn (Java ALICE Environment) is a new Grid middleware framework designed to satisfy the needs of the CERN ALICE Collaboration for the LHC Run 3. It aims to ameliorate past shortcomings of the original AliEn middleware, and to provide a high-performance, and high-scalability service to cope with the increased volumes of collected data. To achieve this, JAliEn comes with many significant architectural changes – something which is not without challenges in a distributed system. Not only are users unfamiliar with how the framework behaves and communicates, it could also potentially introduce system deadlocks.

One possible way of describing and verifying distributed systems is through the use of Petri Nets (PNs), a modelling language designed for such purposes. Through an extension of it, known as Coloured Petri Nets (CPNs), unique "tokens" can be used to describe individual commands and actions as they are sent through the system. This allows to map, model, and possibly verify JAliEn and its architecture, which will be the focus of this contribution.

Keywords: CPN · Grid · ALICE · CERN · LHC · AliEn · JAliEn

1 Introduction

Following multiple upgrades to the ALICE detector at CERN [1], the number of the minimum bias Pb-Pb events to be collected is expected to rise by a factor of 50 during LHC Run 3, compared to the already collected data sample over the past 10 years. To accommodate for this massive increase in data, the current production Grid middleware, AliEn (ALICE Environment) [2], is being replaced with a new middleware stack. Dubbed JAliEn (Java ALICE Environment) [3] for its use of Java as the language of choice, this new middleware is a complete rewrite of the legacy AliEn middleware. It aims to provide a high-performance and high-scalability service, while at the same time avoids inheriting deprecated code and accumulated hardcoded fixes.

L. Bernardinello and L. Petrucci (Eds.): PETRI NETS 2022, LNCS 13288, pp. 195–210, 2022.
https://doi.org/10.1007/978-3-031-06653-5_11

Albeit inspired by the AliEn middleware, JAliEn deviates from it in several ways. There are new database backends, built-in load balancing functionality and a hierarchical approach to configuration – in addition to the new Java codebase. Furthermore, a new authentication scheme is introduced, based around the passing of *token certificates*[1], that allows for assigning fine grained permissions to each required task.

The accommodation of the above changes has resulted in JAliEn having a largely different overall architecture compared to the legacy AliEn, with new system components and means of data exchange. Consequently, there is a loss of familiarity and understanding on how the system is expected to behave – possibly exacerbating the use of the new system for users familiar with the legacy AliEn. Furthermore, AliEn has been successfully used in production for over a decade, giving it high credibility.

To aid users in better understanding the system and to alleviate concerns in regards to its correctness and credibility, a formal software model may thus be used to describe and validate the system. This approach has previously been applied within other experiments at CERN, with promising results [4]. The contribution of this paper is to describe one approach for creating such a model, known as Coloured Petri Nets (CPNs) [5], and consequently present the JAliEn system modelled using this approach. This formal model will in turn be used to study the behaviour of the full system while in-use, performing simulation and state-space exploration through the use of CPN Tools – allowing for not only gauging the correctness of the real system when in use, but also to gain insight to how each individual distributed component comes together to form it.

The rest of this paper is organised as follows: Sect. 2 will discuss initial modelling and abstraction considerations that had to be taken, thereafter providing an overview of the finalised model – a dedicated subsection has been given to each major component. In Sect. 3, the model will be used for simulations, generating a state space report which in turn will be used to examine relevant properties. A conclusion is provided in Sect. 4.

2 Modelling JAliEn Using CPNs

Petri Net colours provide a natural way of modelling the JAliEn middleware: While passing generic PN tokens would be too limited to accurately represent the transitions needed by JAliEn, coloured tokens allow a near 1:1 mapping of relevant data structures. Specifically, a token can be used to represent commands as they are passed through the system, along with JAliEn *token certificates*, which are used to authenticate the actions.

Modelling Considerations. CPNs provide means to verify the correctness of system behaviour. However, Petri nets are also a powerful visual tool for education and documentation. With JAliEn being a new system, and still in active development, changes are major and documentation still scarce. A simulated

[1] This is a JAliEn construct, not to be confused with (C)PN tokens.

CPN model allows users to interact with the "system" in real-time, submit commands and actions, and proceed to monitor how their requests are both handled and forwarded between the various components of the distributed system. This in turn provides valuable insight, allowing users to gain a better understanding of the inner workings of the system, interactions and execution flow. Such a use-case has thus been taken into account during model creation, where additional places/transitions are added for visual aid.

Abstraction Considerations. Selecting an appropriate level of abstraction is essential for a good, usable, model. However, this appropriate level is indeed tightly correlated with its intended use: Reduce the abstraction, gain more detail. Conversely, increase the abstraction, gain a better overview.

The level of abstraction should reflect the intention behind the model. In this case, the model has been devised from the initial idea of utilising CPN tokens as a representation for JAliEn commands and their authentication tokens, which through their interactions enable visualising core exchanges and flow of the real system. Furthermore, there is also the prospect of using the model for documentation and educational purposes in mind. For these reasons, the model should be able to accurately describe the JAliEn token flow, though not needlessly delve into unrelated details.

Model. From a top-down perspective, the JAliEn middleware system can be described as consisting of six distinct (albeit not always unique or sole) actors: The central services, a JBox (JAliEn authentication handler), a VOBox (ALICE front-end for compute clusters), a batch queue (schedules Grid job execution on worker nodes), a worker node (used to execute submitted Grid jobs), and a user – representing actions taken by user(s) from a JAliEn shell.

An overview of the above actors is presented in Fig. 1, using the CPN Tools software [7]. Each actor is represented through a *substitution*, with an associated submodule (see Fig. 2). The communication between each actor is here represented by transitions, and indicated by arcs.

Combined, the actors represented here create a starting point for a full model, though it remains to define the necessary routines and steps for each actor. Furthermore, JAliEn transitions and user interactions require the passing of data structures – something which too needs to be represented. Below we present each of the detailed CPN modules.

2.1 User

While the majority of the actors shown in Fig. 1 represent system components, the submodule labelled "User" consolidates the actions instigated by users interacting with the system.

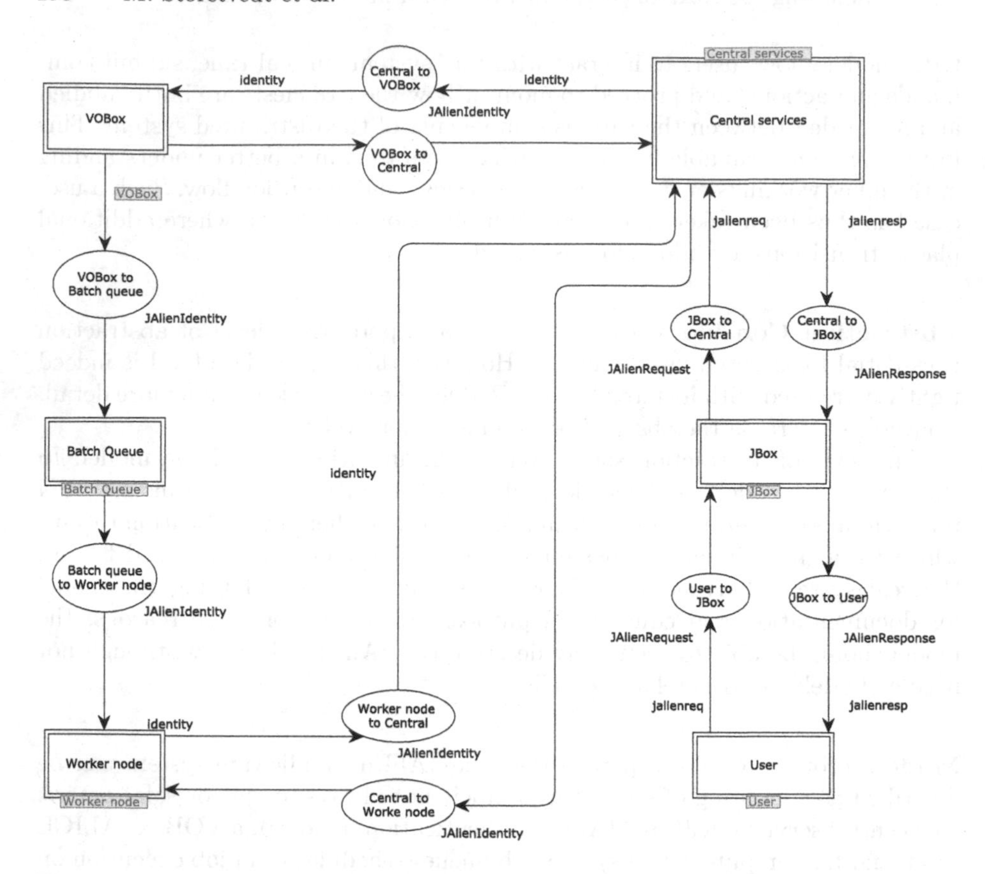

Fig. 1. Overview of the main components (actors) of the JAliEn system modelled in CPN.

Premise. All user-facing interactions are generally done through a JAliEn shell, which provides Unix-like commands for interacting with files on the Grid through an abstracted filesystem, in addition to Grid job submission. A user intending to use the JAliEn system will need to be authenticated, meaning the identity of the user must be included with each request. Identity within JAliEn is handled through the use of authentication tokens, providing sufficient permissions for each intended task/action. Should a user not possess a token, but a common Grid certificate, this certificate will first be exchanged for a token [6].

Modelling Considerations. The user submodule will need to handle the initial stages of the authentication flow, as a valid authentication token is needed for any user commands to be propagated through the system. This is where CPN tokens fit perfectly, as these provide means to simply bundle a command and an authentication token, and have them propagated through the system as a CPN token – as mentioned at the start of Sect. 2.

Model. Each command specified by a user can be represented by a CPN token, and may in this way be passed through the modelled system in a similar manner to the real-world when using a simulator (e.g. CPN Tools). However, as each request must carry an identity, each coloured token must consequently carry two values: The desired command (e.g. `ls`, `ps` or `submit`) and the authentication token. The CPN token is thus defined as a product of `jalienCmd` and `jalienIdentity`, and combined labelled as a `jalienRequest`. An example of a `jalienRequest` token can be found in the bottom left corner of Fig. 2, containing the command `submitJob` and a user certificate for authentication.

Figure 2 provides an overview of the contents found within the User submodule. In its initial state, a coloured token can be seen in the bottom left corner, representing the command to be submitted and the identity. As a user certificate is provided in this example, the following transitions are consequently triggered: The jalienRequest will be forwarded to the central services, but a copy of the coloured token is also put aside, stating "`Identity token missing`". This is used to indicate that the only way to proceed onward from this state, is to receive a proper authentication token – something the forwarded `jalienRequest` will attempt to obtain. Only once a token is received as a response will the true request be forwarded.

2.2 JBox

The JBox generally handles authentication and upstream connections. In other words, it forwards authentication requests centrally when required, or JAliEn commands otherwise. Replies from the central services are similarly forwarded in the opposite direction back to the user/requester. Consequently, the JBox is for this reason simply modelled as an intermediary step for jalienRequests, and the equivalent jalienResponses, between the user and the central services – despite being a core component of the authentication, also maintaining and storing the details of the connection instance.

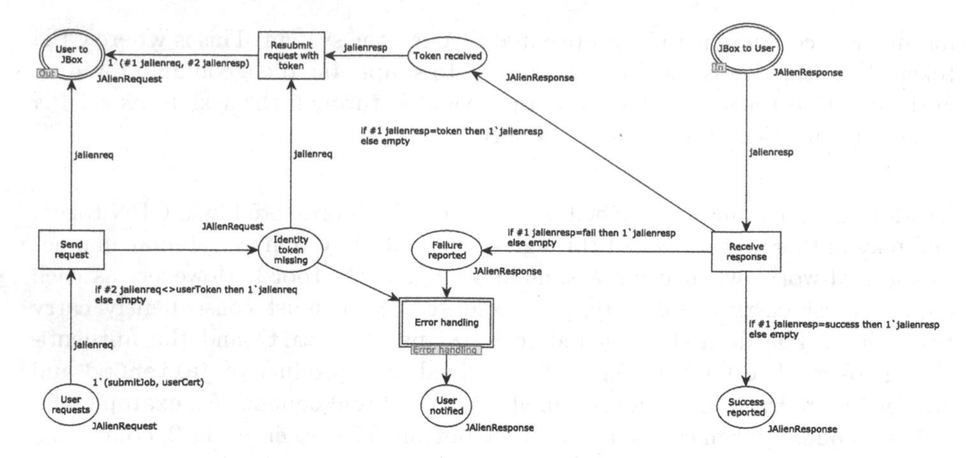

Fig. 2. CPN submodule of the "User". Error handling actions are grouped together in its own submodule, which may be expanded further.

2.3 Central Services

The central services consist of the components required for database connections, lookup in the file catalogue and LDAP (Lightweight Directory Access Protocol), in addition to authentication handling. At its centre is the JCentral – the authoritative source for all JAliEn API requests.

Premise. The JCentral mainly handles API requests, though it also performs calls to LDAP, file catalogue, database and task queue as necessary to satisfy its responsibilities. These include accepting incoming requests, handling them to the extent possible, and providing responses back to caller for each request. All requests must be authenticated before being handled. Conversely, an error response will be generated if this should fail.

Modelling Considerations. Unlike the "User" actor in Sect. 2.1, where most of its logic can be confined to a single submodule, the Central Services are the exact opposite due to their extensive size. Furthermore, they are composed of multiple services and components, many with requests simply being forwarded to them by JAliEn. Taking this into consideration, it becomes sensible to group behaviour based on a response/request flow, as opposed to logical components, so to make the order of passing CPN tokens more evident.

It must also be noted that many components found within the Central Services are irrelevant for the passing of CPN tokens (such as the Task Queue and LDAP). Nevertheless, given how one intended use of the model is for the purpose of documentation, it could make sense to have these added for the sake of overview.

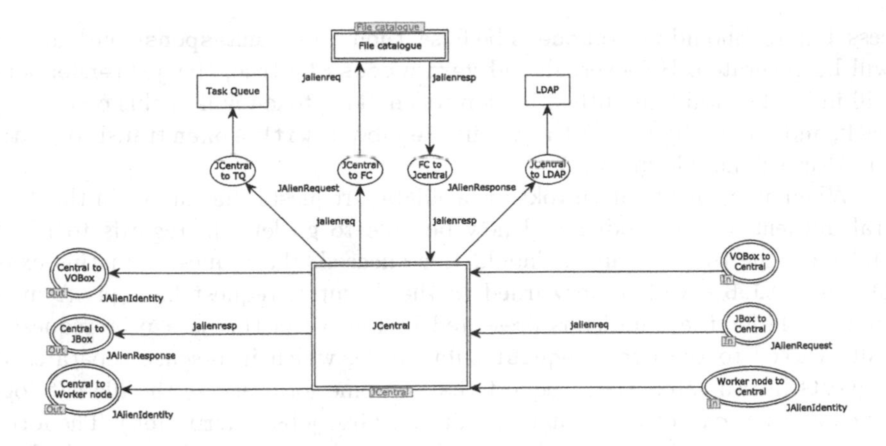

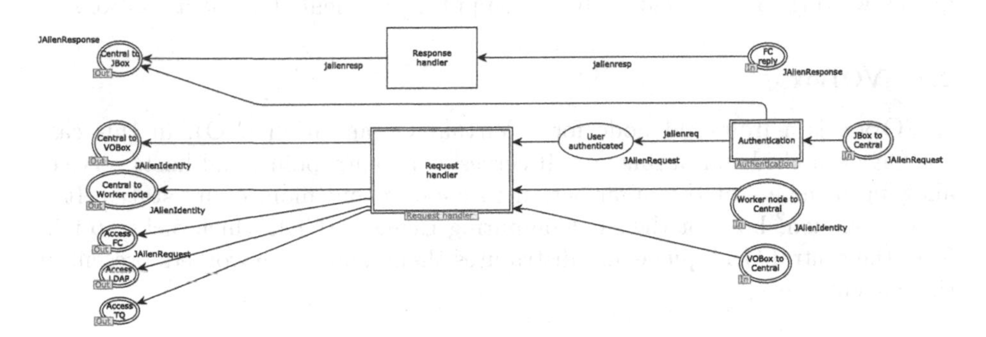

Fig. 3. Initial submodule for the central services (top), and the JCentral submodule when expanded (bottom). Note the addition of 'Task Queue' and 'LDAP', which are solely added for readability.

Model. The first submodule expands to reveal each of the components found in the Central Services and their connection (i.e. JCentral, DB, task queue and LDAP) – though each of these with their own submodule as well. However, the JCentral is again compartmentalised into several submodules for each response/request flow. This is suggested in Fig. 3.

Given how authentication differs greatly between the legacy AliEn and JAliEn, one of the more interesting submodules is the JCentral "Authentication" handler. Continuing on the example discussed in Sect. 2.1, where a jalienRequest was forwarded upstream to exchange a user certificate for a user token, this is also where the CPN token would eventually arrive when simulated.

Figure 4 shows the model of the JCentral authentication handler. As a user certificate is provided in this example, the next transition will go the path indicated by `usercert provided`, in order to authenticate the certificate identity. A Boolean variable is provided in the given model to simulate authentication success/failure: Should `verisuccess` be false, then a `jalienResponse` with an error will be generated. However, should `verisuccess` be true, the `jalienResponse` will instead contain an authentication token. This token will in this case be sent back, and eventually trigger the waiting `Resubmit with token` transition within the User submodule in Fig. 2.

When an authentication token is available, requests that arrive in the JCentral authentication handler will now be able to go left (in regards to Fig. 4), and have the token verified. Should this succeed, the request – in this example `submitJob` – will be forwarded to the JCentral request handler. A model submodule for the handler is presented in Fig. 5. As this is a user request, it will proceed to the `user request` submodule, which in essence separates the requests in two categories: Request that in some form access the file catalogue (such as `ls` or `cd`), or those that submit new Grid jobs (`submitJob`). The former requests are taken by the handler and forwarded appropriately, while the latter (`submitJob`) instigates a separate transition chain: It is submitted to the central queue, waiting to be picked up by a Computing Element (CE) at a VOBox.

2.4 VOBox

A VOBox is a front-end node for a Virtual Organisation (VO), in this case ALICE, at a Grid computing site. It serves as an entrypoint, and hosts services allowing members of the organisation to connect and monitor its status. It is also a common[2] host for the site Computing Element (CE), which picks up jobs from the central Grid queue and distributes them among the workers present at the current site.

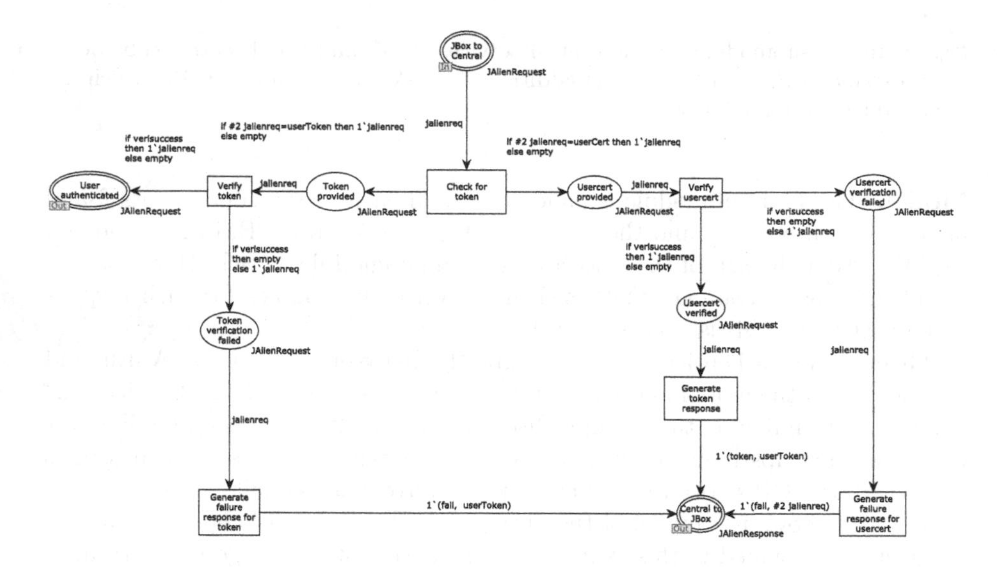

Fig. 4. JCentral authentication handler.

[2] Assumed to be the case throughout the examples within this contribution.

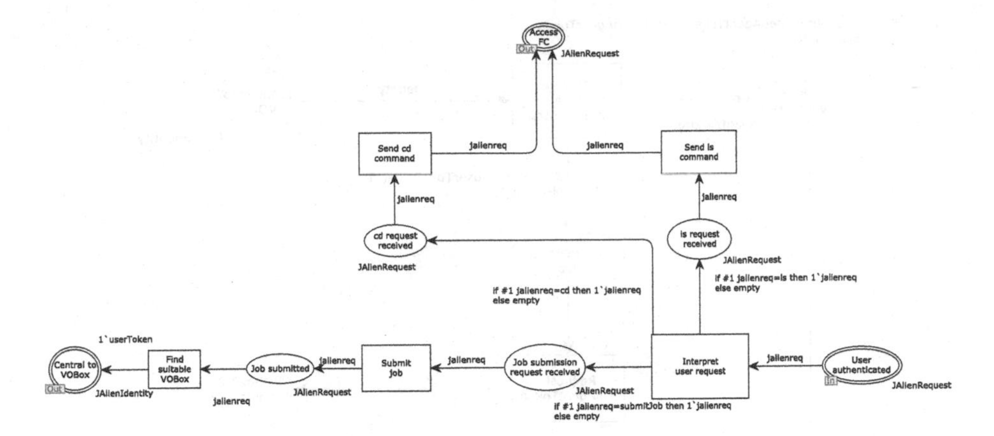

Fig. 5. JCentral user request handler.

Premise. The CE queries the central queue for Grid jobs that may match the site configuration. Should this succeed, a request will be sent for a *JobAgent token.* The JobAgent is a process tasked with executing Grid jobs on each worker node and requires a corresponding token for the permissions required to achieve this. The request for this token is handled by the JCentral, which sends a response containing it back to the CE. The CE will afterwards use this JobAgent token to generate a script that can be used to start a JobAgent process and insert it in the site batch queue – ready to be distributed and executed on a worker node.

Modelling Considerations. The process of requesting a JobAgent token is in many ways reminiscent of the exchange of a user certificate for a user token between User and JCentral: The CE provides a user(host) token, which is thereafter forwarded and exchanged at the JCentral. In this particular case, however, the process is more simplified as there is no error handling.

Model. Given the similarity to the existing User submodule, the corresponding model for the VOBox becomes as shown in Fig. 6. Like in Fig. 2, the CPN token splits in two separate paths, which results in a halt at `Receive JobAgentToken` until a response containing a JobAgent token is also received from the JCentral. Once received, a transition may be used to mimic JobAgent script generation and insertion into the site batch queue.

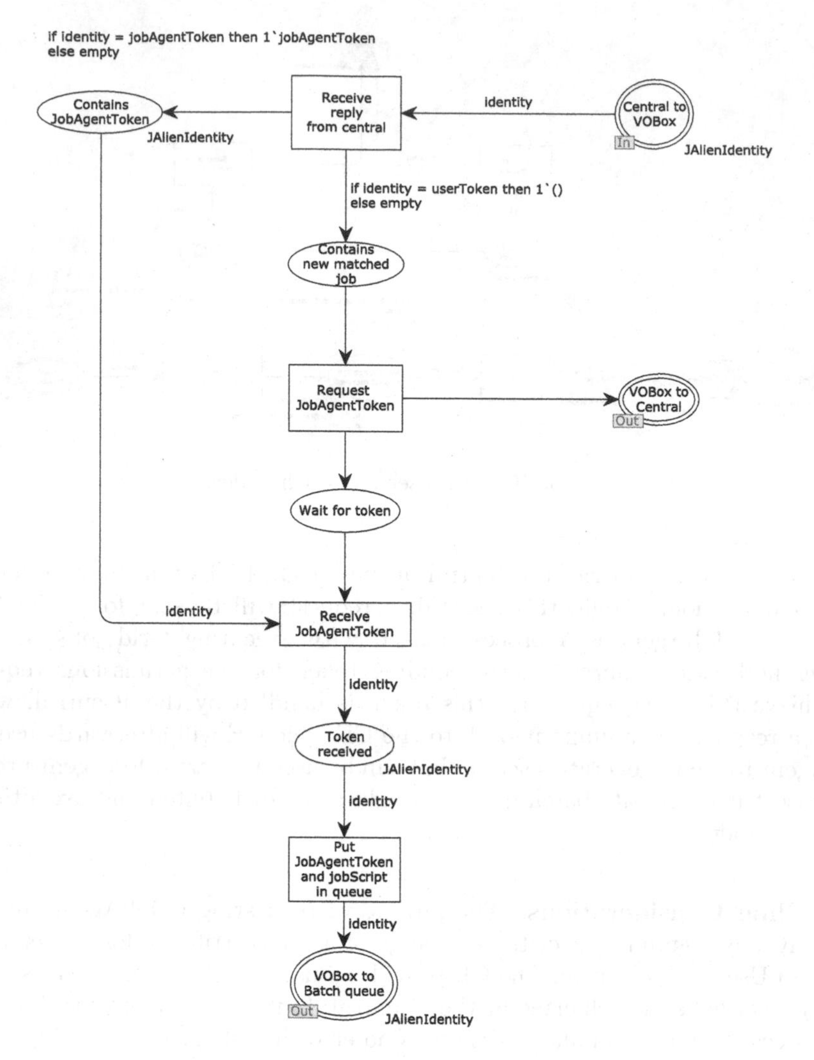

Fig. 6. The CE in a VOBox. Not unlike Fig. 2, the userToken is here used to receive a JobAgentToken from JCentral before continuing.

2.5 Batch Queue

The batch queue maintains a queue of scheduled tasks, in the form of scripts, and distributes these among a set of connected worker nodes. For a JAliEn site, these tasks/scripts are used to start the JobAgent process on each node – which will run and execute scheduled Grid jobs. However, the specific batch queue used differs between Grid sites, with the inner workings of each implementation being unrelated to JAliEn. Consequently, the batch queue has been simply modelled as an intermediary step between generating a JobAgent startup script at the VOBox/CE, and then having it executed at a worker node (visible in Fig. 1).

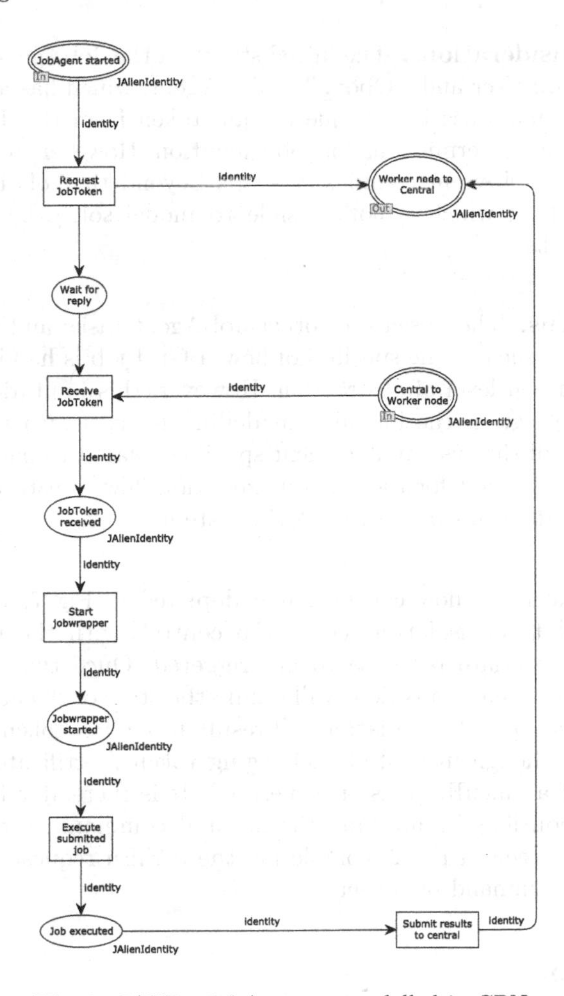

Fig. 7. JAliEn JobAgent as modelled in CPN.

2.6 Worker Node

A worker node is an executing machine at a Grid site, containing a front-end process which pulls scheduled tasks from a site batch queue when there are resources available. These tasks are generally scripts used to start the *JobAgent*, which again pull and execute Grid jobs matching the associated site.

Premise. The JobAgent has two responsibilities: To match the worker node configuration with the requirements of waiting Grid jobs, and to process and execute the jobs that successfully match. The latter is done by launching a separate process known as a *Job Wrapper*, which contains a *JobToken* that allows it to download necessary input files, execute the job, upload the results – and send a response to the JCentral once the job completes.

Modelling Considerations. The initial stages of the JobAgent are not unlike those found within User and VObox. The JobAgent, which has a corresponding JobAgent token, must use it to request a job token from the JCentral, which contains the required permissions for job execution. However, it must be noted that much of the JobAgent complexity spans beyond that of the initial token exchange, something which is not possible to model solely by the passing of authentication tokens.

Modelling Focus. The absence of other JobAgent tasks and responsibilities within the model, such as the specifics of how a Grid job is handled, is a consequence of the chosen level of abstraction. However, this limitation can be considered acceptable given the intended modelling focus, which is aimed at the core interactions of the system. JobAgent specifics related to payload execution are in this case irrelevant for the general execution flow – instead being largely self-contained in its own corner of the Grid system.

Model. The execution flow can be found depicted in Fig. 7, where – and as before – a CPN token is forwarded to the central, with the next transition requiring an appropriate response to be triggered. Once that requirement is satisfied, the subsequent transitions will mimic the start of the JobWrapper, job execution and file upload. The latter will result in a CPN token again arriving at the JCentral, though instead of exchanging tokens/certificates, a new CPN token in form of a jalienResponse is generated. It is thereafter forwarded back to the user responsible for invoking the original command, arriving as either success/failure as seen in Fig. 2, completing the JAliEn request/response circle for the 'submit' command by a user.

3 Simulation

Software toolkits, such as CPN Tools, can be used to run simulations on CPN models: Automatically triggering transitions, moving the CPN tokens through places as indicated by arcs, Booleans and if-statements. This allows us to emulate the execution of a real system in operation, and thus explore properties of interest for the system.

One particular benefit of having an executable model is the possibility of doing state space exploration – by allowing the simulator to run through all possible states of the system model, potential faults and design flaws may be uncovered that are not immediately apparent.

3.1 State Space Insights

```
CPN Tools state space report:
 Statistics
------------------------------------------------------------------------
  State Space
     Nodes:  117
     Arcs:   133
     Secs:   0
     Status: Full

  Scc Graph
     Nodes:  117
     Arcs:   133
     Secs:   0

 Liveness Properties
------------------------------------------------------------------------
  Dead Markings
     9 [29,16,117,116,115,...]

  Dead Transition Instances
     Central_services'LDAP 1
     Central_services'Task_Queue 1
     Handle_user_request'Send_cd_command 1
     Handle_user_request'Send_ls_command 1

  Live Transition Instances
     None

 Fairness Properties
------------------------------------------------------------------------
     No infinite occurrence sequences.
```

Fig. 8. Excerpt from the state space report generated for the command submitJob by CPN Tools. While there are several dead markings and dead transitions present, these are expected for a single command state space report. The exception is the dead transitions for LDAP and Task_Queue, which were solely added to the model for readability.

Deadlocks and unreached end-states are two common types of states/conditions that are generally considered unfavourable in distributed systems. Specifically, given that a distributed system is composed of multiple independent (yet interacting) components, a chain of events may occur that leaves the system in a suspended state, with each and every component waiting for the response of another. Conversely, the system may execute all actions, but still not reach an intended end-state – or simply end up in an infinite loop of repeating actions (fairness). In the context of JAliEn, this result would be disastrous, as it would not only prevent the execution of analysis jobs, but also waste limited computing resources.

To identify the presence of the above states/conditions, a full state space exploration was initiated on the created JAliEn model for each of the modelled commands (`cd`, `ls` and `submitJob`)[3]. The initial results are summarised below:

Liveness. Initially, the model can be found to reach expected end-states for all commands – i.e. 'Success' or 'User notified' (error), the two intended end-states in the model, with no indication of deadlocks being present. However, the state space report also indicates a number of dead markings that do not coincide with the end states, as well as multiple dead transitions. These are however accepted: Each of the modelled commands traverse and instigate different components of the distributed system, and as one state space report was generated for each command, not all components will be interacted with (e.g. calling `cd` will not instigate a submit chain, as needed for `submitJob`). Likewise, this is also reflected in several of the dead transitions. However, the latter is also caused by choices made to make the model more readable for educational purposes (see beginning of Sect. 2). This can be seen in Fig. 8, where there are dead transitions present for `LDAP` and `Task_queue`.

Fairness. Across all state space reports, no infinite sequences were reported, indicating that there are no infinite loops present. While this is only guaranteed to be true for the model, it would coincide with the experience of using the legacy AliEn in production – which never encountered infinite sequences. While not guaranteed, it would suggest that JAliEn may likewise avoid having this issue in production.

Other Properties. While knowing the presence of infinite loops and deadlocks are of high concern, other likewise interesting properties may also be examined within the model. While the high-level model presented in this contribution is unsuitable for identifying intricate edge cases, it may nevertheless be able to reveal other flaws in the model logic.

As the model is tailored to depicting the system flow through the exchange of authentication tokens, with an appropriate abstraction level, it remains the most accurate for this intended use. Within this domain, a crucial safety property is that no tokens are accepted without the proper exchange, or being presented an error – e.g. a user certificate should not be able to do a command without being exchanged for a token. Likewise, each command should not be able to invoke a path unrelated to its own function. In other words, it should not be possible to start with a CPN token representing `ls`, yet execute parts of the system only intended for `submitJob`.

When asserting the correct authentication token, or command, is present in each deviating transition (e.g. through the built-in ML of CPN tools), no issues are detected. That is, no transitions were observed outside of the intended path

[3] Considered representative for possible paths in the system, with each being independent.

for each token/command. For the commands, this can be further confirmed in the above 'Liveness' subsection – the dead markings present when executing the state space for only a single command would indicate that blocked paths exist. This becomes evident in Fig. 8, where there are dead markings in the paths for `ls` and `cd`, for the example command `submitJob`. In turn, this suggests the safety property is satisfied, given that the specified command was unable to interact with the transitions intended for others.

4 Conclusion

This contribution has examined a possible model of the JAliEn Grid middleware system using CPNs. It consists of six main submodules, each representing an actor/component of the system, with CPN tokens carrying commands and identities as indicated by transitions and arcs, mimicking JAliEn requests/responses. Through the use of CPN toolkits, such as CPN Tools, the passing of CPN tokens between valid states is simulated, thus emulating the execution of a real system. From this execution, a state-space exploration was performed, which in turn found no deadlocks, invalid end-states or infinite loops. While the above result is promising, it must be taken in consideration that it only reflects a *model*, and not a real system. Several components are simplified, modelling and interpretations errors may be present, and there is an absence of any external factor. Furthermore, the model is biased to align with the perspective of a user, while in reality a user would not necessarily be the first to instigate actions or exchange certificates. Nevertheless, while this does not guarantee the absence of errors, it helps in alleviating some of the concerns. Furthermore, having an interactive model provides a useful supplement for documentation, allowing users to get an overview of the system as a whole.

References

1. ALICE Collaboration, The ALICE Experiment at the CERN LHC. J. Instrument. **3** (2008). Institute of Physics Publishing/SISSA. https://doi.org/10.1088/1748-0221/3/08/S08002
2. Bagnasco, S., et al.: AliEn: ALICE environment on the GRID. J. Phys. Conf. Ser. **119**, 062012 (2008). https://doi.org/10.1088/1742-6596/119/6/062012
3. Grigoras, A.G., Grigoras, C., Pedreira, M.M., Saiz, P., Schreiner, S.: JAliEn - a new interface between the AliEn jobs and the central services. J. Phys. Conf. Ser. **523**, 012010 (2014). https://doi.org/10.1088/1742-6596/523/1/012010'
4. Hwong, Y.L., Keiren, J.J.A., Kusters, V.J.J., Leemans, S., Willemse, T.A.C.: Formalising and analysing the control software of the compact muon solenoid experiment at the large hadron collider. Sci. Comput. Program. **78**(12), 2435–2452 (2013). https://doi.org/10.1016/j.scico.2012.11.009
5. Jensen, K.: Coloured petri nets – basic concepts, analysis methods and practical use. In: Monographs in Theoretical Computer Science. An EATCS Series, vol. 1, 2nd edn. Springer (1996). https://doi.org/10.1007/978-3-662-03241-1. ISBN 978-3-642-08243-6

6. Pedreira, M.M., Grigoras, C., Yurchenko, V., Storetvedt, M.: The security model of the ALICE next generation grid framework. EPJ Web Conf. **214**, 03042 (2019). https://doi.org/10.1051/epjconf/201921403042
7. Beaudouin-Lafon, M., et al.: CPN/Tools: a tool for editing and simulating coloured petri nets ETAPS tool demonstration related to TACAS. In: Margaria, T., Yi, W. (eds.) TACAS 2001. LNCS, vol. 2031, pp. 574–577. Springer, Heidelberg (2001). https://doi.org/10.1007/3-540-45319-9_39

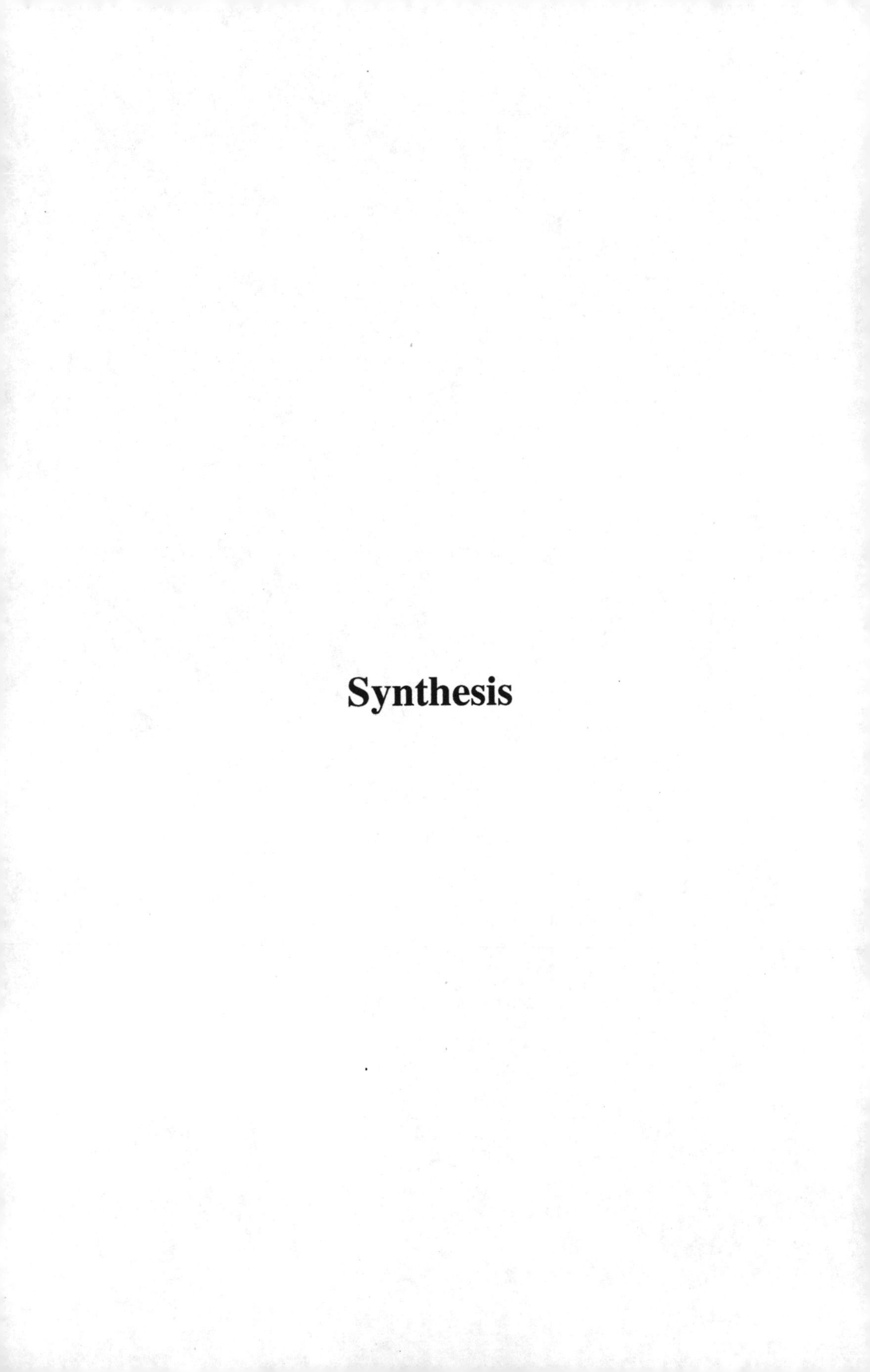

Synthesis

Synthesis of Inhibitor-Reset Petri Nets: Algorithmic and Complexity Issues

Raymond Devillers[1] and Ronny Tredup[2(✉)]

[1] Département d'Informatique, Université Libre de Bruxelles, Boulevard du Triomphe, B1050 Brussels, Belgium
raymond.devillers@ulb.be

[2] Institut Für Informatik - Universität Rostock, Albert-Einstein-Straße 22, D18059 Rostock, Germany
ronny.tredup@uni-rostock.de

Abstract. In this paper, we examine the synthesis problem from a finite labeled transition system when the target is the class of weighted nets with (possibly) inhibitor and/or reset links, or some subclasses of them. We also discuss the intrinsic complexity of some cases; in particular we show that although some subclasses have a polynomial synthesis, most of the time it is NP-complete.

1 Introduction

In order to validate a system, instead of analyzing a model of the latter to check if it satisfies a set of desired properties, the synthesis approach tries to build a model "correct by construction" directly from those properties, and then to implement it. In particular, if the behavior of a system is specified by a finite labeled transition system (LTS for short), more or less efficient algorithms have been developed to build a bounded weighted Petri net with a reachability graph isomorphic to (or close to) the given LTS [3,23]. It is also possible to target some subclasses of Petri nets [8], in particular choice-free nets and some of their specializations [7,9,10,15] which present interesting features.

On the contrary, in order to extend a bit the power of the technique (it may happen that no net of the chosen subclass has an adequate behavior, even for the full class of weighted Petri nets), we may consider superclasses of the classic Petri net subclasses. For instance, in [16], one of us used (weighted) reset arcs. Here, we shall allow both weighted inhibitor and reset arcs, as well as some subcases. Since the analysis of such systems is a bit delicate (some properties even become undecidable [17]), this increases the interest to avoid analysis techniques in favor of synthesis ones.

Petri net synthesis has numerous practical applications, for example, in the field of process discovery to reconstruct a model from its execution traces [1], in supervisory control for discrete event systems [22], and in the design and synthesis of speed-independent circuits [13]. Usually, the synthesized system yields a structural model much smaller than the initial behavioral specification, and allows to build concrete implementations. Moreover, it allows to extract informations about

L. Bernardinello and L. Petrucci (Eds.): PETRI NETS 2022, LNCS 13288, pp. 213–235, 2022.
https://doi.org/10.1007/978-3-031-06653-5_12

concurrency and distributability features from the sequential behavior given by an LTS [5].

The paper is organized as follows. After recalling some basic notions on labeled transition systems and Petri nets, we present an extension of the latter, allowing general inhibitor and reset links. In the next section, we explain how to extend the classical regional approach to synthesize such a net, or a subclass of them, when possible, from a finite transition system. Then, we explain how to characterize the inherent complexity of a synthesis problem, and we delineate a new target class for which the synthesis is polynomial. In Sect. 4.4, we present two (pure) target classes for which the synthesis is NP-complete, and in the next section, we do the same with impure strict reset nets. The last section, as usual, concludes and presents some possible follow up.

2 Preliminaries

Definition 1. TRANSITION SYSTEM
A (deterministic) *labeled transition system* (LTS, for short) $A = (S, E, \delta, \iota)$ consists of two disjoint sets of *states* S and *events* E, a partial *transition function* $\delta : S \times E \longrightarrow S$ and an *initial state* $\iota \in S$. An event e *occurs* at state s, denoted by $s\xrightarrow{e}$, if $\delta(s, e)$ is defined. By $s\xrightarrow{\neg e}$ we denote that $\delta(s, e)$ is not defined. We abridge $\delta(s, e) = s'$ by $s\xrightarrow{e}s'$ and call the latter an *edge* with *source* s and *target* s'. By $s\xrightarrow{e}s' \in A$, we denote that the edge $s\xrightarrow{e}s'$ is present in A.
A sequence $s_0\xrightarrow{e_1}s_1, s_1\xrightarrow{e_2}s_2, \ldots, s_{n-1}\xrightarrow{e_n}s_n$ of edges is called a (directed labeled) *path* (from s_0 to s_n in A). A is called *reachable*, if there is a path from ι to s for every state $s \in S$.
Two LTS $A_1 = (S_1, E, \delta_1, \iota_1)$ and $A_2 = (S_2, E, \delta_2, \iota_2)$ are *isomorphic* if there is a bijection $\zeta\colon S_1 \to S_2$ such that $\zeta(\iota_1) = \iota_2$ and $\delta_1(s, e) = s'$ if and only if $\delta_2(\zeta(s), e) = \zeta(s')$ for all $s, s' \in S_1$ and all $e \in E$.
An LTS $A = (S, E, \delta, \iota)$ is *finite* if so are S and E. □ 1

If an LTS A is not explicitly defined, then we refer to its components by $S(A)$ (states), $E(A)$ (events), δ_A (function), ι_A (initial state). In this paper, we investigate whether a LTS corresponds to the reachability graph of a Petri net. There are various ways to present a Petri net or an extension thereof; here we chose a *link* oriented one, in order to make the definition of (weighted) arcs more uniform:

Definition 2. INHIBITOR-RESET PETRI NET: SPECIFICATION AND SEMANTICS
An *inhibitor-reset Petri net* (IRPN, for short) $N = (P, T, f, \mathfrak{m}_0)$ consists of finite and disjoint sets of *places* P and *transitions* T, a (total) *flow function* $f : P \times T \to \mathbb{N} \times \mathbb{N} \times \{\mathsf{classic}, \mathsf{inhibitor}, \mathsf{reset}\}$ and an *initial marking* $\mathfrak{m}_0 : P \to \mathbb{N}$.
A link (m, n, type) will be said *pure* if $m \cdot n = 0$; it is *k-limited* (for some predefined $k \in \mathbb{N}$) if $m, n \leq k$; it is *plain* if it is 1-limited; it is *strict* if it is 0-limited[1].

[1] I.e., $m = n = 0$: strict inhibitor links correspond to usual inhibitor arcs found in the literature [18,21]; strict reset links correspond to the kind of reset arcs sometimes found in the literature [2]; strict classic links correspond to the fact that some transitions are not connected to some places.

A transition $t \in T$ can *fire* or *occur* in a marking $\mathfrak{m} : P \to \mathbb{N}$, denoted by $\mathfrak{m}\xrightarrow{t}$, and this firing leads to the marking $\mathfrak{m}'$, denoted by $\mathfrak{m}\xrightarrow{t}\mathfrak{m}'$, if, for all places $p \in P$, $f(p,t) = (m, n, \mathsf{type})$ and

1. if $\mathsf{type} = \mathsf{classic}$, then $\mathfrak{m}(p) \geq m$ and $\mathfrak{m}'(p) = \mathfrak{m}(p) - m + n$;
2. if $\mathsf{type} = \mathsf{inhibitor}$, then $\mathfrak{m}(p) \leq m$ and $\mathfrak{m}'(p) = n$;
3. if $\mathsf{type} = \mathsf{reset}$, then $\mathfrak{m}(p) \geq m$ and $\mathfrak{m}'(p) = n$.

This notation extends to sequences $w \in T^*$ and the *reachability set* $RS(N) = \{\mathfrak{m} \mid \exists w \in T^* : \mathfrak{m}_0\xrightarrow{w}\mathfrak{m}\}$ contains all the reachable markings of N. The *reachability graph* of N is the LTS $A_N = (RS(N), T, \delta, \mathfrak{m}_0)$, where, for every reachable marking $\mathfrak{m}$ of N and transition $t \in T$ with $\mathfrak{m}\xrightarrow{t}\mathfrak{m}'$, the transition function δ of A_N is defined by $\delta(\mathfrak{m}, t) = \mathfrak{m}'$. Two IRPNs are *equivalent* if their reachability graphs are isomorphic.

A place in N is said *k-safe* (for some predefined $k \in \mathbb{N}$) if, for each reachable marking $\mathfrak{m}$, $\mathfrak{m}(p) \leq k$; it is *safe* if it is 1-safe; it is *bounded* if it is k-safe for some k not defined a priori. An IRPN N is k-safe|safe|bounded if so are all its places. □2

A classic (and easy) result for classic Petri nets immediately extends to IRPNs:

Corollary 1. Bounded system
An IRPN N is bounded if and only if its reachability graph A_N is finite.

□1

Many subclasses of nets may be defined from this definition. For instance,

Definition 3. Subclasses of Nets
An IRPN $N = (P, T, f, \mathfrak{m}_0)$ is called

- a *strict inhibitor reset (Petri) net*, denoted by SIRPN, if each link is either classic or strict reset or strict inhibitor;
- a (*strict*) *inhibitor (Petri) net*, denoted by (S)IPN, if each link is either classic or (strict) inhibitor;
- a (*strict*) *reset (Petri) net*, denoted by (S)RPN, if each link is either classic or (strict) reset;
- a *(Petri) net*, denoted by PN, if all the links are classic;
- *pure* if so are all the links (note that strict inhibitor and reset links are automatically pure);
- *plain* if so are all the links (note that strict inhibitor and reset links are automatically plain). □ 3

Graphically, the various kinds of links are represented as illustrated in Fig. 1, with the convention that arcs with zero weight are omitted.

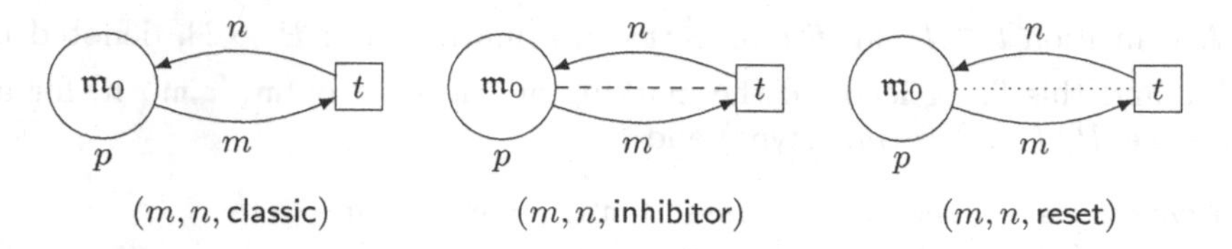

Fig. 1. The three kinds of links between a place p and a transition t (arcs with a null weight are usually omitted in figures).

3 Synthesis

Instead of analyzing a system and (try to) go from a system specification to its behavior (for instance given by the reachability graph), we may go the other way round:

Definition 4. SYNTHESIS
Let $A = (S, E, \delta, \iota)$ be an LTS. An IRPN N synthesizes A if its reachability graph is isomorphic to A. We then say that N *solves* A. □ 4

Synthesis is not exactly the symmetric of analysis, however. Indeed, while a system always has a unique behavior (given by its reachability graph), it may happen that a synthesis fails (then it is interesting to exhibit one or more sources of the failure), and if it is possible, there are infinitely many (behaviorally equivalent) solutions, sometimes with very different structures.

In the following, we shall always assume that the transition system A we start from is finite, so that it may be given explicitly and drawn easily (if not too large), and its synthesis solutions are bounded (see Corollary 1).

Classically [3], synthesis algorithms are related to the construction of *regions*, that we shall here adapt to our context:

Definition 5. IRPN-REGION
Let $A = (S, E, \delta, \iota)$ be a LTS. A pair of mappings $R = (sup, sig)$ that consists of the *sup*port $sup : S \to \mathbb{N}$ and the *sig*nature $sig : E \to \mathbb{N} \times \mathbb{N} \times \{\mathsf{classic}, \mathsf{inhibitor}, \mathsf{reset}\}$ is called a (IRPN-)region of A if, when $\delta(s, e) = s'$, then the following conditions are satisfied:

1. if $sig(e) = (m, n, \mathsf{classic})$, then $sup(s) \geq m$ and $sup(s') = sup(s) - m + n$;
2. if $sig(e) = (m, n, \mathsf{inhibitor})$, then $sup(s) \leq m$ and $sup(s') = n$;
3. if $sig(e) = (m, n, \mathsf{reset})$, then $sup(s) \geq m$ and $sup(s') = n$. □ 5

Remark 1. REGIONS AND PLACES
Intuitively, a region corresponds to a place in an IRPN solving A: sup yields the markings of that place corresponding to the various states of A (hence the initial marking is provided by $sup(\iota)$), and sig yields the links between that place and the various transitions ($E(A)$ must be the same as the set T of transitions of any solution of A).

For convenience, for all $e \in E(A)$, if $sig(e) = (m, n, \mathsf{type})$, then we define $sig^-(e) = m$, $sig^+(e) = n$ and $sig^t = \mathsf{type}$. □ 1

Remark 2. CONSTRUCTION OF THE SUPPORT
If $R = (sup, sig)$ is a region of an LTS $A = (S, E, \delta, \iota)$, then we can reconstruct R already from $sup(\iota)$ and sig, since every state s of A is reachable by a directed labeled path from ι: If $s_0 \xrightarrow{e_1} \ldots \xrightarrow{e_n} s_n$ with $s_0 = \iota$ and $s_n = s$, then, for all $i \in \{0, \ldots, n-1\}$, we get inductively $sup(s_{i+1})$ from $sup(s_i)$ and $sig(e_{i+1})$. Hence, for the sake of simplicity, we shall often present a region $R = (sup, sig)$ only by $sup(\iota)$ and sig (it is necessary however to check that two different paths leading from ι to the same state s yield the same support $sup(s)$, and that the latter is always nonnegative). For an even more compact representation, we shall summarize events with the same signature as follows: $\mathcal{E}^R_{m,n} = \{e \in E \mid sig(e) = (m, n, \mathsf{classic})\}$, and $\mathcal{E}^R_{\mathsf{inhibitor}} = \{e \in E \mid sig(e) = (0, 0, \mathsf{inhibitor})\}$, and $\mathcal{E}^R_{\mathsf{reset}} = \{e \in E \mid sig(e) = (0, 0, \mathsf{reset})\}$. □ 2

Definition 6. SYNTHESIZED NET
If $A = (S, E, \delta, \iota)$ is an LTS and $\mathcal{R}$ a set of regions of A, then the synthesized net $N^{\mathcal{R}}_A$ is defined by $S^{\mathcal{R}}_A = (\mathcal{R}, E, f, \mathfrak{m}_0)$ such that $f(e, R) = sig(e)$, and $\mathfrak{m}_0(R) = sup(\iota)$ for all $R = (sup, sig) \in \mathcal{R}$. □ 6

Definition 7. STATE SEPARATION PROPERTY
Two distinct states $s, s' \in S$ define the *state separation atom*, SSA for short, (s, s') of an LTS $A = (S, E, \delta, \iota)$. A region $R = (sup, sig)$ of A *solves* (s, s') (equivalently: *separates* s and s') if $sup(s) \neq sup(s')$. A state $s \in S$ is called *solvable* if, for every $s' \in S \setminus \{s\}$, there is a region of A that solves the SSA (s, s'). If every state of A is solvable, then A has the *state separation property*, SSP for short. □ 7

Definition 8. EVENT STATE SEPARATION PROPERTY
An event $e \in E$ and a state $s \in S$ of an LTS $A = (S, E, \delta, \iota)$ such that $s \xrightarrow{\neg e}$ define the *event state separation atom*, ESSA for short, (e, s) of A. A region $R = (sup, sig)$ of A *solves* (e, s) (equivalently: *separates* e from s) if $sig^-(e) > sup(s)$ when $sig^t(e) = \mathsf{classic}$ or reset and $sig^-(e) < sup(s)$ when $sig^t(e) = \mathsf{inhibitor}$. An event $e \in E$ is called *solvable* if, for every state $s \in S$ such that $s \xrightarrow{\neg e}$, there is a region of A that solves the ESSA (e, s). If all events of A are solvable, then A has the *event state separation property*, ESSP for short. □ 8

Definition 9. ADMISSIBLE SET
Let $A = (S, E, \delta, \iota)$ be an LTS. A set $\mathcal{R}$ of regions of A is called an *admissible set* if it *witnesses* the SSP and the ESSP of A, i. e., for every SSA, and for every ESSA of A, there is a region in $\mathcal{R}$ that solves it. □ 9

The fundamental characterization [3,14] of synthesizability for classic Petri nets extends immediately to IRPNs:

Theorem 1. SOLVABILITY OF AN LTS
Let $A = (S, E, \delta, \iota)$ be an LTS. A is solvable by an IRPN if and only if there is an admissible set $\mathcal{R}$ of (IRPN-)regions for A, and a possible solution is then $N = N^{\mathcal{R}}_A$. □ *1*

Above, we assumed that the target of a synthesis is the whole set of IRPNs, but it is easy to adapt the discussion to the case where the target is one of the subclasses mentioned in the previous section (or an intersection of them). One simply has to add some constraints to the definition of regions to be considered for the considered LTS:

Definition 10. SUBCLASSES OF REGIONS
A region $R = (sup, sig)$ of an LTS $A = (S, E, \delta, \iota)$ is called

- an *SIRPN*-region if $sig(e) = (m, n, \mathsf{type})$ and $\mathsf{type} \in \{\mathsf{inhibitor}, \mathsf{reset}\}$ imply $m = n = 0$ for all $e \in E$;
- an *(S)IPN*-region if $sig^t(e) \in \{\mathsf{classic}, \mathsf{inhibitor}\}$ for all $e \in E$ (and $sig(e) = (m, n, \mathsf{inhibitor})$ implies $m = n = 0$ for all $e \in E$);
- an *(S)RPN*-region if $sig^t(e) \in \{\mathsf{classic}, \mathsf{reset}\}$ for all $e \in E$ (and $sig(e) = (m, n, \mathsf{reset})$ implies $m = n = 0$ for all $e \in E$);
- a *PN*-region if $sig^t(e) = \mathsf{classic}$ for all $e \in E$;
- a *k-safe region*, if $sup(s) \leq k$ for each $s \in S$ (with $k = 1$ for safeness);
- a *pure* region, if $sig^-(e) = 0$ or $sig^+(e) = 0$ for each $e \in E$;
- a *k-limited* region if $sig^-(e), sig^+(e) \leq k$ for each $e \in E$ (with $k = 1$ for plainness),

where the meaning of the acronyms correspond to Definition 3. □ 10

Then, if we want to restrict our attention to the synthesis of, for example, plain (S)RPNs, we have to look for an admissible set of plain (S)RPN-regions, according to Theorem 1. In the obvious way, we use the corresponding restricted regions for the other net classes (and combinations thereof).

All those separation problems (with possibly additional constraints) may be solved by existing (integer) linear programming tools or SMT-based model checking, but it is necessary to first choose adequately the type of each link. Note however that it is not necessary to solve each separation problem from scratch: we may first check if one of the regions computed previously does not already solve the new separation problem we consider. The result will of course rely on the order in which the various separation problems are considered (besides the fact that each separation problem may sometimes have many possible solutions).

It is possible to also search for solutions with a minimal number of places/regions.

4 Complexity

4.1 Membership in NP

Concerning the inherent complexity of the IRPN synthesis, we may first observe that it is in NP:

Theorem 2. NP-SYNTHESIS
Let $A = (S, E, \delta, \iota)$ be an LTS. Its IRPN solvability is in NP.

Proof: The number of ESSAs and SSAs to be solved is (quadratic, hence) polynomial in the size of A. For each of them, a Turing machine can guess non-deterministically the type of the signature of the events, for instance: $e_{i_1}, \dots, e_{i_p}$ are of type classic, $e_{j_1}, \dots, e_{j_q}$ are of type inhibitor, and $e_{\ell_1}, \dots, e_{\ell_r}$ are of type reset. Then, finding an adequate region amounts to solve a system (of polynomial size) of linear constraints in the integer domain, as follows:

There are $|S| + 2 \cdot |E|$ variables x_i, where $sup(s_i) = x_i$ for all $i \in \{1, \dots, |S|\}$, while $sig^-(e_i) = x_{|S|+i}$, and $sig^+(e_i) = x_{|S|+|E|+i}$ for all $i \in \{1, \dots, |E|\}$; the constraints are then

1. For all $i \in \{1, \dots, |S| + 2 \cdot |E|\}$: add $x_i \geq 0$ (all variables are in $\mathbb{N}$)
2. For all $n \in \{1, \dots, p\}$, and all $i, j \in \{1, \dots, |S|\}$, if $s_i \xrightarrow{e_{i_n}} s_j \in A$, then add
   (a) $x_i - x_{|S|+i_n} \geq 0$, which ensures $sup(s_i) \geq sig^-(e_{i_n})$, and
   (b) $x_j - x_i + x_{|S|+i_n} - x_{|S|+|E|+i_n} = 0$, which ensures $sup(s_j) = sup(s_i) - sig^-(e_{i_n}) + sig^+(e_{i_n})$.
3. For all $n \in \{1, \dots, q\}$, and all $i, j \in \{1, \dots, |S|\}$, if $s_i \xrightarrow{e_{j_n}} s_j \in A$, then add
   (a) $x_i - x_{|S|+j_n} \leq 0$, which this ensures $sup(s_i) \leq sig^-(e_{j_n})$, and
   (b) $x_j - x_{|S|+|E|+j_n} = 0$, this ensures $sup(s_j) = sig^+(e_{j_n})$.
4. For all $n \in \{1, \dots, r\}$, and all $i, j \in \{1, \dots, |S|\}$, if $s_i \xrightarrow{e_{\ell_n}} s_j \in A$, then add
   (a) $x_i - x_{|S|+\ell_n} \geq 0$, which ensures $sup(s_i) \geq sig^-(e_{\ell_n})$, and
   (b) $x_j - x_{|S|+|E|+\ell_n} = 0$, this ensures $sup(s_j) = sig^+(e_{\ell_n})$.
5. In order to solve $\alpha = (s_f, e_g)$, add $x_f - x_{|S|+g} \leq -1$, which ensures $sup(s_f) < sig^-(e_g)$, if e_g corresponds to the types classic or reset, and $x_{|S|+g} - x_f \leq -1$, which ensures $sup(s_f) > sig^-(e_g)$, if e_g corresponds to inhibitor.
6. In order to solve $\alpha = (s_i, s_j)$, create two systems, one where we add $x_i - x_j \leq -1$ and one where we add $x_j - x_i \leq -1$, which ensures $sup(s_i) \neq sup(s_j)$.
7. Other constraints may be added to restrict the target class; for instance, for all $i \in \{1, \dots, |S|\}$: add $x_i \leq k$ if we aim at synthesizing a k-safe net; for all $i \in \{1, \dots, |E|\}$: add $x_{|S|+i} \leq k$ and $x_{|S|+|E|+i} \leq k$ if we aim at synthesizing a k-limited net.

Solving such a system may be assimilated to an integer linear programming problem (without an economic function to optimize, or with a null economic function) (ILP for short), and it is known that ILP belongs to NP [20]. □ 2

4.2 Polynomial Cases

In the procedure described in the previous section to solve a separation property, we first have to fix the type of link for each transition: this is generally exponential, hence we may suspect that (in most cases) the synthesis problem is NP-complete [19]. There are cases however where it is polynomial in the size of the transition system to be solved. For instance, this was shown [3] when the target is the class of classic Petri nets without additional constraints (essentially, in that case, there is no choice to make for the link types and one has to solve a system of homogeneous linear constraints of linear size; one can solve it polynomially in the rational range and then multiply all the found variables by some common factor to get an integer solution, if there is one). The same is true when the target is the class of pure classic Petri nets without additional constraints [4], using slightly different regions.

But there are other interesting cases; for instance we shall now show that the synthesis remains polynomial when the target is the class of strict inhibitor nets (without additional constraints). To do that, we shall first introduce *complementary* places.

Lemma 1. COMPLEMENTARY PLACE

Let $N = (P, T, f, \mathfrak{m}_0)$ *be an IRPN and* $p \in P$ *a place in it bounded by some value* $k \in \mathbb{N}$ *(i.e., for any reachable marking* $\mathfrak{m}$, $\mathfrak{m}(p) \leq k$*). Let also* $l = max_{t \in T}\{n \mid f(p,t)) = (m, n, \textsf{classic})\}$, $h = max_{t \in T}\{m, n \mid f(p,t)) = (m, n, \textsf{type})$ *and* $\textsf{type} \neq \textsf{classic}\}$ *and* $\mu = \max\{k+l, h\}$. *Let* $\widehat{N}$ *be the net obtained from* N *by introducing a fresh place* $\widehat{p}$ *(the complementary of* p*) with initial marking* $\mu - \mathfrak{m}_0(p)$ *and such that, for each* $t \in T$, *the link between* $\widehat{p}$ *and* t *is built as illustrated in Fig. 2. Then* $\widehat{N}$ *and* N *are behaviorally equivalent and, for each marking* $\widehat{\mathfrak{m}}$ *reached in* $\widehat{N}$, $\widehat{\mathfrak{m}}(p) + \widehat{\mathfrak{m}}(\widehat{p}) = \mu$ *(and is thus constant).*

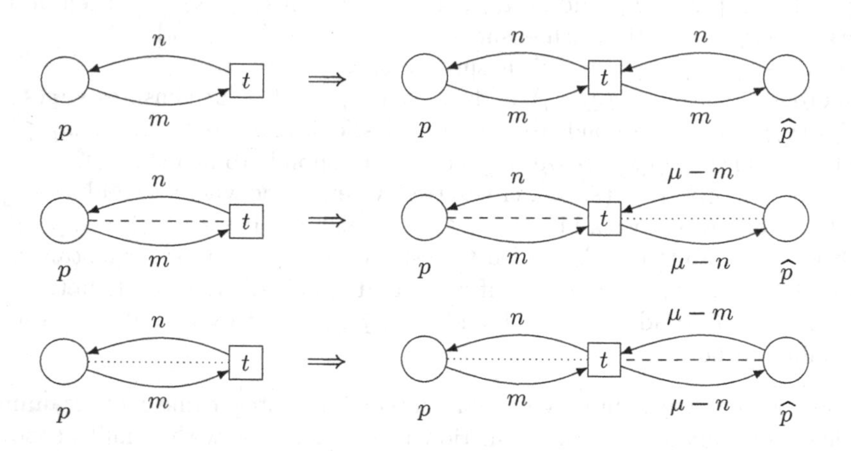

Fig. 2. Construction of a complementary place.

Proof: Since $\widehat{N}$ has one place more than N, and the initial marking of $\widehat{N}$ is the same as N when restricted to P, any evolution of $\widehat{N}$ is also one of N, and the restriction to P of the reached marking in $\widehat{N}$ is the marking reached with the same evolution in N.

If $\mathfrak{m}$ is a marking reachable in N, from the choice of μ we have $\mu - \mathfrak{m}(p) \geq 0$, and if $\widehat{\mathfrak{m}}$ is the marking in $\widehat{N}$ obtained by adding to $\mathfrak{m}$ that $\widehat{\mathfrak{m}}(\widehat{p}) = \mu - \mathfrak{m}(p)$, it may be observed that for any transition $t \in T$, if t is enabled by $\mathfrak{m}$ in N, then it is also enabled by $\widehat{\mathfrak{m}}$ in $\widehat{N}$. Indeed, if $\mathfrak{m}(p) \geq m$, then $\mu - \mathfrak{m}(p) \leq \mu - m$ (reset case), if $\mathfrak{m}(p) \leq m$, then $\mu - \mathfrak{m}(p) \geq \mu - m$ (inhibitor case), and from the choice of μ again $\mu - \mathfrak{m}(p) \geq n$ (classic case). In any case, from the choice of μ again, $\mu - m \geq 0$ and $\mu - n \geq 0$ when they are used. Moreover, if $\mathfrak{m}\xrightarrow{t}\mathfrak{m}'$ in N, and $\widehat{\mathfrak{m}}\xrightarrow{t}\widehat{\mathfrak{m}}'$ in $\widehat{N}$, it is easy to see that the restriction of $\widehat{\mathfrak{m}}'$ to P is $\mathfrak{m}'$, and $\widehat{\mathfrak{m}}'(\widehat{p}) = \mu - \mathfrak{m}'(p)$. Since initially (by construction), $\widehat{\mathfrak{m}}_0(\widehat{p}) = \mu - \mathfrak{m}_0(p)$, by induction we get that the evolutions are the same in N and $\widehat{N}$. Moreover, if $\mathfrak{m}$ is reached after some evolution in N, the marking $\widehat{\mathfrak{m}}$ reached in $\widehat{N}$ after the same evolution only differs from $\mathfrak{m}$ by $\widehat{\mathfrak{m}}(\widehat{p}) = \mu - \mathfrak{m}(p)$.

Finally, from this property, if two evolutions lead to $\mathfrak{m}_1$ and $\mathfrak{m}_2$ in N, and the same evolutions lead to $\widehat{\mathfrak{m}}_1$ and $\widehat{\mathfrak{m}}_2$ in $\widehat{N}$, then $\mathfrak{m}_1 = \mathfrak{m}_2$ if and only if $\widehat{\mathfrak{m}}_1 = \widehat{\mathfrak{m}}_2$. Consequently, we do not only have the same evolutions, but also the same reachability graph (up to isomorphism). □ 1

This exhibits an interesting relationship between inhibitor links and reset ones, but there is also a case where we have an even more interesting relationship between inhibitor links and classic ones.

Lemma 2. Complementary Place and Special Inhibitor Links

In the construction of Lemma 1, if the link between p and t in N is in $\{0\} \times \mathbb{N} \times \{\mathsf{inhibitor}\}$, then in $\widehat{N}$ we may replace the links between t and $\{p, \widehat{p}\}$ by classic ones, as illustrated in Fig. 3.

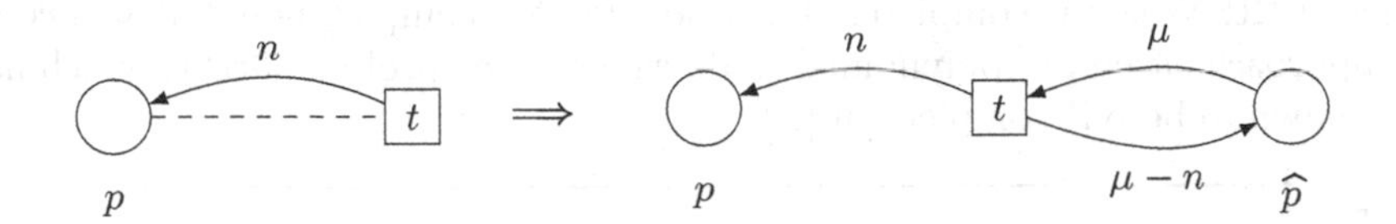

Fig. 3. Construction of a complementary place for special inhibitor links.

Proof: From the construction (and the analysis in the proof of Lemma 1), if $\widehat{\mathfrak{m}}$ is reachable in $\widehat{N}$ and $\widehat{\mathfrak{m}}\xrightarrow{t}$, since $\widehat{\mathfrak{m}}(p) + \widehat{\mathfrak{m}}(\widehat{p}) = \mu$, we must have $\widehat{\mathfrak{m}}(\widehat{p}) = \mu$ and $\widehat{\mathfrak{m}}(p) = \mathfrak{m}(p) = 0$, which indeed allows to fire t from $\mathfrak{m}$ in N. The relationship between the resulting markings in both N and $\widehat{N}$ is as before. □ 2

Corollary 2. Bounded IRPNs With Only Classic and Strongly Guarded Inhibitor Links are Behaviorally Equivalent to Classic Nets

If a bounded IRPN only has links in $\mathbb{N} \times \mathbb{N} \times \{\mathsf{classic}\} \cup \{0\} \times \mathbb{N} \times \{\mathsf{inhibitor}\}$ (i.e., classic or strongly guarded inhibitor links, since m determines how the transition is guarded), then it is behaviorally equivalent to a (bounded) net with only classic links, i.e., a classic Petri net.

Proof: One simply has to apply Lemmata 1 and 2 to each place (or only to the ones having a non-classic link). □ 2

Corollary 3. Bounded SIPNs
Any bounded SIPN is behaviorally equivalent to a bounded net with only classic links, i.e., a classic Petri net. □ 3

Theorem 3. Synthesis of SIPNs
The synthesis problem with the class of SIPNs as target is polynomial.

Proof: Since we only consider finite transition systems, when they have solutions, they are bounded. From Lemma 2, introducing strict inhibitor links does not extend the expressive power of classical links and we only have to search for solutions in the class of (bounded) weighted Petri nets, for which we mentioned before that the problem is polynomial. □ 3

Of course, this result relies on the fact that we consider general weighted Petri nets as target. It may be observed that Lemma 1 and its corollaries are not valid if we require that nets are pure or plain or k-safe. This certainly does not mean that there are no other efficient solvable cases. However, there are important subclasses of IRPNs for which we know they do not have efficient synthesis procedures (unless P = NP), such as (pure) safe SIPN, (pure) safe SRPN, safe SIRPN [24], and (pure) k-safe PN, for any fixed $k \in \mathbb{N}$ [25]. In the following sections, we shall exhibit other NP-complete subclasses.

4.3 General Approach of the NP-completeness Proofs

In Sects. 4.4, and 4.5, we shall show that the synthesis problem for several subclasses of IRPNs is NP-complete. Our proofs for NP-completeness follow a common approach based on (polynomial) reductions of the problem 3SAT, which has been shown to be NP-complete in [19]:

3SAT	
Input:	A pair $(\mathfrak{U}, M)$ with a set of 3-clauses $M = \{M_0, \dots, M_{m-1}\}$ on a finite set $\mathfrak{U}$ of variables.
Question:	Is there an assignment $b : \mathfrak{U} \to \{0, 1\}$ that satisfies all the clauses in M?

It is based on the notion of clause, i.e., a set of literals, where a literal is either a Boolean variable or its negation; a clause is interpreted as a disjunction of its items; and a set of clauses is interpreted as their conjunction. A 3-clause is a clause of size 3 and, if $X \in \mathfrak{U}$ is a variable, then we denote its negation by $\overline{X}$.

Example 1. The instance $(\mathfrak{U}, M)$ with variables $\mathfrak{U} = \{X_0, X_1, X_2, X_3\}$, and clauses $M = \{M_0, M_1, M_2\}$ such that $M_0 = \{X_0, X_1, \overline{X_2}\}$, $M_1 = \{\overline{X_0}, \overline{X_1}, \overline{X_3}\}$, and $M_2 = \{\overline{X_1}, X_2, X_3\}$ allows a positive decision, since $b(X_0) = b(X_2) = 0$, and $b(X_1) = b(X_3) = 1$ defines a truth assignment for M.

In the following, unless explicitly stated otherwise, $(\mathfrak{U}, M)$ is an arbitrary but fixed input of 3SAT, where $\mathfrak{U} = \{X_0, \dots, X_{n-1}\}$, $M = \{M_0, \dots, M_{m-1}\}$ and $M_i = \{L_{i_0}, L_{i_1}, L_{i_2}\}$ for all $i \in \{0, \dots, m-1\}$. By $\mathcal{L} = \bigcup_{i=0}^{m-1}\{L_{i_0}, L_{i_1}, L_{i_2}\}$, we refer to the set of all literals of M. Moreover, by a little abuse of notation, for every assignment b for $(\mathfrak{U}, M)$, we let $b(\overline{X_i}) = 1 - b(X_i)$ for all $i \in \{0, \dots, n-1\}$. Finally, we assume without loss of generality that each variable and its negation occur at least once, but not in the same clause.

The common principle of the proofs for the NP-hardness can be summarized as follows: We reduce $(\mathfrak{U}, M)$ to an LTS A, of size polynomial in n and m, which is the composition of several gadgets, and represents, for every $i \in \{0, \dots, m-1\}$, the clause M_i by a directed path on which the literals of M_i are events. The LTS A has an ESSA α such that if $R = (sup, sig)$ is a region that solves α, and goes along with the addressed net class, then we can extract a truth assignment for $(\mathfrak{U}, M)$ from the signature sig of the literal events of A. Hence, if there is an admissible set $\mathcal{R}$ for A, implying that $\mathcal{R}$ contains a region that solves α, then there is an assignment for $\mathfrak{U}$ that satisfies all clauses of M. Conversely, if there is a truth assignment for $(\mathfrak{U}, M)$, then there is an admissible set of regions for A; in general, we shall construct adequate regions R_i from the construction of sup_i and sig_i. Hence, A is a yes-instance if and only if $(\mathfrak{U}, M)$ is a yes-instance.

4.4 The Synthesis Complexity of Several Plain Subclasses of SIRPN

In this section, we shall prove the following theorem:

Theorem 4. PLAIN SUBCLASSES OF IRPN: NP-COMPLETE CASES
Deciding, for a given LTS A, whether there is

1. *a pure and plain PN, or a pure and plain SRPN, or*
2. *a pure and plain SIPN, or a pure and plain SIRPN, or*
3. *a plain PN, or a plain SIPN, or a plain SRPN, or a plain SIRPN*

whose reachability graph is isomorphic to A is NP-complete. □ 4

Theorem 2 showed that all these synthesis problems are in NP. Hence, it remains to prove the hardness part. For that, we follow our general approach. However, the nets of Item 1, and the ones from Item 2, and 3 need a slightly (but crucially) different construction, which result in LTS A_1, and A_2, respectively. These LTS are defined as follows:

First of all, the LTS A_1 has the edges $h_0 \xrightarrow{k} h_2$ and $h_0 \xrightarrow{u} h_1$, where h_0 is the initial state of A_1. On the other hand, the LTS A_2 has the edge $h_0 \xrightarrow{k} h_1$, and again h_0 is the initial state. Moreover, for every $i \in \{0, \dots, m-1\}$, A_1, and A_2 have the following directed path that represents the clause $M_i = \{L_{i_0}, L_{i_1}, L_{i_2}\}$ by using its literals as events:

$$h_1 \xrightarrow{a_i} t_{i,0} \xrightarrow{k} t_{i,1} \xrightarrow{L_{i_0}} t_{i,2} \xrightarrow{L_{i_1}} t_{i,3} \xrightarrow{L_{i_2}} t_{i,4} \xrightarrow{k} t_{i,5}$$

Finally, for every $i \in \{0, \dots, n-1\}$, the LTSs A_1, and A_2 implement the following gadget G_i that uses the variable X_i and its negation $\overline{X_i}$ as events:

$$G_i = h_0 \xrightarrow{b_i} g_{i,0} \xrightarrow{c_i} g_{i,2}, \quad g_{i,0} \xrightarrow{X_i} g_{i,1} \xrightarrow{\overline{X_i}} g_{i,2}$$

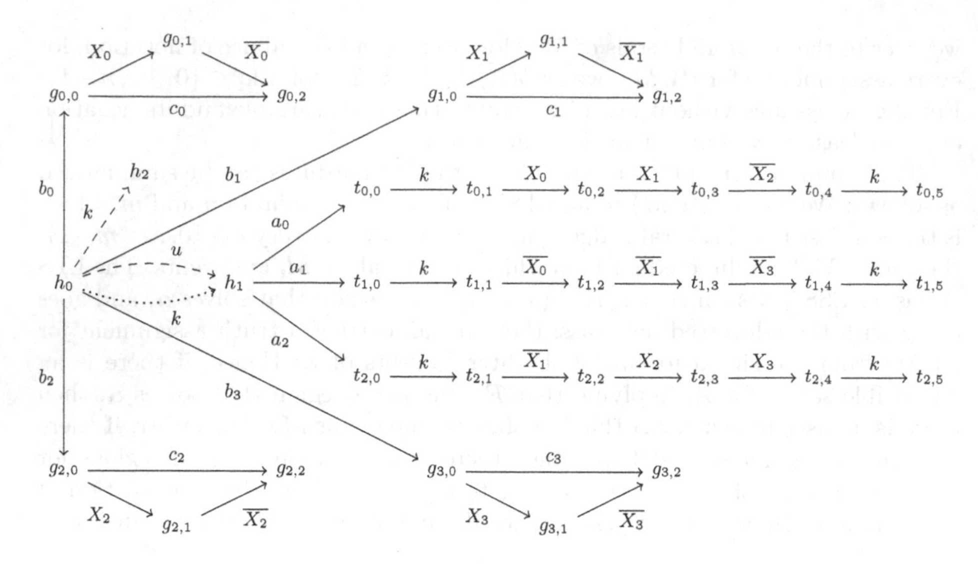

Fig. 4. The reductions based on Example 1: the solid and dashed lines define A_1, and the solid lines and the dotted line define A_2.

In the following, as long as not explicitly stated otherwise, by S_1 and E_1 (S_2 and E_2), we refer to the set of states and the set of events of A_1 (of A_2), respectively. If we apply the reduction to the input of Example 1, then we obtain the LTSs A_1, and A_2 of Fig. 4.

Lemma 3. Plain Solvability of ESSA (k, h_1) Implies Truth Assignment

If the ESSA $\alpha = (k, h_1)$ is solvable (1) by a pure and plain SRPN-region of A_1, respectively (2) by a plain SIRPN-region of A_2, then there is a truth assignment for $(\mathfrak{U}, M)$.

Proof: Let $R = (sup, sig)$ be a region of A_1 that solves α and such that, for all $e \in E_1$, $sig^t(e) \in \{\mathsf{classic}, \mathsf{reset}\}$ and

- if $sig(e) = (x, y, \mathsf{classic})$, then $x = 0$ or $y = 0$, and $x, y \in \{0, 1\}$,
- if $sig(e) = (x, y, \mathsf{reset})$, then $x = 0 = y$,

Since R solves α, we must have $sig(k) = (1, 0, \mathsf{classic})$ and $sup(h_1) = 0$.

Similarly, if $R = (sup, sig)$ is a plain SIRPN-region of A_2 that solves α, then $sig(k) \notin \{(0, 0, \mathsf{reset}), (0, 0, \mathsf{classic}), (0, 1, \mathsf{classic})\}$; moreover, since $h_0 \xrightarrow{k} h_1$, $sig(k) \notin \{(0, 0, \mathsf{inhibitor}), (1, 1, \mathsf{classic})\}$ so that again $sig(k) = (1, 0, \mathsf{classic})$ and $sup(h_1) = 0$.

Let $i \in \{0, \ldots, m-1\}$. Since k occurs at $t_{i,0}$ and at $t_{i,4}$, and $sig(k) = (1, 0, \mathsf{classic})$, we have that $sup(t_{i,0}) \geq 1 \leq sup(t_{i,4})$. Moreover, since $sup(h_1) = 0$, we get that $sig(a_i) = (0, 1, \mathsf{classic})$, $sup(t_{i,0}) = 1$ and $sup(t_{i,1}) = 0$. Hence, there is a $j \in \{0, 1, 2\}$ such that the event L_{i_j} of $\{L_{i_0}, L_{i_1}, L_{i_2}\}$ satisfies

$sig(L_{i_j}) = (0, 1, \mathsf{classic})$. We argue that this implies[2] $sig(\overline{L_{i_j}}) \neq (0, 1, \mathsf{classic})$. Indeed, let $\ell \in \{0, \ldots, m-1\}$ be such that $L_{i_j} \in \{X_\ell, \overline{X_\ell}\}$. If $sig(X_\ell) = sig(\overline{X_\ell}) = (0, 1, \mathsf{classic})$, then we have $sup(g_{\ell,2}) = sup(g_{\ell,0}) + 2$. This implies $sig(c_\ell) = (x, y, \mathsf{classic})$ with $y \geq 2$, which contradicts the plainness of R. Hence, $sig(\overline{L_{i_j}}) \neq (0, 1, \mathsf{classic})$.

Since i was arbitrary, we obtain the following observation: For every $i \in \{0, \ldots, m-1\}$, there is an event $e \in M_i$ such that $sig(e) = (0, 1, \mathsf{classic})$; and for each event $e \in M_i$, if $sig(e) = (0, 1, \mathsf{classic})$, then $sig(\overline{e}) \neq (0, 1, \mathsf{classic})$.

Let $b : \mathfrak{U} \to \{0, 1\}$ be the (well-defined) assignment for $\mathfrak{U}$, which, for all $X \in \mathfrak{U}$, is defined as follows: $b(X) = 1$ if $sig(X) = (0, 1, \mathsf{classic})$, and 0 otherwise.

Needless to say that b is a well-defined assignment, that is, if $b(X) = x$ and $b(X) = y$, then $x = y$. Similarly, if $b(X) = x$, then $b(\overline{X}) = 1 - x$, since both cases are mutually exclusive.

We argue that b satisfies every clause: Let $i \in \{0, \ldots, m-1\}$. As argued above, there is $j \in \{0, 1, 2\}$ such that the literal $L_{i_j} \in M_i$ has signature $sig(L_{i_j}) = (0, 1, \mathsf{classic})$. If $L_{i_j} = X$ for some variable $X \in \mathfrak{U}$, then $b(X) = 1$ and thus M_i is satisfied. Otherwise, $L_{i_j} = \overline{X}$ for some variable $X \in \mathfrak{U}$, which implies $sig(X) \neq (0, 1, \mathsf{classic})$ and thus $b(X) = 0$ according to the definition of b. This, however, implies $b(\overline{X}) = 1$ so that M_i is satisfied. By the arbitrariness of i, we have that b is a truth assignment for $(\mathfrak{U}, M)$. □ 3

By Lemma 3, the existence of an admissible set implies a truth assignment for $(\mathfrak{U}, M)$, since such a set implies the solvability of (k, h_1). Conversely, we have to prove that, if there is a truth assignment for $(\mathfrak{U}, M)$, A_1 and A_2 both allow admissible sets of regions, with signatures corresponding to the wanted target net class. In order to simplify the presentation, we shall only give the support of the initial state and the signatures which are not $(0, 0, \mathsf{classic})$, where we collect events by sets $\mathcal{E}^R_{m,n}$, $\mathcal{E}^R_{\mathsf{inhibitor}}$ and $\mathcal{E}^R_{\mathsf{reset}}$ following Remark 2. This will allow to construct the full region and to check that the construction is sound. Illustrations of our constructions may be found in the appendices.

Fact 1. *All SSAs of A_1 are solvable by pure and plain PN-regions.*

Proof:

- If $sup_0(h_0) = 0$ and $\mathcal{E}^{R_0}_{0,1} = \{e \in E_1 \mid h_0 \xrightarrow{e}\}$, then R_0 solves h_0.
- If $sup_1(h_0) = 0$, $\mathcal{E}^{R_1}_{0,1} = \{u\}$ and $\mathcal{E}^{R_1}_{1,0} = \{a_0, \ldots, a_{m-1}\}$, then R_1 solves h_1.

  Let $i \in \{0, \ldots, n-1\}$.
- If $sup_2(h_0) = 0$ and $\mathcal{E}^{R_2}_{0,1} = \{b_i\}$, then R_2 solves $g_{i,0}, g_{i,1}$ and $g_{i,2}$.
- If $sup_3(h_0) = 0$ and $\mathcal{E}^{R_3}_{0,1} = \{c_i, X_i\}$, then R_3 solves $(g_{i,0}, g_{i,1})$ and $(g_{i,0}, g_{i,2})$.
- If $sup_4(h_0) = 0$ and $\mathcal{E}^{R_4}_{0,1} = \{c_i, \overline{X_i}\}$ then R_4 solves $(g_{i,1}, g_{i,2})$.

As a consequence, since i was arbitrary, all the $g_{i,j}$'s are solvable.

Let $i \in \{0, \ldots, m-1\}$.

- If $sup_5(h_0) = 0$ and $\mathcal{E}^{R_5}_{0,1} = \{a_i\}$, then R_5 solves (t, s) for all $t \in S(T_i)$ and all $s \in S \setminus S(T_i)$.

---

[2] As usual, if $L_{i_j} = \overline{X_\ell}$, we state that $\overline{L_{i_j}} = X_\ell$.

- Let G_{ℓ_0}, G_{ℓ_1} and G_{ℓ_2} be the gadgets containing the events L_{i_0}, L_{i_1} and L_{i_2}, respectively. If $sup_6(h_0) = 0$ and $\mathcal{E}_{0,1}^{R_6} = \{k\} \cup \{L_{i_0}, L_{i_1}, L_{i_2}\} \cup \{c_{\ell_0}, c_{\ell_1}, c_{\ell_2}\}$, then R_6 solves (s, s') for all $s \neq s' \in \{t_{i,0}, \dots, t_{i,5}\}$ (recall that no clause contains both a literal and its negation). As a consequence, since i was arbitrary, all the $t_{i,j}$'s are solvable.

Finally, since, for all $s \in S \setminus \{h_2\}$, we have argued that s is solvable, it follows that h_2 is solvable as well. Altogether, we have witnessed the SSP of A_1. □ 1

Fact 2. *If there is a truth assignment for $(\mathfrak{U}, M)$, then all ESSAs of A_1 are solvable by pure and plain PN-regions.*

Proof: We start with the a_i's: If $sup_0(h_0) = 0$, $\mathcal{E}_{0,1}^{R_0} = \{u\}$ and $\mathcal{E}_{1,0}^{R_0} = \{a_0, \dots, a_{m-1}\}$, then R_0 solves all the a_i's.

We proceed with the c_i's: Let $i \in \{0, \dots, n-1\}$ be arbitrary but fixed and let $i_0, \dots, i_\ell \in \{0, \dots, m-1\}$ be the indices such that $X_i \in M_{i_j}$ for all $j \in \{0, \dots, \ell\}$.

- If $sup_1(h_0) = 0$, $\mathcal{E}_{0,1}^{R_1} = \{b_i, a_{i_0}, \dots, a_{i_\ell}\}$ and $\mathcal{E}_{1,0}^{R_1} = \{X_i, c_i\}$, then R_1 solves (c_i, s) for all necessary states $s \in S \setminus \bigcup_{j=0}^{\ell} S(T_{i_j})$.
- If $sup_2(h_0) = 0$, $\mathcal{E}_{0,1}^{R_2} = \{b_i\} \cup \{a_j \mid j \in \{0, \dots, m-1\} \setminus \{i_0, \dots, i_\ell\}\}$ and $\mathcal{E}_{1,0}^{R_2} = \{\overline{X_i}, c_i\}$, then R_2 solves (c_i, s) for the remaining states s. Since i was arbitrary, all the c_i's are solvable.

We proceed with k and u: Let b be a valid truth assignment for $(\mathfrak{U}, M)$ and let $i \in \{0, \dots, m-1\}$.

- If $sup_3(h_0) = 1$, $\mathcal{E}_{1,0}^{R_3} = \{u, k\}$ and $\mathcal{E}_{0,1}^{R_3} = \{a_0, \dots, a_{m-1}\} \cup \{c_0, \dots, c_{n-1}\} \cup \{X_i, \overline{X_j} \mid b(X_i) = 1, b(X_j) = 0, i, j \in \{0, \dots, n-1\}\}$, then R_3 solves (k, s) and (u, s) for all $s \in \{h_2, h_1\}$.
- If $sup_4(h_0) = 1$, $\mathcal{E}_{1,0}^{R_4} = \{k\} \cup \{b_0, \dots, b_{n-1}\}$ and $\mathcal{E}_{0,1}^{R_4} = \{u\}$, then R_4 solves (k, s) for all $s \in \bigcup_{j=0}^{n-1} \{g_{j,0}, g_{j,1}, g_{j,2}\}$.
- If $sup_5(h_0) = 1$, $\mathcal{E}_{1,0}^{R_5} = \{k\}$ and $\mathcal{E}_{0,1}^{R_5} = \{c_{i_2}, L_{i_2}\} \cup (\{a_0, \dots, a_{m-1}\} \setminus \{a_i\})$, then R_5 solves (k, s) for all $s \in \{t_{i,1}, t_{i,2}, t_{i,3}, t_{i,5}\}$.
- If $sup_6(h_0) = 1$ and $\mathcal{E}_{1,0}^{R_6} = \{u\} \cup \{b_0, \dots, b_{n-1}\}$, then R_6 solves (u, s) for all $s \in S \setminus \{h_2, h_1\}$.

Since i was arbitrary, this completes the separability of k and u.

We proceed with the variable events: Let $i \in \{0, \dots, n-1\}$ be arbitrary but fixed and let $i_0, \dots, i_\ell \in \{0, \dots, m-1\}$ be the indices such that $X_i \in M_{i_j}$ for all $j \in \{0, \dots, \ell\}$.

- If $sup_7(h_0) = 0$, $\mathcal{E}_{1,0}^{R_7} = \{X_i, c_i\}$ and $\mathcal{E}_{0,1}^{R_7} = \{b_i\} \cup \{a_{i_0}, \dots, a_{i_\ell}\}$, then R_7 solves (X_i, s) for all necessary states $s \in S \setminus \bigcup_{j=0}^{\ell} \{t_{i_j,0}, \dots, t_{i_j,4}\}$. (Note that $(X_i, t_{j,5})$ is solved for all $j \in \{0, \dots, m-1\}$).
- If we exchange X_i with $\overline{X_i}$ (according to R_7) and consider $i_0, \dots, i_\ell \in \{0, \dots, m-1\}$ to be the indices such that $\overline{X_i} \in M_{i_j}$ for all $j \in \{0, \dots, \ell\}$, then the resulting region solves $(\overline{X_i}, s)$ for all necessary states $s \in S \setminus \bigcup_{j=0}^{\ell} \{t_{i_j,0}, \dots, t_{i_j,4}\}$ except $s = g_{i,0}$.

- If $sup_8(h_0) = 0$, $\mathcal{E}_{1,0}^{R_8} = \{\overline{X_i}\}$ and $\mathcal{E}_{0,1}^{R_8} = \{X_i\} \cup \{a_0, \ldots, a_{m-1}\}$, then R_8 solves $(\overline{X_i}, g_{i,0})$.

It remains to argue that the literals are separable from the $t_{i,j}$'s, when $j \neq 5$. Let $i \in \{0, \ldots, m-1\}$.

- If $sup_9(h_0) = 0$, $\mathcal{E}_{1,0}^{R_9} = \{L_{i_0}, c_{i_0}\}$ and $\mathcal{E}_{0,1}^{R_9} = \{b_{i_0}, k\}$, then R_9 solves (L_{i_0}, s) for all $s \in \{t_{i,0}, t_{i,2}, t_{i,3}, t_{i,4}\}$.
- If $sup_{10}(h_0) = 0$, $\mathcal{E}_{1,0}^{R_{10}} = \{L_{i_1}, c_{i_1}\}$ and $\mathcal{E}_{0,1}^{R_{10}} = \{c_{i_0}\} \cup \{b_{i_1}\} \cup \{a_0, \ldots, a_{m-1}\} \setminus \{a_i\}$, then R_{10} solves (L_{i_1}, s) for all $s \in \{t_{i,0}, t_{i,1}, t_{i,3}, t_{i,4}\}$. Similarly, one shows that (L_{i_2}, s) can be solved for all $s \in \{t_{i,0}, t_{i,1}, t_{i,2}, t_{i,4}\}$.

Since i was arbitrary, all the literal events are solvable. □ 2

Altogether, we get the following lemma:

Lemma 4. Truth Assignment Implies Suitable Admissible Set for A_1
If there is a truth assignment for $(\mathfrak{U}, M)$, *then* A_1 *has an admissible set of pure and plain PN-regions.*

Let us now consider the synthesizability of A_2. Similarly to the arguments for A_1, there is a set of pure and plain PN-regions that solve all the SSAs of A_2. Hence, we restrict ourselves to the solvability of the ESSAs.

The following fact deals with ESSAs, whose solvability needs possibly impure or inhibitor links:

Fact 3. *There is a pure and plain SIPN-region, as well as an impure and plain PN-region, of* A_2 *that solves* (k, s) *for all* $s \in \bigcup_{i=0}^{n-1} S(G_i)$.

Proof: If $sup_0(h_0) = 0$, $\mathcal{E}_{\text{inhibitor}}^{R_0} = \{k\}$ and $\mathcal{E}_{0,1}^{R_0} = \{b_0, \ldots, b_{m-1}\}$, then R_0 is a suitable pure and plain SIPN-region.
If $sup_1(h_0) = 1$, $\mathcal{E}_{1,1}^{R_1} = \{k\}$ and $\mathcal{E}_{1,0}^{R_1} = \{b_0, \ldots, b_{m-1}\}$, then R_1 is a suitable impure and plain PN-region. □ 3

Fact 4. *If there is a truth assignment* b *for* $(\mathfrak{U}, M)$, *then there is an admissible set of pure and plain PN-regions solving the ESSAs of* A_2 *not addressed by Fact 3.*

Proof: Recall that $\mathcal{L}$ is the set of the literals of M and b is extended to $\mathcal{L}$, i. e., $b(\overline{X}) = 1 - b(X)$ for all $X \in \mathfrak{U}$.

We start with k:

- If $sup_0(h_0) = 1$, $\mathcal{E}_{1,0}^{R_0} = \{k\} \cup \{b_0, \ldots, b_{n-1}\}$ and $\mathcal{E}_{0,1}^{R_0} = \{L \in \mathcal{L} \mid b(L) = 1\} \cup \{a_0, \ldots, a_{m-1}\} \cup \{c_0, \ldots, c_{n-1}\}$, then R_0 solves (k, h_1).

  Let $i \in \{0, \ldots, m-1\}$.

- If $sup_1(h_0) = 2$, $\mathcal{E}_{1,0}^{R_1} = \{k\}$ and $\mathcal{E}_{0,1}^{R_1} = \{L_{i_2}, c_{i_2}\} \cup (\{a_0, \ldots, a_{m-1}\} \setminus \{a_i\})$, then R_1 solves (k, s) for all $s \in \{t_{i,1}, t_{i,2}, t_{i,3}, t_{i,5}\}$.

Since i was arbitrary, the claim follows for k.

We proceed with the a_i's and b_i's: Let $i \in \{0, \ldots, m-1\}$ and $j \in \{0, \ldots, n-1\}$.

- If $sup_2(h_0) = 1$ and $\mathcal{E}_{1,0}^{R_2} = \{a_0, \ldots, a_{m-1}\} \cup \{b_0, \ldots, b_{n-1}\}$, then R_2 solves (a_i, s) and (b_j, s) for all $s \in S \setminus \{h_0, h_1\}$.
- If $sup_3(h_0) = 0$, $\mathcal{E}_{1,0}^{R_3} = \{a_0, \ldots, a_{m-1}\}$ and $\mathcal{E}_{0,1}^{R_3} = \{k\}$, then R_3 solves (a_i, h_0).
- The region R_0 of this proof also solves (b_i, h_1) for all $i \in \{0, \ldots, n-1\}$. Since i and j were arbitrary, this proves the claim for the a_i's and b_i's.

We proceed with the c_i's: Let $i \in \{0, \ldots, n-1\}$.

- If $sup_4(h_0) = 0$, $\mathcal{E}_{1,0}^{R_4} = \{c_i, X_i\}$ and $\mathcal{E}_{0,1}^{R_4} = \{b_i\} \cup \{a_j \mid j \in \{0, \ldots, m-1\} : X_i \in M_j\}$, then R_4 solves (c_i, s) for all $s \in S \setminus (\{t_{j,0}, \ldots, t_{j,5} \mid j \in \{0, \ldots, m-1\} : X_i \in M_j\} \cup \{g_{i,0}\})$.
- If $sup_5(h_0) = 0$, $\mathcal{E}_{1,0}^{R_5} = \{c_i, \overline{X_i}\}$ and $\mathcal{E}_{0,1}^{R_5} = \{b_i\} \cup \{a_j \mid j \in \{0, \ldots, m-1\} : \overline{X_i} \in M_j\}$, then, for all $j \in \{0, \ldots, m-1\}$, R_5 solves (c_i, s) for all $s \in S(T_j)$ if $X_i \in M_j$. Since i was arbitrary, this proves the claim for the c_i's.

It remains to consider the literal events: Let $i \in \{0, \ldots, n-1\}$ be arbitrary but fixed and let $i_0, \ldots, i_\ell$ be the indices such that $X_i \in M_{i_j}$ for all $j \in \{0, \ldots, \ell\}$.

- If $sup_6(h_0) = 0$, $\mathcal{E}_{1,0}^{R_6} = \{c_i, X_i\}$ and $\mathcal{E}_{0,1}^{R_6} = \{a_{i_0}, \ldots, a_{i_\ell}\}$, then R_6 solves (X_i, s) for all $s \in S \setminus (\bigcup_{j=0}^{\ell}\{t_{j,0}, \ldots, t_{j,3}\}$. (We stress that $(X_i, t_{j,4})$ and $(X_i, t_{j,5})$ are solved for every $j \in \{0, \ldots, m-1\}$.)

Similarly, if we interchange X_i with $\overline{X_i}$ and let $i_0, \ldots, i_\ell$ select the clauses that contain $\overline{X_i}$, then the resulting region solves $(\overline{X_i}, s)$ for all $s \in S \setminus (\{g_{i,0}\} \cup \bigcup_{j=0}^{\ell}\{t_{j,0}, \ldots, t_{j,3}\}$.

- If $sup_7(h_0) = 0$, $\mathcal{E}_{1,0}^{R_7} = \{\overline{X_i}\}$ and $\mathcal{E}_{0,1}^{R_7} = \{X_i\} \cup \{a_j \mid j \in \{0, \ldots, m-1\} : \overline{X_i} \in M_j\}$, then R_7 solves $(\overline{X_i}, g_{i,0})$.

It remains to solve the literal events within their gadgets:

- If $sup_8(h_0) = 0$, $\mathcal{E}_{1,0}^{R_8} = \{L_{i_0}, a_i, c_{i_0}\}$ and $\mathcal{E}_{0,1}^{R_8} = \{k, b_{i_0}\}$, then R_8 solves (L_{i_0}, s) for all $s \in \{t_{i,0}, t_{i,2}, t_{i,3}\}$.
- If $sup_9(h_0) = 1$, $\mathcal{E}_{1,0}^{R_9} = \{L_{i_1}, a_i, c_{i_1}\}$ and $\mathcal{E}_{0,1}^{R_9} = \{L_{i_0}, c_{i_0}\}$, then R_9 solves (L_{i_1}, s) for all $s \in \{t_{i,0}, t_{i,1}, t_{i,3}\}$.

Similarly, one shows that (L_{i_2}, s) is solvable for all $s \in \{t_{i,0}, t_{i,1}, t_{i,2}\}$.

Since i was arbitrary, this completes the proof. □ 4

Altogether, we get the following lemma:

Lemma 5. TRUTH ASSIGNMENT IMPLIES SUITABLE ADMISSIBLE SET FOR A_2
If there is a truth assignment for $(\mathfrak{U}, M)$*, then* A_2 *has an admissible set of pure and plain SIPN-regions, and also of impure and plain PN-regions.*

Moreover, gathering Lemmata 3, 4 and 5, we get Theorem 4.

4.5 The Synthesis Complexity of Impure SRPN and SIRPN

The following theorem states the main result of this section:

Theorem 5. SYNTHESIS OF SRPN AND SIRPN IS NP-COMPLETE
Deciding whether there is a SRPN *or a* SIRPN *whose reachability graph is isomorphic to a given LTS is NP-complete.* □ 5

By Theorem 2 the addressed synthesis problems belong to NP. In order to prove the hardness part, we follow again the announced general approach: First of all, the LTS A has the following gadget H_0 that provides the ESSA $\alpha = (k, h_1)$:

$$H_0 = \; k \circlearrowleft h_0 \xrightarrow{y} h_1 \circlearrowright y$$

Moreover, for every $i \in \{0, \dots, m-1\}$, the LTS A has the following gadgets T_i at which the literals of the clause $M_i = \{L_{i_0}, L_{i_1}, L_{i_2}\}$ occur as events:

$$T_i = \; y \circlearrowleft t_{i,0} \xrightarrow{L_{i_0}} t_{i,1} \xrightarrow{L_{i_1}} t_{i,2} \xrightarrow{L_{i_2}} t_{i,3} \circlearrowright k$$

Finally, for every $i \in \{0, \dots, n-1\}$, the LTS has the following three gadgets $G_{i,0}, G_{i,1}$ and $G_{i,2}$, that use the i-th variable and its negation as events:

$$G_{i,0} = g_{i,0,0} \underset{a_i}{\overset{X_i}{\rightrightarrows}} g_{i,0,1} \qquad G_{i,1} = g_{i,1,0} \underset{b_i}{\overset{\overline{X_i}}{\rightrightarrows}} g_{i,1,1} \qquad G_{i,2} = g_{i,2,0} \underset{b_i}{\overset{a_i}{\rightleftarrows}} g_{i,2,1}$$

Let $S_I = \{h_0\} \cup \{t_{0,0}, \dots, t_{m-1,0}\} \cup \bigcup_{i=0}^{n-1} \{g_{i,0}, g_{i,1}, g_{i,2}\}$ be the set of the initial states of A's gadget. Finally, to complete the construction, we use the initial state ι and, for every state $s \in S_I$, a fresh and unambiguous event u_s to connect the gadgets with ι by $\iota \xrightarrow{u_s} s$.

Recall that we here assume all non-classic regions $R = (sup, sig)$ to be strict, that is, if $sig(e) = (m, n, \mathtt{type})$ with $\mathtt{type} \in \{\mathsf{inhibitor}, \mathsf{reset}\}$, then $m = n = 0$. We first observe the following simple facts:

Fact 5. *Let* $i \in \{0, \dots, n-1\}$ *and* $L_i \in \{X_i, \overline{X_i}\}$*, and let* $R = (sup, sig)$ *be a region of* A*. If* $sig(L_i) = (m, n, \mathsf{classic})$ *with* $n > m$*, then* $sig^-(\overline{L_i}) \geq sig^+(\overline{L_i})$*.*

Proof: Let $R = (sup, sig)$ be a region of A, such that $sig(X_i) = (m, n, \mathsf{classic})$ with $n > m$. This implies $sup(g_{i,0,1}) > sup(g_{i,0,0}) \geq 0$ and thus $sig(a_i) = (x, y, \mathsf{classic})$ with $y - x = n - m$. By $g_{i,2,0} \xrightarrow{a_i}$, this implies $sup(g_{i,2,1}) > sup(g_{i,2,0}) \geq 0$, and thus $sig^t(b_i) = \mathsf{reset}$ or $sig(b_i) = (x', y', \mathsf{classic})$ with $x' > y'$. Hence, by the signature of b_i, we have that $sup(g_{i,1,1}) = 0$ or $sup(g_{i,1,0}) > sup(g_{i,1,1})$, which implies that if $sig(\overline{X_i}) = (m', n', \mathsf{classic})$, then $m' > n'$ or $m' = n' = 0$, hence the claim. Analogously, one argues for the other case. □ 5

Fact 6. *Let* $s \neq s' \in S$ *be states and* $e \in E$ *an event such that* $s \xrightarrow{e} s'$ *and* $s' \xrightarrow{e} s'$ *are edges of* A*. If* $R = (sup, sig)$ *is a region of* A *such that* $sup(s) \neq sup(s')$*, then* $sig^t(e) = \mathsf{reset}$ *and* $sup(s') = 0$*.*

Proof: If $sig(e) = (0, 0, \mathsf{inhibitor})$, then we get $sup(s) = sup(s') = 0$, which contradicts $sup(s) \neq sup(s')$. If $sig(e) = (m, n, \mathsf{classic})$, then, by $s' \xrightarrow{e} s'$, we have $sup(s') = sup(s') - m + n$ and thus $m = n$. Moreover, by $s \xrightarrow{e} s'$ and $sup(s') \neq sup(s)$, we get $|-m+n| > 0$ and thus $m \neq n$, which is again a contradiction. Hence, $sig^t(e) = \mathsf{reset}$ and $sup(s') = 0$. □ 6

Lemma 6. STRICT SYNTHESIS IMPLIES TRUTH ASSIGNMENT
If there is an admissible set of SIRPN-regions for A, then there is a truth assignment for $(\mathfrak{U}, M)$.

Proof: Since $\mathcal{R}$ is an admissible set, it contains a region $R = (sup, sig)$ that solves $\alpha = (k, h_1)$. If $sig(k) = (0, 0, \mathsf{inhibitor})$, then $0 = sup(h_0) \neq sup(h_1)$. By Fact 6 and $sup(h_0) \neq sup(h_1)$, we get $sig(y) = \mathsf{reset}$ and thus $sup(h_1) = 0$, which is a contradiction. Moreover, if $sig(k) = (0, 0, \mathsf{reset})$, then R does not solve α. Hence, we have $sig(k) \in \mathbb{N} \times \mathbb{N} \times \{\mathsf{classic}\}$ and $sig^-(k) > sup(h_1) \geq 0$.

Since R is a region, by $h_0 \xrightarrow{k}$, we get $sup(h_0) \geq sig^-(k)$ and thus $sup(h_0) > sup(h_1)$. By Fact 6, this implies $sig^t(y) = \mathsf{reset}$.

Let $i \in \{0, \dots, n-1\}$. By $sig^t(y) = \mathsf{reset}$ and $\xrightarrow{y} t_{i,0}$, we get $sup(t_{i,0}) = 0$. On the other hand, by $t_{i,3} \xrightarrow{k}$, we get $sup(t_{i,3}) \geq sig^-(k) > 0$ and thus $sup(t_{i,3}) > sup(t_{i,0})$. This implies that there is a literal $L \in \{L_{i_0}, L_{i_1}, L_{i_2}\}$ such that $sig^+(L) > sig^-(L)$.

Since i was arbitrary, we obtain the following observation: For every $i \in \{0, \dots, m-1\}$, there is an event $e \in M_i$ such that such that $sig^+(e) > sig^-(e)$, which implies $sig^-(\overline{e}) \geq sig^+(\overline{e})$, by Fact 5.

Let $b : \mathfrak{U} \to \{0, 1\}$ be the assignment for $\mathfrak{U}$, which, for all $X \in \mathfrak{U}$, is defined as follows: $b(X) = 1$ if $sig^+(X) > sig^-(X)$ and 0 otherwise.

The assignment is well-defined, since both cases are mutually exclusive. We argue that b satisfies every clause. Let $i \in \{0, \dots, n-1\}$ be arbitrary but fixed: As argued above, there is $j \in \{0, 1, 2\}$ such that the literal $L_{i_j} \in M_i$ has signature $sig^+(L_{i_j}) > sig^-(L_{i_j})$. If $L_{i_j} = X$ for some variable $X \in \mathfrak{U}$, then $b(X) = 1$ and thus M_i is satisfied. Otherwise, $L_{i_j} = \overline{X}$ for some variable $X \in \mathfrak{U}$, which implies $sig^-(X) \geq sig^+(X)$ (by Fact 5) and thus $b(X) = 0$ according to the definition of b. This, however, implies $b(\overline{X}) = 1$ so that M_i is satisfied. By the arbitrariness of i, we have that b is a truth assignment for $(\mathfrak{U}, M)$. □ 6

For the converse direction, we argue that the existence of a model implies an admissible set of plain SRPN-regions of A. For the presentation of these regions, we shall use the same conventions as for Lemma 4. Moreover, for the sake of simplicity, we often only define (sup, sig) when restricted to $S \setminus \{\iota\}$ and $E \setminus U_s$, which is justified as follows: If $sup : S \setminus \{\iota\}$ and $sig : E \setminus U_s \to \mathbb{N} \times \mathbb{N} \times \{\mathsf{classic}, \mathsf{inhibitor}\}$, such that, for all $s, s' \in S \setminus \{\iota\}$ and for all $e \in E \setminus U_s$, the presence of $s \xrightarrow{e} s'$ implies that (sup, sig) behaves like a region according to Definition 5 and, moreover, for every state $s \in \{h_0\} \cup \{t_{0,0}, \dots, t_{m-1,0}\} \cup \bigcup_{i=0}^{n-1} \{g_{i,0}, g_{i,1}, g_{i,2}\}$ holds $sup(s) \in \{0, 1\}$, then it is easy to see that (sup, sig) can be extended to a region of A, where $sup(\iota) \in \{0, 1\}$ and $sig(u) \in \{(m, n, \mathsf{classic}) \mid m, n \in \{0, 1\}\}$ for all events $u \in U_s$, since the latter occur only once in A. Hence, for an even more compact representation we restrict the representation of *sup* to $s \in \{h_0\} \cup \{t_{0,0}, \ldots, t_{m-1,0}\} \cup \bigcup_{i=0}^{n-1}\{g_{i,0}, g_{i,1}, g_{i,2}\}$, since we can then compute *sup* for all states $S \setminus \{\iota\}$ by Remark 2 (and check its coherence).

Fact 7. *There is a set of plain SRPN-regions of A that solve all SSAs of A.*

Proof: If $sup_0(\iota) = 0$ and $\mathcal{E}_{1,0}^{R_0} = \{e \in E \mid \iota \xrightarrow{e}\}$, then R_0 solves ι.

Let $i \in \{0, \ldots, m-1\}$. If $sup_1(\iota) = 1$ and $\mathcal{E}_{1,0}^{R_1} = \{u_{t_{i,0}}\}$, then R_1 solves (s, s') for all states $s \in S(T_i)$ and all states $s' \in S \setminus S(T_i)$. Similarly, one shows, for any fixed $i \in \{0, \ldots, n-1\}$ and $j \in \{0, 1, 2\}$, that the states $g_{i,j,0}$ and $g_{i,j,1}$ are solvable.

Since i was arbitrary, it only remains to show that different states of the same gadget are separable.

If $sup_2(\iota) = 1$ and $\mathcal{E}_{\text{reset}}^{R_2} = \{y\}$, then R_2 solves (h_0, h_1).

The following implicitly defined regions complete the proof of this fact:

Let $i \in \{0, \ldots, m-1\}$. Let $sup(h_0) = sup_3(t_{0,0}) = \cdots = sup(t_{m-1,0}) = 0$ and, for every $j \in \{0, 1, 2\}$, let $sup_3(g_{i_j,2,0}) = 1$ if $L_{i_j} \notin \mathfrak{U}$ (i.e. L_{i_j} is a negated variable); let $\mathcal{E}_{0,1}^{R_3} = \{L_{i_0}, L_{i_1}, L_{i_2}\} \cup \{a_{i_j} \mid j \in \{0, 1, 2\} \text{ and } L_{i_j} \in \mathfrak{U}\} \cup \{b_{i_j} \mid j \in \{0, 1, 2\} \text{ and } L_{i_j} \notin \mathfrak{U}\}$ and $\mathcal{E}_{\text{reset}}^{R_3} = \{a_{i_j} \mid j \in \{0, 1, 2\} \text{ and } L_{i_j} \notin \mathfrak{U}\} \cup \{b_{i_j} \mid j \in \{0, 1, 2\} \text{ and } L_{i_j} \in \mathfrak{U}\}$. Then R_3 solves (s, s') for all $s \neq s' \in \{t_{i,0}, \ldots, t_{i,4}\}$. Since i was arbitrary, this shows the solvability of the $t_{i,j}$'s.

Let $i \in \{0, \ldots, n-1\}$.

- If, for all $s \in S_I$, $sup(s) = 1$ if $s \in \{g_{i,0,0}, g_{i,1,0}\}$ and otherwise $sup(s) = 0$, and $\mathcal{E}_{\text{reset}}^{R_4} = \{X_i, \overline{X_i}, a_i, b_i\}$, then R_4 solves $(g_{i,j,0}, g_{i,j,1})$ for all $j \in \{0, 1\}$.
- If, for all $s \in S_I$, $sup(s) = 1$ if $s = g_{i,2,0}$ and $sup(s) = 0$ otherwise, $\mathcal{E}_{0,1}^{R_5} = \{\overline{X_i}, b_1\}$ and $\mathcal{E}_{\text{reset}}^{R_5} = \{a_i\}$, then R_5 solves $(g_{i,2,0}, g_{i,2,1})$. Since i was arbitrary, the $g_{i,j,\ell}$'s are solvable. □ 7

Fact 8. *If there is truth assignment for $(\mathfrak{U}, M)$, then there is a set of plain SRPN-regions of A that solves all ESSAs of A.*

Proof: First of all, if $a \in E$ is an event and $q \in S$ a state of a gadget that does not contain a, then (a, q) is solvable by a plain classic region $R = (sup, sig)$: define $sup(s) = 1$ for the states of the gadgets containing a, $sup(s) = 0$ for the other states, $sig(a) = (1, 1, \text{classic})$ and $sig(e) = (0, 1, \text{classic})$ if e goes from ι to any gadget that contains a (and $(0, 0, \text{classic})$ otherwise). Hence, in the following, for every $e \in E$, we only deal explicitly with the gadgets that contain e.

We start with k: Let b be a valid truth assignment for $(\mathfrak{U}, M)$; let $sup_0(h_0) = 1$ and $sup_0(t_{i,0}) = 0$ for all $i \in \{0, \ldots, m-1\}$; for all $i \in \{0, \ldots, n-1\}$, let $sup_0(g_{i,0,0}) = sup_0(g_{i,1,0}) = sup_0(g_{i,2,0}) = 0$ if $b(X_i) = 1$ and let $sup_0(g_{i,2,0}) = 1$ and $sup_0(g_{i,0,0}) = sup_0(g_{i,1,0}) = 0$ if $b(X_i) = 0$; let $\mathcal{E}_{1,1}^{R_0} = \{k\}$, $\mathcal{E}_{\text{reset}}^{R_0} = \{y\} \cup \{a_i \mid i \in \{0, \ldots, n-1\} : b(X_i) = 0\} \cup \{b_i \mid i \in \{0, \ldots, n-1\} : b(X_i) = 1\}$ and $\mathcal{E}_{0,1}^{R_0} = \{X_i, a_i \mid i \in \{0, \ldots, n-1\} : b(X_i) = 1\} \cup \{\overline{X_i}, b_i \mid i \in \{0, \ldots, n-1\} : b(X_i) = 0\}$. Then R_0 solves (k, h_1).

Let $i \in \{0, \dots, m-1\}$.

Let $sup_1(h_0) = sup_1(t_{0,0}) = \dots = sup(t_{m-1,0}) = 1$ and $sup_1(g_{i_2,2,0}) = 1$ if $\overline{L_{i_2}} \in \mathfrak{U}$ (i.e. L_{i_2} is the negation of a variable) and $sup(s) = 0$ for the other states $s \in S_I$; let $\mathcal{E}^{R_1}_{1,1} = \{k\}$, $\mathcal{E}^{R_1}_{0,1} = \{L_{i_2}\} \cup \{b_{i_2} \mid \overline{L_{i_2}} \in \mathfrak{U}\} \cup \{a_{i_2} \mid L_{i_2} \in \mathfrak{U}\}$ and $\mathcal{E}^{R_1}_{\text{reset}} = \cup\{a_{i_2} \mid \overline{L_{i_2}} \in \mathfrak{U}\} \cup \{b_{i_2} \mid L_{i_2} \in \mathfrak{U}\}$. Then R_1 solves (k, s) for all $s \in \{t_{i,0}, t_{i,1}, t_{i,2}\}$. Since i was arbitrary, that completes the solvability of k.

We proceed with y: If, for all $s \in S_I$, $sup_2(s) = 1$ if $s \in \{h_0, t_{0,0}, \dots, t_{m-1,0}\}$ and 0 otherwise, $\mathcal{E}^{R_2}_{1,1} = \{y\}$ and $\mathcal{E}^{R_1}_{\text{reset}} = \bigcup_{i=0}^{m-1} M_i$, then R_2 solves (y, s) for all $s \in \bigcup_{i=0}^{m-1} \{t_{i,1}, t_{i,2}, t_{i,3}\}$. This proves the solvability of y.

We proceed with the a_i's and the b_i's: Let $i \in \{0, \dots, n-1\}$. If, for all $s \in S_I$, $sup(s) = 1$ if $s \in \{g_{i,0,0}, g_{i,2,0}\}$ and 0 otherwise, $\mathcal{E}^{R_3}_{1,0} = \{a_i\}$, $\mathcal{E}^{R_3}_{0,1} = \{b_i, \overline{X_i}\}$ and $\mathcal{E}^{R_3}_{\text{reset}} = \{X_i\}$, then R_3 solves $(a_i, g_{i,0,1})$ and $(a_i, g_{i,2,1})$.

Similarly, one shows the solvability of $(b_i, g_{i,1,1})$ and $(b_i, g_{i,2,0})$. Since i was arbitrary, this shows the solvability of the a_i's and b_i's.

Finally, we argue that the literal events are separable:

Let $i \in \{0, \dots, m-1\}$. If, for all $s \in S_I$, if $s \in \{t_{0,0}, \dots, t_{m-1,0}\} \cup \{g_{i_0,0,0} \mid L_{i_0} \in \mathfrak{U}\} \cup \{g_{i_0,1,0} \mid L_{i_0} \notin \mathfrak{U}\}$, then $sup_4(s) = 1$, and $sup_4(s) = 0$ otherwise, $\mathcal{E}^{R_4}_{1,0} = \{L_{i_0}\}$ and $\mathcal{E}^{R_4}_{\text{reset}} = \{a_{i_0} \mid L_{i_0} \in \mathfrak{U}\} \cup \{b_{i_0} \mid L_{i_0} \notin \mathfrak{U}\}$, then R_4 solves (L_{i_0}, s) for all $s \in \{t_{i,1}, t_{i,2}, t_{i,3}\}$. Moreover, R_4 solves $(L_{i_0}, g_{i_0,0,1})$ if $L_{i_0} \in \mathfrak{U}$, and $(L_{i_0}, g_{i_0,1,1})$ otherwise.

We now argue that (L_{i_1}, s) can be solved for all $s \in \{t_{i,0}, t_{i,1}, t_{i,2}\}$ and $s = g_{i_0,0,1}$ if $L_{i_0} \in \mathfrak{U}$, respectively $s = g_{i_0,1,1}$ if $L_{i_0} \notin \mathfrak{U}$. For space reasons, we consider only the case where $L_{i_1} \in \mathfrak{U}$ and $L_{i_0} \notin \mathfrak{U}$ (i.e. L_{i_0} is a negated variable). The other cases for L_{i_0} and L_{i_1} being a variable or its negation are similar.

If, for all $s \in S_I$, if $s \in \{g_{i_1,0,0}, g_{i_0,2,0}\} \cup \{t_{i,j} \mid j \in \{0, \dots, m-1\} \setminus \{i\}\}$, then $sup_5(s) = 1$, and $sup_5(s) = 0$ otherwise, $\mathcal{E}^{R_5}_{1,0} = \{L_{i_1}\}$, $\mathcal{E}^{R_5}_{0,1} = \{L_{i_0}, b_{i_0}\}$ and $\mathcal{E}^{R_5}_{\text{reset}} = \{a_{i_0}, a_{i_1}\}$, then R_5 solves (L_{i_1}, s) for all $s \in \{t_{i,0}, t_{i,1}, t_{i,2}, g_{i_1,0,1}\}$.

Similarly, one shows that (L_{i_2}, s) is solvable for all $s \in \{t_{i,0}, t_{i,1}, t_{i,3}\}$ (and $g_{i_2,0,1}$ or $g_{i_2,1,1}$, depending on whether $L_{i_2} \in \mathfrak{U}$ or not). Since i was arbitrary, the literal events are solvable. This completes the proof. □ 8

Altogether, we get the following lemma:

Lemma 7. Truth Assignment Implies Suitable Admissible Set for A
If there is a truth assignment for $(\mathfrak{U}, M)$, then there is an admissible set of (plain) SRPN-regions for A.

Since every SRPN-region is a SIRPN-region, gathering Lemmata 6 and 7, we get Theorem 5.

5 Conclusion

We have introduced a rather general class of Petri nets with inhibitor and reset links and showed how to synthesize them. While all the introduced net classes belong to the complexity class NP (Theorem 2), we exhibited some interesting subclasses for which the synthesis is polynomial (Theorem 3), and some for which it is NP-complete (Theorems 4 and 5). Figure 5 provides an overview of our findings, but also shows that there are still some open cases, even if reset and inhibitor links are assumed to be strict.

Net Class	pure	impure	pure + plain	impure + plain
PN	P	P	NPC	NPC
SRPN	open	NPC	NPC	NPC
SIPN	open	P	NPC	NPC
SIRPN	open	NPC	NPC	NPC

Fig. 5. An overview over the complexity of synthesis of PN, SRPN, SIPN and SIRPN, according to whether links are pure or (possibly) impure and/or plain. The results for the pure and impure PN are known from [3,6], the other results were developed here.

As future works, we plan to characterize the synthesis complexity for these open cases pictured by Fig. 5. After that, the next natural step is to characterize the complexity of synthesis of (pure) Petri nets with reset or inhibitor links or both, where these links do not necessarily have to be strict. Moreover, it remains to incorporate the corresponding algorithms into existing synthesis tools like SYNET [3], APT [11], GENET [12], or others.

Acknowledgments. We would like to thank the anonymous reviewers for their detailed comments and helpful suggestions.

References

1. van der Aalst, W.M.P.: Process Mining - Discovery, Conformance and Enhancement of Business Processes. Springer, Heidelberg (2011). https://doi.org/10.1007/978-3-642-19345-3
2. Akshay, S., Chakraborty, S., Das, A., Jagannath, V., Sandeep, S.: On petri nets with hierarchical special arcs. In: Meyer, R., Nestmann, U. (eds.) 28th International Conference on Concurrency Theory, CONCUR 2017, September 5–8, 2017, Berlin, Germany. LIPIcs, vol. 85, pp. 40:1–40:17. Schloss Dagstuhl - Leibniz-Zentrum für Informatik (2017). https://doi.org/10.4230/LIPIcs.CONCUR.2017.40
3. Badouel, E., Bernardinello, L., Darondeau, P.: Polynomial algorithms for the synthesis of bounded nets. In: Mosses, P.D., Nielsen, M., Schwartzbach, M.I. (eds.) CAAP 1995. LNCS, vol. 915, pp. 364–378. Springer, Heidelberg (1995). https://doi.org/10.1007/3-540-59293-8_207
4. Badouel, E., Bernardinello, L., Darondeau, P.: Petri Net Synthesis. TTCSAES, Springer, Heidelberg (2015). https://doi.org/10.1007/978-3-662-47967-4
5. Badouel, E., Caillaud, B., Darondeau, P.: Distributing finite automata through Petri net synthesis. Formal Asp. Comput. **13**(6), 447–470 (2002). https://doi.org/10.1007/s001650200022
6. Badouel, E., Darondeau, P.: On the Synthesis of General Petri Nets. Research Report RR-3025, INRIA (1996). https://hal.inria.fr/inria-00073668

7. Best, E., Devillers, R.: Characterisation of the state spaces of live and bounded marked graph petri nets. In: Dediu, A.-H., Martín-Vide, C., Sierra-Rodríguez, J.-L., Truthe, B. (eds.) LATA 2014. LNCS, vol. 8370, pp. 161–172. Springer, Cham (2014). https://doi.org/10.1007/978-3-319-04921-2_13
8. Best, E., Devillers, R., Erofeev, E., Wimmel, H.: Target-oriented Petri net synthesis. Fundamenta Informaticae **175**, 97–122 (2020). https://doi.org/10.3233/FI-2020-1949
9. Best, E., Devillers, R., Schlachter, U.: Bounded choice-free Petri net synthesis: algorithmic issues. Acta Informatica **55**(7), 575–611 (2017). https://doi.org/10.1007/s00236-017-0310-9
10. Best, E., Devillers, R.R., Erofeev, E.: A new property of choice-free Petri net systems. In: Application and Theory of Petri Nets and Concurrency - 41st International Conference, PETRI NETS 2020, Paris, France, 24–25 June 2020, Proceedings, pp. 89–108 (2020). https://doi.org/10.1007/978-3-030-51831-8_5
11. Best, E., Schlachter, U.: Analysis of petri nets and transition systems. In: Proceedings 8th Interaction and Concurrency Experience, ICE 2015, Grenoble, France, 4–5 June 2015, pp. 53–67 (2015). https://doi.org/10.4204/EPTCS.189.6
12. Carmona, J., Cortadella, J., Kishinevsky, M.: Genet: a tool for the synthesis and mining of petri nets. In: Ninth International Conference on Application of Concurrency to System Design, ACSD 2009, Augsburg, Germany, 1–3 July 2009, pp. 181–185 (2009). https://doi.org/10.1109/ACSD.2009.6
13. Cortadella, J., Kishinevsky, M., Kondratyev, A., Lavagno, L., Yakovlev, A.: A region-based theory for state assignment in speed-independent circuits. IEEE Trans. CAD Integr. Circ. Syst. **16**(8), 793–812 (1997). https://doi.org/10.1109/43.644602
14. Desel, J., Reisig, W.: The synthesis problem of Petri nets. Acta Inf. **33**(4), 297–315 (1996). https://doi.org/10.1007/s002360050046
15. Devillers, R., Hujsa, T.: Analysis and synthesis of weighted marked graph Petri nets. In: Application and Theory of Petri Nets and Concurrency - 39th International Conference, PETRI NETS 2018, Bratislava, Slovakia, 24–29 June 2018, Proceedings, pp. 19–39 (2018). https://doi.org/10.1007/978-3-319-91268-4_2
16. Devillers, R.R.: Synthesis of (choice-free) reset nets. In: Application and Theory of Petri Nets and Concurrency - 42nd International Conference, PETRI NETS 2021, Virtual Event, 23–25 June 2021, Proceedings, pp. 274–291 (2021). https://doi.org/10.1007/978-3-030-76983-3_14
17. Dufourd, C., Finkel, A., Schnoebelen, P.: Reset nets between decidability and undecidability. In: Automata, Languages and Programming, 25th International Colloquium, ICALP 1998, Aalborg, Denmark, 13–17 July 1998, Proceedings, pp. 103–115 (1998). https://doi.org/10.1007/BFb0055044, https://doi.org/10.1007/BFb0055044
18. Flynn, M.J., Agerwala, T.: Comments on capabilities, limitations and correctness of petri nets. In: Lipovski, G.J., Szygenda, S.A. (eds.) Proceedings of the 1st Annual Symposium on Computer Architecture, Gainesville, FL, USA, December 1973, pp. 81–86. ACM (1973). https://doi.org/10.1145/800123.803973
19. Garey, M.R., Johnson, D.S.: Computers and Intractability: A Guide to the Theory of NP-Completeness. Freeman, W. H, New York City (1979)
20. Gathen, J., Sieveking, M.: A bound on solutions of linear integer equalities and inequalities. Proc. Am. Math. Soc. **72**, 155–155 (1978). https://doi.org/10.2307/2042554
21. Hack, M.: Petri net language. In: MIT Technical Report 159 (1976)

22. Holloway, L.E., Krogh, B.H., Giua, A.: A survey of Petri net methods for controlled discrete event systems. Disc. Event Dyn. Syst. **7**(2), 151–190 (1997). https://doi.org/10.1023/A:1008271916548
23. Schlachter, U.: Over-approximative Petri net synthesis for restricted subclasses of nets. In: Language and Automata Theory and Applications - 12th International Conference, LATA 2018, Ramat Gan, Israel, 9–11 April 2018, Proceedings, pp. 296–307 (2018). https://doi.org/10.1007/978-3-319-77313-1_23
24. Tredup, R.: The complexity of synthesizing sf nop-equipped boolean Petri nets from g-bounded inputs. Trans. Petri Nets Other Model. Concurr. **15**, 101–125 (2021)
25. Tredup, R.: The complexity of synthesis of *b*-bounded petri nets. Fundamenta Informaticae **183**(1–2), 2019 (2021). https://fi.episciences.org/8821

The Synthesis Problem for Repeatedly Communicating Petri Games

Paul Hannibal(✉) and Ernst-Rüdiger Olderog

Carl von Ossietzky University Oldenburg, 26129 Oldenburg, Germany
paul.jonathan.hannibal1@uni-oldenburg.de,
ernst.ruediger.olderog@informatik.uni-oldenburg.de

Abstract. Petri games are a multi-player game model for the automatic synthesis of distributed systems, where the players are represented as tokens on a Petri net and grouped into environment players and system players. As long as the players move in independent parts of the net, they do not know of each other; when they synchronize at a joint transition, each player gets informed of the entire causal history of the other players.

We present a subclass of Petri games, for which the synthesis problem is decidable, with finitely many sources of nondeterminism, which are caused by the finitely many environment players, and with finitely many system players. All players satisfy a synchronisation condition guaranteeing that they know within a bounded number of own moves what each other player's next (non)deterministic move has been. This differs from existing approaches that limit the number of the system players or environment players. We show that for Petri games in this subclass deciding the existence of a winning strategy for the system players with a global safety condition is in EXPTIME.

Keywords: Synthesis · Distributed systems · Concurrent systems · Petri nets · Petri games · Unfolding

1 Introduction

A game can be interpreted as a formal specification of a reactive system. If the system is distributed over several processes, a multi-player game is appropriate for its specification. In a multi-player game one distinguishes between environment and system players. A system player can control or choose which move it takes next. An environment player is uncontrollable for the system players; they have to react to all options of the environment players. A strategy for the system players decides all choices that they have to make during a play, which is a possibly infinite sequences of moves. A strategy is *winning* if it fulfills a given winning condition against all behaviors of the environment. Thus, an implementation of a winning strategy can be seen as a correct implementation of a reactive system. The *synthesis problem* asks whether there exists a winning strategy for the system players and calls for the automatic generation of such a strategy if it

L. Bernardinello and L. Petrucci (Eds.): PETRI NETS 2022, LNCS 13288, pp. 236–257, 2022.
https://doi.org/10.1007/978-3-031-06653-5_13

exists. Such an automatic generation is useful for implementation tasks, which are prone to errors.

In this paper, we consider Petri games as a formalisation of multi-player games. A Petri game extends a Petri net by dividing its places into system and environment places. A token on a system place represents a system player and a token on an environment place represents an environment player. Specific to Petri games is the notion of informedness of the players. As long as the players move in independent parts of the net, they do not know of each other; when they synchronize at a joint transition, each player gets informed of the entire causal history of the other players. Petri games are equipped with a global safety condition, which is formalized as a set of 'bad' markings, i.e., sets of places which must never be reached simultaneously by the players.

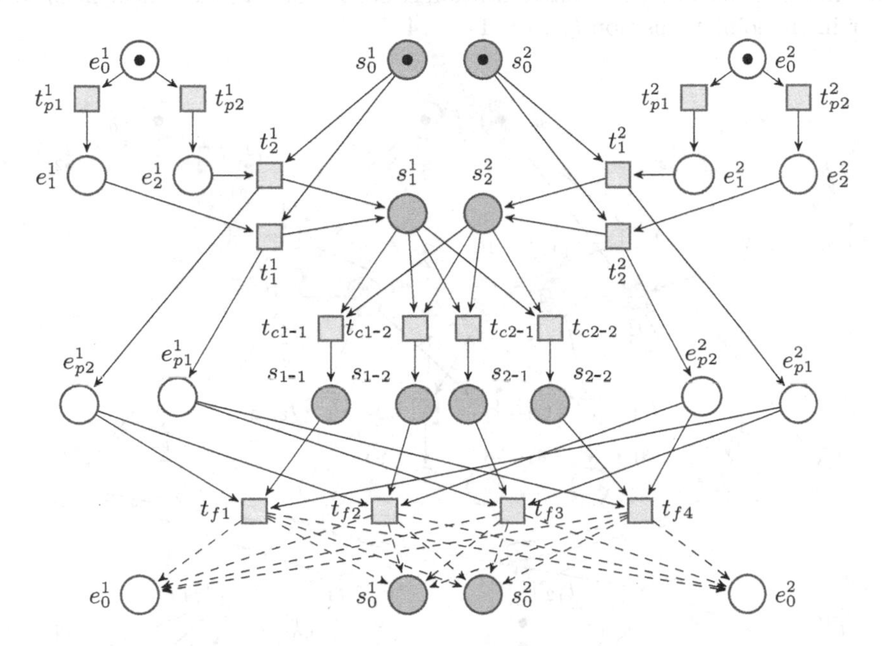

Fig. 1. A Petri game: the grey places belong to the system players and the white places to the environment players. The two environment players have two ports each to choose via t_{p1}^i and t_{p2}^i, $i = 1, 2$. The goal for the system players is to connect the chosen ports such that the environment players can communicate; every marking containing a place $s_{i\text{-}j}$ and a place $e_{p_k}^1$ with $k \neq i$ or a place $e_{p_l}^2$ with $l \neq j$ is a bad marking. The outgoing edges of the transitions t_{fi}, $i = 1, \ldots, 4$, putting the tokens back on the initial places are shown with dotted lines to keep the Petri net overseeable; those are the same places.

Petri games have been introduced in [13], where it has been shown that, limiting the environment to one player, the synthesis problem is EXPTIME-complete [13]. The dual case, limiting the system to one player, is also EXPTIME-complete [12]. For Petri games with unboundedly many players the synthesis problem is undecidable in general [13]. An approach of limiting the size of the winning strategies to search for is presented as *bounded synthesis* in [9].

In this paper, we introduce a subclass of Petri games with finitely many system players and finitely many environment players that satisfy a synchronisation condition which ensures that all players hear from each other directly or indirectly within a bounded number of own transitions, or that the Petri net is acyclic. Games with an arbitrary number of system and environment players are key to specifying many realistic distributed reactive systems. We show that for Petri games in this subclass the synthesis problem is decidable in EXPTIME.

Figure 1 shows a Petri game in an abstract communication setting, where two environment players have two ports each to connect to via transitions. The system players have to link the correct ports after knowing the choice of each environment player. If done so, the environment players communicate over the connected channel and every token is put back to the initial places. The Petri game in Fig. 1 satisfies the synchronisation condition as all tokens hear from each other in the joint transition t_{fi}, $i = 1, \ldots, 4$.

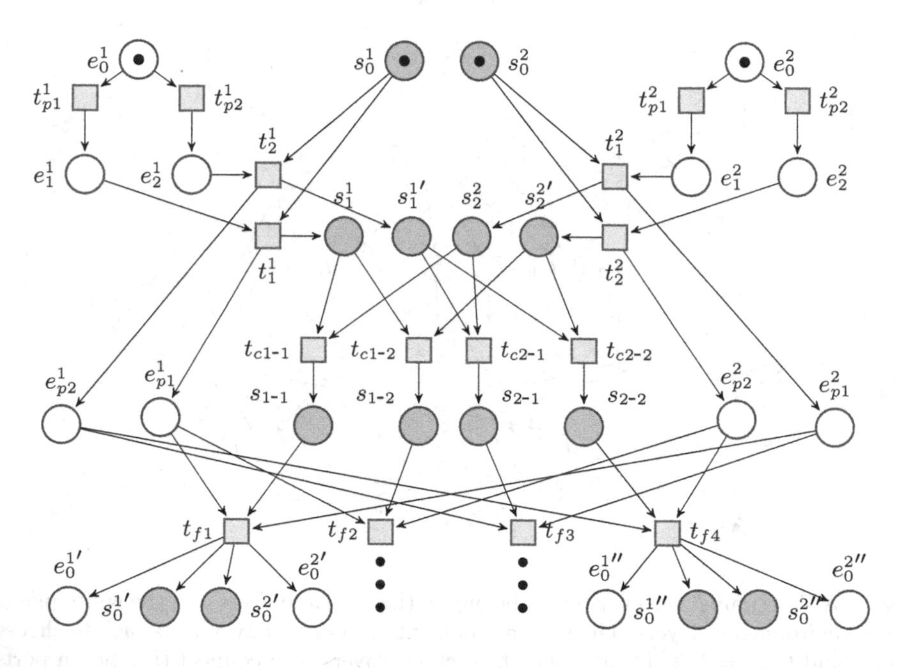

Fig. 2. An initial part of a winning strategy of the Petri game in Fig. 1. This is a part of the unfolding of the Petri net. The places $e_0^{1'}$, $s_0^{1'}$, $s_0^{1'}$, $s_0^{1'}$ are new instances of the initial places.

Figure 2 shows an initial part of the described winning strategy. The winning strategy is a part of the unfolding of the Petri net, which is itself a Petri net, where every flow of tokens through the net is represented by distinct new places and transitions: the token initially on place s_0^1 can distinguish its causality on the places s_1^1 and $s_1^{1'}$: on place s_1^1 the system players knows that the environment player is connected to port one and in $s_1^{1'}$ to port two. It does not know the port the other environment player connected to. From there on, the system players can choose from four transitions to connect the ports. In the winning strategy, only the correct option is chosen based on the causality.

The presented solution of the synthesis problem of the described subclass of Petri games is a reduction to a two-player graph game with a safety condition. The idea for the system player is to plan so far ahead that no information, that is not yet available in the Petri net unfolding, is used for the decisions made in the winning strategy.

The remainder of this paper is organized as follows. In Sect. 2 we introduce the notions of Petri games and graph games. In Sect. 3 we define the reduction to graph games and show its correctness. Related work and conclusions are presented in Sect. 4 and Sect. 5.

2 Foundations

In this section, we define branching processes and unfoldings as they are defined in [7]. Also, we define Petri games and their winning strategies.

The *power set* for a set A is denoted as $2^A = \{B \mid B \subseteq A\}$. The *set of nonempty finite subsets* for a set A is denoted as $2^A_{nf} = \{B \mid B \subseteq A \wedge B \text{ is nonempty and finite}\}$ and the *set of finite subsets* as 2^A_f. A *Petri net* is a 5-tuple $(\mathcal{P}, \mathcal{T}, pre, post, In)$, where $\mathcal{P}$ is the set of *places*, $\mathcal{T}$ is the set of *transitions*, *pre* and *post* are flow mappings, *In* is the *initial marking* and the following properties hold: $\mathcal{P} \cap \mathcal{T} = \emptyset$, $pre :\mapsto 2^{\mathcal{P}}_{nf}$, $post : \mathcal{T} \mapsto 2^{\mathcal{P}}_f$, and $In \subseteq \mathcal{P}$ is the initial marking. A Petri net is called finite if $\mathcal{P} \cup \mathcal{T}$ is finite. The flow mappings *pre* and *post* are extended for places as usual: $\forall p \in \mathcal{P} : pre(p) = \{t \in \mathcal{T} \mid p \in post(t)\}$ and $\forall p \in \mathcal{P} : post(p) = \{t \in \mathcal{T} \mid p \in pre(t)\}$. The *flow relation* $\mathcal{F}$ is defined as $\mathcal{F} = \{(p,t) \in \mathcal{P} \times \mathcal{T} \mid p \in pre(t)\} \cup \{(t,p) \in \mathcal{T} \times \mathcal{P} \mid p \in post(t)\}$. A *marking* M of a Petri net $\mathcal{N}$ is a multi-set over $\mathcal{P}$. In particular, *In* is a marking. A Petri net N is called *safe*, if for all reachable markings $M(p) \leq 1$ for all $p \in \mathcal{P}$ holds. Then, M is a subset of $\mathcal{P}$.

A transition $t \in \mathcal{T}$ is *enabled* in marking M, if $pre(t) \subseteq M$. If t is enabled, the transition t can be *fired*, such that the new marking is $M' = M \setminus pre(t) \cup post(t)$. This is denoted as $M|t\rangle M'$. The marking M' is also denoted by $M|t\rangle$. This notation is extended to sequences of enabled transitions $M|t_1 \dots t_n\rangle M'$ and $M|t_1 \dots t_n\rangle$, respectively A marking M is *reachable*, if there exists a sequence of enabled transitions $(t_k)_{k=\{1,\dots,n\}}$ and $In|t_1 \dots t_n\rangle M$. This sequence can be empty. The set of all reachable markings is denoted as $\mathcal{R}(N)$, where N is a Petri net. We call a *node* x a place or a transition $x \in \mathcal{P} \cup \mathcal{T}$.

The following definitions are essential for understanding this paper and are also taken from [7]. A node $x \in \mathcal{P} \cup \mathcal{T}$ is a *causal predecessor* of y, denoted as $x \leq y$, if there is a sequence $(x_i, y_i)_{i=1,\dots,n}$, $(x_i, y_i) \in \mathcal{F}$, where $x_1 = x$, $y_n = y$ and $y_i = x_{i+1}$ forall $i = 1, \dots, n-1$. Furthermore, $x \leq x$ holds for all $x \in \mathcal{P} \cup \mathcal{T}$. Two nodes $x, y \in \mathcal{P} \cup \mathcal{T}$ are *causally related*, if and only if $x \leq y$ or $y \leq x$ holds. We say x is a *causal successor* of y, if and only if $y \leq x$ holds. We define the *past* of a node in an occurrence net as the set of all causal predecessors $Past(x) = \{y \in \mathcal{P} \cup \mathcal{T} \mid y \leq x\}$ as this is used in later definitions.

Two nodes $x_1, x_2 \in \mathcal{P} \cup \mathcal{T}$ are *in conflict*, denoted $x_1 \# x_2$, if there exist two transitions $t_1, t_2 \in \mathcal{T}$, $t_1 \neq t_2$ with $pre(t_1) \cap pre(t_2) \neq \emptyset$ and $t_i \leq x_i$, $i = 1, 2$. A node $x \in \mathcal{P} \cup \mathcal{T}$ is in self-conflict, if $x \# x$. Informally speaking, two nodes are in conflict if two transitions exist that share some place in their presets and each node is a causal successor of one of those transitions. Two nodes $x, y \in \mathcal{P} \cup \mathcal{T}$ are *concurrent*, denoted $x || y$, if they are neither causally related nor in conflict.

A Petri net N is *finitely preceded*, if for every node $x \in \mathcal{P} \cup \mathcal{T}$ the set $Past(x)$ is finite. A Petri net N is *acyclic*, if the directed graph $(\mathcal{P} \cup \mathcal{T}, \mathcal{F})$ is acyclic. The two following definitions lead to the definition of a branching process, which represents several runs of the underlying Petri net taken together. A run is represented by a (possibly infinite) firing sequence of transitions. An *occurrence net* is a Petri net $N = (\mathcal{P}, \mathcal{T}, pre, post, In)$ with the following properties: N is acyclic, finitely preceded, $\forall p \in \mathcal{P} : |pre(p)| \leq 1$, no transition $t \in \mathcal{T}$ is in self-conflict, and $In = \{p \in \mathcal{P} | \, pre(p) = \emptyset\}$.

A homomorphism from one Petri net to another maps each node to a node such that the preset and postset relations are preserved including the initial marking. Let $N_1 = (\mathcal{P}_1, \mathcal{T}_1, pre_1, post_1, In_1)$ and $N_2 = (\mathcal{P}_2, \mathcal{T}_2, pre_2, post_2, In_2)$ be two Petri nets. A *homomorphism* from N_1 to N_2 is a mapping $h : \mathcal{P}_1 \cup \mathcal{T}_1 \rightarrow \mathcal{P}_2 \cup \mathcal{T}_2$ with following properties: $h(\mathcal{P}_1) \subseteq \mathcal{P}_2$ and $h(\mathcal{T}_1) \subseteq \mathcal{T}_2$, for all transitions $t \in \mathcal{T}_1$, h restricted to $pre_1(t)$ is a bijection between $pre_1(t)$ and $pre_2(h(t))$, for all transitions $t \in \mathcal{T}_1$, h restricted to $post_1(t)$ is a bijection between $post_1(t)$ and $post_2(h(t))$, and the restriction of h to In_1 is a bijection between In_1 and In_2. An *isomorphism* is an bijective homomorphism.

The previous definitions now lead to the definition of a branching process of a Petri net resembling multiple runs of the Petri net.

Branching Process. Let $N_0 = (\mathcal{P}_0, \mathcal{T}_0, pre_0, post_0, In_0)$ be a Petri net. A *branching process* of N_0 is a pair $B = (N, \pi)$, where $N = (\mathcal{P}, \mathcal{T}, pre, post, In)$ is an occurrence net and π a homomorphism from N to N_0 such that:

(*) For all $t_1, t_2 \in \mathcal{T}$: if $pre(t_1) = pre(t_2)$ and $\pi(t_1) = \pi(t_2)$, then $t_1 = t_2$.

The notion of the set of all reachable markings of a branching process $B = (N, \pi)$ is extended to $\mathcal{R}(B) = \mathcal{R}(N)$.

The property $(*)$ of the definition of a branching process ensures that every run of the Petri net is represented at most once. Informally speaking, a run only consists of concurrent and causally related nodes and a node can be part of multiple runs. Nodes that are in conflict, cannot belong to the same run.

Homomorphism on Branching Processes. Given two branching processes $B_1 = (N_1, \pi_1)$ and $B_2 = (N_2, \pi_2)$ of a Petri net N_0. A *homomorphism* from B_1 to B_2 is a homomorphism h from N_1 to N_2 such that $\pi_2 \circ h = \pi_1$. It is called an *isomorphism* if h is an isomorphism. The branching processes B_1 and B_2 are *isomorphic* if there exists an isomorphism from B_1 to B_2.

As the names of the nodes of isomorphic branching processes may differ we define the set of canonical names such that every branching process is isomorph to a branching process with canonical names.

Canonical Names. Let $N = (\mathcal{P}, \mathcal{T}, pre, post, In)$ be a Petri net, the set of *canonical Names* CAN is the smallest set such that, if $x \in \mathcal{P} \cup \mathcal{T}$ and A is a finite subset of CAN, then $(x, A) \in CAN$.

A canonical name of a node in a branching process is composed of the label of the node and the set of the canonical names of all nodes in its preset. The label of a node x is the name of the node in the underlying Petri net, denoted as $\pi(x)$.

Canonical Coding. Let $B = (N, \pi)$ be a branching process of N_0 and $N = (\mathcal{P}, \mathcal{T}, pre, post, In)$. Then the *canonical coding* of B is a mapping $cod_B : \mathcal{P} \cup \mathcal{T} \to CAN$ with $cod_B(x) = (\pi(x), cod_B(pre(x)))$. Note that $pre(x)$ is a set and thus $cod_B(pre(x)) = \{cod_B(y) | \, y \in pre(x)\}$. Consequently, a branching process is called *canonical* if all nodes have canonical names.

Canonical Branching Process. A branching process $B = (N, \pi)$ is called *canonical*, if $\mathcal{P} \cup \mathcal{T} \subseteq CAN$ and $cod_B(x) = x$ for all $x \in \mathcal{P} \cup \mathcal{T}$.

Note that the definition of a canonical branching process requires the initial marking of the branching process to be a set rather than a multi set.

A natural partial order on branching processes is defined in the following.

Subprocess Relation of Branching Processes. Let B_1 and B_2 be two branching processes of a Petri net $N = (\mathcal{P}, \mathcal{T}, pre, post, In)$. Then B_1 approximates B_2, denoted $B_1 \leq B_2$, if there exists an injective homomorphism denoted $h_{\leq}$ from B_1 to B_2.

Note that this partial order is independent of a branching processes being canonical. Restricting this partial order to canonical branching processes results in a partial order, too. Now we define the maximal canonical branching process as the unfolding of a Petri net.

Unfolding. A branching process $B = (N, \pi)$ with $N = (\mathcal{P}, \mathcal{T}, pre, post, In)$ is isomorphic to the unfolding $unf(N_0)$ of the underlying Petri net $N_0 = (\mathcal{P}_0, \mathcal{T}_0, pre_0, post_0, In_0)$ if and only if the following holds: For all transitions $t_0 \in \mathcal{T}_0$ and all sets $C \subseteq \mathcal{P}$ of pairwise concurrent places exists $t \in \mathcal{T}$ with $pre(t) = C$ and $\pi(t) = t_0$, if the restriction of π to C is a bijection between C and $pre_0(t_0)$. The notation $unf(N_0)$ denotes the canonical unfolding. We refer to the components of the unfolding as $\mathcal{T}_{unf(N_0)}$, $\mathcal{P}_{unf(N_0)}$, $pre_{unf(N_0)}$, $post_{unf(N_0)}$, and $In_{unf(N_0)}$.

As we only consider Petri games of finite and safe Petri nets in this paper, these properties are part of the definition of a Petri game. Note that we allow tokens to be generated or deleted. Also, we allow tokens to transition from a system place to an environment place and vice versa.

Definition 1 (Petri game). *A* Petri-game *of an underlying finite and safe Petri net N is a tuple $G = (\mathcal{P}^S, \mathcal{P}^E, \mathcal{T}, pre, post, In, \mathcal{B})$, where the places are two disjoint sets $\mathcal{P}^S$, called the* system places *and $\mathcal{P}^E$, called the* environment places *with* $\mathcal{P}^S \cup \mathcal{P}^E = P$*. The sets* $\mathcal{P}$ *and* $\mathcal{T}$ *are finite.* $\mathcal{B}$ *is the set of* bad markings.

We assume the underlying Petri net of a Petri game to be safe for formal simplicity of the canonical branching processes. Note that a finite and safe Petri net has a bounded number of places in every reachable marking.

Now we can define a winning strategy as a branching process of the underlying Petri net, that satisfies four properties.

Definition 2 (Winning strategy). *A* winning strategy *of a Petri-game* $G_P = (\mathcal{P}_0^S, \mathcal{P}_0^E, \mathcal{T}_0, pre_0, post_0, In_0, \mathcal{B})$ *with underlying Petri-net* $N_0 = (\mathcal{P}_0, \mathcal{T}_0, pre_0, post_0, In_0)$ *is a branching process* $B = (N, \pi)$ *of* N_0 *with* $N = (\mathcal{P}, \mathcal{T}, pre, post, In)$ *and the following properties.*

1. ***Justified refusal:*** *Let* $C \subseteq \mathcal{P}$ *be a set of pairwise concurrent places and* $t \in \mathcal{T}_0$ *a transition with* $\pi(C) = pre_0(t)$*. If no* $t' \in \mathcal{T}$ *with* $\pi(t') = t$ *and* $pre(t') = C$ *exists, then there exists a place* $p \in C$ *with* $\pi(p) \in \mathcal{P}_0^S$*, such that* $t \notin \pi(post(p))$*.*
2. ***Safety:*** *For all reachable markings* M *in* N *holds* $\pi(M) \notin \mathcal{B}$*.*
3. ***Determinism:*** *For all* $p \in \mathcal{P}$ *with* $\pi(p) \in \mathcal{P}_0^S$ *and for all reachable markings* M *in* N *with* $p \in M$ *exists at most one transition* $t \in post(p)$*, which is enabled in* M*.*
4. ***Deadlock avoiding:*** *For all reachable markings* M *in* N *exists an enabled transition, if a transition is enabled in* $\pi(M)$ *in the underlying Petri-net* N_0*.*

We fix the notations $G_P = (\mathcal{P}_0^S, \mathcal{P}_0^E, \mathcal{T}_0, pre_0, post_0, In_0, \mathcal{B})$ *of the Petri game* G_P *and* $N_0 = (\mathcal{P}_0, \mathcal{T}_0, pre_0, post_0, In_0)$ *of the underlying Petri net* N_0*.*

The four properties of a winning strategy can be interpreted as follows: The *justified refusal* property forces the system player in each place to allow all instances of an outgoing transition t or no instance at all. This enables the representation of the decisions of the system player as commitment sets in its places: each transition is allowed for every possible instance or it is forbidden at all. The *safety* property ensures that no bad markings are reachable. The *determinism* property ensures that for each system place at most one transition is enabled in every reachable marking. The *deadlock avoiding* property ensures that the system allows at least one transition in every reachable marking if an enabled transition exists in that marking.

2.1 Graph Games

A graph game is a two player game with perfect information played on a directed graph, called an arena. The vertices in the arena divide into system vertices and environment vertices. The system player, who chooses the next move in system vertices, is referred to as player 0 and the environment player, who chooses the next move in environment vertices, is player 1.

An *arena* $A = (V, V_0, V_1, E)$ consists of a finite set V of vertices, disjoint subsets $V_0, V_1 \subseteq V$ with $V = V_0 \cup V_1$ denoting the vertices of player 0 and player

1, a set $E \subseteq V \times V$ of (directed) edges such that every vertex has at least one outgoing edge, i.e., $\{v_0 \mid (v, v_0) \in E\}$ is non-empty for every $v \in V$. The size of A, denoted by $|A|$, is defined to be $|V|$. A *play* in an arena $A = (V, V_0, V_1, E)$ is an infinite sequence of vertices $\mu = v_1 v_2 v_3 \ldots \in V^\omega$ such that $(v_n, v_{n+1}) \in E$ holds for every $n \in \mathbb{N}$. We say μ starts in the vertex v_1. The set of plays in A is denoted by $Plays(A)$, the set of all plays starting in v by $Plays(A, v)$, and we define $Plays(A, V') = \bigcup_{v \in V'} Plays(A, v)$ for every $V' \subseteq V$.

A strategy of a player determines the next vertex of a play, if the current vertex belongs to the player. A *strategy* for Player $i \in \{0, 1\}$ in an arena (V, V_0, V_1, E) is a function $\sigma : V^* V_i \mapsto V$ such that $\sigma(wv) = v'$ implies $(v, v') \in E$ for every $w \in V^*$ and every $v \in V_i$. The set of plays obtained by following a strategy is the set of consistent plays. A play $v_1 v_2 v_3 \ldots$ in an arena $A = (V, V_0, V_1, E)$ is consistent with a strategy σ for Player i in A if $v_{n+1} = \sigma(v_1 \ldots v_n)$ for every $n \in \mathbb{N}$ with $v_n \in V_i$. Given a vertex v, we denote the set of plays that are consistent with σ and start in v with $Plays(A, v, \sigma)$. Finally, we define $Plays(A, V', \sigma)$ for $V' \subseteq V$ by $Plays(A, V', \sigma) = \bigcup_{v \in V'} Plays(A, v, \sigma)$.

The winning condition in a safety graph game is to remain in safe vertices. The set of safe vertices is a subset of all vertices. A *safety graph game* $G = (A, S)$ consists of an arena A with vertex set V and a set of safe vertices $S \subseteq V$. We call a sequence μ winning for Player 0 if, and only if $Occ(\mu) \subseteq S$. $Occ(\mu)$ denotes all vertices occurring in μ: $Occ(\mu) := \{v \in V \mid \exists n \in \mathbb{N} : v_n = v\}$. A strategy is winning, if all consistent plays remain in safe vertices at all time: A strategy σ of a safety graph game $G = (A, S)$ is called winning in a vertex $v \in V$, if all consistent plays $\mu \in Plays(A, v, \sigma)$ are winning.

A strategy σ for Player i in an arena (V, V_0, V_1, E) is *positional* if $\sigma(wv) = \sigma(v)$ for all $w \in V^*$ and $v \in V_i$. Safety graph games are determined with positional winning strategies. These strategies are called positional because they do not need any memory of the vertices visited so far.

Safety graph games can be solved with the standard attractor construction in linear time in the number of edges of the underlying arena [2].

3 Reduction of Petri Games to Graph Games

In this section, we reduce Petri games, where the underlying Petri net satisfies a synchronisation condition, to safety graph games. We call the synchronisation condition non-simultaneous synchronisation condition as the players do not need to take one joint transition. They do need to be causally dependent from each other directly or indirectly after they take a bounded number of transitions. The only exception occurs, if there are only finite firing sequences of transitions in the Petri net meaning the Petri net is acyclic.

Informally speaking the non-simultaneous synchronisation condition defined in the following expresses that every token in the Petri net has to take a transition within firing at most n transitions or no transition is enabled anymore after $n - 1$ transitions. This ensures that every token hears from every other token directly or indirectly after firing finitely many transitions, if there are still enabled transitions in the Petri net.

Definition 3 (Non-simultaneous synchronisation condition). *A Petri net N satisfies the non-simultaneous synchronisation condition if and only if there exists a bound $n \in \mathbb{N}$ such that*

$$\forall M \in \mathcal{R}(N) : \forall s \in M : \neg \exists t_1 \dots t_n : M[t_1 \dots t_n\rangle M' \wedge \forall k = 1, \dots n : s \notin \mathit{pre}(t_k).$$

This bound is unrelated to the number of tokens in the Petri net. An equivalent characterisation of Petri nets that satisfy this synchronisation condition is to bound the number of concurrent transitions for every place in the unfolding of a Petri net. A Petri net satisfies the non-simultaneous synchronisation if and only if there exists a bound $m \in \mathbb{N}$ such that for all places $p \in \mathit{unf}(N)$ the set of concurrent transitions in the unfolding is bounded by m, i.e. $|\{t \in \mathcal{T}_{\mathit{unf}(N)} \mid p||t\}| \leq m$.

The remainder of this section is structured as follows: in Subsect. 3.1 we introduce further definitions for Petri games such that we can express a winning strategy in a Petri game to be planned part by part in a graph game. In Subsect. 3.2 we define the graph game to which a Petri game is reduced. In Subsect. 3.3 we show how to construct a winning Petri game strategy if the graph game has a winning strategy and vice versa.

3.1 Extended Petri Game Semantics

In the following, we introduce another synchronisation condition, the local synchronisation condition, that is defined for the nodes in a branching process as opposed to the non-simultaneous synchronisation condition that is defined for a Petri net.

The reduction from Petri games to graph games, where both players have perfect information, needs to ensure that the decisions of the system player in the graph game do not rely on information it would not have in the Petri game. Therefore, we introduce a local synchronisation condition that ensures that the system player has sufficient information in the graph game. The idea is that the system player plans so far ahead until it meets the local synchronisation condition for all tokens such that it cannot abuse information in the graph game it would not have in the Petri game. We define the set of enabled transitions in a marking of a Petri net and the local synchronisation condition dependent on this set as follows. The set of the enabled transitions in a marking M of a Petri net is defined as $\mathcal{T}_M^{En} = \{t \in \mathcal{T} \mid t \text{ is enabled in } M\}$. The notation $\mathcal{T}_B^{En}$ is used for a branching process B. This is the set of enabled transitions in the initial marking of B.

Definition 4 (Local synchronisation condition (abbreviated LSC)). *For a branching process $B = (N, \pi)$ of a Petri net N_0 the* local synchronisation condition, *abbreviated LSC, is defined for a node $x \in \mathcal{P} \cup \mathcal{T}$ as follows: x satisfies the LSC if*

$$\forall t \in \mathcal{T}_B^{En} : t \in \mathit{Past}(x) \vee x \# t.$$

The local synchronisation condition is called local because the causal past of a node determines if it is satisfied: a node satisfies the local synchronisation condition if it has information of all enabled transitions in the initial marking whether they have been fired or not. The information that one of those transitions was not fired is equal to having a node in conflict to that transition in the past of the node that satisfies the local synchronisation condition.

In the graph game, the system player and environment player take alternate turns. The system player plans a branching process, i.e. the system player determines its strategy part by part. Then, the environment player chooses a set of enabled and pairwise concurrent transitions that are fired. Afterwards, the system player has to plan further ahead, and so on. We define firing transitions in a branching process which results in a branching process containing all remaining nodes that were not in conflict to a fired transition or a causal predecessor of fired transition. As the initial marking of this new branching process changed and therefore it might not be a branching process by definition, we extend the set of branching process to those starting in an arbitrary reachable marking of the underlying Petri net.

Definition 5 (Extended set of branching processes of a Petri net). *An extended branching process $B = (N, \pi)$ of a safe Petri net $N_0 = (\mathcal{P}_0, \mathcal{T}_0, pre_0, post_0, In_0)$ is a branching process of a Petri net $N_0' = (\mathcal{P}_0, \mathcal{T}_0, pre_0, post_0, M)$, where $M \in \mathcal{R}(N_0)$. The set of all extended branching processes of a Petri net N_0 or a Petri Game G_P is denoted as $EB(N_0)$ and $EB(G_P)$, respectively. The notion of the unfolding of N_0' as an extended branching process is added as $unf_M(N_0)$.*

In the following, we define firing a set of pairwise concurrent transitions in an extended branching process.

Definition 6 (Firing transitions in extended branching processes). *Let $B = (N, \pi)$ be an extended branching process with $N = (\mathcal{P}, \mathcal{T}, pre, post, In)$ and $T_f \subseteq T_B^{En}$ a set of pairwise concurrent transitions. Then $B[T_f\rangle B'$ denotes the firing of all transitions in T_f in an arbitrary order, resulting in the extended branching process*

$$\begin{aligned} B' = &((\mathcal{P} \setminus \{s \in \mathcal{P} \mid s \in pre(T_f) \vee \exists t' \in T_f : t' \# s\}, \\ &\mathcal{T} \setminus \{t \in \mathcal{T} \mid t \in T_f \vee \exists t' \in T_f : t' \# t\}, pre\restriction_{\mathcal{T}'}, post\restriction_{\mathcal{T}'}, M[T_f\rangle)), \pi\restriction_{\mathcal{T}' \cup \mathcal{P}'}), \end{aligned}$$

where the components of B' are referred to as $\mathcal{P}'$, $\mathcal{T}'$, pre', $post'$, In' and π', and $pre\restriction_{\mathcal{T}'}$ and $post\restriction_{\mathcal{T}'}$ are the restrictions of pre and post to the transitions $\mathcal{T}'$ of B', and $\pi\restriction_{\mathcal{T}' \cup \mathcal{P}'}$ the restriction of π to $\mathcal{T}' \cup \mathcal{P}'$.

The branching process B' is also denoted as $B[T_f\rangle$. We extend these notations for sequences of sets of concurrent transitions as $B[T_{f_1} \dots T_{f_n}\rangle B'$ and $B[T_{f_1} \dots T_{f_n}\rangle$ respectively. Here, $M[T_f\rangle$ denotes the marking reached after firing all transitions in T_f in an arbitrary order. Let $\tilde{B}' \cong B'$ denote the canonical branching process, which is isomorph to B' via an isomorphism $\Phi : B' \mapsto \tilde{B}'$.

To ensure that the decisions of the system player which transitions are allowed to be fired in each place are final, such that the system does not change its decisions when it has more information, we define a commitment set mapping for an extended branching process, which states for every place the set of transitions which are allowed to be fired. This ensures the *justified refusal* property of a winning strategy in a Petri game. For environment places the allowed transitions are not restricted. In the graph game, the system player has to choose a commitment set for every system place, when it is added, and this commitment set is kept the same from there on.

Definition 7 (Commitment set mapping). *Given a Petri game G_P, a commitment set mapping CS of an extended branching process $B = (N, \pi) \in EB(G_P)$, where $N = (\mathcal{P}, \mathcal{T}, pre, post, M)$, is a mapping with the following properties:*

$$\begin{aligned} CS :& \mathcal{P} \rightarrow 2^{\mathcal{T}_0} \\ & \forall s \in \mathcal{P} : CS(s) \subseteq post_0(\pi(s)) \\ & \wedge (\pi(s) \in \mathcal{P}_0^E \Rightarrow CS(s) = post_0(\pi(s))) \end{aligned}$$

For example, the commitment set of the system place s_1^1 in Fig. 2 would be $\{t_{\text{c1-1}}, t_{\text{c1-2}}\}$ as it needs to connect port one of the first environment player to either port one or port two of the other environment player.

In the following definition we extend the subprocess relation $\leq$ to a subprocess relation $\leq_{CS'}^{CS}$ respecting given commitment set mappings of extended branching processes. This means that the commitment sets are preserved under the subprocess homomorphism $h_{\leq}$.

Definition 8 (Subprocess relation of extended branching processes with commitment set mapping). *Let B_1 and B_2 be extended branching processes of a safe Petri net N_0 with commitment set mappings CS_1 and CS_2. B_1 approximates B_2 with respect to the commitment set mappings CS_1 and CS_2, denoted $B_1 \leq_{CS_2}^{CS_1} B_2$, if and only if $B_1 \leq B_2$ and for all $p \in \mathcal{P}_1$ for the commitment set $CS_1(p) = CS_2(h_{\leq}(p))$ holds.*

Note that this defines a partial order on tuples of extended branching processes and their commitment set mappings.

Now, we can define planning segments as tuples of extended branching processes and a commitment set mapping, such that every node necessary to satisfy the local synchronisation condition for every maximally progressed place is added and no more nodes are included. We call a place maximally progressed, if and only if it satisfies the LSC or has an empty postset. Thus, either the LSC is satisfied or no further transitions are allowed in the commitment set mapping. Exactly those transitions which are allowed in the commitments set are added until the LSC is satisfied.

Definition 9 (Planning segment). *A tuple (B, CS) consisting of an extended branching process $B = (N, \pi)$ with $N = (\mathcal{P}, \mathcal{T}, pre, post, M)$ of a Petri game G_P*

with underlying Petri net N_0 *and a commitment set mapping* CS *of* B *is a* planning segment, *if and only if the following holds:*

$$\forall t \in \mathcal{T} : \forall p \in pre(t) : \pi(t) \in CS(p) \\ \wedge \exists t' \in \mathcal{T} : (t \leq t' \wedge \exists s \in pre(t') : s \textit{ does not satisfy the LSC})) \quad (1)$$

$$\neg \exists (B', CS') : (B', CS') \textit{ satisfies } (1) \wedge B \leq^{CS}_{CS'} B' \quad (2)$$

Let $PSeg(G)$ *denote the set of all planning segments of a Petri game* G*. The notation* T^{En}_v *is used for a planning segment* $v \in PSeg(G)$*. This is the set of enabled transitions in the initial marking of its branching process* B*.*

Note: The second part of property (1): $\exists t' \in \mathcal{T} : (t \leq t' \wedge \exists s \in pre(t') : s$ does not satisfy LSC) ensures that every added transitions is necessary to satisfy the LSC. Thereby, it allows transitions, that are causal predecessors of other transitions, where the preset does not satisfy the local synchronisation condition, to be added despite the possibility that all places in the preset of the transition itself already satisfy the local synchronisation condition. The property (2) of such a tuple ensures that all such transitions are added by requiring the tuple to be maximal.

3.2 Corresponding Graph Game

The set of all planning segments is used as the set of vertices of the environment player in the graph game. In the following, we define decision sets that are the possible decisions for the environment player in one of its vertices. This means that every decision set corresponds to an outgoing edge from an environment vertex in the graph game. A decision set consists of a set of concurrent transitions that are enabled in the current marking. The environment player chooses to fire these transitions.

Definition 10 (Decision sets). *For a Petri game* G_P *and a planning segment* $(B, CS) \in PSeg(G)$ *with initial marking* In *of* B*, the set of decision sets for the environment player is defined as follows:*

$$DSets((B, CS)) = \{DS \subseteq T^{En}_{In} \mid \forall t_i, t_j \in DS, t_i \neq t_j : t_i || t_j\}$$

An example of a branching process of a planning segment is shown in Fig. 2, if the places in the post sets of the transitions t_{f2} and t_{f3} are added, which are indicated with vertical dots. The places after firing a transition t_{fi}, $i = 1, \ldots, 4$, satisfy the local synchronisation condition as they know which port each environment player chose. The branching process in Fig. 2 is a planning segment, if a suitable commitment set mapping is added.

In the following, we define deterministic and deadlock avoiding planning segments. The intuition here is quite similar to the determinism and deadlock avoidance of a winning strategy in a Petri game: eventually, assuming exactly those transitions allowed in the commitment sets are added to the branching process the definition is equal for the places and transitions that are actually in the branching process. The definitions are not only for planning segments but for extended branching processes with a commitment set mapping.

Definition 11 (Deterministic and deadlock avoiding planning segments). *Let G_P be a Petri game. An extended branching process with a commitment set mapping (B, CS), where $B = (N, \pi)$ and $N = (\mathcal{P}, \mathcal{T}, pre, post, In)$, is called* deterministic *in a reachable marking $M \in \mathcal{R}(B)$, if for all $p \in M$ with $\pi(p) \in \mathcal{P}^S$ exist at most one transition $t \in CS(p)$, where $pre_0(t) \subseteq \pi(M)$ and for all $p \in \pi\restriction_M^{-1}(pre_0(t)) : t \in CS(p)$ holds.*

An extended branching process with a commitment set mapping (B, CS) is called deadlock avoiding *in a reachable marking $M \in \mathcal{R}(B)$ if and only if the following holds: If a transition $t \in \mathcal{T}_0$ is enabled in $\pi(M)$, then there exists a transition $t' \in \mathcal{T}_0$, where for all $p \in \pi\restriction_M^{-1}(pre_0(t')) : t' \in CS(p)$ holds.*

The set of bad planning segments is defined as expected.

Definition 12 (Bad planning segments). *Let G_P be a Petri game and let $PSeg(G_P)$ denote the set of all planning segments. The set of* bad planning segments *is defined as:*

$$\begin{aligned} PSeg_{bad} = \{ & (B, CS) \in PSeg \mid \exists M \in \mathcal{R}(B) : \pi(M) \in \mathcal{B} \\ & \vee (B, CS) \text{ is not deterministic or not deadlock avoiding in } M.\} \end{aligned}$$

Now, turn to the graph game. We define the starting vertices of the system player in the graph game. A winning strategy has to be winning starting in one of those vertices to construct a winning strategy in the Petri game later on. A starting vertex of the graph game consists of the branching process with just the places of the initial marking, their arbitrary commitment sets and no transitions. Note the pre-image of *pre* and *post* are empty.

Definition 13 (Starting vertices). *The set of starting vertices in the graph game for a Petri game $G_P = (\mathcal{P}_0^S, \mathcal{P}_0^E, \mathcal{T}_0, pre_0, post_0, In_0, \mathcal{B})$ is defined as:*

$$\begin{aligned} Start = \{ & (B = ((In_0, \emptyset, pre, post, In_0), \pi), CS) \mid \\ & CS \text{ is a commitment set mapping of } B\} \end{aligned}$$

The reduction of a Petri game to a graph game is now defined as follows: the set of environment vertices V_1 is the set of all planning segments. The set of system vertices V_0 is the union of the set of starting vertices and the set of extended branching processes with commitment set mapping, that are reached by firing a decision set of an arbitrary planning segment. The commitment sets remain the same for the remaining places. The system vertices are denoted as pairs with a 0 in the second component to distinguish them from the environment vertices; the first component of a system vertex might be a planning segment. The set of directed edges consists of three sets: First, firing a decision set resembles an edge from the planning segment to a system vertex. Second, the system player has to plan further ahead to meet the local synchronisation condition resulting in an edge from a system vertex to a planning segment. Third, the graph game loops if no transition is enabled. All edges preserve a subprocess relation in their direction. The commitment sets for all those places kept due to the subprocess relation remain the same. The set of all safe vertices are all vertices that are not bad planning segments.

Definition 14 (Corresponding graph game). *For a Petri game $G_P = (\mathcal{P}_0^S, \mathcal{P}_0^E, \mathcal{T}_0, pre_0, post_0, In_0, \mathcal{B})$ we define a safety graph game $G_{Graph} = (\mathcal{A}, S)$, $\mathcal{A} = (V, V_0, V_1, E)$ as follows:*

$$\boldsymbol{V_1} = PSeg(G_p)$$

$$\boldsymbol{V_0} = \{((\tilde{B}', CS'), 0) \mid \exists (B, CS) \in V_1 : \exists DS \in DSets((B, CS)) : \tilde{B}' = \Phi(B|DS\rangle) \wedge CS' = CS \restriction_{\mathcal{P}'} \circ \Phi^{-1}\} \cup Start \times \{0\}$$

$$\boldsymbol{E} = \{(((B, CS), 0), (B', CS') \in V_0 \times V_1 \mid B \leq_{CS'}^{CS} B'\}$$
$$\cup \{((B, CS), ((\tilde{B}', CS'), 0)) \in V_1 \times V_0 \mid \exists DS \in DSets((B, CS)) : \tilde{B}' = \Phi(B|DS\rangle) \wedge CS' = CS \restriction_{\mathcal{P}'} \circ \Phi^{-1}\}$$
$$\cup \{((B, CS), (B, CS) \in V_1 \times V_1 \mid T_B^{En} = \emptyset\}$$

$$\boldsymbol{S} = V_1 \cup V_0 \setminus PSeg_{bad}(G_p)$$

The sets $\mathcal{P}$ and $\mathcal{P}'$ denote the places of the branching processes B and B', respectively. Φ denotes the isomorphism from the branching process B' to the canonical branching process $\tilde{B}'$. We refer to this graph game as the corresponding *graph game of G.*

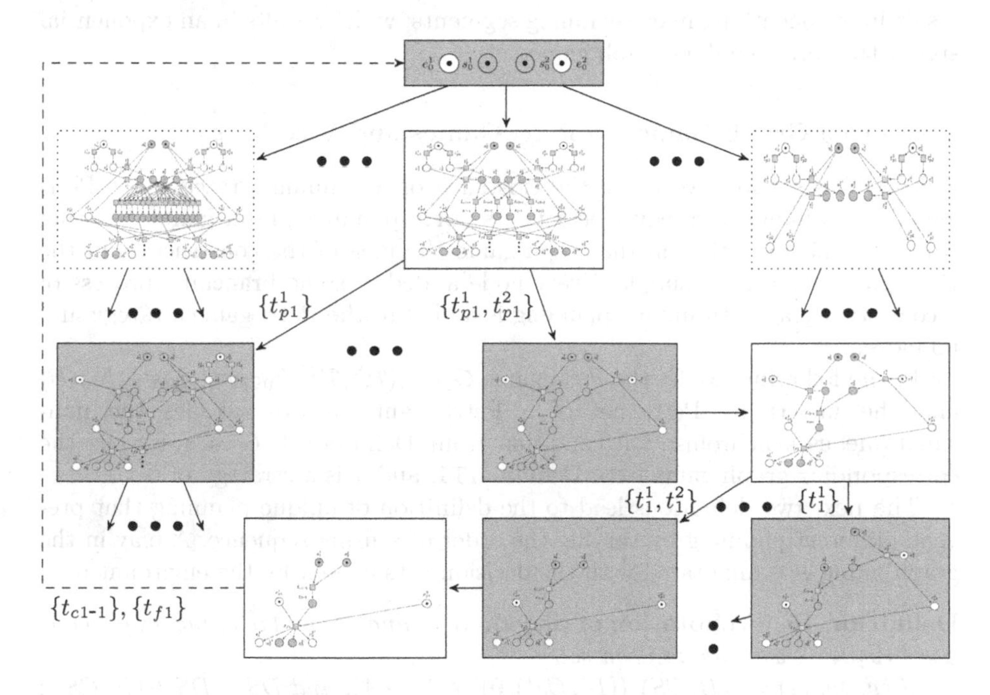

Fig. 3. A part of the corresponding safety graph game of the Petri game in Fig. 1 starting in the vertex, where the commitment set mapping allows all transitions. The commitment set mappings are implicit within the branching processes. White vertices belong to the environment player and gray to the system player. The dotted white vertices are bad planning segments. Edges are annotated with their decision sets. Decision sets with unique planning are blue, later defined in Definition 17.

The graph game is almost bipartite: As long as $\mathcal{T}_{In}^{En} \neq \emptyset$, vertices of the system player and the environment player alternate. This is not necessary when the extended branching process with its commitment set mapping of a vertex of the system player is already a planning segment again. The graph game is kept bipartite in that case for formal simplicity reasons.

In Fig. 3, a part of the corresponding graph game starting in the vertex, that allows all transitions in the initial marking, is shown. The planning segments represent all possible combinations of commitment sets, from the unfolding on the left over the initial part of a winning strategy in Fig. 2 to not allowing any transition on the right. The edges are annotated with decision sets. A blue decision set belongs to a play with unique planning, later defined in Definition 17. The play following the blue decision sets is winning. Note that the *LSC* is always satisfied after a transition t_{fi}.

Lemma 1 (Corresponding graph game has exponential size). *The corresponding graph game of a Petri game has exponential size.*

Proof (Proof sketch).[1] The corresponding graph game has exponential size as the *LSC* ensures that every token has to hear from every other token within taking n own transitions. The combinations of commitment sets in the system places result in exponentially many planning segments, which results in an exponential size of the corresponding graph game.

3.3 From Graph Games to Petri Games and Vice Versa

First in this section, we prove the existence of a winning strategy in a Petri game if a winning strategy exists in its corresponding graph game. Starting with a winning strategy in the graph game the idea of the construction of the Petri game strategy is simple: Every node added in some branching process of a consistent play with unique planning is added to the Petri game strategy and no more.

In the following, we fix the Petri game $G_P = (\mathcal{P}_0^S, \mathcal{P}_0^E, \mathcal{T}_0, pre_0, post_0, In_0, \mathcal{B})$ and the underlying Petri net of a Petri game always satisfies the non-simultaneous synchronisation condition from Definition 3, $G = (\mathcal{A}, S)$ is the corresponding graph game from Definition 14, and σ is a strategy of G.

The next two definitions lead to the definition of unique planning that prevents different planning by varying the order of a firing sequence. A play in the graph game gets annotated with the decision sets chosen by the environment:

Definition 15 (Annotation of plays). *We annotate a play of the graph game* $\mu = v_1 v_2 v_3 \dots$ *with its decision sets:*

If $(v_n, v_{n+1}) = ((B, CS), ((\tilde{B}', CS'), 0)) \in V_1 \times V_0$ *and* $DS \in DSets(B, CS)$: $\tilde{B}' = \Phi(B|DS\rangle) \wedge CS' = CS\restriction_{\mathcal{P}'} \circ \Phi^{-1}$ *we write* $v_n \xrightarrow{DS} v_{n+1}$.

[1] Full proofs will appear in an extended version of this paper on arXiv.

We define the canonical branching process of a play in the graph game as the branching process yielded by sequentially attaching the planning segments of the play. As every planning segment itself has canonical names, the names in the branching process of a play are not the same as in the planning segments in general. A subprocess homomorphism $h_{\leq}$ for each planning segment maps its places and transitions to its part of the branching process of the play. The commitment set mapping of the branching process of a play is compounded by the commitment set mappings of the planning segments.

Definition 16 (Branching process and commitment set mapping of a play). *Let $\mu = v_1 v_2 v_3 \ldots$ be a play or a prefix of a play of G and $DS_1 DS_2 \ldots$ the sequence of annotated decision sets. The branching process of the play μ, $B_\mu = (N_\mu, \pi_\mu)$, is defined as the smallest canonical branching process with respect to $\leq$ such that the sequence of annotated decision sets can be fired in B_μ. We refer to the components of N_μ as $\mathcal{P}_\mu, \mathcal{T}_\mu, pre_\mu, post_\mu$, and In_μ.*

For all prefixes $v_1 v_2 \ldots v_n$ with $v_n = (B_{v_n}, CS_{v_n}) \in V_1$ and its sequence of annotated decision sets $DS_1 DS_2 \ldots DS_m$, we define a part of the planning segment v_n in B_μ, denoted $\tilde{v}_n$, as follows: $\tilde{v}_n = (B_{\tilde{v}_n}, CS_{\tilde{v}_n})$, where $B_{\tilde{v}_n} = h_{\leq}(B)$ and B is the maximal canonical branching process with respect to $\leq$ such that $B \leq B_\mu | DS_1 DS_2 \ldots DS_m \rangle$ with the subprocess homomorphism $h_{\leq} : B \mapsto B_\mu | DS_1 DS_2 \ldots DS_m \rangle$ and $B \leq B_{v_n}$. The components of $B_{\tilde{v}_n}$ are denoted as $\mathcal{P}_{\tilde{v}_n}, \mathcal{T}_{\tilde{v}_n}, pre_{\tilde{v}_n}, post_{\tilde{v}_n}$ and $In_{\tilde{v}_n}$. For all $p \in \mathcal{P}_{\tilde{v}_n}$, the commitment set mapping $CS_{\tilde{v}_n}(p)$ is defined as $CS_{v_n}(p_n)$ if $h_{\leq}(p_n) = p$. The commitment set mapping $CS_\mu(p)$ of B_μ for all $p \in \mathcal{P}_\mu$ is defined as $CS_\mu(p) = CS_{\tilde{v}_n}(p)$, if $p \in \mathcal{P}_{\tilde{v}_n}$.

Note that the compound commitment set mapping CS_μ is well defined as already existing commitment sets are preserved by the edges in the graph game.

Now we define the set of plays with unique planning. Firing the same set of transitions in a different order can result in differently planned branching processes in the graph game. Those different branching processes might not result in a winning strategy for the Petri game when they are added together. We only consider those plays in the graph game, where every transition is fired as soon as possible or never. This means that every transition chosen in a decision set must not be enabled in an earlier planning segment.

Definition 17 (Set of plays with unique planning). *Let $Plays(G, \sigma)$ be the set of consistent plays and let $DS_1^\mu DS_2^\mu \ldots$ denote the sequence of annotated decision sets of a play $\mu \in Plays(G, \sigma)$. The set of consistent plays with unique planning is defined as*

$$\begin{aligned} Plays_u(G, \sigma) = \{&\mu \in Plays(G, \sigma) \mid \forall t \in DS_j^\mu, j \geq 2 : t \notin \mathcal{T}^{En}_{v_{DS_{j-1}^\mu}} \\ &\vee \exists t' \in DS_{j-1}^\mu : pre_{v_{DS_{j-1}^\mu}}(t') \cap pre_{v_{DS_j^\mu}}(t) \neq \emptyset\}, \end{aligned}$$

where $v_{DS_{j-1}^\mu}$ is the vertex of the play μ with outgoing annotation DS_{j-1}^μ and $pre_{v_{DS_i^\mu}}$ the preset mapping of the branching process of $v_{DS_i^\mu}$.

This definition makes use of the canonical names. An enabled transition in the initial marking has the same name for all instances of that transition. If a transition is enabled in the initial marking of a previous planning segment, we do not allow that transition in the decision set. But, we have to consider that the transition in the current planning segment can be another instance of the transition of the previous planning segment that just has the same name. This is the case, if a transition t_0 of the Petri net is enabled right again after firing itself or after firing a transition in conflict to the previous instance of t_0 such that an instance with the same name is in $\mathcal{T}^{En}_{v_{DS^{\mu}_{j-1}}}$, but the second instance cannot be fired earlier. The condition $\exists t' \in DS^{\mu}_{j-1} : pre_{v_{DS^{\mu}_{j-1}}}(t') \cap pre_{v_{DS^{\mu}_{j}}}(t) \neq \emptyset$ ensures that a second instance with the same name can be chosen in the decision set.

The idea of the construction of the winning strategy in the Petri game is to merge the branching processes of every possible and to the graph game strategy consistent unique play. We define the union of two branching processes componentwise. As the proof of the construction of a winning strategy in the Petri game is by induction, we define a winning prefix and its union naturally. The branching processes of prefixes of winning plays in the graph game are winning prefixes.

In the following we show three crucial properties of the branching process of a single play consistent to the winning strategy. The first property is that every transition has to be allowed in the commitment sets of its preset.

Lemma 2 (All transitions are allowed in commitment sets). *For the branching process B_μ of a play $\mu \in Plays(G, \sigma)$ it holds that for every transition $t \in \mathcal{T}_\mu$ for all $p \in pre_\mu(t) : t \in CS_\mu(p)$.*

Proof. As in the definition of a planning segment, every transition has to be in the commitment set of every place in its preset.

The second property is that every reachable marking in the branching process of a play is reachable in one of its planning segments.

Lemma 3 (Every reachable marking is reachable in a planning segment). *For the branching process B_μ of a play $\mu \in Plays(G, \sigma)$ it holds that for every reachable marking $M \in \mathcal{R}(B_\mu)$ exists a planning segment $v \in V_1$ in μ, where M is reachable in its part $B_{\tilde{v}}$ of the branching process B_μ.*

Proof (Proof sketch). We need to distinguish two cases. In the first case, a transition $t \in \mathcal{T}_\mu$ is enabled in M. Then we can take the prefix $v_1 v_2 \dots v_n$ of μ, where $v_n \in V_1$ is the first vertex, where In_{v_n} enables a transition $t' \in \mathcal{T}_\mu$, which is enabled in M. Every reachable marking in a branching process is a set of pairwise concurrent places and the maximally progressed places in v_n either satisfy the *LSC* or no further transitions are allowed. Thus, as those places are either not concurrent to the places in $pre(t')$ or no concurrent places can be added later on, M is reachable in $B_{\tilde{v}_n}$.

In the second case, a planning segment v_i is reached eventually, where M is the initial marking $In_{\tilde{v}_i}$.

The third property is that every transition allowed in the commitment sets of its preset is added in a branching process of some play with unique planning.

Lemma 4 (All allowed transitions are added). *Let $\mu = v_1 v_2 \ldots \in Plays_u(G, \sigma)$ be a consistent play with unique planning. If M is a reachable marking in v_i such that no transition in the past of a place of M could have been fired in a previous planning segment of μ and $\exists t_0 \in \mathcal{T}_0 : pre_0(t_0) \subseteq \pi_{v_i}(M) \wedge \forall p \in \pi_{v_i} \restriction_M^{-1} (pre_0(t_0)) : t_0 \in CS_{v_i}(p)$, then there exists a unique play $\mu' \in Plays_u(G, \sigma)$ with a same prefix $\mu^{|i|} = \mu'^{|i|}$, and a vertex v'_j that is reached in μ' with $j \geq i$ via the sequence of annotated decision sets $DS_1 \ldots DS_n$ from $v'_i \ldots v'_j$ such that there exists $t \in \mathcal{T}_{v'_j}$ with $pre_{v'_j}(t) = \Phi(\pi_{v_i} \restriction_M^{-1} (pre_0(t_0)))$ and $\pi_{v'_j}(t) = t_0$, where Φ is the isomorphism from $B|DS_1 \ldots DS_n\rangle$ to its canonical branching process.*

Proof. Let $t_0 \in \mathcal{T}_0$ be such a transition in v_i. Since M is a reachable marking in v_i, we construct a play $\mu' \in Plays_u(G, \sigma)$, where the environment player chooses its decision sets in a way that every transition in the past of any place of M is chosen as soon as possible and no transition in conflict to one of the places in M is chosen. Since B_μ is finitely preceded it follows that a planning segment v'_j, $j \geq i$, is reached, where a place $p \in \Phi(\pi_{v_i} \restriction_M^{-1} (pre_0(t_0)))$ is in the initial marking of v'_j, and the commitment sets of the places remained the same due to the construction of the arena A. The place p does not satisfy the LSC of v'_j and since v'_j is maximal with respect to its commitment set mapping, we have $t \in \mathcal{T}_{v'_j}$ with $pre_{v'_j}(t) = \Phi(\pi_{v_i} \restriction_M^{-1} (pre_0(t_0)))$ and $\pi_{v'_j}(t) = t_0$.

Lemma 5 (From graph games to Petri games). *If a winning strategy $\sigma : V^* V_0 \to V$ for player 0 exists in G starting in a vertex $v_1 \in Start$, then there exists a winning strategy σ' in the Petri game G_P.*

Proof (Proof sketch). We show that the union of the branching processes of all plays with unique planning is winning in the Petri game. For the maximally progressed places in a planning segment that satisfy the LSC, all decision sets, which are allowed in plays with unique planning, are distinguishable. This means that each of those places refers to exactly one such decision set. Thus, for two different decision sets the maximally progressed places are disjoint and further transitions added later in the play have disjoint presets, such that every node is added only in one branching process of a prefix of a play with unique planning. For the maximally progressed places, that do not satisfy the LSC, no transition is allowed to be added due to the commitment sets. From there on, the commitment set is kept the same in all plays with that prefix.

Now, the three shown properties for branching processes of a play with unique planning together with σ being a winning strategy ensure the winning properties of the Petri game strategy: Lemma 2 and Lemma 4 ensure the justified refusal property; Lemma 3 ensures the safety property; Lemma 2 and Lemma 3 ensure the determinism property; Lemma 3 and Lemma 4 ensure the deadlock avoiding property.

Now, we show the existence the other way round.

Lemma 6 (From Petri games to graph games). *Let G_P be a Petri game and σ a winning strategy for G_P. Then there exists a winning strategy σ' for player 0 in the safety graph game $G = (\mathcal{A}, S)$ from Definition 14.*

Proof (Proof sketch). We choose the planning segments according to the given Petri game strategy, where the commitment sets are equal to the postsets of the places in the Petri game strategy.

We conclude with the theorem that states that the synthesis problem for Petri games of the presented subclass is decidable in EXPTIME.

Theorem 1 (Synthesis). *Let G_P be a Petri game, where the underlying Petri net satisfies the non-simultaneous synchronisation condition from Definition 3, then the existence of a winning strategy, and deriving it if existent, is decidable in EXPTIME.*

Proof. The equivalence of the existence of a winning strategy follows from the two implications in Lemma 5 and Lemma 6. The corresponding graph game has exponential size dependent on the size of the Petri game. The graph game takes linear time to be solved dependent on the number of edges, which are at most square as many as vertices. This results in exponential complexity overall.

4 Related Work

There have been quite a few approaches to the synthesis problem. We distinguish between works that address a single-process synthesis problem, where the system consists of one process that has all its information accessible, and works that address a multi-process synthesis problem, where multiple processes exist that are all partially informed. One of the latter is presented in this paper. The works on Petri nets in [5,23] belong to the former.

The synthesis problem was first introduced in [6]. Pnueli and Rosner introduced a setting of synchronous processes that communicate via shared variables [22]. For a single process, this setting is known to be decidable [4,21]. For multiple processes, this setting is known to be undecidable [22]. In particular, information forks have been found to be a necessary and sufficient criterion for the undecidability in that distributed setting [14]. There have been positive decidability results on specific architectures with multiple processes, including pipelines [24], rings [16], and acyclic architectures [14]. However, all the positive results for multiple processes have non-elementary complexity. A general game model in this type of setting is introduced by Walukiewicz and Mohalik [19]. Another line of work concerns the alternating-time temporal logics, which are interpreted over concurrent game structures [1].

Petri nets are conceptually connected to event structures by their unfoldings [18,20]. As an application example, unfoldings are used to determine the set of all reachable markings in a Petri net [8]. We use net unfoldings to define strategies on Petri games. The causal past of a node is the only available information; concurrent and future actions are invisible.

Zielonka automata are another distributed setting introduced in [25]. These are weakly bisimilar on their winning strategies to Petri games that have a local safety condition, with an exponential blow-up [3]. The synthesis problem is decidable for games on acyclic Zielonka automata [15] and on Zielonka automata with a strong synchronisation condition together with constraints on the winning condition that allow no distinction of differently ordered executions of the same trace [17]. We allow such distinctions, but the synchronisation condition in this paper is similar to the one in [17] and restricts the Petri nets to those that only allow loops including all tokens or that are acyclic.

5 Conclusions

We have presented a subclass of Petri games with an arbitrary mixture of system and environment players for which the synthesis problem is decidable in EXPTIME. Petri games use the tokens as carriers of information and link their information flow to their causality. This makes Petri games a suitable formalism to reason about distributed applications. The presented approach might seem anti-intuitive as it does not use the causal past of the players; it uses the causality to plan ahead. This subclass allows us to model distributed systems with a hierarchic communication structure, where every part has to check on its subordinated parts within a bound. Every part may consist of multiple processes itself, that communicate repeatedly within a bound. For example, we can model a control for several traffic lights geared to each other which react to the current traffic situation. A failure of a traffic light could be modeled within the Petri game. Another possible modeling is the communication structure of a round robin protocol. We cannot express exact timing constraints in Petri games.

This is the first work approaching the synthesis problem for distributed systems that allows finitely many system players and finitely many environment players while the global safety condition allows to distinguish between different interleavings of the same trace. Also, we allow tokens to be generated or deleted, which makes these Petri games a convenient way to model resource allocations and situations where processes are generated or deleted. This is not possible in the setting of Zielonka automata.

In future work, we will investigate weaker synchronisation conditions. We will also implement the presented decision procedure and compare it to existing ones in [9,11]. Another challenge is to extend the winning condition to reachability, Büchi, or parity conditions. A reachability condition is harder to check as we need to ensure that for all plays a designated marking is reached in every possible order of firing its transitions, which is not possible to check within the presented planning segments. In [10], it is shown that Petri games with a global reachability condition are undecidable.

References

1. Alur, R., Henzinger, T.A., Kupferman, O.: Alternating-time temporal logic. J. ACM **49**(5), 672–713 (2002)
2. Alur, R., Madhusudan, P., Nam, W.: Symbolic computational techniques for solving games. Int. J. Softw. Tools Technol. Transf. **7**(2), 118–128 (2005)
3. Beutner, R., Finkbeiner, B., Hecking-Harbusch, J.: Translating asynchronous games for distributed synthesis. In: Fokkink, W.J., van Glabbeek, R. (eds.) 30th International Conference on Concurrency Theory, CONCUR 2019, Amsterdam, The Netherlands, 27–30 August 2019. LIPIcs, vol. 140, pp. 26:1–26:16. Schloss Dagstuhl - Leibniz-Zentrum für Informatik (2019). https://doi.org/10.4230/LIPIcs.CONCUR.2019.26
4. Bloem, R., Schewe, S., Khalimov, A.: CTL* synthesis via LTL synthesis. In: Fisman, D., Jacobs, S. (eds.) Proceedings Sixth Workshop on Synthesis, SYNT@CAV 2017, Heidelberg, Germany, 22nd July 2017. EPTCS, vol. 260, pp. 4–22 (2017). https://doi.org/10.4204/EPTCS.260.4
5. Buy, U.A., Darabi, H., Lehene, M., Venepally, V.: Supervisory control of time Petri nets using net unfolding. In: 29th Annual International Computer Software and Applications Conference, COMPSAC 2005, Edinburgh, Scotland, UK, 25–28 July 2005, vol. 2. pp. 97–100. IEEE Computer Society (2005). https://doi.org/10.1109/COMPSAC.2005.148
6. Church, A.: Applications of recursive arithmetic to the problem of circuit synthesis. Summ. Summer Inst. Symb. Logic **1**, 3–50 (1957)
7. Engelfriet, J.: Branching processes of Petri nets. Acta Inf. **28**(6), 575–591 (1991)
8. Esparza, J.: Model checking using net unfoldings. Sci. Comput. Program. **23**(2), 151–195 (1994)
9. Finkbeiner, B.: Bounded synthesis for petri games. In: Meyer, R., Platzer, A., Wehrheim, H. (eds.) Correct System Design. LNCS, vol. 9360, pp. 223–237. Springer, Cham (2015). https://doi.org/10.1007/978-3-319-23506-6_15
10. Finkbeiner, B., Gieseking, M., Hecking-Harbusch, J., Olderog, E.: Global winning conditions in synthesis of distributed systems with causal memory. CoRR abs/2107.09280 (2021). https://arxiv.org/abs/2107.09280
11. Finkbeiner, B., Gieseking, M., Olderog, E.-R.: ADAM: causality-based synthesis of distributed systems. In: Kroening, D., Păsăreanu, C.S. (eds.) CAV 2015. LNCS, vol. 9206, pp. 433–439. Springer, Cham (2015). https://doi.org/10.1007/978-3-319-21690-4_25
12. Finkbeiner, B., Gölz, P.: Synthesis in distributed environments. In: Lokam, S., Ramanujam, R. (eds.) 37th IARCS Annual Conference on Foundations of Software Technology and Theoretical Computer Science (FSTTCS 2017). Leibniz International Proceedings in Informatics (LIPIcs), vol. 93, pp. 28:1–28:14. Schloss Dagstuhl-Leibniz-Zentrum fuer Informatik, Dagstuhl, Germany (2018). https://doi.org/10.4230/LIPIcs.FSTTCS.2017.28. http://drops.dagstuhl.de/opus/volltexte/2018/8406
13. Finkbeiner, B., Olderog, E.R.: Petri games: synthesis of distributed systems with causal memory. Inf. Comput. **253**, 181–203 (2017)
14. Finkbeiner, B., Schewe, S.: Uniform distributed synthesis. In: Proceedings of the 20th IEEE Symposium on Logic in Computer Science (LICS 2005), Chicago, IL, USA, 26–29 June 2005, pp. 321–330 (2005). https://doi.org/10.1109/LICS.2005.53

15. Genest, B., Gimbert, H., Muscholl, A., Walukiewicz, I.: Asynchronous games over tree architectures. In: Fomin, F.V., Freivalds, R., Kwiatkowska, M., Peleg, D. (eds.) ICALP 2013. LNCS, vol. 7966, pp. 275–286. Springer, Heidelberg (2013). https://doi.org/10.1007/978-3-642-39212-2_26
16. Kupferman, O., Vardi, M.: Synthesizing distributed systems. In: Proceedings - Symposium on Logic in Computer Science, pp. 389–398 (2001). https://doi.org/10.1109/LICS.2001.932514. https://www.scopus.com/inward/record.uri?eid=2-s2.0-0034873871&doi=10.1109
17. Madhusudan, P., Thiagarajan, P.S., Yang, S.: The MSO theory of connectedly communicating processes. In: Sarukkai, S., Sen, S. (eds.) FSTTCS 2005. LNCS, vol. 3821, pp. 201–212. Springer, Heidelberg (2005). https://doi.org/10.1007/11590156_16
18. Meseguer, J., Montanari, U., Sassone, V.: Process versus unfolding semantics for place/transition petri nets. Theoret. Comput. Sci. **153**(1), 171–210 (1996)
19. Mohalik, S., Walukiewicz, I.: Distributed games. In: Pandya, P.K., Radhakrishnan, J. (eds.) FSTTCS 2003. LNCS, vol. 2914, pp. 338–351. Springer, Heidelberg (2003). https://doi.org/10.1007/978-3-540-24597-1_29
20. Nielsen, M., Plotkin, G., Winskel, G.: Petri nets, event structures and domains, part I. Theoret. Comput. Sci. **13**(1), 85–108 (1981)
21. Pnueli, A., Rosner, R.: On the synthesis of an asynchronous reactive module. In: Ausiello, G., Dezani-Ciancaglini, M., Della Rocca, S.R. (eds.) ICALP 1989. LNCS, vol. 372, pp. 652–671. Springer, Heidelberg (1989). https://doi.org/10.1007/BFb0035790
22. Pnueli, A., Rosner, R.: Distributed reactive systems are hard to synthesize. In: 31st Annual Symposium on Foundations of Computer Science, St. Louis, Missouri, USA, 22–24 October 1990, vol. II, pp. 746–757 (1990). https://doi.org/10.1109/FSCS.1990.89597
23. Raskin, J., Samuelides, M., Begin, L.V.: Petri games are monotone but difficult to decide. Technical report, Université Libre De Bruxelles (2003)
24. Rosner, R.: Modular synthesis of reactive systems. Ph.D. thesis, Weizmann Institute of Science, Rehovot, Israel (1992)
25. Zielonka, W.: Notes in finite asynchronous automata. RAIRO - Theoret. Inform. Appl. - Informatique Théorique et Applications **21**(2), 99–135 (1987)

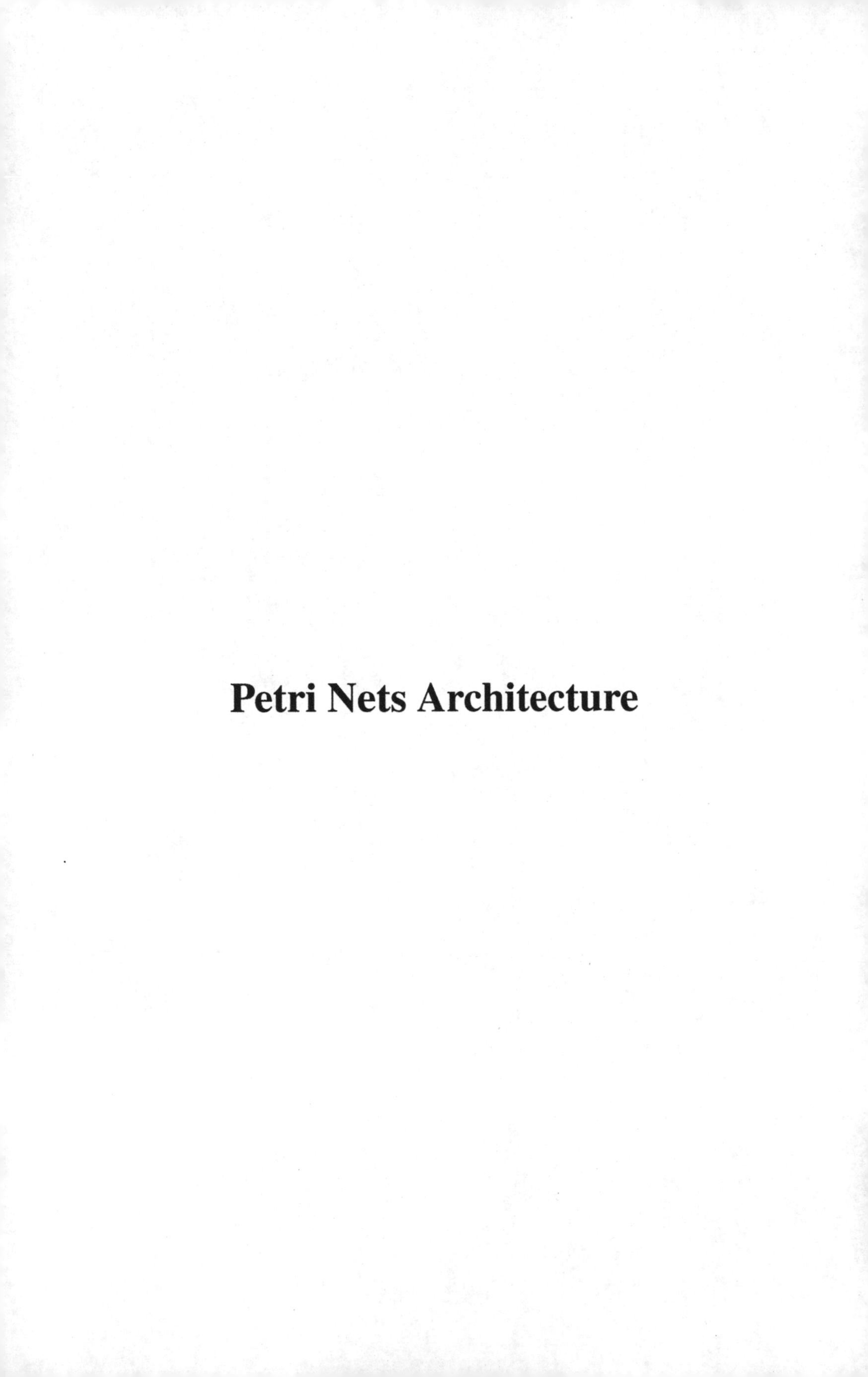

Petri Nets Architecture

Avoiding Exponential Explosion in Petri Net Models of Control Flows

Victor Khomenko[1(✉)], Maciej Koutny[1], and Alex Yakovlev[2]

[1] School of Computing, Newcastle University,
1 Science Square, Newcastle upon Tyne NE4 5TG, UK
{victor.khomenko,maciej.koutny}@ncl.ac.uk
[2] School of Engineering, Newcastle University,
Merz Court, Newcastle upon Tyne NE1 7RU, UK
alex.yakovlev@ncl.ac.uk

Abstract. We look at modelling of a choice between several 'bursts' of concurrent actions in a Petri net. If 'silent' transitions are disallowed, a construction based on Cartesian product is traditionally used, resulting in an exponential explosion in the model size.

We demonstrate that this exponential explosion can be avoided. We show the equivalence between this modelling problem and the problem of finding an edge clique cover of a complete multipartite graph, which gives major insights into the former problem as well as linking it to the existing results from graph theory.

It turns out that the exponential number of places created by the Cartesian product construction can be improved down to polynomial (quadratic) in the worst case, and down to logarithmic in the best (non-degraded) case. For example, to express a choice between 10 pairs of concurrent transitions, the Cartesian product construction creates 1024 places, even though 6 places are sufficient. We also derive several lower and upper bounds on the numbers of places and arcs.

As these results affect the 'core' modelling techniques based on Petri nets, eliminating a source of an exponential explosion, we hope they will have applications in Petri net modelling and translations of various formalisms to Petri nets. As an example, applying them to translate Burst Automata to Petri nets reduces the size of the resulting Petri net from exponential down to polynomial.

Keywords: Petri net · Complete multipartite graph · Edge clique cover · Control flow · Burst Automata

1 Introduction

Petri nets (PN) are often used as a modelling formalism, and their advantages include the simplicity of semantics, the intuitive graphical notation, the abundance of software tools, and the possibility of capturing behaviours concisely without making subsequent processing (e.g., formal verification or synthesis) undecidable. In particular, the possibility to create concise models is often the

L. Bernardinello and L. Petrucci (Eds.): PETRI NETS 2022, LNCS 13288, pp. 261–277, 2022.
https://doi.org/10.1007/978-3-031-06653-5_14

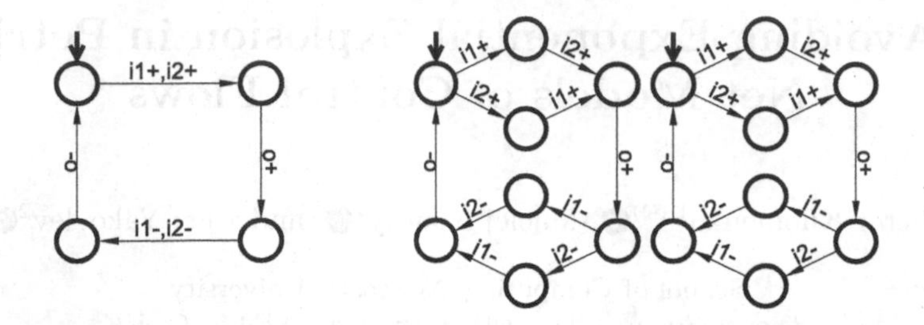

Fig. 1. A BA specification of the C-element and an FSM expressing its interleaving semantics. A C-element waits for both inputs to switch to 1 (actions i_1^+ and i_2^+) before switching its output to 1 (action o^+), and then waits for both inputs to switch to 0 (actions i_1^- and i_2^-) before switching its output to 0 (action o^-). It is assumed that the environment fulfills its part of the contract, i.e. each input switches only once before the output switches.

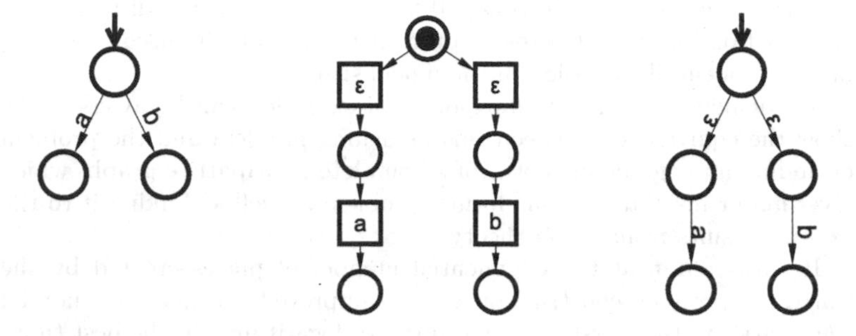

Fig. 2. A BA with singleton bursts, so coinciding with the FSM expressing its interleaving semantics (left); its PN translation prefixing each burst with a silent transition (middle); and the reachability graph (FSM) of this PN (right). Note that the two FSMs are language-equivalent but not weakly bisimilar.

key advantage of PNs over simpler formalisms like Finite State Machine (FSMs). Indeed, it is generally accepted that one is likely to encounter the exponential *state space explosion* [9] during, e.g., formal verification—this problem is believed to be fundamental (unless P=PSPACE), and mitigating this explosion using heuristics has been a hot research topic for many years. However, encountering an exponential explosion already during the modelling stage is both regrettable and indicative of problems in modelling techniques or even the formalism itself.

Unfortunately, some common PNs modelling techniques can indeed result in exponentially large models, even in common cases of simple control flows. As a motivating example we consider *Burst Automata* (BA) [3]—a more general version of Burst Mode [8] formalism from the area of asynchronous circuits design. Intuitively, BAs are similar to FSMs, except that their arcs are labelled not by single actions but by sets of actions ('bursts') which fire concurrently. One can define the interleaving semantics of BAs by allowing the actions in a burst to fire in any order, which results in the usual FSM, see the example in Fig. 1. For the purposes of formal verification and circuit synthesis, it would be advantageous to develop a translation from BAs to PNs, in order to be able to use existing PN software. This means some kind of behavioural equivalence between the FSM expressing the interleaving semantics and the reachability graph of the resulting PN is required, e.g., language equivalence or bisimulation. As BAs are a very simple FSM-like formalism, it would be reasonable to expect that such a translation would be quite simple and efficient.

However, developing a compact translation from BAs to PNs is more complicated than one might expect. In particular, efficiently expressing a choice between several bursts of concurrent transitions is not trivial in PNs. In [3], a language-preserving linear size translation is proposed, that prefixes each burst with a silent 'fork' transition and then uses another silent 'join' transition after the burst to detect completion. Unfortunately, there are situations when this translation is unacceptable. First of all, silent transitions turn a deterministic model into a non-deterministic one which is often undesirable (e.g., non-determinism cannot be directly implemented physically, say in an asynchronous logic circuit [4]). Second, language equivalence may be too weak (e.g., it does not preserve branching time temporal properties or even deadlocks), and prefixing bursts with silent transitions breaks not only strong but also weak bisimulation, see Fig. 2.

To preserve strong bisimulation, the following *Cartesian Product Construction* ($\times$-construction) is traditionally used, see e.g., [2]. To express a choice between several bursts (i.e., sets of concurrent transitions) $B_1, B_2, \ldots, B_n$, this construction would create a set of places corresponding to tuples in the Cartesian product $B_1 \times B_2 \times \cdots \times B_n$, so that a place corresponding to a tuple $(b_1, \ldots, b_n)$ is connected to each transition b_i occurring in the tuple. This means that the number of created places is $|B_1| \cdot |B_2| \cdot \ldots \cdot |B_n|$, i.e., the PN size is exponential in the number of bursts.

In this paper we focus on efficiently expressing a choice between several bursts without using silent transitions. We demonstrate that the $\times$-construction is often sub-optimal, in particular one can always avoid an exponential explosion—a polynomial (quadratic) number of places is sufficient even in the worst case. Moreover, in the case of each burst containing two transitions, the $\times$-construction requires 2^n places while the construction proposed in this paper needs only $\log_2 n$ places (asymptotically)—a dramatic double-exponential reduction.

We believe that the proposed construction, which is the main contribution of the paper, will have many applications, as it affects the 'core' modelling techniques for PNs. In particular, translations from various formalisms to PNs relying on the $\times$-construction can be significantly improved by using the proposed construction instead, thus eliminating a source of exponential explosion. In Sect. 8, we show how to improve the bisimulation-preserving translation from BAs to PNs described in [3], from exponential down to polynomial. Note that the developed translation is just a simple example of applying the proposed construction rather than the focus of the paper, and there are more applications, e.g., in our forthcoming paper [7] we develop a polynomial PN translation for arbitrary control flows built from atomic actions using sequencing, parallel composition, and choice.

The proposed construction is based on the observation that the problem of implementing a choice between concurrent bursts in a PN using k places is equivalent to finding an *edge clique cover* of a complete multipartite graph with k cliques. Hence, the minimal possible number of places is equivalent to the *edge clique cover number* (a.k.a. *intersection number* [5]) of a certain complete multipartite graph. The latter is a problem investigated for decades. Even though it is not completely solved, there are many useful published results, and we prove several new results helping to improve some upper bounds on the number of PN places and arcs. In practice, the optimality is usually not required, and one can use simple approximations with useful lower and upper bounds.

2 Basic Notions

In this section, we provide some basic notions related to Petri nets, complete multipartite graphs, and edge clique covers.

Petri Nets
We focus on *safe* (i.e., at most one token per place) PNs, which are often used for modelling control flows. For a safe PN, the total number of tokens in its initial marking cannot exceed the number of places, so we can define its size as the total number of places, transitions, and arcs, disregarding the initial marking. Note that the size of a PN is dominated by its arcs, except the uninteresting degraded case when there are many isolated nodes.

In this paper, the set of transitions is usually given (e.g., when translating a model from some other formalism to PNs, the transitions often correspond to the occurrences of actions in that model), and the objective is to express the intended behaviour using small numbers of places and arcs. Note that having a small number of places is often desirable for formal verification as they correspond to state variables, and having a small number of arcs is desirable as they dominate the PN size.

Complete Multipartite Graphs
We consider undirected graphs with no parallel edges and no self-loops. A graph is called *multipartite* if its vertices are partitioned into several sets in such a way that there are no edges between vertices in the same partition. A multipartite graph is *complete* if, for every pair of vertices from different partitions, there is an edge connecting them. A complete multipartite graph with the partitions of sizes $t_1 \leq t_2 \leq \cdots \leq t_n$ will be denoted $K_{t_1,t_2,\ldots,t_n}$. Note that for the purposes of this paper one can assume that multipartite graphs have at least two partitions and that each partition contains at least two vertices—violating these assumptions leads to simple degraded cases. If the sizes of all partitions in $K_{t_1,t_2,\ldots,t_n}$ are equal, $t_1 = t_2 = \cdots = t_n = t$, the graph is called *balanced* and will be denoted $K_n(t)$.

Edge Clique Covers
A *clique* in a graph is a set of vertices which are pairwise connected by edges. A clique is called *maximal* if it is not a subset of any other clique. Note that for a complete multipartite graph, every maximal clique contains exactly one vertex

from each partition, and vice versa, by picking one vertex from each partition one always obtains a maximal clique.

A set of cliques in a graph form an *edge clique cover* (ECC) if for every edge in the graph there is at least one clique that contains both endpoints of this edge. The number of cliques in an ECC is called its *size*. Note that, given an ECC, one can expand each clique in it to some maximal one, without increasing the size of the ECC. The minimum possible size of an ECC of a graph G is called the *edge clique cover number* (a.k.a. *intersection number*) of G, and will be denoted $ecc(G)$.

3 Equivalence Between the Problems of Modelling a Choice Between Bursts in a PN and Finding an ECC of a Complete Multipartite Graph

Suppose that we have pairwise disjoint[1] bursts $B_1, B_2, \ldots, B_n$ where each burst is a non-empty set of transitions, and the intention is to create a choice between these bursts. Hence, the problem is to add some places and connect them to these transitions (by place→transition arcs) so that the transitions within each burst must be concurrent,[2] but any pair of transitions from different bursts must be in conflict. More precisely, the following requirements must be satisfied:

ReqChoice For every pair of transitions from different bursts, there is a place connected to both of them (this creates choices between transitions from different bursts).

ReqConc A place cannot be connected to more than one transition from the same burst (otherwise these transitions would not be concurrent). An alternative and stricter formulation of this requirement is that a place must be connected to exactly one transition from each burst—this in addition ensures that no tokens are left behind after a burst fires, which is essential for control flows containing cycles, a common case in practice. We will denote this strict formulation by **ReqConc**(strict). Note that if **ReqConc**(strict) is satisfied then the number of arcs can be determined from the number of places by multiplying the latter by the number of bursts, i.e., by n.

ReqNoDups No two places are connected to the same set of transitions (this would create redundancy – one of such places can be removed without affecting the PN's behaviour).

Furthermore, we define the problem size as $|B_1| + \cdots + |B_n|$—the upper and lower bounds on the number of required places and arcs will be relative to it.

[1] In practice, transitions are often labelled by actions, and there is no requirement for the sets of labels corresponding to bursts to be disjoint, e.g., it is possible to have bursts $B_1 = \{t_1\}$ and $B_2 = \{t_2\}$ where t_1 and t_2 are distinct transitions labelled by the same action a. In particular, BAs allow non-determinism, and the PN translation in Sect. 8 expresses it by creating distinct transitions with the same label.

[2] Note that the $|B_i|$-ary concurrency relation intended here is not reducible to pairwise concurrency.

One can see that the $\times$-construction satisfies the above conditions, including **ReqConc**(strict). However, it generates $|B_1| \cdot |B_2| \cdot \ldots \cdot |B_n|$ places and $n \cdot |B_1| \cdot |B_2| \cdot \ldots \cdot |B_n|$ arcs, i.e., the PN size is exponential.

The natural questions now are whether the above requirements are possible to achieve with fewer places and arcs—in particular, whether the PN size can be polynomial in the problem size, what would be the minimal number of places, and whether it is possible to derive some useful lower and upper bounds on the size of the smallest PN. We show that these problems can be reformulated in terms of finding ECCs of a complete multipartite graph, which provides revealing insights and helps one to find positive answers to these questions.

Consider the conflict (i.e., choice) relation between the transitions. It is symmetric and irreflexive, and so can be represented by an undirected graph without self-loops such that there is an edge between two vertices iff the transitions corresponding to these vertices belong to different bursts. Thus, the graph is a complete multipartite graph with its partitions corresponding to the bursts, i.e., $K_{|B_1|,|B_2|,\ldots,|B_n|}$. Furthermore, any complete multipartite graph represents the conflict relation for some family of bursts.

Now consider a place connected to several transitions picked from different bursts. The vertices corresponding to these transitions form a clique in the graph representing the conflict relation. Thus, given any ECC of this graph, one can create a place for each clique in the ECC and connect it to transitions corresponding to the vertices occurring in the clique, and the resulting set of places will satisfy the above requirements. Furthermore, one can in addition satisfy **ReqConc**(strict) by extending every clique in the ECC to a maximal one. Hence, there is a 1-to-1 correspondence between cliques and places, or between maximal cliques and places in the **ReqConc**(strict) case. This simple observation allows one to answer some of the posed questions:

- The *trivial* ECC where each edge is covered by a clique with two vertices (which can then be extended to a maximal clique if necessary) has only a quadratic number of cliques at most, which yields a polynomial translation with a quadratic number of places and either quadratic (for **ReqConc**) or cubic (for **ReqConc**(strict)) number of arcs—already a huge improvement on the exponential $\times$-construction.
- In bipartite graphs there are no cliques with more than two vertices, and so the minimal ECC coincides with the trivial one. Consider the case of two bursts of equal size, $|B_1| = |B_2| = k$. One can see that the graph of the conflict relation is the bipartite graph $K_{k,k}$, i.e., the number of PN places cannot be smaller than k^2. This observation yields a quadratic worst case lower bound on the number of places, that matches the above upper bound. Furthermore, this gives a quadratic worst case lower bound on the number of arcs, that matches the above upper bound for the **ReqConc** case, though there is still a gap between this lower bound and the upper bound for **ReqConc**(strict).

These simple observations, though giving matching worst case upper and lower bounds on the number of places as well as the number of arcs in the **ReqConc** case, do not provide the full picture. It turns out that one can often do much better than the above quadratic worst case lower bound suggests, e.g., in the next section, we give an example where a logarithmic number of places is sufficient.

Furthermore, the **ReqConc**(strict) case is more important in practice, and given that the size of a PN is normally dominated not by places but by arcs, the gap between quadratic lower and cubic upper bounds on the number of arcs needs narrowing and ideally closing. In what follows, we derive some improved upper bounds in the **ReqConc**(strict) case, including a quadratic upper bound for the cases where the sizes of bursts are the same or at least not too different, i.e., the graph of the conflict relation is balanced or almost balanced.

4 A Logarithmic Case

Above we derived a polynomial (quadratic) worst-case bound on the number of places, which is an exponential improvement over the $\times$-construction. In this section, we consider the best-case scenario, excluding the degraded cases of a single burst, and the situation when some of the bursts are singletons. The former is trivial, and the latter can be reduced to a smaller non-degraded case by removing all singleton bursts. It turns out that in the best case a logarithmic number of places is both necessary and sufficient, yielding a dramatic double-exponential improvement on the $\times$-construction.

Consider the case of each burst containing two transitions. The $\times$-construction would create 2^n places, but it turns out that $\log_2 n$ places are sufficient (asymptotically). The conflict relation graph in this case is $K_n(2)$, and the problem boils down to finding a small ECC of this graph. This problem was solved in [6], which proved that $ecc(K_n(2)) \sim log_2 n$.

The idea of the construction is as follows. For simplicity, we assume that $n = \binom{k-1}{k/2}$ for some even k. One can check (e.g., using wolframalpha.com) that

$$\lim_{k\to+\infty} \frac{\log_2 n}{k} = \lim_{k\to+\infty} \frac{\log_2 \binom{k-1}{k/2}}{k} = 1 ,$$

and so $k \sim log_2 n$. Consider the family of subsets of size $k/2$ of $\{1,\dots,k\}$. One can build a multipartite graph with these subsets as vertices, and with two vertices being connected iff their intersection is non-empty (this construction is called the *intersection graph* [5] of a family of sets). In our case each subset has a non-empty intersection with all but one other subset in the family, viz. its complement, and so $K_n(2)$ is the intersection graph of this family of subsets. One can then observe that the vertices corresponding to subsets sharing some element i are pairwise connected and so form a clique. Moreover, the cliques corresponding to the elements of $\{1,\dots,k\}$ form an ECC of size k.

For example, let $n = 10 = \binom{6-1}{6/2}$, i.e., $k = 6$ places are sufficient to express all the conflicts between 10 binary bursts (compared to $2^{10} = 1024$ places created by the $\times$-construction). We consider all 3-element subsets of the set $\{1,\dots,6\}$ and pair subsets with their complements:

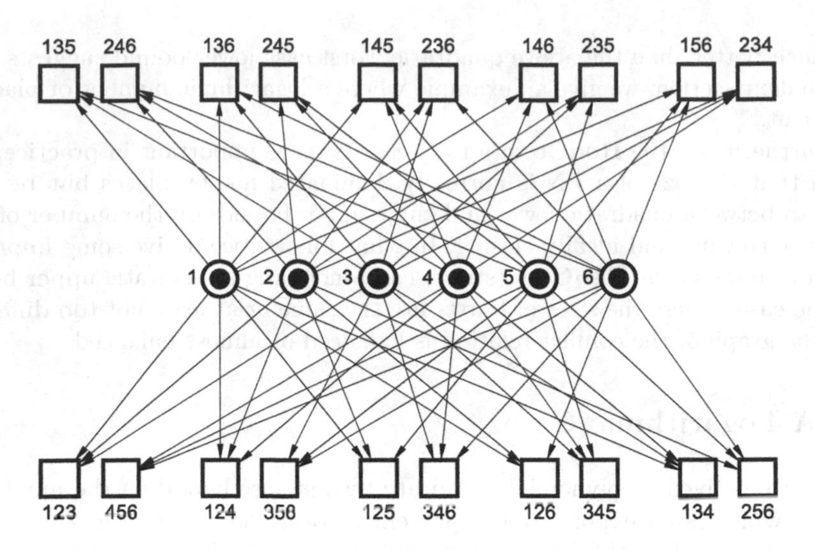

Fig. 3. A PN model with 6 places expressing a choice between 10 binary bursts. The transitions are labelled by 3-element subsets of $\{1,\ldots,6\}$, e.g., '123' corresponds to $\{1,2,3\}$.

$$\begin{array}{ll}
\{1,2,3\} & \{4,5,6\} \\
\{1,2,4\} & \{3,5,6\} \\
\{1,2,5\} & \{3,4,6\} \\
\{1,2,6\} & \{3,4,5\} \\
\{1,3,4\} & \{2,5,6\} \\
\{1,3,5\} & \{2,4,6\} \\
\{1,3,6\} & \{2,4,5\} \\
\{1,4,5\} & \{2,3,6\} \\
\{1,4,6\} & \{2,3,5\} \\
\{1,5,6\} & \{2,3,4\}
\end{array}$$

One can see that the intersection graph of this family of subsets is $K_{10}(2)$ where the vertices in each of the 10 partitions correspond to these 10 pairs of complimentary sets. Furthermore, one can cover all the edges of this graph by 6 cliques, where the i-th clique comprises vertices corresponding to the subsets containing i, for each $i = 1,\ldots,6$.

In the corresponding PN, the pairs of transitions in these 10 bursts can be labelled by the above pairs of subsets, the places be labelled by the numbers 1,..., 6, and the connection be such that there is an arc from place i to a transition labelled by subset S iff $i \in S$, see Fig. 3.

5 Upper Bounds for the Balanced Case

In this section, we provide two different upper bounds on the number of places and arcs for the balanced case, i.e., when all the bursts have the same size. It turns out one can improve the trivial upper bound on the number of arcs derived in Sect. 3.

The idea of the first bound is based on the observation that, by generating a sufficient number k of random cliques, one can cover every edge with high probability (in the sense that the expected number of uncovered edges is <1), which implies that there is an ECC of size k. Furthermore, the probability of an edge being uncovered falls exponentially with k, so k does not have to be large. The following result was inspired by [1, Lemma 3.2]. Using the specifics of balanced complete multipartite graphs, we reformulated that result to avoid references to graph's complement, streamlined the proof, and obtained a better multiplicative constant.

Proposition 1. *$ecc(K_n(t)) \leq \left\lceil 2t^2 \ln \frac{tn}{\sqrt{2}} \right\rceil$ if $n, t \geq 2$.*

Proof. One can pick a random maximal clique in $K_n(t)$ by randomly and uniformly picking a vertex from each of the n parts. Suppose that k (to be chosen appropriately below) such cliques are picked independently.

Given a random clique as above, an edge (u, v) is covered by it iff both u and v were picked from their partitions, i.e., the probability that a particular edge is covered by a random maximal clique is $1/t^2$. Hence the probability that an edge is not covered by any of the k chosen random cliques is

$$\left(1 - \frac{1}{t^2}\right)^k \leq e^{-k/t^2},$$

where the inequality follows from $1 - x \leq e^{-x}$ for all real x.

There are $\frac{t^2 n(n-1)}{2} < \frac{t^2 n^2}{2}$ edges in $K_n(t)$, so the expected number of edges not covered by any of the k cliques does not exceed

$$\frac{t^2 n^2}{2} e^{-k/t^2}.$$

We now choose $k = \left\lceil t^2 \ln \frac{t^2 n^2}{2} \right\rceil > t^2 \ln \frac{t^2 n^2}{2}$ (note that the inequality here is indeed strict because $n, t \geq 2$ are integers and the natural logarithm of an integer distinct from 1 is never an integer). Substituting this value for k into the above formula for the expected number of edges not covered by any of the k cliques, we have

$$\frac{t^2 n^2}{2} e^{-\frac{\left\lceil t^2 \ln \frac{t^2 n^2}{2} \right\rceil}{t^2}} < \frac{t^2 n^2}{2} e^{-\frac{t^2 \ln \frac{t^2 n^2}{2}}{t^2}} = 1,$$

i.e., it is strictly below 1. Hence, it is possible to choose

$$k = \left\lceil t^2 \ln \frac{t^2 n^2}{2} \right\rceil = \left\lceil 2t^2 \ln \frac{tn}{\sqrt{2}} \right\rceil$$

cliques forming an ECC of $K_n(t)$. □

One can see that the proof of the above proposition yields a randomised algorithm for generating a small ECC, and the derived upper bound is usually good in practice. However, it is expressed in terms of the number n of partitions in $K_n(t)$ and their (common) size t rather than the number $v = t \cdot n$ of vertices $K_n(t)$ (note that v is the problem size that was defined as the total size of all bursts). Reformulating this bound in terms of v yields

$$\left\lceil 2\frac{v^2}{n^2} \ln \frac{v}{\sqrt{2}} \right\rceil = O\left(\frac{v^2}{n^2} \log v\right),$$

which means that the number of places in the PN is $O\left(v^2 \log v\right)$ if n is fixed. This means that in some cases this upper bound can be worse than the trivial quadratic one derived in Sect. 3. However, the above expression allows one to improve on the trivial cubic upper bound on the number of arcs in the **Req-Conc**(strict) case derived in Sect. 3. Recall that in this case the number of arcs equals to the number of places multiplied by n, i.e.,

$$n\left\lceil 2\frac{v^2}{n^2} \ln \frac{v}{\sqrt{2}} \right\rceil = O\left(\frac{v^2}{n} \log v\right).$$

This bound is quite good, and becomes $O\left(v \log v\right)$ if n is linear in v. The worst case is when n is fixed—the bound then becomes $O\left(v^2 \log v\right)$, which almost (but not quite) matches the quadratic worst case lower bound.

The natural question now is whether one can achieve quadratic (in v) upper bounds on the numbers of both places and arcs, thus completely closing the gap between the worst case lower and upper bounds at least in the balanced case. The following simple result provides a new (to our knowledge) upper bound on the edge clique cover number of $K_n(t)$ that allows one to answer this question affirmatively.

Proposition 2. *$ecc(K_n(t)) \leq nt^2/2$, where $t, n \geq 2$.*

Proof. We construct an ECC comprised of the cliques:

- C_{pij}, where $p = 1, \ldots, n$, $i = 2, \ldots, t$ and $j = 1, \ldots, i-1$, such that C_{pij} comprises the i-th vertex from p-th partition and the j-th vertex from every other partition.
- C_i, where $i = 1, \ldots, t$, containing the i-th vertex from every partition.

Note that the edge from the i-th vertex from p-th partition to j-th vertex in some other partition p' is covered by either C_{pij} (if $i > j$) or $C_{p'ji}$ (if $j > i$) or C_i (if $i = j$), so it is indeed an ECC. The size of this ECC is

$$nt(t-1)/2 + t = nt^2/2 - nt/2 + t \leq nt^2/2\ ,$$

and so $ecc(K_n(t)) \leq nt^2/2$. □

The above result can be used to derive quadratic upper bounds on the numbers of both places and arcs. Indeed, when expressed in terms of v, this bound becomes $\frac{v^2}{2n}$ and so cannot be worse than $O(v^2)$ even if n is fixed. Furthermore, the number of arcs in **ReqConc**(strict) case can be obtained by multiplying by n, which yields $\frac{v^2}{2}$.

Hence, in the balanced case we have obtained quadratic worst case upper bounds on both places and arcs, which match the quadratic worst case lower bound derived in Sect. 3.

6 Upper Bounds for the Almost Balanced Case

In this section, we show that the upper bounds from the previous sections can be transferred (up to a multiplication by a constant) to the unbalanced case provided it is not 'too unbalanced'. The idea is based on the following observations:

- $K_{t_1,t_2,\ldots,t_n}$ is an induced sub-graph in $K_n(t_n)$, and so $ecc(K_{t_1,t_2,\ldots,t_n}) \leq ecc(K_n(t_n))$, as any ECC of the latter can be turned into an ECC of the former by deleting the vertices which are not in $K_{t_1,t_2,\ldots,t_n}$ from each clique (and then extending each clique to a maximal one if required).
- The number of vertices in $K_n(t_n)$ is not much greater than that in $K_{t_1,t_2,\ldots,t_n}$ provided that the latter is not 'too unbalanced', so the upper bounds derived in the previous section do not become too large w.r.t. the size of $K_{t_1,t_2,\ldots,t_n}$.

Formally, let $b \geq 1$ be some fixed real number. A complete multipartite graph $K_{t_1,t_2,\ldots,t_n}$ is called *b-balanced* if $t_n \leq \frac{b}{n}\sum_{i=1}^{n} t_i = \frac{b}{n}v$, i.e., the size of the biggest partition is within the factor b of the average partition size. One can observe that the number of vertices in $K_n(t_n)$ is then

$$nt_n \leq n\left(\frac{b}{n}v\right) = bv,$$

i.e., within the factor b of the number v of vertices in $K_{t_1,t_2,\ldots,t_n}$. Hence the bounds derived in Sect. 5 can be lifted to 'almost balanced' graphs as follows.

Proposition 3 (Lifting Propositions 1 and 2 to almost balanced case). *Suppose that $K_{t_1,t_2,\ldots,t_n}$ is b-balanced and $n, t_i \geq 2$. Then*

$$ecc(K_{t_1,t_2,\ldots,t_n}) \leq \left\lceil 2\frac{(bv)^2}{n^2} \ln \frac{bv}{\sqrt{2}} \right\rceil$$

and

$$ecc(K_{t_1,t_2,\ldots,t_n}) \leq \frac{(bv)^2}{2n},$$

where $v = \sum_{i=1}^{n} t_i$ is the number of vertices in $K_{t_1,t_2,\ldots,t_n}$.

Proof. Follows directly from the $\left\lceil 2\frac{v^2}{n^2} \ln \frac{v}{\sqrt{2}} \right\rceil$ and $\frac{v^2}{2n}$ bounds for the balanced case derived in Sect. 5. □

Hence, in the almost balanced case the numbers of places and arcs are quadratic in the worst case. Furthermore, Proposition 5 below improves the former bound down to $\left\lceil 2t_n t_{n-1} \ln \frac{v}{\sqrt{2}} \right\rceil$. Note that, for a b-balanced graph, $t_{n-1} \leq t_n \leq \frac{bv}{n}$ and, moreover, b is removed from under the logarithm.

7 Upper Bounds for the Unbalanced Case

In this section, we consider the case of complete multipartite graphs $K_{t_1,t_2,\ldots,t_n}$ which are 'very unbalanced'. We slightly improve on the trivial upper bound on the number of arcs. However, this may still be cubic in the worst case, e.g., if n, t_n and t_{n-1} are linear in v.

First, we recall the following result from [5]. It refers to *edge clique partitions*, which are a special case of ECCs with every edge covered exactly by one clique. We denote the *edge clique partition number* $ecp(G)$ of a graph G as the smallest possible number of cliques in an edge clique partition of G. Trivially, $ecc(G) \leq ecp(G)$.

Proposition 4 (adapted from [5]). *Let G be a graph with v vertices. Then G has an edge clique partition of size at most $\lfloor v^2/4 \rfloor$ consisting of edges and triangles. Moreover, $ecp(G) = \lfloor v^2/4 \rfloor$ if and only if G is $K_{\lfloor v/2 \rfloor, \lceil v/2 \rceil}$.*

This translates into a $\lfloor v^2/4 \rfloor$ upper bound on the number of places and $n \lfloor v^2/4 \rfloor$ upper bound on the number of arcs in **ReqConc**(strict) case, which may be $O(v^3)$ if n (the number of bursts) grows linearly in v.

It should be noted that this bound is rather pessimistic: Edge clique partition is a very special case of ECC, and moreover only edges and triangles are used in this partition. Hence one can hope that the bound on arcs could be improved:

- The worst case for Proposition 4 is a balanced bipartite graph, in which case $n = 2$ and so the number of arcs is quadratic (and it remains quadratic for any fixed n).
- When n grows linearly, much larger cliques than edges and triangles can be formed.

As a result, we hope that a sub-cubic or even a quadratic bound on the number of arcs can be obtained, and leave this question for future research.

We now generalise Proposition 1 to the unbalanced case, which gives a slightly better bound than that derived in Sect. 6.

Proposition 5. *$ecc(K_{t_1,t_2,\ldots,t_n}) \leq \left\lceil 2t_n t_{n-1} \ln \frac{v}{\sqrt{2}} \right\rceil$ where v is the number of vertices in $K_{t_1,t_2,\ldots,t_n}$ and $n, t_i \geq 2$.*

Proof. One can pick a random maximal clique in $K_{t_1,t_2,\ldots,t_n}$ by randomly and uniformly picking a vertex from each of the n partitions. Suppose that k (to be chosen appropriately below) such cliques are picked independently.

Given a random clique as above, an edge (u, v) is covered by it iff both u and v were picked from their partitions, i.e., the probability that an edge (u, v) is covered by a random maximal clique is $1/(t_u t_v)$, where t_u and t_v are the sizes of partitions u and v are coming from. Hence the probability that an edge (u, v) is not covered by any of the k chosen random cliques is

$$\left(1 - \frac{1}{t_u t_v}\right)^k \leq e^{-k/(t_u t_v)} \leq e^{-k/t_n t_{n-1}},$$

where the former inequality follows from $1 - x \leq e^{-x}$ for all real x.

There are $\frac{1}{2}\left(v^2 - \sum_{i=1}^{s} t_i^2\right) < \frac{v^2}{2}$ edges in the graph, so the expected number of edges not covered by any of the k cliques does not exceed

$$\frac{v^2}{2} e^{-k/t_n t_{n-1}}.$$

We now choose $k = \left\lceil t_n t_{n-1} \ln \frac{v^2}{2} \right\rceil > t_n t_{n-1} \ln \frac{v^2}{2}$ (note that the inequality here is indeed strict because $v \geq 4$ and $t_i \geq 2$ are integers and the natural logarithm of an integer distinct from 1 is never an integer). Substituting this value for k into the above formula for the expected number of edges not covered by any of the k cliques, we have

$$\frac{v^2}{2} e^{-\frac{\left\lceil t_n t_{n-1} \ln \frac{v^2}{2} \right\rceil}{t_n t_{n-1}}} < \frac{v^2}{2} e^{-\frac{t_n t_{n-1} \ln \frac{v^2}{2}}{t_n t_{n-1}}} = 1,$$

i.e., it is strictly below 1. Hence, it is possible to choose

$$k = \left\lceil t_n t_{n-1} \ln \frac{v^2}{2} \right\rceil = \left\lceil 2 t_n t_{n-1} \ln \frac{v}{\sqrt{2}} \right\rceil$$

cliques forming an ECC of $K_{t_1, t_2, \ldots, t_n}$. □

8 A Polynomial Bisimulation-Preserving Translation from BAs to PNs

In [3], three translations from BAs to PNs were developed. One of them is linear—but it uses silent 'fork' and 'join' transitions for each burst and so preserves only language equivalence but not bisimulation (not even weak bisimulation)—as illustrated in Fig. 2. The second translation uses 'join' (but not 'fork') transitions and the ×-construction. As a result, it preserves weak bisimulation but is exponential. The third translation uses neither 'fork' nor 'join' transitions, and preserves strong bisimulation, but it also depends on the ×-construction and thus is exponential (and often larger than the second translation).

We now show how to eliminate a source of exponential explosion in the latter construction by replacing the ×-construction by the ECC-based one, thus obtaining a polynomial bisimulation-preserving translation from BAs to PNs. Note that the developed translation is just a simple example of applying the proposed ECC-based construction rather than the focus of the paper, and there are more applications, e.g., in our forthcoming paper [7] we develop a polynomial PN translation for arbitrary control flows built from atomic actions using sequencing, parallel composition, and choice.

The improved translation is illustrated in Fig. 4 and works as follows.

- Each burst B is represented by $|B|$ transitions corresponding to occurrences of actions in B and labelled by the corresponding actions. (Empty bursts are interpreted as ε-transitions in FSMs, and so for the purposes of this translation are replaced by singleton bursts $\{\varepsilon\}$.) No other transitions are created by the translation. Then each state s of the BA is considered in turn, together with its incoming and outgoing bursts.
- For the incoming bursts of s, a set P_s^{in} of new places is created, so that $|P_s^{in}|$ is the maximal input burst cardinality (and hence of linear size). The transitions in the input bursts are then connected by transition→place arcs to the places in P_s^{in}, so that the i-th transition in each burst is connected to the i-th place in P_s^{in}. Moreover, for bursts with fewer than $|P_s^{in}|$ transitions, extra arcs are added so that each place in P_s^{in} is connected to exactly one transition in the burst (e.g., one can connect the first transition in the burst to all the unmatched places in P_s^{in}).
- For the outgoing bursts of s, the **ReqConc**(strict) variant of the ECC-based construction presented in this paper is applied, yielding a set P_s^{out} of places. Note that $|P_s^{out}|$ is at most quadratic, and the number of created arcs is at most cubic (in the total size of all output bursts).
- To enforce the causality between the input and output bursts, a set of places $P_s^{(in,out)} = P_s^{in} \times P_s^{out}$ is created,[3] where each place $(p_{in}, p_{out}) \in P_s^{(in,out)}$ inherits its incoming arcs from p_{in} and its outgoing arcs from p_{out}. After that, the places in P_s^{in} and P_s^{out} are removed from the PN, together with their arcs.
- Finally, if s is the initial state of the BA, all the places in $P_s^{(in,out)}$ are initially marked.

Clearly, the resulting PN is strongly bisimilar with the original BA (in fact, its reachability graph is isomorphic to the FSM expressing the interleaving semantics of BA, which is an even stronger equivalence), and its size is polynomial in the size of BA, improving thus the exponential translation of [3].

It should be noted that self-loops with non-singleton bursts in BAs may cause the resulting PN to be unsafe (2-bounded). If this is undesirable, the problem can be easily avoided by replicating BA states with self-loops before the translation, as follows. If s is a state with a self-loop, a new state s' is created that inherits all

[3] This aspect of the translation can be improved as explained in our forthcoming paper [7]. In fact, the number of places in the Petri net in Fig. 4(bottom) can be reduced from 12 down to 6. However, the given simple construction is already sufficient for the translation to be polynomial.

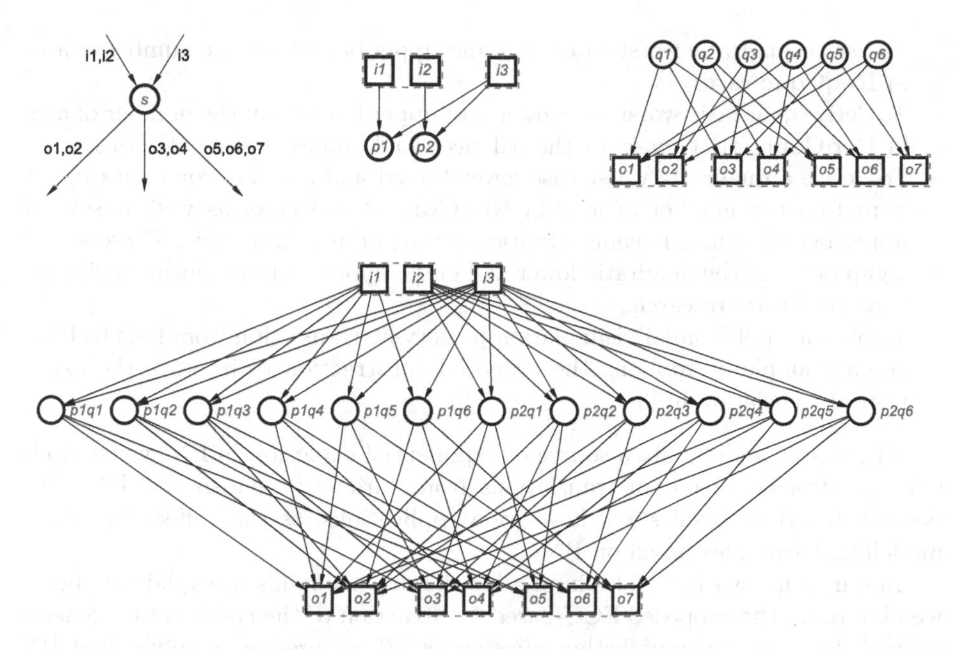

Fig. 4. An example of bisimulation-preserving BA to PN translation: (top-left) A BA state with its incoming and outgoing bursts; (top-middle) PN translations of the incoming bursts—the maximal incoming burst size is two, so two places are created; (top-right) PN translations of the outgoing bursts—$ecc(K_{2,2,3}) = 6$ places are created; (bottom) combined PN—the places corresponds to pairs in $\{p_1, p_2\} \times \{q_1, \dots, q_6\}$.

the incoming and outgoing arcs of s, except the self-loops. Then each self-loop at s labelled by a burst B is replaced by two B-labelled arcs, $s \to s'$ and $s' \to s$. This transformation at most doubles the size of the BA.

9 Conclusions

In this paper, we observed that the $\times$-construction often used for the modelling of a choice between concurrent bursts is sub-optimal and causes an exponential explosion in the size of PNs that can be avoided by better modelling. We showed equivalence between this modelling problem, and the problem of finding an ECC of a complete multipartite graph. It provided helpful insights into the former problem as well as linking it to existing results from graph theory. This enabled us:

- To show that the exponential number of places created by the $\times$-construction can be improved down to polynomial (quadratic) even in the worst case, and down to logarithmic in the best (non-degraded) case.
- To derive quadratic worst case lower and upper bounds on the number of places.

- To derive quadratic worst case lower and upper bounds on the number of arcs in **ReqConc** case.
- To derive quadratic worst case lower and upper bounds on the number of arcs in **ReqConc**(strict) case for the balanced and 'almost balanced' cases.
- To derive a quadratic worst case lower bound and a cubic worst case upper bound on the number of arcs in **ReqConc**(strict) case, as well as several upper bounds which in some situation can be better than cubic. There is still a gap between the quadratic lower and cubic upper bounds, closing which we leave for future research.
- To obtain a polynomial bisimulation-preserving translation from BAs to PNs, as an example of applying the proposed construction to improve the exponential translation in [3].

These results eliminate a source of exponential explosion in PNs when modelling control flows and in translations from various formalisms to PNs. We believe that these results will have wide applications, as they affect the 'core' modelling techniques based on PNs.

In our future work, besides improving the upper bounds as explained above, we plan to lift the proposed ECC-based construction to the case of more general control flows, and in our forthcoming paper [7] we present a polynomial PN translation for arbitrary control flows built from atomic actions using sequencing, parallel composition, and choice.

Acknowledgements. We are grateful to the anonymous reviewers for constructive comments and suggestions. A. Yakovlev acknowledges support from EPSRC (EP/N031768/1).

References

1. Alon, N.: Covering graphs by the minimum number of equivalence relations. Combinatorica **6**(3), 201–206 (1986)
2. Best, E., Devillers, R.R., Koutny, M.: Petri Net Algebra. Monographs in Theoretical Computer Science. An EATCS Series. Springer, Heidelberg (2001). https://doi.org/10.1007/978-3-662-04457-5
3. Chan, A., Sokolov, D., Khomenko, V., Lloyd, D., Yakovlev, A.: Burst automaton: framework for speed-independent synthesis using burst-mode specifications (2021, submitted paper)
4. Cortadella, J., Kishinevsky, M., Kondratyev, A., Lavagno, L., Yakovlev, A.: Logic Synthesis for Asynchronous Controllers and Interfaces. Springer, Heidelberg (2002). https://doi.org/10.1007/978-3-642-55989-1
5. Erdös, P., Goodman, A.W., Pósa, L.: The representation of a graph by set intersections. Can. J. Math. **18**, 106–112 (1966)
6. Gregory, D.A., Pullman, N.J.: On a clique covering problem of Orlin. Discrete Math. **41**(1), 97–99 (1982)
7. Khomenko, V., Koutny, M., Yakovlev, A.: Slimming down Petri boxes: Compact Petri net models of control flows (2022, in preparation)

8. Nowick, S.M., Dill, D.L.: Synthesis of asynchronous state machines using a local clock. In: Proceedings 1991 IEEE International Conference on Computer Design: VLSI in Computer & Processors, ICCD 1991, Cambridge, MA, USA, 14–16 October 1991, pp. 192–197. IEEE Computer Society (1991)
9. Valmari, A.: The state explosion problem. In: Reisig, W., Rozenberg, G. (eds.) ACPN 1996. LNCS, vol. 1491, pp. 429–528. Springer, Heidelberg (1998). https://doi.org/10.1007/3-540-65306-6_21

The Ins and Outs of Petri Net Composition

Elvio G. Amparore(✉) and Susanna Donatelli

Università degli Studi di Torino, Torino, Italy
{amparore,susi}@di.unito.it

Abstract. Composition is a key issue in Petri net modelling. It is a topic that has been studied for a long time, and that finds practical application in many Petri net frameworks and tools. Multiple approaches to composition exist, based on place or transition refinement, place superposition, transition synchronization, or sub-net substitution to cite a few. In this paper we revisit the peculiarities and technicalities (the ins and outs) of net composition based on the labelling of the net elements. We shall express general composition of nets through a combinatorial operator, that, instantiated with different policies, and completed with operators for parallel composition, label rewriting and restriction, allows us to define different forms of place-based and transition-based compositions. The use of this composition framework for model construction is also examined. For composition based on multisets of labels, we also provide an algorithm for the construction of the composed net that uses a modified version of the Farkas algorithm for the computation of semiflows.

Keywords: Petri net composition · Petri Box Calculus · CCS · CSP

1 Introduction

Composition has attracted the interest of Petri net researchers from the very early stages of the research in the field. Composition has been studied as an algebra for building nets from smaller "basic" blocks, as in the seminal work on Petri Box Calculus (PBC) [6,7] or as a way to compose existing models, independently from how they have been built (as in many tools). The definition of a "well-thought" algebra typically allows to exploit the composition also at the solution level, possibly at the price of some rigidity in the modelling process. On the other side, composing arbitrary Petri nets provides a lot of flexibility, but typically it is not as strong in terms of compositional properties and analyses. When a full algebra, with operators and associated properties is not available, it is left to the modeller to compose "reasonable" models in a "reasonable" way.

There exist multiple ways to perform composition. Models can be composed based on place or transition superposition, more rarely on both, and on place, transition or subnet substitution (also known as *refinement*). Composition rules can be based on place and transition names or on labels associated to the net elements. Especially for what concerns composition based on transitions, different

L. Bernardinello and L. Petrucci (Eds.): PETRI NETS 2022, LNCS 13288, pp. 278–299, 2022.
https://doi.org/10.1007/978-3-031-06653-5_15

interpretations are present: we shall call *CCS-like* the CCS [26] inspired transition composition (like in PBC), and *CSP-like* the CSP [19] inspired transition superposition (like in [5]).

While net algebras and their operators have been studied in-depth [6,7,23], less attention has been devoted to the composition of arbitrary nets. The research questions that motivated our work were to understand the ins and outs of net composition, and how different forms of composition can be described, and later implemented, in a single framework, and whether the standard duality principle of place and transitions carries over in this context. As a result we have defined a framework for net composition that encompasses different composition rules, whether based on net elements' label, set of labels, or multisets of labels, and whether rooted on place superposition or transition synchronization, or both at the same time, and considering different form of transition-based synchronization.

In this framework, composition of nets is expressed through a *combinatorial operator*, that, instantiated with different *policies*, allows us to define different forms of place-based and transition-based compositions. Composition policy rules are defined over the labelling of the net elements using multisets of labels. The framework is completed by operators for parallel composition, label rewriting and restriction, all implemented in the GreatSPN [1] software. The proposed framework does not include explicit operators for recursion, nor for place, transition or subnet substitution/refinement.

We can summarize the paper's contributions as follows:

- A new framework for Petri net composition in which place- and transition-based composition are treated uniformly, with a new generic composition technique, controlled by an input composition *policy instance.*
- Two composition policies (unary conjugated and n-ary structured, inspired by CCS, CSP, and PBC) and the algorithms to compute the associated policy instances.
- Examples of how known operators of other languages and various modelling patterns can be defined in the proposed framework.

1.1 A Few Examples of Net Composition

Before proceeding to the main part of the paper, it is worth to set the ground by examining a few examples of net composition. We limit these first examples to nets in which at each element is associated at most one label.

Figure 1 shows an example of *CCS-like* parallel composition, which is based on actions and co-actions (*conjugate* actions). The co-action of a is named $\hat{a}$. Actions are the labels, and are depicted on top of the transition names. In CCS when two processes are put in parallel, each action of one process synchronizes with the co-actions of the other process and vice-versa. The joint action and co-action leads to a new transition labeled τ, and no further synchronization can occur. In the figure the label τ is omitted, assuming that each net element that does not have an associated label is labelled with τ. Moreover each action and co-action are still executable in the composed process, unless a *restriction*

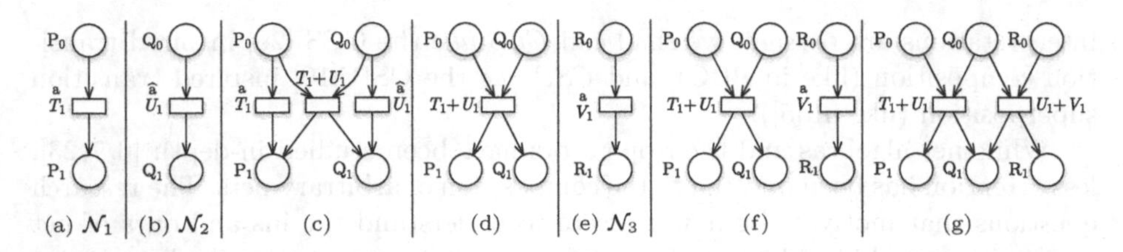

Fig. 1. CCS-like conjugate synchronization.

is specified. Figure 1(c) shows the composition of $\mathcal{N}_1$ with $\mathcal{N}_2$: it features a synchronized transition labelled τ, while still allowing both a and $\hat{a}$ to be executed. Figure 1(d) shows the composition of $\mathcal{N}_1$ with $\mathcal{N}_2$ with restriction over a: it features only the τ labelled synchronized transition. If the net in Fig. 1(d) is further composed with $\mathcal{N}_3$ no synchronization is possible, resulting in the net in Fig. 1(f).

If instead $\mathcal{N}_1$ and $\mathcal{N}_3$ are composed first, since all transitions are labelled with a, no synchronization occurs. If the resulting net is then composed with $\mathcal{N}_2$, it results in the net in Fig. 1(g), with U_1 composing with both T_1 and V_1. *From a modelling point of view* this can be seen as two processes ($\mathcal{N}_1$ and $\mathcal{N}_3$) that access at their will, but in mutual exclusion, the same resource, where label a can be interpreted as "providing" a resource and $\hat{a}$ as "requesting" it.

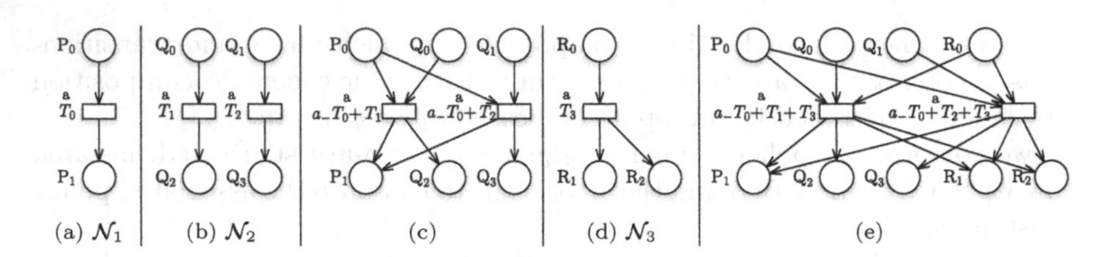

Fig. 2. CSP-like synchronization.

Figure 2 shows an example of a *CSP-like* parallel composition on a synchronization set S made of the single action a. In CSP there is no notion of co-actions. When two processes synchronize over S, each action $a \in S$ of one process synchronizes with every other action a in the other process. Actions that are not in S can still be freely executed. There is no restriction on the execution of actions that are not in S. The transition that represents the synchronization is also labelled with a, so that further synchronization can occur, permitting a straightforward implementation of the synchronization among any number of processes (multi-way synchronization).

Figure 2(c) is the result of the composition of $\mathcal{N}_1$ and $\mathcal{N}_2$ over the synchronization set $S = \{a\}$. All transitions labeled with a from $\mathcal{N}_1$ (i.e. T_0) are composed with every transition labeled with a from $\mathcal{N}_2$ (i.e. T_1 and T_2), resulting in two new transitions $a\_T_0{+}T_1$ and $a\_T_0{+}T_2$. Once synchronized, the merged transitions T_0, T_1 and T_2 are not preserved, and do not appear in (c). When (c) is composed with $\mathcal{N}_3$ resulting in net (e) that can be interpreted as a multi-way synchronization, modelling the case in which three processes need to reach a shared barrier, but may choose two different ways to do so.

From a modelling point of view also the use of CSP-like synchronization may require some cautions. With the same example of Fig. 2, if we interpret $\mathcal{N}_1$ as a resource, and $\mathcal{N}_2$ as two requests for that resource, than the net in Fig. 2(c) correctly represents the acquisition of the resource. If the obtained model is later composed with a net like $\mathcal{N}_3$, which feature *another* request of the resource, the resulting model, shown again in Fig. 2(e), does not correctly represent the resource acquisition. Indeed in this case the order of composition is important: to get the intended behaviour all resource requests have to be composed first (by composing $\mathcal{N}_2$ with $\mathcal{N}_3$ with an empty synchronization set) and then they may be composed with the resource model of $\mathcal{N}_1$.

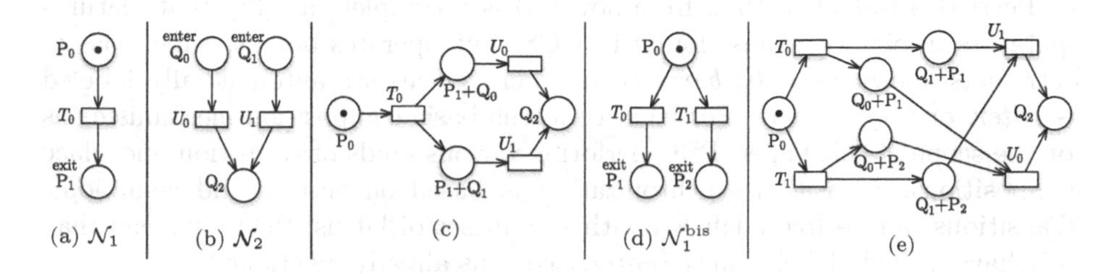

Fig. 3. Place-based composition.

Figure 3 shows an example of place composition. The net in Fig. 3(c) is the result of a simple sequential composition of $\mathcal{N}_1$ and $\mathcal{N}_2$: places are labelled as *enter*, *exit*, or τ, and exit places of net $\mathcal{N}_1$ are superposed to the enter places of net $\mathcal{N}_2$. The net in Fig. 3(e) is instead the result of the composition of net $\mathcal{N}_1^{\mathrm{bis}}$ with net $\mathcal{N}_2$: the two exit places combined with the two enter places produce four places in the resulting net, which induces a synchronization over U_0 and U_1 among two "control flows" (the one coming through P_1 and the one from P_2) that are in mutual exclusion in $\mathcal{N}_1^{\mathrm{bis}}$, leading to a deadlock. This is somehow counter-intuitive and shows that the PBC choice of having only two labels for places (*enter*, *exit*) may not be always adequate from a modelling point of view.

2 Previous Work on Composition

This section reviews some of the most relevant net algebras and examine which composition support is provided by a number of well-known Petri net tools. This review is certainly non-exhaustive, but it is meant to overview the large variety of composition rules present in the literature and provide pointers to them. Note that most of the literature presents nonuniform criteria to compose places and transitions. The framework proposed in this paper (Sect. 4) will follow instead the classical duality of Petri nets, and place- and transition-based composition will be treated using uniform rules. Since most net composition is based on *labels*, the type of labelling function is also a relevant aspect.

2.1 Net Algebras and Composition Frameworks

An early proposal for an algebra of Petri nets was provided in [23]. Starting from *simple nets* (a sequence of "head" place - transition - "tail" place), larger nets are generated through tail over head place-based composition operators. The list of operators includes superposition, merging, joining, exclusion, and other operations, with a focus on preserving structure in the composition formula. Transition-based refinement (a transition is substituted by an expression of nets) is also defined.

An early proposal for CSP-like transition synchronization can be found in [13].

Petri Box Calculus (PBC from now on) is a complete algebra that features operators inspired by those found in CCS, but operates on a specific class of Petri nets known as *Petri boxes* (nets where places are automatically labeled as *enter*, *exit* or internal). For what concerns basic composition mechanisms as for the scope of this paper, PBC performs various kinds of transition and place compositions. Transition synchronization is based on actions and co-actions. Transitions can be freely labeled with a multiset of labels. Places instead may only have a single label among $\{enter, exit\}$, as already mentioned.

Place and transition composition operators are separated. Places can be composed as *sequences* or *choices*. Transitions are composed by *synchronization*, which is a unary operator. Performing $\mathcal{N}\,\mathrm{sy}\{a\}$ synchronizes label a over net $\mathcal{N}$, which leads to the addition of a new transition for each pair of $a, \hat{a}$ transitions that can be "merged". Unary synchronization can also lead to unexpected consequences, see [7, p. 23]. Multiple nets can be composed by *parallel composition*, followed by a synchronization.

In PBC transition labels are *multisets*, which is needed to ensure that the synchronization of multiple labels is order-independent. This choice is relevant (see [7, p. 21]), and for this reason in this paper we also consider multisets of labels for each net element.

A limitation of PBC is that it does not have an explicit "multi-handshake" (or multi-way) synchronization. This type of synchronization occurs in practice [7, Sec. 2.8 and Chap. 9], for example every time a process needs to perform atomic operations on multiple variables. The *box algebra* [6, Sec. 4] is a generalization of PBC, a more abstract and general algebra that allows for multi-way synchronization, which facilitate the definition of a box algebra semantics of a process algebra like TCSP [18], as illustrated in [7, Sec. 8.2.2]. A second limitation of PBC is that its algorithm for the computation of the synchronized transitions may not terminate, generating infinite synchronizations [6, Sec. 4.5].

The work in [2] proposes a formal composition model for synchronizing multilabeled transitions in a similar way to PBC but, unlike PBC, this approach is guaranteed to always terminate. This synchronization is shown to be equivalent to the *semiflows* computation problem (i.e. finding the anullers of a matrix).

The use of semiflows, however, hides some technical details related to minimality, which result in generating only a subset of the possible interactions. We shall review this approach to overcome its limits in Sect. 5.1.

The Petri net standard PNML [29] provides support for *multi-page nets* (files containing multiple nets), but does not include a compositional specification. An examination of how to add modules in PNML, and how to construct nets from instances of such modules is given in [22], in an high level Petri net context.

Driving the modeller in the use of composition to build large models of computing systems was the objective of the PSR methodology [14], that organizes models into three layers (Processes, Services, and Resources). Each layer is defined in isolation, and then composed through CSP-like transition superposition. Transitions may have a set of associated labels, but in a well-defined manner: only single labels in the Resource layer (to model the "offer" of a resource), sets of labels for the Service layer (to model a service that acquires two or more resources at the same time) and single labels for the Process layer. Multisets of labels are not allowed, which means, for example, that a service cannot acquire two copies of the same resource at the same time.

Composition has also been considered for colored and high level nets: here the additional complexity is to appropriately define how to deal with all the extra information associated to places, transitions and arcs. High-level Petri net composition using the BPN^2 framework was introduced in [8]. Such composition is shown to be consistent with the unfolding and the operators of PBC. In CPN, a model [21] can be organised as a set of hierarchically related modules. "Substitution" transitions are replaced by subnets with well specified place-oriented input and output ports. Component aggregation of CPNs is described in [20], based on communicating modules. Hierarchical composition of Generalized Colored Stochastic Petri Nets (GCSPN) was first defined in [9], while transition superposition for the colored class of *well-formed nets* [10] was defined in [4]. A colored extension of the PSR methodology is given in [4].

2.2 Composition in Tools

Many tools supports some form of compositionality. Snoopy [17] implements hierarchical nets through P/T refinements. Nodes can be abstracted by a macro node, and a fine/coarse hierarchy can be visualized [16].

A CPN model in the CPN-Tools framework [27] can be defined hierarchically using pages and subpages [21]. Special substitution transitions connect superpages with subpages, and special tags (in/out) in the subpage allow to define the inner behaviour.

In Möebius [11] submodels are composed through superposition of places (shared state variables) [24]. It has two state-sharing formalisms: *Replicate/Join* composition and *Graph* composition. Fused places/transition have the same name (that must be unique in every composed model), hence it reduces to a simple merge of the same-name elements, and no complex combination of multiple net elements is possible.

In the ITS-tools framework [28] multiple GAL (*Guarded Action Language*) instances can be composed over synchronized events, but not over shared variables [25]. Events are labeled with symbols which guide the synchronization [3].

The GreatSPN [1] tool supports binary composition over labelled places and transitions, with some restricted form of multiset labelling and some support for colored net composition, following the rules defined in [4,5].

3 Definitions

Let Σ be a set of tags (also called symbols, actions, etc.). Tags will be used to label the places and transitions of a Petri net. Since we also label places, we prefer the use of the term *tag* instead of the more broadly used term *action*. Given a tag $a \in \Sigma$, let $\hat{a}$ be its *conjugate* (or complementary) tag. By convention $\hat{a} \neq a$ and $\hat{\hat{a}} = a$. Let $\hat{\Sigma} = \Sigma \cup \{\hat{a} \mid \forall a \in \Sigma\}$ be the set of all tags including their conjugated counterparts and $\mathcal{M}(\hat{\Sigma})$ be the set of all natural multisets of tags (including their conjugates). Elements of $\mathcal{M}(\hat{\Sigma})$ are indicated by formal sums and τ denotes the empty multiset, so, given the set of tags $\Sigma = \{a, b, c\}$, $a + 2 \cdot \hat{a} + 2b$, $\hat{a} + c$ and τ are examples of multisets of tags. A multiset of tags is *canonical* if it does not include both a tag and its conjugate (therefore $a+2\cdot\hat{a}+2b$ is not canonical).

Given $\sigma \in \mathcal{M}(\hat{\Sigma})$ and $A \subseteq \Sigma$, we indicate with $\sigma \setminus A$, the multiset obtained by removing all tags in A, and their conjugates. Notation $\sigma[a]$ denotes the multiplicity of a in σ.

Definition 1 (Labeled Petri net). *It is a tuple $\mathcal{N} = \langle P, T, I, O, \mathbf{m}_0, lab \rangle$, where P is the set of places, T is the set of transitions, $I : P \times T \to \mathbb{N}$ is the input function, $O : T \times P \to \mathbb{N}$ is the output function, $\mathbf{m}_0 : P \to \mathbb{N}$ is the initial marking, and $lab : (P \cup T) \to \mathcal{M}(\hat{\Sigma})$ is the net element labeling function.*

We use the term *net element* to identify elements in $(P \cup T)$ and *label* (of a net element) to indicate the multiset of tags associated to the net element by the labelling function *lab*. We consider only labels that are canonical. Let $\hat{\Sigma}_P$ and $\hat{\Sigma}_T$ be the subsets of $\hat{\Sigma}$ that appear on the labels of the place set P and on the transition set T, respectively. With $\mathcal{M}(P)$ we denote the set of the natural multisets of places, which can be represented as a weighted sum of elements of P, like $P_1 + 3{\cdot}P_4 + P_5$. Similarly $\mathcal{M}(T)$ is used for the transitions.

For notational convenience, we also use a matrix-oriented representation of a Petri net. Let $\mathbf{I} : |P| \times |T|$ and $\mathbf{O} : |T| \times |P|$ be the input and the output matrix of $\mathcal{N}$, respectively, with $\mathbf{I}[p,t] = I(p,t)$ and $\mathbf{O}[p,t] = O(p,t)$. Let $\mathbf{L}_\mathrm{P} : |P| \times |\Sigma_P|$ be the place labeling matrix, where $\mathbf{L}_\mathrm{P}[p,a]$ is the multiplicity of tag a in $lab(p)$ and it is negative if a appears conjugated, positive otherwise. Similarly, let $\mathbf{L}_\mathrm{T} : |T| \times |\Sigma_T|$ be the transition labeling matrix. A full example of the net matrices will be given at the end of Sect. 4.

Definition 2 (Semiflows). *Given an integer matrix $\mathbf{A}$, a flow $\mathbf{f}$ is an integer vector s.t. $\mathbf{f} \cdot \mathbf{A} = 0$, i.e. $\mathbf{f}$ is a* left anuller *of $\mathbf{A}$. A semiflow is a non-negative* *flow. The support* $[\![\mathbf{f}]\!]$ *of a flow* $\mathbf{f}$ *is the set of indices of the non-zero values, i.e.* $[\![\mathbf{f}]\!] = \{i \mid \mathbf{f}[i] \neq 0\}$*. A semiflow is canonical iff the g.c.d. of its non-zero entries is* 1*. A semiflow is minimal iff it is canonical and its support does not strictly contain the support of any other semiflow of* $\mathbf{A}$*. The set of all minimal semiflows is finite and unique, and let* $\mathbf{F}$ *be the matrix of the minimal semiflows [12, p. 82].*

4 A Framework for Net Composition

We proceed by defining a framework for composing labelled Petri nets that allows us to define, among others, the cases discussed in Sect. 1. We define the basic operations for net composition in terms of four basic operations. Three operations perform basic transformations on labels (tag rewriting, restriction) and merge multiple nets together without combining the elements (parallel composition). The fourth operation (combinatorial composition) combines net elements together by applying a policy.

Tag Rewriting. A *tag rewriting function* is a function $\lambda : \hat{\Sigma} \to \hat{\Sigma}$ that transforms tags. By extension, given a multiset of tags $\phi = w_1 \cdot a_1 + \ldots + w_n \cdot a_n$, let $\lambda(\phi)$ be the canonical multiset resulting from the application of λ to every tag, i.e. the canonical form of $w_1 \cdot \lambda(a_1) + \ldots + w_n \cdot \lambda(a_n)$. We define the *tag rewriting* operation on a net $\mathcal{N}$, denoted as $\lambda(\mathcal{N})$, as an operation that builds a new net $\mathcal{N}'$ where labels have been rewritten, i.e. $lab' = \lambda \circ lab$.

Parallel Composition. This operation juxtaposes multiple independent nets together into a single net. Given $\mathcal{N}_1 \ldots \mathcal{N}_n$ nets, let $\mathcal{N}_1 \parallel \ldots \parallel \mathcal{N}_n$ be a new net $\mathcal{N}'$ defined as:

- $P' = \cup_{i=1}^n P_i$ and $T' = \cup_{i=1}^n T_i$;
- $I'(p', t') = I_{\theta(p')}(p', t')$ if $\theta(p') = \theta(t')$, and 0 otherwise;
- $O'(t', p') = I_{\theta(p')}(t', p')$ if $\theta(p') = \theta(t')$, and 0 otherwise;
- $\mathbf{m}'_0(p') = (\mathbf{m}_0)_{\theta(p')}(p')$;
- $lab'(p') = lab_{\theta(p')}(p')$ and $lab'(t') = lab_{\theta(t')}(t')$;

where the function $\theta : P \times T \to \mathbb{N}$ is defined to associate each P/T elements of $\mathcal{N}$ to the index of the original net $\mathcal{N}_i$.

Restriction. This operation removes from a net $\mathcal{N}$ all elements whose label includes any of the tags in the set of *restriction tags* $A \subseteq \Sigma$ or their conjugates. The new net is indicated as $\mathcal{N}' = \mathcal{N} \setminus A$ and it is defined by:

- $P' = \{p \in P \mid \forall a \in A : lab(p)[a] = lab(p)[\hat{a}] = 0\}$;
- $T' = \{t \in T \mid \forall a \in A : lab(t)[a] = lab(t)[\hat{a}] = 0\}$;
- $I'(p, t) = I(p, t)$ and $O'(t, p) = O(t, p)$, for all $p \in P', t \in T'$;
- $\mathbf{m}'_0(p) = \mathbf{m}_0(p)$, for all $p \in P'$.
- $lab'(x) = lab(x)$, for all $x \in P' \cup T'$.

Combinatorial Composition. This operation alters the behaviour of a net $\mathcal{N}$ by defining a new set of places and transitions made as combinations of the net elements of the original net. Each new place (resp. transition) that is being composed is identified by a multiset of *composing places (resp. transitions)* from the original net. We divide the net composition into two tasks:

1. Identifying which *multisets* of places (transitions) will be composed together to form each new place (transition). These are described by a *composition instance* π, which can be generated by *composition policy* (defined in the next section).
2. Defining a new net with the new net elements, connected according to a composition of the original input and output functions.

A *composition instance* π is a pair $\pi = \langle C_P, C_T \rangle$, with $C_P \subseteq \mathcal{M}(P) \times \mathcal{M}(\hat{\Sigma}_P)$ and $C_T \subseteq \mathcal{M}(T) \times \mathcal{M}(\hat{\Sigma}_T)$. We use the notation $\langle \phi, \sigma \rangle$ to denote tuples in C_P, and $\langle \psi, \varsigma \rangle$ to denote tuples in C_T.

Given a net $\mathcal{N}$ and a composition instance π, the *combined net* $\mathcal{N}' = \mathcal{N} * \pi$ is obtained in the following way. Each tuple $\langle \phi, \sigma \rangle \in C_P$ defines a new place p' of $\mathcal{N}'$, s.t. the multiset ϕ tells the weighted combination of places of $\mathcal{N}$ that are combined together to form p', while σ is the label of p' Transitions follow a similar schema from C_T.

The combined net $\mathcal{N}'$ is defined as

- $P' = \{\text{new place } p' \text{ for each } \langle \phi, \sigma \rangle \in C_P\}$;
- $T' = \{\text{new transition } t' \text{ for each } \langle \psi, \varsigma \rangle \in C_T\}$;
- $I'(p', t') = \sum_{p \in P} \sum_{t \in T} \phi_{p'}[p] \cdot \psi_{t'}[t] \cdot I(p, t)$;
- $O'(p', t') = \sum_{p \in P} \sum_{t \in T} \phi_{p'}[p] \cdot \psi_{t'}[t] \cdot O(p, t)$;
- $\mathbf{m}_0'(p') = \sum_{p \in P} \phi_{p'}[p] \cdot \mathbf{m}_0(p)$;
- $lab(p') = \sigma_{p'}$ and $lab(t') = \varsigma_{t'}$.

with $\langle \phi_{p'}, \sigma_{p'} \rangle$ and $\langle \psi_{t'}, \varsigma_{t'} \rangle$ the tuples that originated p' and t', respectively.

Figure 4 shows an example of a composed net, where π is

$$C_P = \left\{ \begin{array}{c} \langle P_0, \tau \rangle, \\ \langle P_1, \tau \rangle, \\ \langle P_2, e \rangle, \\ \langle P_3, 2\hat{e} \rangle, \\ \langle \{2 \cdot P_2 + P_3\}, \tau \rangle \end{array} \right\}, \quad C_T = \left\{ \begin{array}{c} \langle T_0, a + b + c \rangle, \\ \langle T_1, \hat{a} + \hat{b} + 2d \rangle, \\ \langle T_0 + T_1, c + 2d \rangle \end{array} \right\}$$

i.e. two new net elements are added, $2 \cdot P_2 + P_3$ and $T_0 + T_1$, and all the other net elements are preserved. In the figure, drawn with the GreatSPN tool, multisets of tags are represented as tags separated by bars, so for instance $\hat{a} + \hat{b} + 2d$ is depicted as $\widehat{a}|\widehat{b}|2d$. Observe that the new arcs connecting the new nodes have the sum of the multiplicities. For instance, $O(T_0 + T_1, 2 \cdot P_2 + P_3) = 3$ because it is $2 \cdot O(T_0, P_2) + O(T_1, P3)$.

Alternatively, we can view the combinatorial composition as a matrix operation over the net elements. Let $\mathbf{F}_\mathrm{P}$ be a $|P'| \times |P|$ matrix that encodes the places

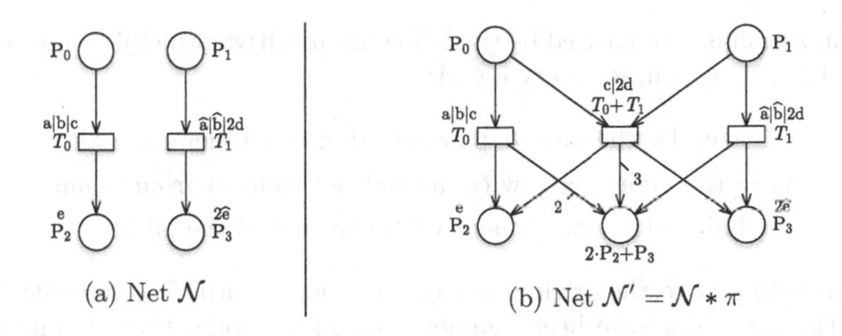

Fig. 4. Example of combinatorial composition for both places and transitions.

in C_P and $\mathbf{L}'_\mathrm{P}$ be a $|P'| \times |\Sigma_P|$ matrix that encodes the place labels, where conjugated labels are represented as negative values. Thus $[\mathbf{F}_\mathrm{P}|\mathbf{L}'_\mathrm{P}]$ encodes C_P in matrix form. Similarly, $\mathbf{F}_\mathrm{T} : |T'| \times |T|$ and $\mathbf{L}'_\mathrm{T} : |T'| \times |\Sigma_T|$ are used for C_T. For the example of Fig. 4, we have:

$$[\mathbf{F}_\mathrm{P}|\mathbf{L}'_\mathrm{P}] = \begin{array}{c} \begin{matrix} P_0 & P_1 & P_2 & P_3 & e \end{matrix} \\ \left[\begin{array}{cccc|c} 1 & 0 & 0 & 0 & 0 \\ 0 & 1 & 0 & 0 & 0 \\ 0 & 0 & 1 & 0 & 1 \\ 0 & 0 & 0 & 1 & -2 \\ 0 & 0 & 2 & 1 & 0 \end{array}\right] \end{array} \qquad [\mathbf{F}_\mathrm{T}|\mathbf{L}'_\mathrm{T}] = \begin{array}{c} \begin{matrix} T_0 & T_1 & a & b & c & d \end{matrix} \\ \left[\begin{array}{cc|cccc} 1 & 0 & 1 & 1 & 1 & 0 \\ 0 & 1 & -1 & -1 & 0 & 2 \\ 1 & 1 & 0 & 0 & 1 & 2 \end{array}\right] \end{array}$$

Then, we can write the net composition in terms of matrix operations:

- The input matrix is $\mathbf{I}' = \mathbf{F}_\mathrm{P} \times \mathbf{I} \times \mathbf{F}_\mathrm{T}^T$;
- The output matrix is $\mathbf{O}' = \mathbf{F}_\mathrm{T} \times \mathbf{O} \times \mathbf{F}_\mathrm{P}^T$;
- The initial marking $\mathbf{m}'_0 = \mathbf{F}_\mathrm{P} \times \mathbf{m}_0$.

If we indicate with **Id** the identity matrix, whenever $[\mathbf{Id}|\mathbf{L}_\mathrm{P}]$ is a submatrix of $[\mathbf{F}_\mathrm{P}|\mathbf{L}'_\mathrm{P}]$ under some row permutation, the transformation is a *place extension*, since all places of $\mathcal{N}$ are preserved in $\mathcal{N}'$. A similar notion of *transition extension* can be defined on $[\mathbf{F}_\mathrm{T}|\mathbf{L}'_\mathrm{T}]$.

5 Composition Policies

We now define two *composition policies* to generate composition instances according to two paradigms inspired by PBC and of CSP.

5.1 Unary Conjugated Composition

We start by considering a composition policy for places and transitions which is based on the merging of conjugated tags, as in PBC transition synchronization. Let $A \subseteq \Sigma$ be the set of tags considered for the operation. The PBC transition synchronization is summarized by the following intuitive principle (adapted from [6, sec. 4.5]), by which, for every $a \in A$:

$$\begin{array}{c}\text{Repeatedly choose } a, \hat{a}\text{–pairs of labeled net elements, and}\\ \text{each time create a new composed net element from them.}\\ \text{Label this new element with the sum of the labels.}\end{array} \quad (1)$$

The idea is to add to the original net new elements until all combinations that reduce the $a, \hat{a}$–pairs have been enumerated. This implies that (1) enumerates both the combinations that reach a τ-label, as well as all intermediate steps which may still have tags of Σ.

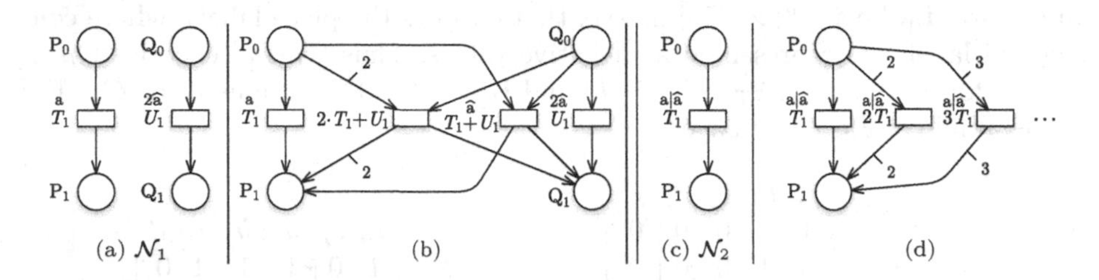

Fig. 5. Multitag conjugate composition and infinite PBC synchronization.

Figure 5(b) shows an example of the application of (1) to $\mathcal{N}_1$: T_1 and U_1 are retained, synchronization of T_1 and U_1 leads to transition $T_1 + U_1$, labelled with $\hat{a}$, which can be synchronized with T_1, leading to transition $2 \cdot T_1 + U_1$, labelled τ. There are no other composition of transitions that satisfies (1).

Note that this approach, which is the one employed by PBC for transition composition [6, Sec. 4.5], may repeatedly choose the same transition as a pair if it is labelled with both a tag a and its conjugate $\hat{a}$. Figure 5(d) shows an example of such PBC synchronization, applied to $\mathcal{N}_2$. In that case, transition T_1 is composed infinitely many times with itself, since each composition adds the labels $a + \hat{a}$ and then removes a single pair of $a, \hat{a}$ tags. Such scenario may happen because PBC labels may be multisets of tags that are not canonical. That's why we restrict our work to canonical labels.

The computation of all transitions pairs to be added can be non trivial, and Anisimov proposes in [2] an algorithm that is based on the computation of *minimal P/T-semiflows*. The Anisimov algorithm works on transitions with canonical labels, and it is based on the Farkas algorithm [12,15] for computing the *minimal P/T-semiflows*: therefore it is proved to always terminate. The intuition is that the goal of finding which combination of net elements reduces the sum of their labels to τ is equivalent to finding a linear combinations $\mathbf{f}$ of labels that cancel the tags in A, i.e. $\mathbf{f} \cdot \mathbf{L} = 0$. Vector $\mathbf{f}$ is, by Definition 2, a semiflow of $\mathbf{L}$. The Anisimov algorithm although, by considering only minimal semiflows, does not correspond exactly to (1) because:

- compositions that result in a transition with a label different from τ, like transition $T_1 + U_1$ in Fig. 5(b), are not generated;
- the algorithm does not either generates all composed transitions labelled with τ. Indeed the algorithm only considers minimal semi-flows, while these transitions corresponds to semi-flows that are non-minimal, although canonical (as explained later with reference to the example in Fig. 6).

Algorithm 1. Modified Farkas algorithm for unary conjugated composition.

```
1: procedure ConjugateCompSet(L, A) // L is a N × M matrix
2:     [D|A] ← [Id|L]
3:     for j between 1 and M do
4:         if j corresponds to a column of A then
5:             for each r1 ≠ r2 with A[r1, j] > 0 ∧ A[r2, j] < 0 do
6:                 [d|a] = [D|A][r1, ·] + [D|A][r2, ·]
7:                 [d|a] ← [d|a] / gcd([d|a])
8:                 if [d|a] does not appear in [D|A] then
9:                     [D|A] ← AppendRows([D|A], [d|a])
10:                end if
11:            end for
12:        end if
13:    end for
14:    return [D|A]
15: end procedure
```

We therefore propose a modified Farkas algorithm for determining the instance of a composition policy that follows (1) and that overcomes the limitations of the Anisimov algorithm listed above. The pseudocode is shown in Algorithm 1. To compute the composition instance π, the unary conjugated composition policy extends both places and transitions simultaneously. To do so, the method is used twice, once for the places and once for the transitions, i.e.

$$[\mathbf{F}_{\mathrm{P}}|\mathbf{L}'_{\mathrm{P}}] \leftarrow \textsc{ConjugateCompSet}\mathbf{L}_{\mathrm{P}}, A$$
$$[\mathbf{F}_{\mathrm{T}}|\mathbf{L}'_{\mathrm{T}}] \leftarrow \textsc{ConjugateCompSet}\mathbf{L}_{\mathrm{T}}, A$$

Assume that we want to compute the composition of transitions. The method takes in input a $|T| \times |\Sigma_T|$ matrix $\mathbf{L}$, where $\mathbf{L}[t, s]$ is the multiplicity of tag s in label $lab(t)$, and Σ_T is the set of tags appearing on transitions. Conjugated tags appear as negative numbers in $\mathbf{L}$. The objective of Algorithm 1 is to find all linear combinations of labels that combine a–$\hat{a}$ pairs, until all τ combinations are generated. The loop at lines 3–13 considers one tag at a time. The tags in A are used to generate the combination rows. The inner loop 5–11 identifies all candidate combinations of net elements with a–$\hat{a}$ pairs in their labels. A combination is obtained by summing row r_1 with r_2, assuming that r_1 has tag a and r_2 has tag $\hat{a}$. The algorithm generates all such combinations until a fixed point is reached. Convergence is guaranteed since each new row $[\mathbf{d}|\mathbf{a}]$ generated at line 6 is such that $|\mathbf{a}[j]| < |\mathbf{A}[r_1, j]|$ and $|\mathbf{a}[j]| < |\mathbf{A}[r_2, j]|$, and the algorithm stops when all possible rows with $\mathbf{a} = 0$ are generated. However, Algorithm 1 may require an exponential number of steps to terminate.

While generating all the combinations that reduce the tag pairs, the algorithm may reach a point where all A–tags are zeroed for a row. In that case, the vector $\mathbf{d}$ for that row is a semiflow of the initial system. Unlike the Farkas algorithm, there is no check of *minimality* of such semiflows. Therefore all canonical semiflows are found. A second difference from the Farkas algorithm is the selection of the candidate vector at line 6. For semiflow computation, the vector $[\mathbf{d}|\mathbf{a}]$ would be computed as:

$$m_2 \cdot [\mathbf{D}|\mathbf{A}][r_1, \cdot] \; + \; m_1 \cdot [\mathbf{D}|\mathbf{A}][r_2, \cdot], \quad m_1 = |\mathbf{A}[r_1, j]|, \; m_2 = |\mathbf{A}[r_2, j]|$$

which would zero the value of $\mathbf{a}[j]$. By not multiplying by m_1 and m_2, all intermediate steps to reach the zero for $\mathbf{a}[j]$ are stored as rows in $[\mathbf{D}|\mathbf{A}]$. Each intermediate step can be seen as a new pair of $a, \hat{a}$ tags being cancelled from two groups of net elements, therefore implementing the principle (1).

Consider the net in Fig. 6(a) and the set $A = \{a, b\}$. The initial $[\mathbf{Id}|\mathbf{L}]$ matrix for transitions is shown in (2a). Row operations combine progressively row pairs, until a fixpoint is reached. All rows are kept.

$$\begin{array}{r} \\ row_1: \\ row_2: \\ row_3: \\ row_4: \end{array} \begin{array}{c} \begin{array}{cccccc} T_1 & T_2 & T_3 & T_4 & a & b \end{array} \\ \left[\begin{array}{cccc|cc} 1 & 0 & 0 & 0 & 2 & 0 \\ 0 & 1 & 0 & 0 & 0 & -1 \\ 0 & 0 & 1 & 0 & -1 & 1 \\ 0 & 0 & 0 & 1 & -1 & 0 \end{array}\right] \end{array}$$

(2a) Matrix $[\mathbf{Id}|\mathbf{L}]$

$$\begin{array}{r} \\ row_5: \\ row_6: \\ row_7: \\ row_8: \\ row_9: \\ row_{10}: \\ row_{11}: \\ row_{12}: \\ row_{13}: \\ row_{14}: \end{array} \begin{array}{c} \begin{array}{cccccc} T_1 & T_2 & T_3 & T_4 & a & b \end{array} \\ \left[\begin{array}{cccc|cc} 1 & 0 & 1 & 0 & 1 & 1 \\ 1 & 0 & 0 & 1 & 1 & 0 \\ 1 & 0 & 2 & 0 & 0 & 2 \\ 1 & 0 & 1 & 1 & 0 & 1 \\ 1 & 0 & 0 & 2 & 0 & 0 \\ 0 & 1 & 1 & 0 & -1 & 0 \\ 1 & 1 & 1 & 0 & 1 & 0 \\ 1 & 1 & 2 & 0 & 0 & 1 \\ 1 & 1 & 1 & 1 & 0 & 0 \\ 1 & 2 & 2 & 0 & 0 & 0 \end{array}\right] \end{array} \begin{array}{l} \\ \\ \\ \\ \\ \text{minimal s.f.} \\ \\ \\ \\ \text{s.f.} \\ \text{minimal s.f.} \end{array} \qquad (2)$$

(2b) Rows appended to matrix $[\mathbf{D}|\mathbf{A}]$

The final matrix $[\mathbf{D}|\mathbf{A}]$ is made by all the initial rows of (2a) together with the rows in (2b), which contain the canonical semiflows (minimal and not), if they exists, and all the intermediate pairwise combinations.

The Anisimov algorithm generates the composition elements based exclusively on the minimal semiflows, i.e. Fig. 6(b).[1] The full unary conjugated composition policy, denoted as $\mathcal{N} *_C A$ with $A = \{a, b\}$, corresponds to generating the 14 transitions resulting from the rows of $\mathbf{D}$ in (2b). Each row $[\mathbf{d}|\mathbf{a}]$ results

[1] In (2b) row_{13} is not minimal because its support (all four transitions) contains the supports of both row_9 and row_{14}.

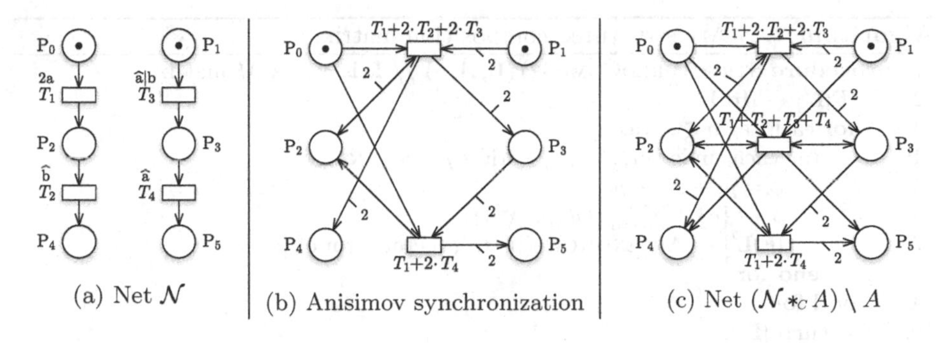

(a) Net $\mathcal{N}$
(b) Anisimov synchronization
(c) Net $(\mathcal{N} *_C A) \setminus A$

Fig. 6. Unary conjugated composition of $\mathcal{N}$, with $A = \{a, b\}$.

in a tuple $\langle \psi, \varsigma \rangle \in C_T$. Note that every initial transition is also preserved, since they appear as rows in $\mathbf{D}$. If we further restrict to A, for sake of readability, we obtain the net $(\mathcal{N} *_C A) \setminus A$ depicted in Fig. 6(c), where only the elements corresponding to semiflows are added.

The unary conjugated composition policy is a *place/transition extension*, since new net elements are added and no net element is removed.

5.2 N-Ary Structured Composition

The second composition policy that we consider is defined over a parallel composition of $n > 1$ nets $\mathcal{N} = (\mathcal{N}_1 \parallel \ldots \parallel \mathcal{N}_n)$. Again, a set of tags $A \subseteq \Sigma$ is defined to guide the policy. For each tag $a \in A$, new net elements result from composing one net element from every subnet $\mathcal{N}_1 \ldots \mathcal{N}_n$ that is labeled with a. For this policy there is no notion of conjugated tags. Moreover, the resulting multiplicity is 1 independently of the input tag multiplicities. When focusing on transitions only, this composition is similar to the parallel composition of CSP [19].

The sets C_P and C_T are computed independently. Consider the problem of identifying the places that will be composed together. For each tag $a \in A$, for every tuple of places $\langle p_1, \ldots p_n \rangle$ with $\forall i \geq n : \theta(p_i) = i \wedge a \in lab(p_i)$, then the tuple $\langle \phi, \sigma \rangle$ belongs to C_P, with:

- $\phi = p_1 + \ldots + p_n$, with all weights being one;
- $\sigma = \{a\} + \sum_{i=1}^{n} \big(lab(p_i) \setminus A\big)$.

An equivalent definition applies for the transitions.

Algorithm 2 shows the pseudo-code of the structured composition policy. To compute C_P and C_T, the algorithm is applied twice, i.e.

$$[\mathbf{F}_\mathrm{P} | \mathbf{L}'_\mathrm{P}] \leftarrow \textsc{StructuredCompSet}\mathbf{L}_\mathrm{P}, A, \theta$$
$$[\mathbf{F}_\mathrm{T} | \mathbf{L}'_\mathrm{T}] \leftarrow \textsc{StructuredCompSet}\mathbf{L}_\mathrm{T}, A, \theta$$

We define two variations of the structured composition:

Algorithm 2. N-Ary structured composition matrix.

1: **procedure** StructuredCompSet($\mathbf{L}, A, \theta$) // $\mathbf{L}$ is a $N \times M$ matrix
2: $[\mathbf{F}|\mathbf{L}'] \leftarrow [\mathbf{Id}|\mathbf{L}]$
3: **for** each tag $a \in A$ **do**
4: **for** each tuple $\langle p_1, \dots, p_n \rangle$ with $\theta(p_i) = i$ **do**
5: $\phi = p_1 + \ldots + p_n$
6: $\sigma \leftarrow \{a\} + \sum_{i=1}^{n} \big(lab(p_i) \setminus A\big)$
7: $[\mathbf{F}|\mathbf{L}'] \leftarrow$ AppendRows($[\mathbf{F}|\mathbf{L}']$, vector form of $[\phi|\sigma]$)
8: **end for**
9: **end for**
10: **return** $[\mathbf{F}|\mathbf{L}']$
11: **end procedure**

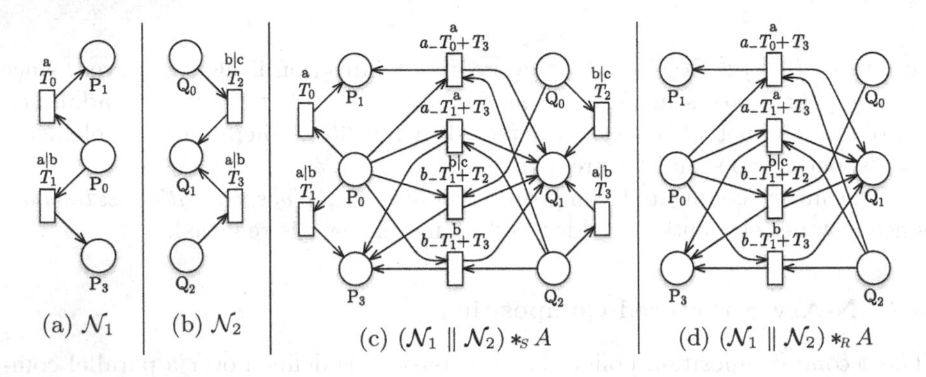

(a) $\mathcal{N}_1$ (b) $\mathcal{N}_2$ (c) $(\mathcal{N}_1 \parallel \mathcal{N}_2) *_S A$ (d) $(\mathcal{N}_1 \parallel \mathcal{N}_2) *_R A$

Fig. 7. Structured composition example, with $A = \{a, b\}$.

- *Structured extension:* $(\mathcal{N}_1 \parallel \ldots \parallel \mathcal{N}_n) *_S A$ extends $\mathcal{N}$ with the new elements. All existing net elements are kept.
- *Restricted structured composition:* $(\mathcal{N}_1 \parallel \ldots \parallel \mathcal{N}_n) *_R A$ first extends $\mathcal{N}$ with the new elements, and then removes all the elements of $(\mathcal{N}_1 \parallel \ldots \parallel \mathcal{N}_n)$ that were used to generate the new elements.

Figure 7 shows an example of both a structured and a restricted extensions of a parallel composition of two nets, with $A = \{a, b\}$. Only transitions are composed in this example. Four new transitions are added to C_T, corresponding to $\langle T_0 + T_3, a\rangle$, $\langle T_1 + T_3, a\rangle$, $\langle T_1 + T_2, b + c\rangle$, and $\langle T_1 + T_3, b\rangle$. Observe that $T_1 + T_3$ is composed twice, once for tag a and once for tag b. The *restricted* structured composition results in Fig. 7(d).

Notes on Operations. Tag rewriting and *restriction* could also be defined as composition policies.

6 Modelling Using the Composition Framework

We now focus on the usefulness of the framework for generating new models from existing ones.

6.1 Place-Based Composition

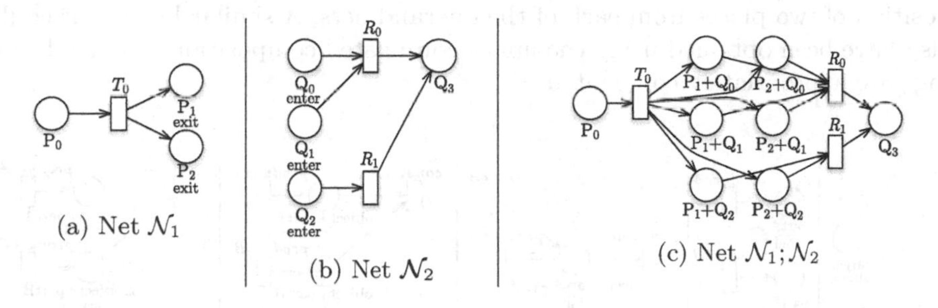

Fig. 8. Sequence as: $(\mathcal{N}_1 \parallel \lambda_{enter \rightarrow \widehat{exit}}(\mathcal{N}_2)) *_C A \setminus A$, with $A = \{exit\}$.

Place Sequence. Sequential composition is a typical composition pattern found in several algebras (CCS, CSP, PBC, and others). In the following we show how to use the framework to provide the sequence operator of PBC, that is based on the notion of entry and exit places. Assume a net $\mathcal{N}_1$ has some places labeled with a tag *exit* and a net $\mathcal{N}_2$ has some places labeled as *enter*. We can connect the exit places of $\mathcal{N}_1$ with the enter places of $\mathcal{N}_2$ by means of tag rewriting and combinatorial composition. Both the unary conjugated composition and the n-ary structured composition can be adopted. Figure 8 shows an example of sequential composition on places performed using unary conjugated composition, that leads to the formula: $(\mathcal{N}_1 \parallel \lambda_{enter \rightarrow \widehat{exit}}(\mathcal{N}_2)) *_C A \setminus A$, with $A = \{exit\}$. Tag rewriting is particularly useful in these situations, since it allows to identify pairs of conjugated tags that do not need to have the same name in the operand nets.

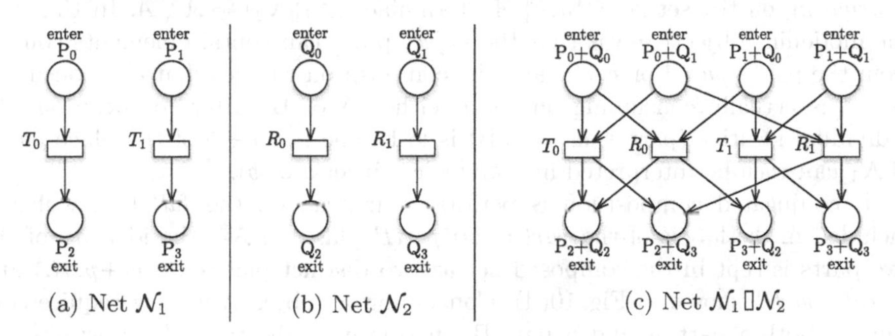

Fig. 9. Choice as $(\mathcal{N}_1 \parallel \mathcal{N}_2) *_R A$, with $A = \{enter, exit\}$.

Place Choice. Another common compositional pattern is *choice*. Again, our example consider the choice operator of PBC. In this setting, *enter* places of the two nets are multiplied together, the same for *exit* places, to split and then merge the control flows of the nets. Figure 9 shows an example of choice composition, taken from [6, Fig. 4]. The resulting net is obtained using restricted n-ary structured composition on the two tags $\{enter, exit\}$ leading to $(\mathcal{N}_1 \parallel \mathcal{N}_2) *_R A$, with $A = \{enter, exit\}$. Note that each place in the final net results from a composition of two places from each of the operand nets. A similar behaviour could also have been obtained using the unary conjugated composition, provided that tags are appropriately conjugated.

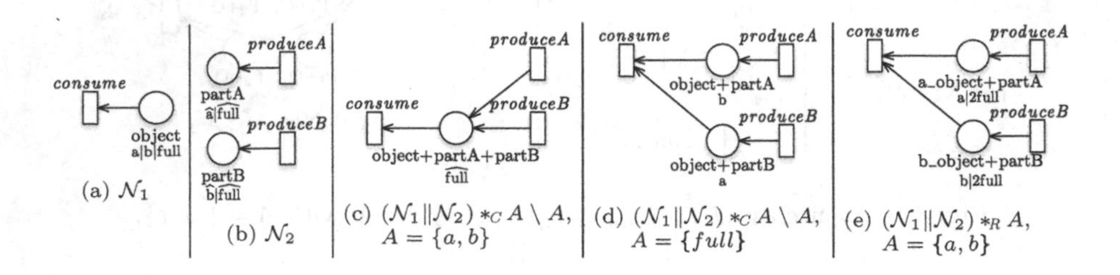

Fig. 10. Examples of place composition.

General Place Composition. Figure 10 shows different types of place-based composition for two nets $\mathcal{N}_1$ and $\mathcal{N}_2$, that induce a different interpretation of the multiset of tags associated to places. Indeed using different type of tags and different composition policies we can achieve rather diverse interpretations: this can be an advantage, but it requires a certain modelling expertise to appropriately master the composition process. Nets $\mathcal{N}_1$ and $\mathcal{N}_2$ in Fig. 10(a) and Fig. 10(b) can be interpreted as a very simple consumer and producer models. The net in Fig. 10(c) is obtained through unary conjugated composition, followed by restriction, on the set $A = \{a, b\}$. In formulae: $(\mathcal{N}_1 \| \mathcal{N}_2) *_C A \setminus A$. In this case the modelling objective was that the *object* place can contain elements coming from the places *partA* or *partB* and the composition ensures that any consumed object is actually consuming one part, either A or B. After composition, the individual identity (part A or part B) is lost. The tag $a + b$ of the place *object* of $\mathcal{N}_1$ can then be interpreted in a *or*-logic (either a or b).

If conjugated composition is performed instead on the *full* tag, which is included in the labels of the *partA* and *partB* places of $\mathcal{N}_2$, the identity of the two parts is kept in the composed net as two distinct places *object+partA* and *object+partB*, shown in Fig. 10(d). Consuming an object will now require consuming both a part A and a part B. In this way, the tag *full* interprets the composition using an *and*-logic (one token from every *full* place).

When place composition is realized through the n-ary structured composition policy, the result is similar to the *and*-logic. If composition is performed on tags $\{a, b\}$, each tag will result in an individual place, as in Fig. 10(e). Similarly, if composition is performed on $\{full\}$, this will again lead to two places. To have a 3-way composition resulting in a single place, the net $\mathcal{N}_2$ has to be separated further into two subnets, one for part A and one for part B, such that each subnet has only a single place tagged with *full*.

6.2 Transition-Based Composition

Composition of concurrent events as a composition of the transitions sharing the same tags (*synchronization*) is an important feature of any net algebra. We shall first consider how the proposed framework can express the CCS-like and CSP-like parallel operators, through which synchronization can be achieved, to then show how to express multi-way synchronization in two different forms.

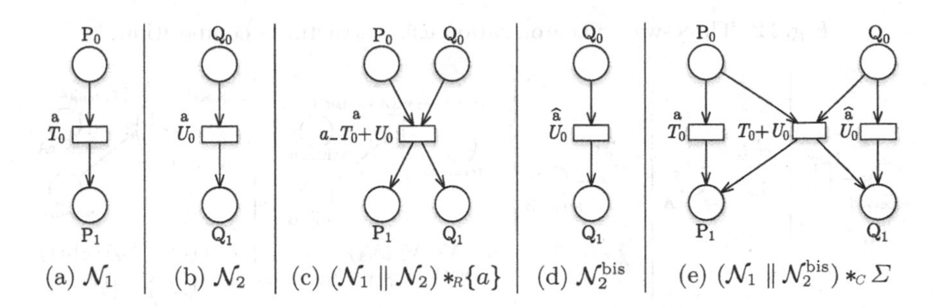

Fig. 11. CSP and CCS parallel operators.

CSP and CCS Parallel Operators. Nets $\mathcal{N}_1$ and $\mathcal{N}_2$ in Fig. 11 show two simple processes, let's say P and Q, that can both execute a. The net equivalent to the CSP process $P \parallel_{\{a\}} Q$ (parallel composition of P and Q with synchronization over action a) can be obtained as $(\mathcal{N}_1 \parallel \mathcal{N}_2) *_R \{a\}$, and it is depicted in Fig. 11(c).

If we now consider for process Q the net $\mathcal{N}_2^{\text{bis}}$ in Fig. 11(d), the net equivalent to the CCS process $P \parallel Q$ (parallel composition of P and Q over conjugate actions) can be obtained as $(\mathcal{N}_1 \parallel \mathcal{N}_2) *_C \{\Sigma\}$, and it is depicted in Fig. 11(e). Note that the composition is over the whole set Σ of actions (tags) as in CCS (in its original form): there is no way to limit the set of actions on which synchronization takes place and the resulting net can correctly execute independently also action a and $\hat{a}$, while the synchronized action is labelled τ. Moreover the two actions a and $\ddot{a}$ are still executable.

Multi-way Synchronization requires a different approach, depending on whether we have a single common tag or conjugate tags. In the former case we can use structured composition, while in the latter one we need to use multiple tags.

Figure 12 shows how to realize a three-way synchronization with a single common tag. The three operand nets $\mathcal{N}_1$, $\mathcal{N}_2$ and $\mathcal{N}_3$ all have a transition with tag a. Restricted structured composition merges these transitions into a single transition, that is also connected to all input and output places.

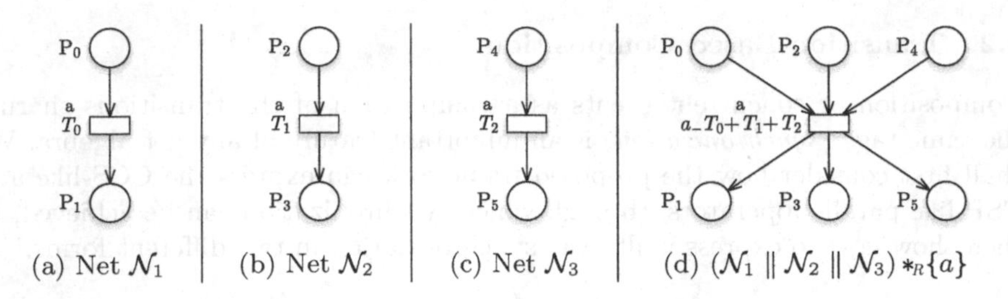

Fig. 12. Three-way synchronization using structured composition.

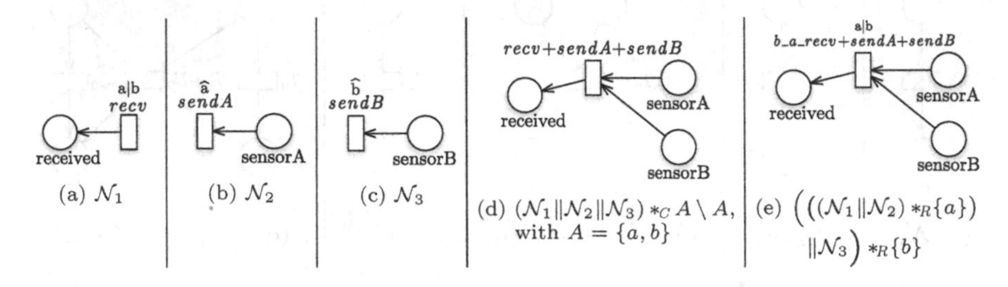

Fig. 13. Three-way synchronization using conjugated tags.

Figure 13 shows a three-way synchronization realized using conjugate tags. Nets $\mathcal{N}_1$, $\mathcal{N}_2$, and $\mathcal{N}_3$ depicts the component of a system in which a token is placed into place *received* only when the two sensors' values are read in a single moment. The synchronization can be achieved through unary conjugated composition (merging all tags), as in Fig. 13(d), or by subsequently merging one tag after the other as multiple nested restricted structured composition, as in Fig. 13(e).

Multitag Synchronization. The other important aspect that is covered by the composition over multitags is that all transitions that fully complement the synchronized tags are generated (if possible). Figure 6(c) is an example of this behaviour. In this way, complex dependencies among the tags can be expressed. An example of application is a transition that needs to acquire n resources of type a, that are provided and locked by another transition with a $\hat{a}$ transition. The resulting synchronization consists in a single acquisition and n lock events into a single transition. When complementarity of the tags is implicit in the structure of the net, structured composition is also an option, as in Fig. 7(c).

Prototype. The proposed framework has been implemented as a prototype inside the GreatSPN software (https://github.com/greatspn/SOURCES). All figures used in the paper were generated using the proposed composition framework, with the exception of Fig. 5(d, e), that was generated manually. The framework is developed inside the graphical editor, where multiple nets can be composed together using the unary conjugated composition, or the n-ary structured composition. Restriction is optional and can be applied after each composition.

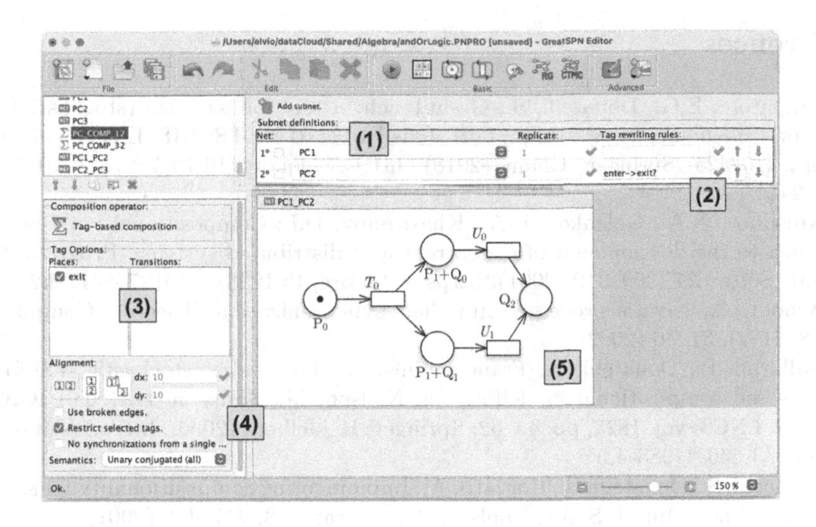

Fig. 14. Prototype implemented inside the GreatSPN graphical interface.

Figure 14 shows how the composition interface looks like in the tool. Composition pages are special subnets, and are defined following this schema. In (1) the composed nets are selected, and optional tag rewriting rules are specified (2). The composition tags for places and transitions are then selected (3), together with the composition policy, the optional tag restriction and other parameters (4). The composed net is then shown in the central pane (5).

7 Conclusions

In this paper we propose a novel framework for net composition that is focused on the simultaneous joint combination of places and transitions into a single policy-based combinatorial operator. Multiple policies can be defined for composition, and we provided a CCS-like unary conjugated composition, as well as a CSP-like n-ary structured composition. Conjugated composition is based on the intuitions of [2], but it is modified to follow the synchronization rules of the Petri box calculus. We have reviewed several common cases for Petri net composition and various modeling patterns, showing the effectiveness of the proposed approach in modeling terms. The operators have been defined in net syntactical terms. While this definition is enough to proceed, for example, to an implementation, it lacks a formal semantic interpretation of the composed net behaviour in terms of the possible executions of the composed nets. While for transition composition defined by the policies of Sect. 5.1 and 5.2 this may be an attainable goal, it is less clear how this can be achieved when the policies are applied to place-based composition, a topic that has received less attention in the literature and that certainly deserves more investigation.

References

1. Amparore, E.G., Donatelli, S.: GreatTeach: a tool for teaching (stochastic) petri nets. In: Khomenko, V., Roux, O.H. (eds.) PETRI NETS 2018. LNCS, vol. 10877, pp. 416–425. Springer, Cham (2018). https://doi.org/10.1007/978-3-319-91268-4_24
2. Anisimov, N.A., Golenkov, E.A., Kharitonov, D.I.: Compositional petri net approach to the development of concurrent and distributed systems. Program. Comput. Softw. **27**, 309–319 (2001). https://doi.org/10.1023/A:1012758417962
3. Arnold, A.: Nivat's processes and their synchronization. Theoret. Comput. Sci. **281**(1–2), 31–36 (2002)
4. Ballarini, P., Donatelli, S., Franceschinis, G.: Parametric stochastic well-formed nets and compositional modelling. In: Nielsen, M., Simpson, D. (eds.) ICATPN 2000. LNCS, vol. 1825, pp. 43–62. Springer, Heidelberg (2000). https://doi.org/10.1007/3-540-44988-4_5
5. Bernardi, S., Donatelli, S., Horvath, A.: Implementing compositionality for stochastic petri nets. Int. J. Softw. Tools Technol. Transf. **3**, 417–430 (2001)
6. Best, E., Devillers, R., Hall, J.G.: The box calculus: a new causal algebra with multi-label communication. In: Rozenberg, G. (ed.) Advances in Petri Nets 1992. LNCS, vol. 609, pp. 21–69. Springer, Heidelberg (1992). https://doi.org/10.1007/3-540-55610-9_167
7. Best, E., Devillers, R., Koutny, M.: Petri Net Algebra. Springer, Heidelberg (2013). https://doi.org/10.1007/978-3-662-04457-5
8. Best, E., Fleischhack, H., Fraczak, W., Hopkins, R.P., Klaudel, H., Pelz, E.: A class of composable high level Petri nets. In: De Michelis, G., Diaz, M. (eds.) ICATPN 1995. LNCS, vol. 935, pp. 103–120. Springer, Heidelberg (1995). https://doi.org/10.1007/3-540-60029-9_36
9. Buchholz, P.: Hierarchies in colored GSPNs. In: Ajmone Marsan, M. (ed.) ICATPN 1993. LNCS, vol. 691, pp. 106–125. Springer, Heidelberg (1993). https://doi.org/10.1007/3-540-56863-8_43
10. Chiola, G., Dutheillet, C., Franceschinis, G., Haddad, S.: On well-formed coloured nets and their symbolic reachability graph. In: Jensen, K., Rozenberg, G. (eds.) High-level Petri Nets, pp. 373–396. Springer, Heidelberg (1991). https://doi.org/10.1007/978-3-642-84524-6_13
11. Clark, G., et al.: The Möbius modeling tool. In: 9th International Workshop on Petri Nets and Performance Models, pp. 241–250 (2001)
12. Colom, J.M., Silva, M.: Convex geometry and semiflows in P/T nets. A comparative study of algorithms for computation of minimal p-semiflows. In: Rozenberg, G. (ed.) ICATPN 1989. LNCS, vol. 483, pp. 79–112. Springer, Heidelberg (1991). https://doi.org/10.1007/3-540-53863-1_22
13. De Cindio, F., et al.: A Petri net model for CSP. In: Proceedings of Convención Informática Latina (CIL 1981), Barcelona, vol. 81, pp. 392–406 (1981)
14. Donatelli, S., Franceschinis, G.: The PSR methodology: integrating hardware and software models. In: Billington, J., Reisig, W. (eds.) ICATPN 1996. LNCS, vol. 1091, pp. 133–152. Springer, Heidelberg (1996). https://doi.org/10.1007/3-540-61363-3_8
15. Farkas, J.: Theorie der einfachen Ungleichungen. Journal für die reine und angewandte Mathematik (Crelles Journal) **1902**(124), 1–27 (1902)
16. Heiner, M.: How to draw a hierarchical Petri net? https://www-dssz.informatik.tu-cottbus.de/DSSZ/Software/FAQ#QQuestions5. Accessed Jan 2022

17. Heiner, M., Herajy, M., Liu, F., Rohr, C., Schwarick, M.: Snoopy - a unifying petri net tool. In: Haddad, S., Pomello, L. (eds.) PETRI NETS 2012. LNCS, vol. 7347, pp. 398–407. Springer, Heidelberg (2012). https://doi.org/10.1007/978-3-642-31131-4_22
18. Hoare, C.A.R.: Communicating Sequential Processes. Prentice Hall, Hoboken (1985)
19. Hoare, C.A.R.: Communicating sequential processes. Commun. ACM **21**(8), 666–677 (1978)
20. Jensen, K.: Coloured Petri nets. In: Brauer, W., Reisig, W., Rozenberg, G. (eds.) Petri Nets: Central Models and Their Properties. LNCS, vol. 254, pp. 248–299. Springer, Heidelberg (1987). https://doi.org/10.1007/BFb0046842
21. Jensen, K., Kristensen, L.M., Wells, L.: Coloured Petri nets and CPN tools for modelling and validation of concurrent systems. Int. J. Software Tools Tech. Transf. **9**(3), 213–254 (2007)
22. Kindler, E.: Modular PNML revisited: some ideas for strict typing. Arbeitsberichte aus dem Arbeitsberichte aus dem Fachbereich Informatik 20 (2007)
23. Kotov, V.E.: An algebra for parallelism based on petri nets. In: Winkowski, J. (ed.) MFCS 1978. LNCS, vol. 64, pp. 39–55. Springer, Heidelberg (1978). https://doi.org/10.1007/3-540-08921-7_55
24. Lampka, K., Siegle, M.: Symbolic composition within the Möbius framework. In: Proceedings of the 2nd MMB Workshop, pp. 63–74 (2002)
25. Mieg, Y.T.: From Symbolic Verification To Domain Specific Languages. Ph.D. thesis. Sorbonne Université, UPMC; Laboratoire d'informatique de Paris 6 [LIP6] (2016)
26. Milner, R. (ed.): A Calculus of Communicating Systems. LNCS, vol. 92. Springer, Heidelberg (1980). https://doi.org/10.1007/3-540-10235-3
27. Ratzer, A.V., et al.: CPN tools for editing, simulating, and analysing coloured petri nets. In: van der Aalst, W.M.P., Best, E. (eds.) ICATPN 2003. LNCS, vol. 2679, pp. 450–462. Springer, Heidelberg (2003). https://doi.org/10.1007/3-540-44919-1_28
28. Thierry-Mieg, Y.: Symbolic model-checking using ITS-tools. In: Baier, C., Tinelli, C. (eds.) TACAS 2015. LNCS, vol. 9035, pp. 231–237. Springer, Heidelberg (2015). https://doi.org/10.1007/978-3-662-46681-0_20
29. Weber, M., Kindler, E.: The petri net markup language. In: Ehrig, H., Reisig, W., Rozenberg, G., Weber, H. (eds.) Petri Net Technology for Communication-Based Systems. LNCS, vol. 2472, pp. 124–144. Springer, Heidelberg (2003). https://doi.org/10.1007/978-3-540-40022-6_7

Process Mining

Discovering Process Models with Long-Term Dependencies While Providing Guarantees and Handling Infrequent Behavior

Lisa L. Mannel(✉) and Wil M. P. van der Aalst

Process and Data Science (PADS), RWTH Aachen University, Aachen, Germany
{mannel,wvdaalst}@pads.rwth-aachen.de

Abstract. In process discovery, the goal is to find, for a given event log, the model describing the underlying process. While process models can be represented in a variety of ways, Petri nets form a theoretically well-explored description language. In this paper, we present an extension of the eST-Miner process discovery algorithm. This approach computes a set of places which are considered to be fitting with respect to a user-definable fraction of the behavior described by the given event log, by evaluating all possible candidate places using token-based replay. The set of replayable traces is determined for each place in isolation, i.e., they do not need to be consistent. When combining these places into a Petri net by connecting them to the corresponding transitions, which are uniquely labeled for each activity in the event log, the resulting net can replay exactly those traces that can be replayed by each of the inserted places. Thus, inserting places without further checks may results in deadlocks and thus low fitness of the Petri net. In this paper, we explore a variant of the eST-Miner, that aims to select a subset of the discovered places such that the resulting Petri net guarantees a definable minimal fitness while maintaining high precision with respect to the input event log. Various place selection strategies are proposed and their impact on the returned Petri net is evaluated by experiments using both real and artificial event logs.

Keywords: Process discovery · Petri nets · eST-Miner

1 Introduction and Related Work

More and more corporations and organizations support their processes using information systems, which record the occurring behavior and represent this data in the form of *event logs*. Each event in such a log has a name identifying the executed activity (activity name), an identification mapping the event to some execution instance (case id), a time stamp showing when the event was observed, and often extended meta-data of the activity or process instance. In the field of *process discovery*, we utilize the event log to identify relations between the activities (e.g. pre-conditions, choices, concurrency), which are then expressed within a process model, for example a Petri net [1–4]. This is non-trivial for various reasons. We cannot assume that the given event log is complete, as some possible behavior might be yet unobserved. Also, real-life event logs often contain noise in the form of incorrectly recorded data or deviant behavior, which

L. Bernardinello and L. Petrucci (Eds.): PETRI NETS 2022, LNCS 13288, pp. 303–324, 2022.
https://doi.org/10.1007/978-3-031-06653-5_16

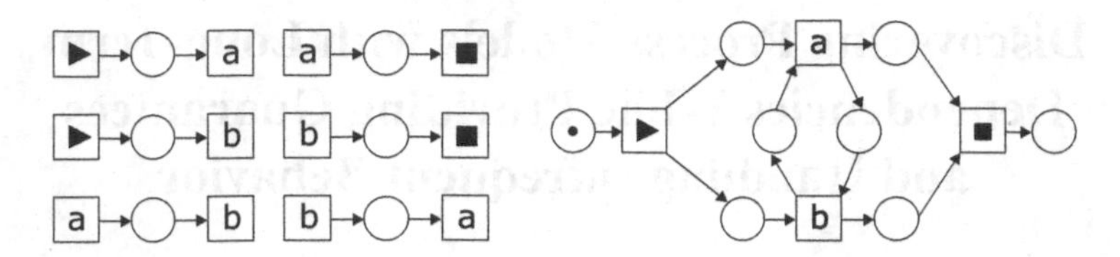

Fig. 1. Consider the event log $L = [\langle \blacktriangleright, a, b, \blacksquare \rangle^{40}, \langle \blacktriangleright, b, a, \blacksquare \rangle^{60}]$ and the set of candidate places on the left. Assuming, that we set the eST-Miner to accept all places that can replay at least 35% of the event log, it would add all those places and return the Petri net on the right. Although each individual place has at least 40 fitting traces, the whole model cannot replay any trace.

is not desired to be reflected in the process model. Correctly classifying behavior as noise can be hard to impossible. An ideal process model can reproduce all behavior contained in an event log, while not allowing for unobserved behavior. It should represent all dependencies between events and at the same time be simple enough to be understandable by a human interpreter. Computation should be fast and robust to noise. Usually, it is impossible to fulfill all these requirements at the same time. Thus, different algorithms focus on different quality criteria, while neglecting others. As a result, the models returned for a given event log can differ significantly.

Many existing discovery algorithms abstract from the full information given in a log and/or generate places heuristically, in order to decrease computation time and complexity of the returned process models. While this is convenient in many applied settings, the resulting models are often underfitting, in particular when processes are complex. Examples are the Alpha Miner variants [5], the Inductive Mining family [6], genetic algorithms or Heuristic Miner. In contrast to these approaches, which are not able to (reliably) discover complex model structures, algorithms based on region theory [7–17] discover models whose behavior is the minimal behavior representing the input event log. On the downside, these approaches are known to be rather time-consuming, cannot handle noise, and tend to produce complex, overfitting models which can be hard to interpret. A combination of strategies has been introduced in [18], which aims to circumvent performance issues by limiting the application of region theory to small fragments of a pre-descovered Petri net.

In [19] we introduced the discovery algorithm eST-Miner. This approach aims to combine the capability of finding complex control-flow structures like longterm-dependencies with an inherent ability to handle low-frequent behavior while exploiting the token-game to increase efficiency. The basic idea is to evaluate all possible places, defined by all possible combinations of uniquely labeled transitions, to discover a set of fitting ones. Efficiency is significantly increased by skipping uninteresting parts of the search space. This may decrease computation time immensely compared to the brute-force approach evaluating every single candidate place, while still providing guarantees with regard to fitness and precision.

While traditional region-theory uses a global perspective to find a set of feasible places, the eST-Miner evaluates each place separately, that is from a local perspective. This allows us to easily enforce all kinds of constraints definable on the place level, e.g., constraints on the number or type of connected transitions, token throughput or similar. In particular, we are able to filter infrequent behavior locally, by requiring each place to be able to replay only a certain fraction of the traces in the event log. A candidate place will be accepted, if the event log contains sufficient support for the relation between the activities as defined by the place. In contrast to common noise filtering techniques which loose information by removing infrequent trace variants or infrequent activities from the event log, this approach can also consider infrequent information to discover relations between activities.

The local perspective of the eST-Miner ensures that all occurrences of activities within a log can contribute to the discovered model. However, when a set of discovered fitting places is combined into a Petri net, this Petri net allows only for the behavior in the intersection of the behaviors allowed by all inserted places. Thus, the Petri net may include deadlocks or dead parts, resulting in a much lower overall fitness than the fitness of each individual place and an overly complicated model. In extreme cases, the constructed net cannot replay any trace at all as illustrated by the small example in Fig. 1. Assuming we decide to add places that replay only a fraction of 0.35 of the traces, the Petri net discovered for the given event log cannot fire any transition after the start transition.

In this paper, we aim to remedy this issue by selecting a subset of the discovered places which can be combined into a Petri net with definable minimal fitness, while simultaneously striving for high precision and simplicity, without loosing the desirable properties of the eST-Miner. Thus, we require the algorithm to maintain its ability to discover and model non-local dependencies, to deal with infrequent behavior and to provide guarantees without over- or underfitting. Additionally, the time and space consumption should remain reasonable, in particular more scalable than classic region theory approaches.

Section 2 provides basic notation and definitions. In Sect. 3, we briefly review the basics of the standard eST-Miner. Our new concepts are introduced in Sects. 4 and 5, and their experimental evaluation is presented in Sect. 6. Finally, Sect. 7 concludes this work by summarizing our findings and suggesting possibilities for future work.

2 Basic Notations, Event Logs, and Process Models

A set, e.g. $\{a, b, c\}$, does not contain any element more than once, while a multiset, e.g. $[a, a, b, a] = [a^3, b]$, may contain multiples of the same element. The intersection of two sets contains only elements that occur in both sets, i.e., $\{x, y\} \cap \{y, z\} = \{y\}$, while the intersection of two multisets contains each element with its minimum frequency, i.e., $[x, y^2, z] \Cap [y^5, z^2] = [y^2, z]$. By $\mathbb{P}(X)$ we refer to the power set of the set X, and $\mathbb{M}(X)$ is the set of all multisets over this set. In contrast to sets and multisets, where the order of elements is irrelevant, in sequences the elements are given in a certain order, e.g., $\langle a, b, a, b\rangle \neq \langle a, a, b, b\rangle$. The size of a set, multiset or sequence X, that is $|X|$, is defined to be the number of elements in X.

We define activities, traces, and logs as usual, except that we require each trace to begin with a designated start activity (▶) and end with a designated end activity (■). Note that this is a reasonable assumption in the context of processes, and that any log can easily be transformed accordingly.

Definition 1 (Activity, Trace, Log). *Let $\mathcal{A}$ be the universe of all possible* activities *(e.g., actions or operations), let* $\blacktriangleright \in \mathcal{A}$ *be a designated start activity and let* $\blacksquare \in \mathcal{A}$ *be a designated end activity. A* trace *is a sequence containing* $\blacktriangleright$ *as the first element,* $\blacksquare$ *as the last element and in-between elements of* $\mathcal{A} \setminus \{\blacktriangleright, \blacksquare\}$. *Let* $\mathcal{T}$ *be the set of all such traces. A* log $L \in \mathbb{M}(\mathcal{T})$ *is a multiset of traces.*

In this paper, we use an alternative definition for Petri nets. We only allow for places connecting transitions, called here activities, that are initially empty (without tokens), because we allow only for traces starting with $\blacktriangleright$ and ending with $\blacksquare$. These places are uniquely identified by the non-empty sets of input activities I and output activities O. Each activity corresponds to exactly one uniquely labeled transition, therefore, this paper refers to transitions as activities.

Definition 2 (Petri nets). *A Petri net is a pair* $N = (A, P)$, *where* $A \subseteq \mathcal{A}$ *is the set of activities including start and end* ($\{\blacktriangleright, \blacksquare\} \subseteq A$) *and* $P \subseteq \{(I|O) \mid I \subseteq A \wedge I \neq \emptyset \wedge O \subseteq A \wedge O \neq \emptyset\}$ *is the set of places. We call* I *the set of* ingoing *activities of a place and* O *the set of* outgoing *activities.*

Note that if $p = (I|O)$, then $\bullet p = I$ and $p\bullet = O$ using standard notation.

A place is *fitting* if it can replay (parts of) the event log without missing or remaining tokens. Otherwise, it is *unfitting*.

Definition 3 (Fitting and Unfitting Places, compare [20]**).** *Let* $N = (A, P)$ *be a Petri net, let* $p = (I|O) \in P$ *be a place, and let* σ *be a trace. With respect to the given trace* σ, p *is called*

- unfitting, *denoted by* $\boxtimes_\sigma(p)$, *if and only if* $\exists k \in \{1, 2, ..., |\sigma|\}$ *such that* $|\{i \mid i \in \{1, 2, ...k-1\} \wedge \sigma(i) \in I\}| < |\{i \mid i \in \{1, 2, ...k\} \wedge \sigma(i) \in O\}|$ *or* $|\{i \mid i \in \{1, 2, ...|\sigma|\} \wedge \sigma(i) \in I\}| > |\{i \mid i \in \{1, 2, ...|\sigma|\} \wedge \sigma(i) \in O\}|$,
- fitting, *denoted by* $\square_\sigma(p)$, *if and only if not* $\boxtimes_\sigma(p)$.

We extend these notions to the whole log using the noise parameter: with respect to a log L *and parameter* $\tau \in [0, 1]$, p *is called* fitting, *denoted by* $\square_L^\tau(p)$, *if and only if* $|\{\sigma \in L \mid \square_\sigma(p)\}|/|L| \geq \tau$, *and* unfitting *otherwise.*

Definition 4 (Behavior of a Petri net). *We define the* behavior *of the Petri net* (A, P) *to be the set of all fitting traces, that is* $\{\sigma \in \mathcal{T} \mid \forall p \in P \colon \square_\sigma(p)\}$.

Note that we only allow for behaviors of the form $\langle \blacktriangleright, a_1, a_2, \ldots a_n, \blacksquare \rangle$ (Definition 1) such that places are empty at the end of the trace and never have a negative number of tokens.

We are often interested in the traces of the event log which are replayable by certain (sets of) places.

Definition 5 (Multisets of Fitting Traces). *For an event log* L *and a place* p, *the multiset of log traces replayable by* p *is*

$$fit(L, p) = [\sigma \in L \mid \square_\sigma(p)].$$

For an event log L *and a Petri net* $N = (A, P)$, *the multiset of log traces replayable by* N *is the intersection of all log traces replayable by the places in* P, *i.e.,*

$$fit(L, N) = [\sigma \in L \mid \forall p \in P \colon \square_\sigma(p)] = \bigcap_{p \in P} fit(L, p).$$

3 Introducing the eST-Miner

Several variants and extensions of the eST-Miner have been proposed in the past years. In the following, we briefly introduce the eST-Miner variant used as the basis of this work. For further details, we refer the reader to the respective papers.

As input, the algorithm takes a log L and a parameter $\tau \in [0,1]$, and returns a Petri net as output. A place is considered *fitting*, if it allows to replay at least a fraction τ of traces in the event log. Inspired by language-based regions, the basic strategy of the approach is to begin with a Petri net whose transitions correspond exactly to the activities used in the given log. From the finite set of unmarked, intermediate places, the subset of all fitting places is computed and inserted. To facilitate further computations and human readability, *implicit* places are identified and removed [21–23]. A place is implicit if its removal does not increase the behavior of the Petri net. Implicit places can be detected based on the structure of the Petri net as proposed for the first eST-Miner variant [19], or by using the faster replay-based implicit place removal strategy introduced in [24]. The latter one is applied in the experimentation of this paper.

The algorithm uses token-based replay to evaluate the candidate places. To avoid replaying the log on the exponential number of candidates (i.e., all pairs of subsets of activities, $(2^{|A|}-1)^2$), it organizes the potential places as a set of trees, such that certain properties hold. When traversing the trees, these properties allow to cut off subtrees, and thus candidates, based on the replay result of their parent [19]. This greatly increases efficiency, while still guaranteeing that all fitting places are found.

An example of such a tree-structured candidate space is shown in Fig. 2. Note the incremental structure of the trees, i.e., the increase in distance from the roots corresponds to the increase of input (red edges) and output (blue edges) activities. However, the organization of candidates within the same depth and their connections to other candidates is not fixed, but defined by the order of ingoing activities ($>_i$) and outgoing activities ($>_o$).

Definition 6 (Complete Candidate Tree). *Let A be a set of activities and let $>_i, >_o$ be two total orderings on this set of activities. A* complete candidate tree *is a pair $CT = (N, F)$ with $N = \{(I|O) \mid I \subseteq A \setminus \{\blacksquare\} \wedge O \subseteq A \setminus \{\blacktriangleright\} \wedge I \neq \emptyset \wedge O \neq \emptyset\}$. We have that $F = F_{red} \cup F_{blue}$, with*

$$F_{red} = \{((I_1|O_1),(I_2|O_2)) \in N \times N \mid |O_2| = 1 \wedge O_1 = O_2$$
$$\wedge \exists a \in I_1 \colon \big(I_2 \cup \{a\} = I_1 \wedge \forall a' \in I_2 \colon a >_i a'\big)\} \textit{ (red edges)}$$
$$F_{blue} = \{((I_1|O_1),(I_2|O_2)) \in N \times N \mid I_1 = I_2$$
$$\wedge \exists a \in O_1 \colon \big(O_2 \cup \{a\} = O_1 \wedge \forall a' \in O_2 \colon a >_o a'\big)\} \textit{ (blue edges)}.$$

If $((I_1|O_1),(I_2|O_2)) \in F$, we call the candidate $(I_1|O_1)$ the child *of its* parent $(I_2|O_2)$.

The purpose of the tree structured candidate space is to enable skipping of sets of uninteresting candidates to improve time and space efficiency. The runtime of the eST-Miner strongly depends on the number of candidate places skipped during the search for fitting places.

When a candidate place is evaluated to be fitting, i.e., it can replay a fraction of τ traces in the event log, existing variants of the eST-Miner simply insert the place into the

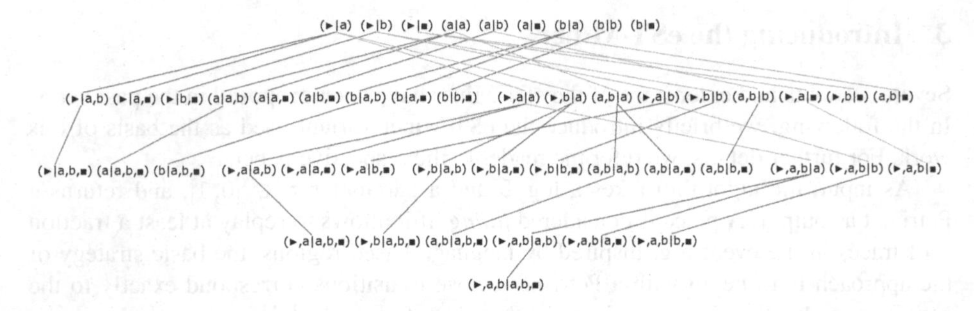

Fig. 2. Example of a tree-structured candidate space for the set of activities $\{\blacktriangleright, a, b, \blacksquare\}$, with orderings $\blacksquare >_i b >_i a >_i \blacktriangleright$ and $\blacksquare >_o b >_o a >_o \blacktriangleright$.

ID	Traces in L	p_1 $(\blacktriangleright\|a)$	p_2 $(a\|c)$	p_3 $(a\|b)$	p_4 $(c\|e)$	p_5 $(b\|e)$	p_6 $(e\|\blacksquare)$	p_7 $(b\|c,d)$	p_8 $(d,e\|\blacksquare)$	N
1	$\langle \blacktriangleright, a, b, c, e, \blacksquare \rangle^{60}$	✓	✓	✓	✓	✓	✓	✓	✓	✓
2	$\langle \blacktriangleright, a, b, d, \blacksquare \rangle^{20}$	✓	✗	✓	✓	✗	✗	✓	✓	✗
3	$\langle \blacktriangleright, a, c, b, e, \blacksquare \rangle^{15}$	✓	✓	✓	✓	✓	✓	✗	✓	✗
4	$\langle \blacktriangleright, a, b, d, e, \blacksquare \rangle^{5}$	✓	✗	✓	✗	✓	✓	✓	✗	✗

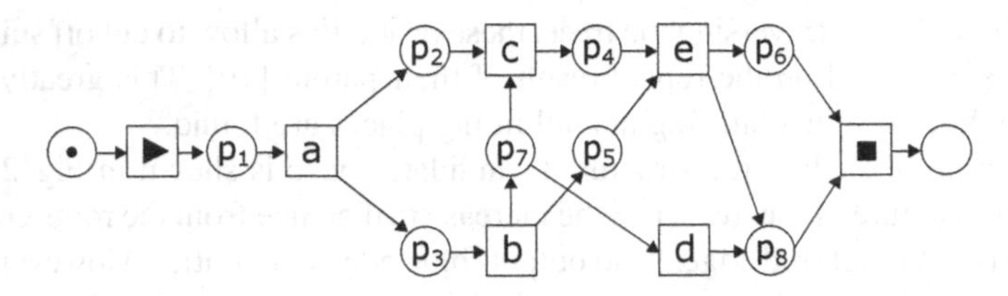

Fig. 3. The table indicates for each of the given trace variants and candidate places whether the place can replay that trace variant. The Petri net N is created by inserting all places which can replay at least $0.75 \cdot |L| = 75$ traces. However, N can replay only the first trace variant, i.e., $0.6 \cdot |L| = 60$ traces.

Petri net by connecting it to its uniquely labeled ingoing and outgoing transitions. Consider the example event log and subset of candidate places in Fig. 3. Of the (incomplete) subset of candidate places, the places p_1 to p_8 are fitting the event log for $\tau = 0.75$. Inserting these places results in the given Petri net N, which can replay only the first trace variant corresponding to 60% of traces. The introductory example in Fig. 1 illustrates, that the fraction of replayable traces may even decrease to 0. Such a result is undesirable, since it is unnecessarily complex with respect to the behavior it represents, not free of dead parts and likely to disappoint user expectations with respect to fitness. This work explores strategies of maintaining the fitness threshold τ as a minimal fitness threshold of the returned, deadlock-free Petri net by inserting only a selection of the discovered fitting places.

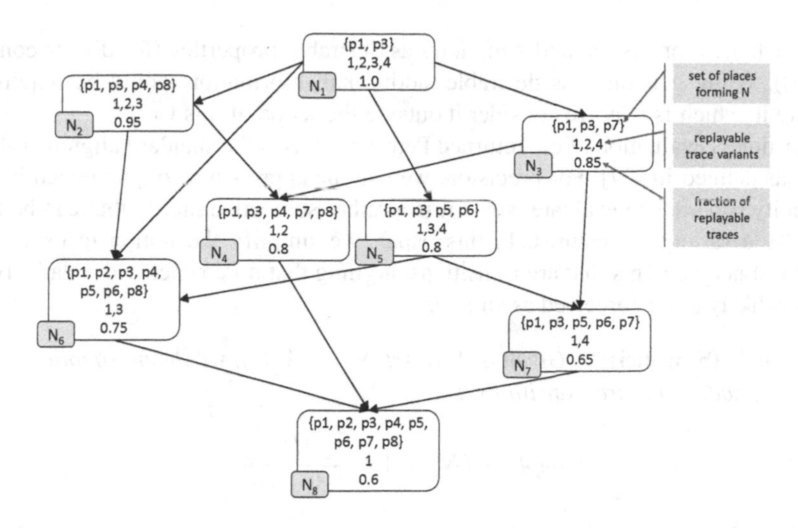

Fig. 4. Consider the set of places given in Fig. 3. This figure shows all possible combinations of these places such that adding any other place to the corresponding Petri net would decrease the number of replayable log traces. Each set of places, i.e., Petri net, is annotated with the list of trace variants it can replay and the corresponding fraction of log traces. Note, that N_8 corresponds to the Petri net shown in Fig. 3.

4 Place Selection

Consider the set of all fitting places discovered for a certain noise threshold τ. The selection of an adequate subset of these places, such that also the resulting Petri net can replay a fraction of at least τ traces, is non-trivial for a variety of reasons. First of all, the definition of an optimal solution is not straightforward. Several maximal subsets of places satisfying this requirement may exist, which differ, for example, in size, fraction of replayable traces, place complexity (number of connected activities) or subjective 'interestingness' measures for the places retained. Figure 4 illustrates all maximal sets of places that can be built from the example places given in Fig. 3. These sets are maximal in the sense that adding any of the other places would decrease the number of replayable log traces. Depending on the choice of the minimal fitness threshold τ, the optimal solution is not clear.

Furthermore, even if we have somehow obtained a notion of optimality, first collecting all fitting places and then computing an optimal solution can quickly become unfeasible, both in terms of time complexity and memory requirements. This is due to the very large number of fitting but potentially implicit places discovered by the eST-Miner. Unfortunately, knowledge of which places are contained in the Petri net is required to identify implicit places reliably.

To circumvent the issue of time and space complexity, we combine the eST-Miners sequential place evaluation procedure with a guided greedy place selection approach, which is decsribed in detail in Subsects. 4.1 and 4.2. In the absence of a clear notion of optimality, we propose and investigate several heuristic selection strategies and evaluate their impact on different quality aspects of the returned Petri net. In this paper, we consider fitness, precision, and simplicity as desirable properties (for details compare [25,26]). While generality is desirable, additional information would be required to evaluate it, which is why we consider it outside the scope of this work.

Our fitness evaluation of the returned Petri net N uses the standard alignment-based fitness as defined in [27]. For precision, we use the approach as implemented in [28]. Simplicity is harder to evaluate, since it is a rather subjective metric that can be influenced by a variety of features. In this paper, we simplify the notion to express the fraction of net elements that are transitions, arguing that a Petri net with relatively few places is likely to be perceived as simple.

Definition 7 (Simplicity). *Given a Petri net $N = (A, P)$, we define simplicity as the fraction of nodes that are transitions, i.e.,*

$$\mathit{Simplicity}(N) = 1 - \frac{|P|}{|P| + |A|}.$$

All used quality metrics return values between 0 and 1, where a value close to 1 indicates high quality in general. However, note that for simplicity a value around 0.5 indicates a model with roughly as many places as transitions, for example a simple sequence, while a higher value would arise from a Petri net with extremely few places. Therefore, we consider a value close to 0.5 to be rather optimal in terms of simplicity.

4.1 Place Classification

When making the decision to insert a place into the model, this reduces the possible choices we can make later on: the place constrains the behavior of the model and only places with a sufficiently large intersection of replayable traces can be added to the model at a later point. Consider the example place combinations in Fig. 4 with a fitness threshold of $\tau = 0.75$ and assume that the model already contains the places p_1 and p_3. If the next fitting place we discover is p_7, and we immediately insert it into the Petri net, we can no longer discover a Petri net including, for example, p_6, without violating our global fitness constraint. Such choices may prevent us from discovering a more desirable solution. Therefore, we aim to capture the main behavior of the log by using heuristics to postpone, or even disallow, the addition of very restrictive places.

To this end, we introduce a new parameter δ which is our main tool to guide the choice of places while balancing fitness, precision and simplicity. This δ specifies the largest acceptable reduction in replayable traces when adding a place to the model. Optionally, δ can be adapted for each place individually using an adaption function *adapt* to favor certain places over others, according to the users preferences. Favored places can be added earlier, despite being rather restrictive, while other places will be added only if they do not constrain the behavior too much. Examples for such strategies are presented in Sect. 5.

Definition 8 formalizes the use of τ, δ and *adapt* to decide for the discovered fitting places, whether they should be added, kept for later re-evaluation or discarded.

Definition 8 (Place Classification Using τ, δ and *adapt*). *Consider a set of activities A, a set of places $P \subseteq \mathbb{P}(A) \times \mathbb{P}(A)$, a place $p \in \mathbb{P}(A) \times \mathbb{P}(A)$ and an event log L. We* *use parameters* $\tau \in [0,1]$ *and* $\delta \in [0,1]$, *and a function adapt*: $([0,1], \mathbb{P}(A) \times \mathbb{P}(A)) \rightarrow [0,1]$ *to categorize p as follows:*

$$\begin{aligned} \mathtt{keep}_{L,A,\tau}(P,p) &= |\mathit{fit}(L(A,P)) \cap \mathit{fit}(L,p)| \geq \tau \cdot |L| \\ \mathtt{add}_{L,A,\tau,\delta}(P,p) &= \mathtt{keep}_{L,A,\tau}(P,p) \\ &\quad \wedge |\mathit{fit}(L(A,P))| - |\mathit{fit}(L,(A,P)) \cap \mathit{fit}(L,p)| \leq \mathit{adapt}(\delta,p) \cdot |L| \end{aligned}$$

If $\mathtt{keep}_{L,A,\tau}(P,p)$ *does not hold, p will be discarded.*

In the following subsection, we give an overview of the complete approach.

4.2 Selection Framework

An overview of our approach, indicating inputs, outputs and use of parameters, is given in Fig. 5. Since we consider simplicity to be a desirable property, we set the eST-Miner to traverse the *complete candidate tree* using BFS rather than DFS. Thus, places with few connected activities are evaluated first and can therefore be inserted into the model at an earlier stage. Furthermore, we limit the traversal depth to places with d_{cut} activities, arguing that places with many transitions are generally not desirable - such places are usually devastating to simplicity while their constraints can be sufficiently approximated by much simpler places.

After the eST-Miner framework evaluates a candidate place p to be fitting with respect to a fraction τ of traces (Definition 3), we use the *classification functions* given in Definition 8 to decide whether the place should immediately be added to the output Petri net, discarded forever or kept for re-evaluation. In the latter case it is added to a queue Q of potential places which is sorted according to how interesting a place is. In our case, we sort by place simplicity (few transitions are better) and place fitness (number of replayable log traces). Optionally, the length of Q can be limited, trading an improvement in time and space complexity for potentially lowered model quality.

Whenever the BFS candidate traversal reaches a new level in the *complete candidate tree*, we revisit the potential places queue Q and re-evaluate its places using the classification functions before proceeding with the traversal of more complex places. This makes sense to promote simplicity in particular together with the delta adaption functions proposed in Sect. 5, which give preference to places less complex than indicated by the current tree level. After reaching the lowest tree level, the approach can either terminate immediately, or iterate over the potential places queue while artificially increasing the tree depth d^+ times. This can be relevant for delta adaption functions depending on place complexity, as exemplified in Sect. 5. Here, the artificial tree depth allows for increased leniency also for the most complex places evaluated.

Finally, the resulting Petri net N may contain dead parts: in particular infrequent activities with erratic behavior are likely to occur only in those traces that are no longer replayable on N. Therefore, as a final step, we detect and remove all activities that do not occur in $\mathit{fit}(L,N)$ together with their connected arcs.

Before returning this Petri net as final output, the eST-Miner framework removes implicit places, merges places that are equal except for self-loops, i.e., $(I \cup X_1 | O \cup X_1)$ and $(I \cup X_2 | O \cup X_2)$ are merged into $(I \cup X1 \cup X_2 | O \cup X_1 \cup X_2)$, and adds start and end places.

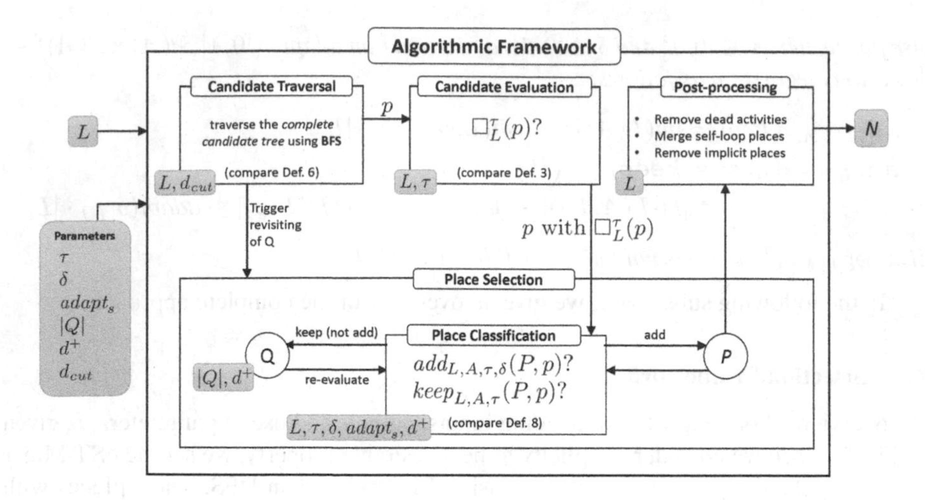

Fig. 5. Overview of the presented approach, including input, output and parameter use.

The approach returns a Petri net N satisfying the following guarantees.

Theorem 1 (Guarantees). *Given a set of activities A, event log L over A, an adaption function adapt*: $([0,1], \mathbb{P}(A) \times \mathbb{P}(A)) \rightarrow [0,1]$ *and parameters* $\tau \in [0,1], \delta \in [0,1], s \in \mathbb{N}, |Q| \in \mathbb{N}, d^+ \in \mathbb{N}, d_{cut} \in \mathbb{N}$, *the presented approach computes a Petri net* $N = (A', P)$ *with* $A' \subseteq A$, *such that N can replay at least* $\tau \cdot |L|$ *traces from L and every transition in A' can be fired at least once.*

Furthermore, if the length of Q is not limited, and thus a place p is discarded only if it does not satisfy $keep_{L,A,\tau}(P,p)$, the set of places P is maximal in the sense that no place from the set of evaluated candidate places can be added without violating the fitness constraints imposed by the chosen heuristics.

5 Selection Strategies

As illustrated by the example place combinations in Fig. 4, the order of places added can have a significant impact on the selected subset of places and thus the behavior of the returned Petri net. The presented framework allows for a wide range of heuristic functions, optimizing the place selection individually towards a variety of possible user interests. Thus, obviously, the examples presented in the following are by far not exhaustive and entirely different choices are possible, but they can serve as a starting point to an investigation of the impact and suitability of our approach.

The *linear* and *sigmoid* delta adaption functions both aim to promote fitness and simplicity. The *constant* and *no delta* delta adaption functions are introduced to be used as a baseline in our experiments, towards which the effect of the other strategies can be compared.

No Delta. As a baseline to compare to, we introduce a function that ignores delta and simply adds every fitting place to the Petri net as soon as it is discovered. Within the framework, this can be formalized to

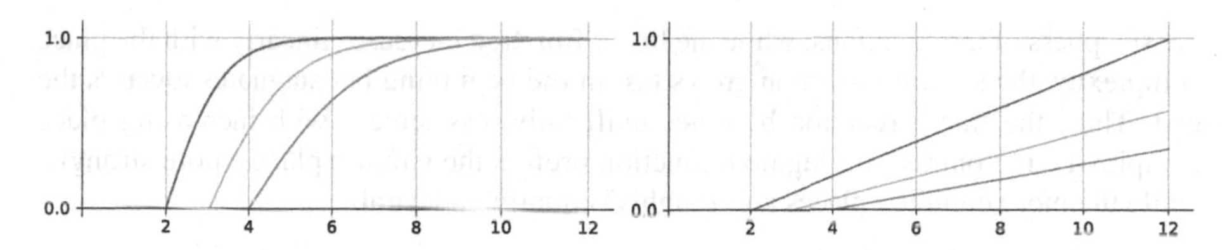

Fig. 6. Example behavior of the delta adaption modifiers $mod_{sigmoid}(\delta)$ (left) and $mod_{linear}(\delta)$ (right) for three places with 2, 3 and 4 activities, respectively. The x-axis indicates the current tree depth d, with $d_{max} = 12$, while the y-axis indicates the modifier to be multiplied with δ.

$$adapt_{noDelta}(\delta, p) = 1.$$

Constant Delta. Trivially, we can choose not to adapt delta at all. We simply add every fitting, non-discarded place that does not reduce the replayable traces from the log by a fraction of more than delta. Formally, this resembles the identity function:

$$adapt_{constant}(\delta, p) = \delta$$

Linear and Sigmoid Delta Adaption. For a set of activities A, let $d_{max} = 2|A|$ be the maximum depth of the complete candidate tree, and let $d \in [2, 3, ..., d_{max}]$ be the current depth of the candidate tree traversal. We call $s \in \mathbb{N}\backslash\{0\}$ the *steepness modifier*. Consider a place $p = (I|O)$.

The *linear delta adaption function* computes the adapted δ as follows:

$$\begin{aligned} adapt_{linear}(\delta, (I|O)) &= \delta \cdot mod_{linear}((I|O)) \\ &= \delta \cdot \frac{s}{(|I| + |O|)} \cdot \frac{d - (|I| + |O|)}{d_{max} - 2} \end{aligned}$$

We define the *sigmoid delta adaption function* as follows:

$$\begin{aligned} adapt_{sigmoid}(\delta, (I|O)) &= \delta \cdot mod_{sigmoid}((I|O)) \\ &= \delta \cdot \left(\frac{2}{1 + \exp\left((-1) \cdot \frac{s}{(|I|+|O|)} \cdot (d - (|I| + |O|))\right)} - 1 \right) \end{aligned}$$

Figure 6 illustrates the behavior of the modifier each adaption function multiplies with the parameter δ for three example places of varying complexity. Both, the *linear* and *sigmoid* delta adaption are designed to prefer simple places. When a place originates from the currently traversed level of the *complete candidate tree*, i.e., it is among the most complex places currently available, both functions will evaluate to 0, meaning that only a perfectly fitting place can be added. The simpler the evaluated place is compared to the current tree level, the larger the result of the function and the more unfitting traces are allowed, with δ marking the maximal returnable value. The only difference is the steepness of the functions: while the linear function increases linearly with the place complexity, the sigmoid function grows fast in the beginning but stagnates towards the end. Thus, the linear function becomes uniformly less lenient with increasing place complexity. In contrast, the sigmoid function prefers the simpler places more strongly, while the more complex places are (roughly) equally undesirable.

6 Experimental Results and Evaluation

We perform several introductory experiments where we run the proposed algorithm with a wide variation of combinations of possible parameter settings on several event logs with different properties. To investigate the impact of the proposed heuristics, we use a lexicographical ordering of the activities, thus fixing the order of candidate evaluation. The purpose is to focus on the effect of the different parameters, and possibly derive which of them are the most relevant for the discovery of certain models and whether certain (combinations of) settings are preferable.

6.1 Experimental Setup

Table 1 provides an overview of the event logs used in our experimentation. Sepsis has a relatively high number of different trace variants, all of which have comparable frequencies with the most frequent trace making up only 3.33% of the event log. Activities are repeated often within a trace, which must lead to looping behavior within a Petri net with uniquely labeled transitions. RTFM is rather large, with a moderate variety of trace variants and activities. Both for variants and activities some are very frequent while others are quite infrequent. Teleclaims is an established artificial log useful for testing discovery of various control-flow structures. With Orders we can demonstrate the algorithms ability to discover complex control flow structures, as well as the option to abstract from rare behavior. For each event log we perform 4200 runs of the algorithm with varying combinations of the different parameters, as specified in Table 2. Note, that we keep the order of place candidate traversal fixed for all runs.

6.2 Results and Evaluation

For each model discovered we compute alignment-based fitness, precision and simplicity as described in Sect. 4. Based on alignment-based fitness and precision, we also present the F_1-Score, i.e., the harmonic mean of alignment-based fitness and precision.

Table 1. List of logs used for the evaluation. The upper part lists real-life logs while the lower part shows artificial logs. Logs are referred to by their abbreviations.

Log name	Abbreviation	Activities	Trace variants	Reference
Sepsis	Sepsis	16	846	[29]
Road Traffic Fine Management	RTFM	11	231	[30]
Teleclaims	Teleclaims	11	12	[25]
Order-Handling	Orders	8	9	[31]

Table 2. Overview of the parameters settings used in our experimentation. The combinations result in 4200 runs for each event log. The values ranges were chosen based on a smaller set of preliminary experiments, aiming to investigate a wide range of parameter settings on the one hand, while on the other hand avoiding unnecessary complexity resulting from variation without notable impact. For example, for our inputs no places were discarded for $|Q| \geq 10000$. For d^+ we chose a very low and a very high value to evaluate whether it had any impact at all. Finally, for the chosen event logs $d_{cut} = 5$ has shown to be sufficient to find complex structures with the standard eST-Miner, i.e., increasing the traversed tree depth increases computation time but has no strong impact on model quality.

Parameter	Used values	Purpose
τ	0.3, 0.4, 0.5, 0.6, 0.7, 0.8, 0.9	Defines the minimal fraction of log traces that every place, as well as the final Petri net, must be able to replay.
δ	0.05, 0.1, 0.15, 0.2, 0.25	Used to define the allowed reduction in log traces replayable by N when adding a place.
adapt	$adapt_{noDelta}$, $adapt_{constant}$, $adapt_{linear}$, $adapt_{sigmoid}$	The delta adaption function used to guide the heuristics.
s	1, 2, 3, 4, 5	The steepness of the increase of the adaption function (relevant for $adapt_{linear}$ and $adapt_{sigmoid}$ only).
$\|Q\|$	100, 1000, 10000	The maximal number of places stored in Q.
d^+	0, 10	Artificial tree depth to re-evaluate places in Q after end of tree traversal (relevant for $adapt_{linear}$ and $adapt_{sigmoid}$ only).
d_{cut}	5	Stop candidate traversal after the specified tree level.

In Fig. 7 an overview of the quality results of the 4200 models generated for each log is given. Fitness and simplicity remain rather stable, with fitness being generally high and simplicity values clustering around 0.5 (which we consider a good value, recall Definition 7). On the other hand, precision, and by extension the F_1-score, vary a lot for the discovered models. This clearly indicates that the choice of parameters has a strong impact on this quality aspect.

While we discovered only 7 unique models for `Orders`, there were 19 unique models found for `RTFM`, 27 for `Teleclaims` and 140 for `Sepsis`. The quality results and frequencies of the 10 most frequently discovered Petri nets are given in Fig. 8 (Model IDs 1 to 10). Additionally, we provide the same results for the models discovered by the Inductive Miner infrequent (IMf) with default settings as implemented in ProM [32], the models discovered by the eST-Miner with $\tau = 1.0$ (comparable to region theory results), as well as the model with the highest F_1-Score discovered over all runs.

In Figs. 9, 10, 11 and 12 we present a selection of models for each log: the model discovered by IMf, the model discovered by the eST-Miner with $\tau = 1.0$, the most frequently discovered model and the model with the highest F_1-score discovered by the experiments with our proposed approach.

All models shown in Fig. 9 were discovered for the `Orders` log. For this rather simple event log, all models achieve relatively high scores with respect to the quality metrics. However, some notable differences in the expressed behavior can be observed in particular with respect to the activities *send invoice, send reminder, pay* and *cancel*

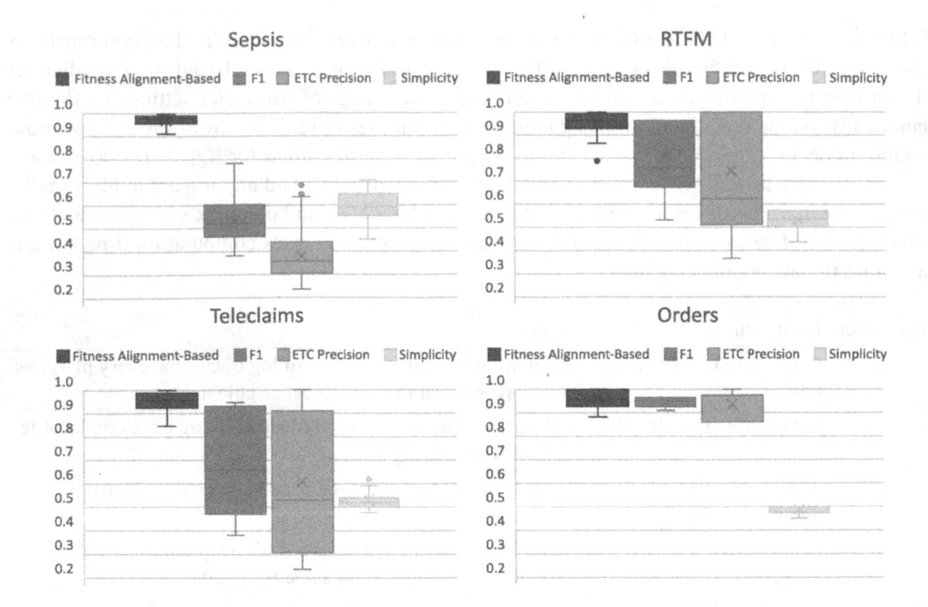

Fig. 7. Overview of model quality results for all 4200 runs with varying parameters but fixed candidate traversal order.

order. According to the event log, in most cases execution of *send invoice* is eventually followed either by *pay* (and then delivery) or by *cancel order*, but never both. In rare cases, payment occurs before sending the invoice. After sending the invoice, reminders can be sent repeatedly, until payment is received or the order is canceled. This behavior is fully expressed only by the model discovered using the eST-Miner with $\tau = 1.0$, which is comparable to results produced by region-based approaches. Since payment before sending the invoice is rare, users may prefer the other models which focus on behavior where payment arrives after sending the invoice. The model discovered by IMf further deviates from the log by not allowing for repeated reminders (occurring in 25% of the traces), and enabling the cancellation of orders after payment. In contrast to the model discovered most frequently by our extended eST-Miner, the model with the highest F_1-Score does not contain the activity *cancel order* (occurring in 13.03% of traces) at all, resulting in slightly lower fitness but notably increased precision.

The `Sepsis` event log exhibits many repetitions of activities and a comparatively high control-flow variance, with 846 trace variants in 1050 traces, the most frequent of which occurs only 35 times. Thus, the discovery of a model with simultaneously high fitness and precision is challenging. Figure 10 presents a selection of discovered Petri nets. The IMf manages to discover groups of activities that occur in sequence, however, within these groups the activities are in parallel and mostly skipable, resulting a very low precision. The eST-Miner with $\tau = 1.0$ illustrates a disadvantage of requiring perfect fitness: the resulting model allows for nearly all possible behaviors. For the most frequent model discovered by our extension, this problem becomes less severe and is significantly reduced for the discovered model with the highest F_1-Score. This model manages to capture the main behavior hidden in the traces while ignoring infrequent activity behavior, achieving comparatively high precision.

Log	Metric	IMf (default settings)	eST (τ = 0.1)	Model ID										highest F1
				1	2	3	4	5	6	7	8	9	10	
RTFM (discovered 19 unique models)	F1	0.6432	0.6531	0.6763	0.9638	0.6794	0.9371	0.7570	0.8816	0.9270	0.8540	0.7792	0.6757	Model 2
	Fitness	0.9820	1.0000	0.9940	0.9301	0.9642	0.9642	0.9603	0.7882	0.8640	0.8707	0.9809	0.9642	
	Precision	0.4782	0.4849	0.5125	1.0000	0.5245	0.9114	0.6248	1.0000	1.0000	0.8379	0.6463	0.5201	
	Simplicity	0.5526	0.6842	0.5714	0.5000	0.5714	0.5333	0.5200	0.4667	0.5000	0.5000	0.5333	0.5333	
	Frequency	-	-	1271	1107	383	367	312	243	150	150	34	33	
Sepsis (discovered 140 unique models)	F1	0.3215	0.3266	0.4264	0.4639	0.5205	0.5752	0.6086	0.4849	0.4672	0.6893	0.6036	0.4582	0.7836
	Fitness	0.9060	1.0000	0.9942	0.9846	0.9603	0.9681	0.9560	0.9679	0.9781	0.9230	0.9635	0.9942	0.9115
	Precision	0.1954	0.1952	0.2714	0.3034	0.3570	0.4091	0.4464	0.3235	0.3069	0.5500	0.4395	0.2977	0.6871
	Simplicity	0.5106	0.7826	0.6667	0.5926	0.5926	0.5909	0.5385	0.5806	0.6087	0.4516	0.6154	0.6522	0.52
	Frequency	-	-	408	354	198	187	186	180	180	150	150	120	52
Teleclaims (discovered 27 unique models)	F1	0.9496	0.5993	0.9316	0.4697	0.3895	0.4536	0.6272	0.9469	0.9328	0.8621	0.8202	0.8825	Model 6
	Fitness	0.9490	1.0000	0.9538	1.0000	0.9889	0.9889	0.9733	0.9244	0.8740	0.9423	0.9572	0.8461	
	Precision	0.9503	0.4279	0.9105	0.3069	0.2425	0.2943	0.4627	0.9706	1.0000	0.7945	0.7175	0.9222	
	Simplicity	0.5172	0.5200	0.5000	0.5417	0.6190	0.5909	0.5200	0.5200	0.4815	0.5200	0.5000	0.4783	
	Frequency	-	-	650	507	474	440	374	356	210	182	182	150	
Orders (discovered 7 unique models)	F1	0.9340	0.9319	0.9258	0.9664	0.9502	0.9374	0.9319	0.9096	0.9296				Model 2
	Fitness	0.9600	1.0000	0.9996	0.9562	0.9279	0.8822	1.0000	0.9279	0.9562				
	Precision	0.9094	0.8725	0.8622	0.9768	0.9735	1.0000	0.8725	0.8921	0.9044				
	Simplicity	0.5000	0.4762	0.5000	0.4737	0.5000	0.4706	0.4762	0.4737	0.4500				
	Frequency	-	-	1472	1424	1000	225	64	11	4				

Fig. 8. Overview of the qualitative evaluation results of the IMf (default settings), the eST-Miner with $\tau = 0$ and the 10 most frequently discovered models by the presented approach including their frequencies. The final column indicates the model with the best F_1-Score (only for Sepsis this is not contained in the most frequent models). A selection of models is presented for each event log in Figs. 9, 10, 11 and 12.

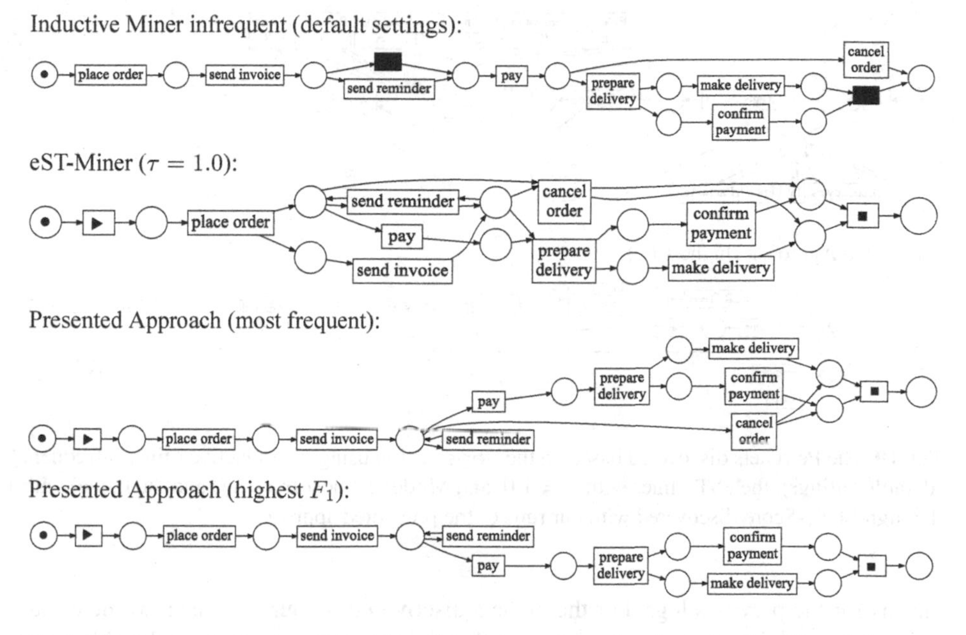

Fig. 9. The Petri nets discovered based on the Orders log using the Inductive Miner infrequent (default settings), the eST-Miner with $\tau = 1.0$, and Model 1 (most frequent) and Model 2 (highest F_1-Score) discovered with our runs of the presented approach.

Figure 11 shows Petri nets discovered from the RTFM log. Considering the models discovered by IMf and eST-Miner with $\tau = 1.0$, we observe the same general tenden-

Inductive Miner infrequent (default settings):

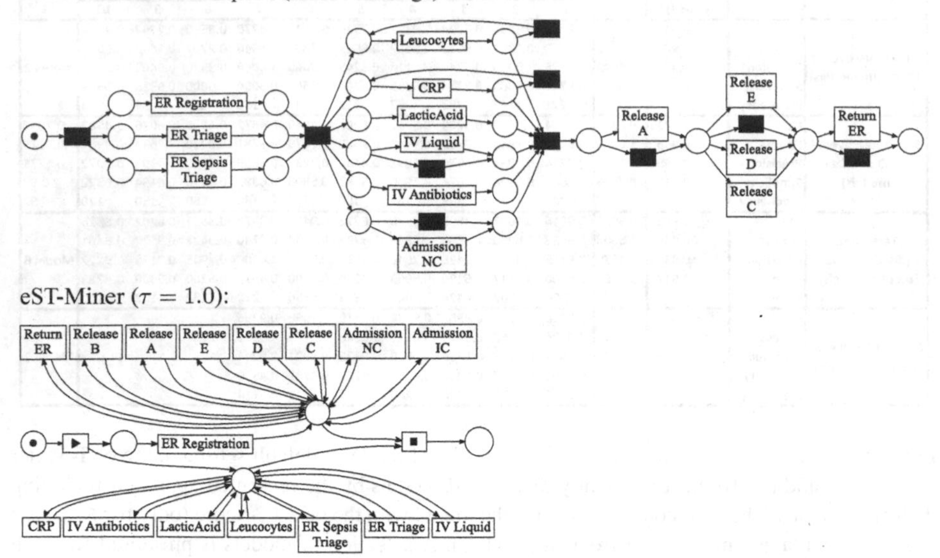

Presented Approach (most frequent):

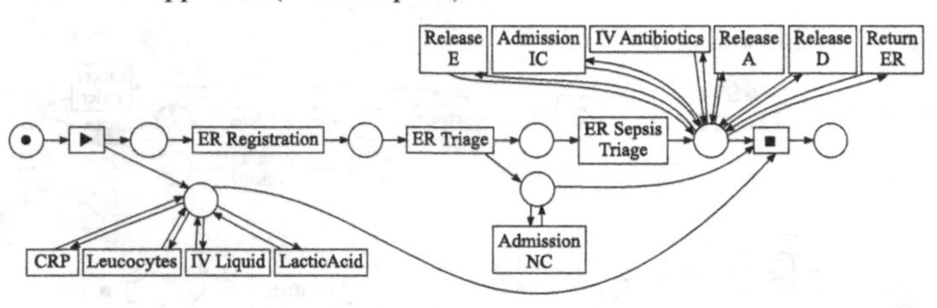

Presented Approach (highest F_1):

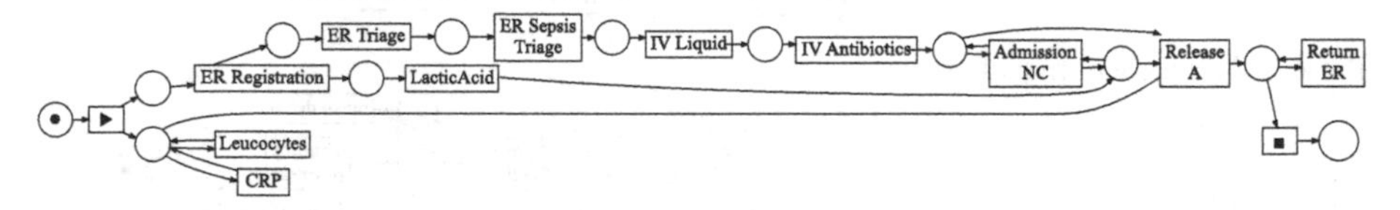

Fig. 10. The Petri nets discovered based on the `Sepsis` log using the Inductive Miner infrequent (default settings), the eST-Miner with $\tau = 1.0$, and Model 1 (most frequent) and the model with the highest F_1-Score discovered with our runs of the presented approach.

cies as for the previous logs. For the models discovered by our approach, we note that quite many activities are missing, meaning that they are not part of any replayable trace from the event log. The reason can be found by investigation of this particular event log, which describes two very distinct sub-processes, the more frequent of which consists of the activities still contained in the model. The activities of the infrequent sub-process related to the appeals have been filtered to focus on the main process.

Inductive Miner infrequent (default settings):

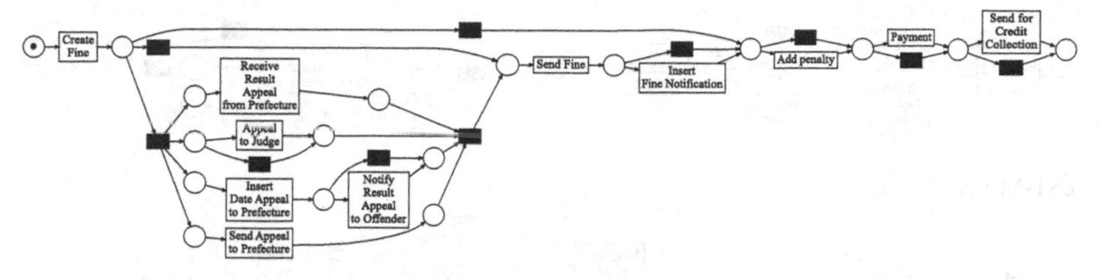

eST-Miner ($\tau = 1.0$):

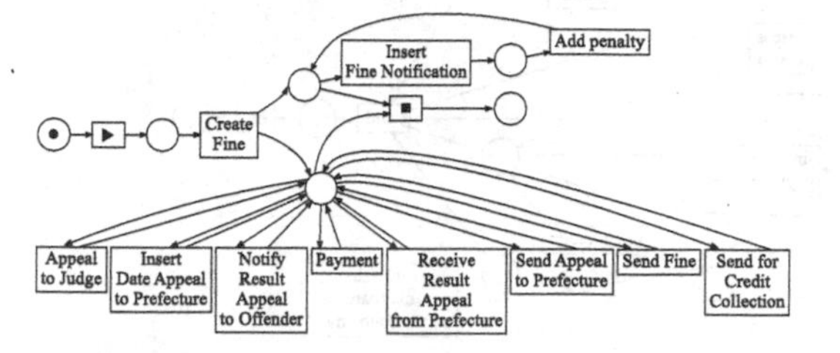

Presented Approach (most frequent):

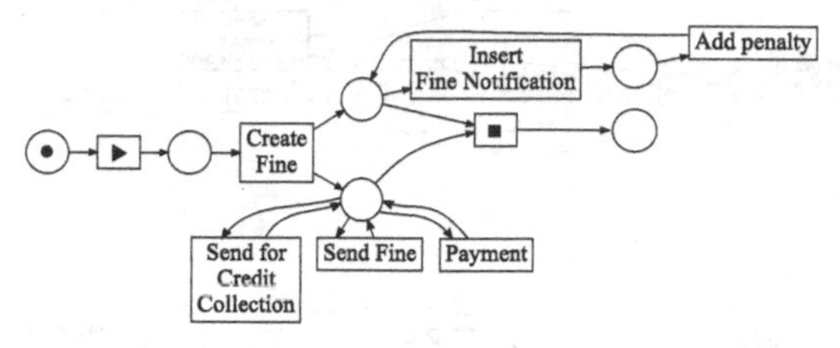

Presented Approach (highest F_1):

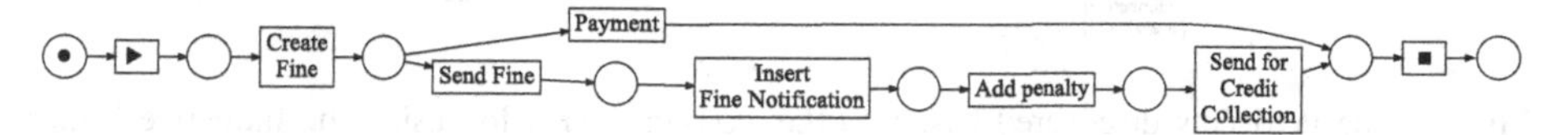

Fig. 11. The Petri nets discovered based on the RTFM log using the Inductive Miner infrequent (default settings), the eST-Miner with $\tau = 1.0$, and Model 1 (most frequent) and Model 2 (highest F_1-Score) discovered with our runs of the presented approach.

A set of process models discovered from the `Teleclaims` log is presented in Fig. 12. The models discovered by the IMf and our approach express similar behavior, with the main difference being the representation of skipable activities: with all transitions being uniquely labeled, our approach has to rely on loop constructs rather than silent activities. The eST-Miner with $\tau = 1.0$ does not abstract from infrequent behavior, which in this case results in a perfectly fitting but quite complex model.

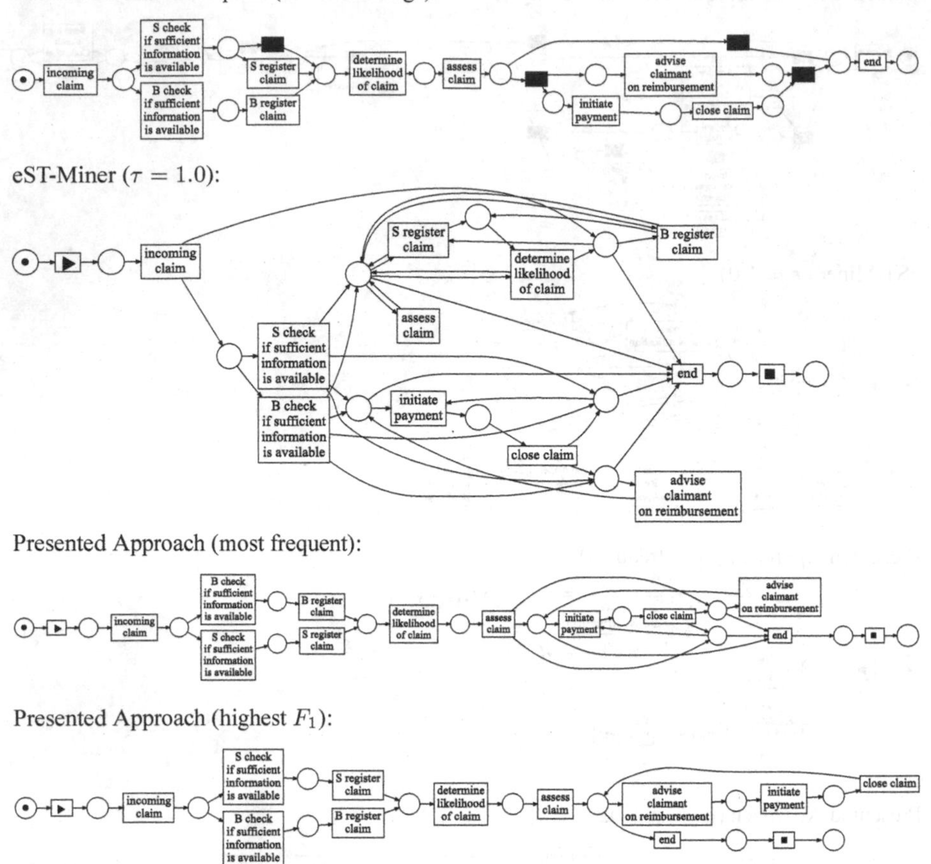

Fig. 12. The Petri nets discovered based on the `Teleclaims` log using the Inductive Miner infrequent (default settings), the eST-Miner with $\tau = 1.0$, and Model 1 (most frequent) and Model 6 (highest F_1-Score) discovered with our runs of the presented approach.

Our results indicate that even minor gains in fitness are usually accompanied by a major drop in precision. The models with the best F_1-Score are usually those with the highest precision value. From Figs. 9, 10, 11 and 12 we can observe that these models seem to abstract well from infrequent activity behavior, giving a clear representation of the main process. However, models with higher fitness may reveal complex control-flow structures and interesting variations. The presented approach is able to return models anywhere on this scale between fitness and precision based on the choice of parameters.

While the quality metrics clearly indicate that our approach is able to discover models balancing fitness and precision while maintaining reasonable simplicity, the choice of parameters has a significant impact that requires further investigation. We used decision tree analysis to search for certain parameter settings that would result in the highest quality models as indicated by the F_1-Score. The results of this analysis are shown in Table 3, where each line represents a set of parameter combinations that leads to the discovery of the best model.

For the four event logs investigated in this paper, the most important parameters seem to be τ and δ. This is not surprising, since τ has a direct impact on which places are available for addition to the Petri net and δ is limiting the range of the adaption strategies, which include the use of s. Notably, the artificial tree depth d^+ as well as $|Q|$ have had no major impact on the discovery of any of the examined models and the sigmoid and linear delta adaption strategies are often interchangeable.

Some interdependencies between the parameters are expected, and seem to be confirmed by the results in Table 3. For example, when using the delta adaption functions $adapt_{constant}$ and $adapt_{noDelta}$, the steepness modifier s has no impact. For the `Orders` log there is an indication of an exceptionally low s value working well with a high artificial tree depth modifier d^+, and for low τ values requiring the use of δ to discover the best model. For event logs like the `RTFM` log and the `Orders` log, which have a few very dominant trace variants, we seem to generally achieve good results for rather high values of τ. In contrast, for event logs with a high variety of traces as for example `Sepsis` log, a low τ-value seems mandatory. Most likely, the large variety of fitting places allows for obtaining high precision, while our heuristics seems to successfully ensure the focus on the main behavior. However, more results are needed to validate such speculations.

Interestingly, the results from the `Sepsis` log, which contains a high variety of traces, seem to confirm our algorithms ability to discover the main behavior hidden in an event log even in the absence of main trace variants: for a low value of τ, e.g. $\tau = 0.3$, the fraction of log traces replayable by the return Petri net is indeed close to 0.3, however, the alignment-based fitness reliably remains above 0.9, indicating that most of the traces are close to being replayable. We can conclude that the returned model successfully expresses the core behavior of the process.

To summarize, the results clearly show that high quality models balancing the different quality aspects can be discovered. There is a significant variance in some of the metrics, particularly precision, indicating that the settings of the algorithm have a notable impact. Our preliminary investigation using decision trees shows, that certain parameter choices result in high quality models. It gives a first indication about which parameters have a more notable impact and whether certain settings are more suitable for logs with certain properties. Further experimentation needs to be performed to investigate to which degree a generalization is possible. Note, that the impact of the candidate traversal order has not been investigated yet, and may allow for further improvements.

Table 3. Overview of the parameter choices resulting in the discovery of the model with the highest F_1-Score. For each log we indicate how often this model has been discovered in our experimental runs and in which figure to find it. For each parameter that our decision tree analysis has revealed to be impactful, the possible values are indicated. Each line corresponds to a set of parameter combinations, with the frequency of the model being discovered using these combinations given to the right.

Log and Model	τ	δ	**Strategy**	s	d^+	#Combinations
Sepsis (Fig. 10) (discovered 52 times)	0.3	0.15	$adapt_{constant}$	–	–	30
	0.3	0.25	$adapt_{sigmoid}$	[3, 5]	–	18
RTFM (Fig. 11) (discovered 1107 times)	[0.3, 0.6]	[0.15, 0.25]	$adapt_{constant}$	–	–	360
	[0.4, 0.6]	[0.2, 0.25]	$adapt_{sigmoid}$, $adapt_{linear}$	[2, 5]	–	288
	0.6	–	$adapt_{noDelta}$	–	–	150
	[0.4, 0.6]	0.15	$adapt_{linear}$	[3, 5]	–	54
Teleclaims (Fig. 12) (discovered 356 times)	[0.3, 0.4]	[0.15, 0.25]	$adapt_{constant}$	–	–	180
	[0.3, 0.4]	[0.2, 0.25]	$adapt_{sigmoid}$	[4, 5]	–	48
	[0.3, 0.4]	0.15	$adapt_{linear}$	[3, 4]	–	24
	[0.3, 0.4]	0.25	$adapt_{linear}$	[4, 5]	–	24
Orders (Fig. 9) (discovered 1424 times)	[0.7, 0.8]	[0.15, 0.25]	–	[2, 5]	–	576
	[0.3, 0.6]	[0.15, 0.2]	$adapt_{sigmoid}$, $adapt_{constant}$	[2, 5]	-	384
	[0.7, 0.8]	–	$adapt_{noDelta}$	–	–	120
	[0.7, 0.8]	[0.15, 0.25]	–	1	10	72

7 Conclusion

In this paper, we introduced an extension to the eST-Miner that returns a Petri net which satisfies user-definable minimal fitness requirements. The presented approach employs heuristics to efficiently select a suitable subset of the discovered places, while aiming towards high precision and simplicity. The algorithm is capable to discover complex control-flow structures such as non-local dependencies, to deal with noise in the event log and to provide guarantees without over- or underfitting.

Our first experiments, using four different event logs, clearly show that not only is it possible to discover high-quality models using the introduced approach, but also the heuristics applied have a significant impact on the obtained Petri net. Based on the parameter settings, models with a very different focus with respect to fitness, precision and the handling of infrequent behavior can be discovered. Some parameters have a stronger effect than others and some parameter choices seem to be more suitable for logs with certain properties, which should be verified by further experimentation.

Next to an analysis of the running-time, future work includes further experimentation to explore the generalization the preliminary results, as well as the impact of the candidate place traversal order and its interaction with the heuristics used. Improvements or variations of the strategies are likely possible. It would be particularly interesting to investigate to which degree the approach can be used to prioritize non-standard quality aspects, for example related to user interests such as compliance or performance.

The dead transitions removed from the model because they are no longer part of the replayable event log give rise to further possible extensions of the eST-Miner. When detected early on, they can be used to identify and cut off candidate subtrees consisting of dead places to improve the running time. Further investigation into the cause of their removal my lead to better noise handling strategies to improve the quality of discovered models. Finally, it would be interesting to investigate whether the presented place selection strategies can be adapted to improve other algorithms as well.

Acknowledgments. Special thanks goes to Tobias Brockhoff for supporting the implementation and evaluation of the presented experiments. We thank the Ministry of Culture and Science of the German State of North Rhine-Westphalia (MKW) and the Excellence Strategy of the Federal Government and the Länder for supporting our research.

References

1. Reisig, W.: Understanding Petri Nets: Modeling Techniques, Analysis Methods, Case Studies. Springer, Berlin Heidelberg (2013). https://doi.org/10.1007/978-3-642-33278-4
2. Desel, J., Oberweis, A., Reisig, W., Rozenberg, G.: Petri Nets and Business Process Management. Geschäftsstelle Schloss Dagstuhl, Saarbrücken (1998)
3. Desel, J., Esparza, J.: Free-Choice Petri Nets. Cambridge Tracts in Theoretical Computer Science. Cambridge University Press, Cambridge (1995)
4. Berthelot, G.: Transformations and decompositions of nets. In: Brauer, W., Reisig, W., Rozenberg, G. (eds.) Petri Nets: Central Models and Their Properties, pp. 359–376. Springer, Berlin, Heidelberg (1987). https://doi.org/10.1007/978-3-540-47919-2_13
5. Wen, L., van der Aalst, W.M.P., Wang, J., Sun, J.: Mining process models with non-free-choice constructs. Data Min. Knowl. Disc. **15**(2), 145–180 (2007)
6. Leemans, S.J.J., Fahland, D., van der Aalst, W.M.P.: Discovering block-structured process models from event logs - a constructive approach. In: Colom, J.-M., Desel, J. (eds.) PETRI NETS 2013. LNCS, vol. 7927, pp. 311–329. Springer, Heidelberg (2013). https://doi.org/10.1007/978-3-642-38697-8_17
7. Badouel, E., Bernardinello, L., Darondeau, P.: Petri Net Synthesis. Text in Theoretical Computer Science, EATCS Series. Springer, Heidelberg (2015). https://doi.org/10.1007/978-3-662-47967-4
8. Lorenz, R., Mauser, S., Juhás, G.: How to synthesize nets from languages: a survey. In: Proceedings of the 39th Conference on Winter Simulation: 40 Years! The Best is Yet to Come, WSC 2007, pp. 637–647. IEEE Press, Piscataway, NJ, USA (2007)
9. Bergenthum, R., Desel, J., Lorenz, R., Mauser, S.: Process mining based on regions of languages. In: Alonso, G., Dadam, P., Rosemann, M. (eds.) BPM 2007. LNCS, vol. 4714, pp. 375–383. Springer, Heidelberg (2007). https://doi.org/10.1007/978-3-540-75183-0_27
10. van der Werf, J.M.E.M., van Dongen, B.F., Hurkens, C.A.J., Serebrenik, A.: Process discovery using integer linear programming. In: van Hee, K.M., Valk, R. (eds.) PETRI NETS 2008. LNCS, vol. 5062, pp. 368–387. Springer, Heidelberg (2008). https://doi.org/10.1007/978-3-540-68746-7_24
11. van Zelst, S.J., van Dongen, B.F., van der Aalst, W.M.P.: Avoiding over-fitting in ILP-based process discovery. In: Motahari-Nezhad, H.R., Recker, J., Weidlich, M. (eds.) BPM 2015. LNCS, vol. 9253, pp. 163–171. Springer, Cham (2015). https://doi.org/10.1007/978-3-319-23063-4_10
12. van Zelst, S.J., van Dongen, B.F., van der Aalst, W.M.P.: ILP-based process discovery using hybrid regions. In: ATAED@Petri Nets/ACSD (2015)
13. Carmona, J., Cortadella, J., Kishinevsky, M.: A region-based algorithm for discovering PETRI nets from event logs. In: Dumas, M., Reichert, M., Shan, M.-C. (eds.) BPM 2008. LNCS, vol. 5240, pp. 358–373. Springer, Heidelberg (2008). https://doi.org/10.1007/978-3-540-85758-7_26

14. Darondeau, P.: Deriving unbounded Petri nets from formal languages. In: Sangiorgi, D., de Simone, R. (eds.) CONCUR 1998. LNCS, vol. 1466, pp. 533–548. Springer, Heidelberg (1998). https://doi.org/10.1007/BFb0055646
15. Bergenthum, R., Desel, J., Lorenz, R., Mauser, S.: Synthesis of Petri nets from finite partial languages. Fundam. Inf. **88**(4), 437–468 (2008)
16. Ehrenfeucht, A., Rozenberg, G.: Partial (set) 2-structures. Acta Informatica **27**(4), 343–368 (1990)
17. Carmona, J., Cortadella, J., Kishinevsky, M., Kondratyev, A., Lavagno, L., Yakovlev, A.: A symbolic algorithm for the synthesis of bounded PETRI nets. In: van Hee, K.M., Valk, R. (eds.) PETRI NETS 2008. LNCS, vol. 5062, pp. 92–111. Springer, Heidelberg (2008). https://doi.org/10.1007/978-3-540-68746-7_10
18. Kalenkova, A., Carmona, J., Polyvyanyy, A., La Rosa, M.: Automated repair of process models using non-local constraints. In: Janicki, R., Sidorova, N., Chatain, T. (eds.) PETRI NETS 2020. LNCS, vol. 12152, pp. 280–300. Springer, Cham (2020). https://doi.org/10.1007/978-3-030-51831-8_14
19. Mannel, L.L., van der Aalst, W.M.P.: Finding complex process-structures by exploiting the token-game. In: Donatelli, S., Haar, S. (eds.) PETRI NETS 2019. LNCS, vol. 11522, pp. 258–278. Springer, Cham (2019). https://doi.org/10.1007/978-3-030-21571-2_15
20. van der Aalst, W.M.P.: Discovering the glue connecting activities. In: de Boer, F., Bonsangue, M., Rutten, J. (eds.) It's All About Coordination. LNCS, vol. 10865, pp. 1–20. Springer, Cham (2018). https://doi.org/10.1007/978-3-319-90089-6_1
21. Garcia-Valles, F., Colom, J.: Implicit places in net systems. In: Proceedings of the 8th International Workshop on Petri Nets and Performance Models, pp. 104–113 (1999)
22. Berthomieu, B., Botlan, D.L., Dal-Zilio, S.: Petri net reductions for counting markings. CoRR abs/1807.02973 (2018)
23. Colom, J.M., Silva, M.: Improving the linearly based characterization of P/T nets. In: Rozenberg, G. (ed.) ICATPN 1989. LNCS, vol. 483, pp. 113–145. Springer, Heidelberg (1991). https://doi.org/10.1007/3-540-53863-1_23
24. Mannel, L.L., Bergenthum, R., van der Aalst, W.M.P.: Removing implicit places using regions for process discovery. In: Proceedings of the International Workshop on Algorithms & Theories for the Analysis of Event Data (ATAED) 2020, vol. 2625, pp. 20–32 (2020). CEUR-WS.org
25. van der Aalst, W.M.P.: Process Mining: Data Science in Action, 2nd edn. Springer, Heidelberg (2016). https://doi.org/10.1007/978-3-662-49851-4
26. Carmona, J., van Dongen, B.F., Solti, A., Weidlich, M.: Conformance Checking - Relating Processes and Models. Springer, Cham (2018). https://doi.org/10.1007/978-3-319-99414-7
27. Adriansyah, A.: Aligning observed and modeled behavior. Ph.D. thesis, Mathematics and Computer Science (2014)
28. Muñoz-Gama, J., Carmona, J.: A fresh look at precision in process conformance. In: Hull, R., Mendling, J., Tai, S. (eds.) BPM 2010. LNCS, vol. 6336, pp. 211–226. Springer, Heidelberg (2010). https://doi.org/10.1007/978-3-642-15618-2_16
29. Mannhardt, F.: Sepsis cases-event log (2016)
30. De Leoni, M., Mannhardt, F.: Road traffic fine management process (2015)
31. van der Aalst, W.M.P.: Spreadsheets for BPM. Bus. Process. Manag. J. **24**, 105–127 (2010)
32. van Dongen, B.F., de Medeiros, A.K.A., Verbeek, H.M.W., Weijters, A.J.M.M., van der Aalst, W.M.P.: The ProM framework: a new era in process mining tool support. In: Ciardo, G., Darondeau, P. (eds.) ICATPN 2005. LNCS, vol. 3536, pp. 444–454. Springer, Heidelberg (2005). https://doi.org/10.1007/11494744_25

Aligning Event Logs to Resource-Constrained ν-Petri Nets

Dominique Sommers(✉), Natalia Sidorova, and Boudewijn van Dongen

Department of Mathematics and Computer Science, Eindhoven University of Technology, Eindhoven, The Netherlands
{d.sommers,n.sidorova,b.f.v.dongen}@tue.nl

Abstract. Systems with shared resources can be modeled and analyzed using high-level Petri nets in a natural way. Choosing a model type suitable for the use in conformance checking introduces challenges related to constraints the model should put on resource types and resource instances. In this paper, we propose a model for systems with shared resources based on resource-constrained Petri nets and ν-Petri nets that can be used in the context of conformance checking. Our model allows for case and resource isolation, allowing for proper simulation of multiple cases involving shared resources.

With this minimal extension, we show that we can use existing state-of-the-art conformance checking techniques to compute alignments on complete event logs rather than on individual case instances. We show that previously undetected deviations caused by inter-case dependencies can now be exposed, providing valuable information regarding the exhaustive workflow in the process.

Keywords: Petri nets · Shared resources · Conformance checking · Inter-case dependencies

1 Introduction

Process models often include descriptions of resources executing activities within the process, since the availability of resources puts constraints on the process execution. Event logs, recording process executions, also often include indications which resource executed which activity and when, usually mentioning the exact person(s) or machine(s) that were involved in the activity. Such event logs can be used for conformance checking, i.e. checking whether and where the actual process behavior recorded in an event log deviates from the behavior prescribed by a process model.

Various types of conformance checking techniques that exist so far are primarily focused on the control flow of a process, without taking into account

This work is done within the project "Certification of production process quality through Artificial Intelligence (CERTIF-AI)", funded by NWO (project number: 17998).

L. Bernardinello and L. Petrucci (Eds.): PETRI NETS 2022, LNCS 13288, pp. 325–345, 2022.
https://doi.org/10.1007/978-3-031-06653-5_17

the information about resources, and do so by looking at individual cases going through the process separately. In the context of resources, it is crucial to consider all the cases going through the system at the same time, since these cases share resources available in the system.

Another challenge in the conformance checking of processes with shared resources is the necessity to consider not only resource types (e.g. whether it is a doctor or a nurse who has to perform a particular activity), but also the resource identity (e.g. which doctor performed a surgery). This information is critical for checking resource related constraints that should be imposed by the model, e.g. that the patient had a follow-up appointment with the doctor who conducted the surgery, or that the second-opinion appointment is not planned with the same doctor whom the patient already met.

In this paper, we first address the question how to model resource-constrained processes in order to enable conformance checking and then adapt a conformance checking method to dealing with resource-constrained processes. We build our model on basis of resource-constrained Petri nets [31] and ν-Petri nets [22]. The model allows to specify resource types by using resource places, and case and resource identities by using case ids and resource ids as token colors. We use the alignment mechanism introduced in [1] as basis for our alignment method to do conformance checking on resource-constrained ν-Petri nets.

Related Work. In [3] and [31], Petri nets are extended with resources to model availability of durable resources, as well as their claims and releases by cases running through the system. ν-Petri nets [22] allow for case isolation as a minimal extension to classical Petri nets via name creation and name management. An advantage as opposed to more advanced Petri net extensions is that coverability and termination are decidable for ν-Petri nets.

Other extensions such as Catalog Petri nets [10], synchronizing proclet models [9], resource and instance-aware workflow nets (RIAW-nets) [17], and DB-nets [18] inherit the functionality of ν-Petri nets. Additionally, these extensions implement concepts from databases, shared resources, and proclet channels. We show that we do not require such additional functionality and aim for a minimal extension on Petri nets.

Many conformance checking techniques use alignments to directly connect the behavior of a system recorded in a log with the behavior allowed by a process model. Alignments can expose exactly where the recorded behavior and the model agree, which activities prescribed by the model are missing in the log and which log activities should not be performed according to the model [6, 28]. Rule checking techniques [14, 25] are conformance checking techniques that check if specific business rules are respected, and they can be useful in case the process model does not describe the whole process behavior. Case-replay techniques [4, 24, 28, 29] aim to identify specific deviations between modeled and observed behavior. We choose alignments as basis for our conformance checking method since they are designed for fully-specified processes (potentially with invisible transitions) and target at discovering a broad range of deviations in the process behavior.

Conformance checking usually targets isolated cases from the workflow perspective. More advanced techniques consider resources and data on top of the control flow; in [7], the control flow is considered first, after which other perspectives are checked. This method can provide misleading results in case of shared resources, since resources put additional constraints on the control flow. In [15], this is partially mitigated by balancing the different perspectives in a customizable manner. More recently, a technique was proposed to consider all perspectives at once [16], but cases are still considered individually, also when they are run in parallel, and tokens are uncolored, making it impossible to capture resource ids in the model.

Alignments, as well as the other techniques, are computed primarily focusing on the detection of workflow deviations for individual cases. Work has been done to take into account multiple perspectives like data attributes and resources to check, besides the workflow, whether the correct data attributes and resources were involved [2,7,15,16]. However, they still operate on a case-by-case basis. With resource-constrained Petri nets, violations regarding inter-case dependencies remain undetected.

Outline. This paper is organized as follows. Section 2 presents basic definitions related to Petri nets and event logs. In Sect. 3, we focus on processes with shared resources, introduce the notion of resource-constrained ν-Petri nets and some modeling patterns. In Sect. 4, we steer towards the problem of conformance checking and investigate the missing link with inter-case dependencies caused by shared resources. In Sect. 5, we propose a solution exploiting ν-Petri nets to compute alignments which allows for exposing violations. We conclude in Sect. 6 by discussing our contributions and directions for future work.

2 Preliminaries

In this section we present the notations that we will use throughout the paper.

2.1 Petri Nets

Petri nets can be used as a tool for the representation, validation and verification of workflow processes to provide insights in how the process behaves [21].

Definition 1 *(Multiset). A multiset m over a set X is $m : X \rightarrow \mathbb{N}$, denoted as $X^{\oplus}$. The support supp(m) of a multiset m is the set $\{x \in X \mid m(x) > 0\}$.*

For $m_1, m_2 \in \mathbb{N}$, we write $m_1 \leq m_2$ if $\forall_{x \in X} : m_1(x) \leq m_2(x)$, and $m_1 < m_2$ if $m_1 \leq m_2 \wedge m_1 \neq m_2$. We define $m_1 + m_2$ as $(m_1 + m_2)(x) = m_1(x) + m_2(x)$ for all $x \in X$. For $m_1 \geq m_2$, we define $m_1 - m_2$ as $(m_1 - m_2)(x) = m_1(x) - m_2(x)$ for all $x \in X$.

In some cases we consider multisets over a set X as vectors of length $|X|$, where we assume arbitrary but fixed orderings of elements of X.

Definition 2 *(Petri net). A* Petri Net [19] *is a 3-tuple* $N = \langle P, T, \mathcal{F} \rangle$*, where* P *is the set of places,* T *is the set of transitions,* $P \cap T = \emptyset$*,* $\mathcal{F} : (P \times T) \cup (T \times P) \to \mathbb{N}$ *is the flow of the net. The incidence matrix* F *of a Petri net* N *is a matrix with a row for each place* $p \in P$ *and a column for each transition* $t \in T$ *and it is defined by* $F(p, t) = \mathcal{F}(t, p) - \mathcal{F}(p, t)$*.*

We write $P(N)$*,* $T(N)$ *and* $\mathcal{F}(N)$ *to indicate that we refer to the set of places, the set of transitions and the flow relation of a net* N*.*

$N_1 \cap N_2$*,* $N_1 \cup N_2$*, and* $N_1 \subseteq N_2$ *denote intersection, union, and subsets of nets, respectively, defined on the sets of nodes and arcs of* N_1 *and* N_2*.*

A labeled *Petri net* $N = \langle P, T, \mathcal{F}, \ell \rangle$ *additionally defines a labeling* $\ell : T \to \Sigma^{\perp} = \Sigma \cup \{\tau\}$ *assigning each transition* t *a label* $\ell(t)$ *from alphabet* Σ *or* $\ell(t) = \tau$ *for silent transitions. We assume that the intersection, union and subsets are only defined for two labeled Petri nets* N_1*,* N_2 *where* $\ell_1(t) = \ell_2(t)$ *for any transition* $t \in T_1 \cap T_2$*.*

Definition 3 *(Post-set, Pre-set). Given a transition* $t \in T$*, its* pre-set $^\bullet t$ *and* post-set $t^\bullet$ *are multisets defined as follows:* $^\bullet t(p) = \mathcal{F}(p, t)$ *and* $t^\bullet(p) = \mathcal{F}(t, p)$ *for* $p \in P$*. Correspondingly, for a place* $p \in P$ *we have* $^\bullet p(t) = \mathcal{F}(t, p)$ *and* $p^\bullet(t) = \mathcal{F}(p, t)$ *for* $t \in T$*.*

Definition 4 *(Marking). A* marking $m : P \to \mathbb{N}$ *of Petri net* $N = \langle P, T, \mathcal{F} \rangle$ *assigns how many tokens each place contains. A marking defines the* state *of* N*.*

Definition 5 *(Enabling and firing of transitions, Reachable markings). A transition* $t \in T$ *is* enabled *for firing if and only if* $m \geq {}^\bullet t$*. We denote the* firing *of* t *by* $m \xrightarrow{t} m'$*, where* m' *is the resulting marking after firing* t *and is defined by* $m' = m - {}^\bullet t + t^\bullet$*. For a transition sequence* $\sigma = \langle t_1, \ldots, t_m \rangle$ *we write* $m \xrightarrow{\sigma} m'$ *to denote the consecutive firing of transitions* t_1 *to* t_m*. We also write* $m \xrightarrow{*} m'$ *if there is some* $\sigma \in T^*$ *such that* $m \xrightarrow{\sigma} m'$*.*

The set of reachable markings $\mathcal{R}(N, m)$ *from marking* m *in a Petri net* N *is the set* $\{m' \mid m \xrightarrow{*} m'\}$*.*

Definition 6 *(Place invariant). A place invariant* [12] *is a row vector* $I : P \to \mathbb{Q}$ *such that* $I \cdot F = 0$*. We denote the set of all place invariants as* $\mathcal{I}_N$*, which is a linear subspace of* $\mathbb{Q}^P$*.*

The main property of place invariants is that for any two markings m_1, m_2 such that $m_1 \xrightarrow{*} m_2$ and any place invariant I holds: $I \cdot m_1 = I \cdot m_2$.

Definition 7 *(Distributed run). A distributed run describes a partial order of transition occurrences represented as an acyclic occurrence net* π [20]*. An occurrence net* $\pi = \langle B, E, G \rangle$ *is a Petri net where each place* $b \in B$ *is called a condition, each* $e \in E$ *is called an event, the transitive closure* G^+ *is acyclic. Each* $b \in B$ *has at most one pre-event and at most one post-event, i.e.* $|^\bullet b| \leq 1$ *and*

$|b^\bullet| \leq 1$. *A labeled occurrence net* $\pi = \langle B, E, G, \ell \rangle$ *is an unfolding of a Petri net* N *where each condition (event) is labeled with a set of labels of the form* (x, id) *where* x *refers to a place (transition) of* N*, and* id *is an instance identifier.*

Definition 8 *(Net system, Language, Execution sequence). A* Net system *is a tuple* $SN = (N, m_i, m_f)$*, where* N *is a Petri Net and* m_i *and* m_f *are respectively the initial and final marking. The* language *of* SN *is the set* $\mathcal{L}(SN) = \{\sigma \in T^* \mid m_i \xrightarrow{\sigma} m_f\}$ *of all full firing sequences of* SN.

An execution sequence *in a net system* $SN = (N, m_i, m_f)$ *is a distributed run of steps, starting at the initial marking* m_i *and ending at the final marking* m_f.

2.2 Event Logs

An event log records action executions as events where each event records at least the action that occurred, the time of occurrence and the case identifier of the case in which the action occurred. Often resources are also recorded as event attributes, e.g. the actors executing the action. Typically, there are several types of resources, and it is generally known beforehand which resources of which types are involved in which actions.

Definition 9 *(Cases, Resources).* Id_c *denotes the* set of case identifiers. *An identifier of case* c *is denoted as* id_c.

$R = \{r_1, \ldots, r_m\}$ *is the* set of resource types. *Each resource instance with an identifier* id_r *belongs to some resource type* $r \in R$. Id_r *denotes the set of resource instances of type* $r \in R$. *We assume that* $Id_r \cap Id_{r'} = \emptyset$ *for any* $r \neq r'$. $Id_R = \uplus_{r \in R} Id_r$ *denotes the set of the resource instances of all types.*

Note that if one would want to capture resource instances with multiple types $R' \subseteq R$, a new type should be constructed containing all types from R'.

With the notation on cases, resources and resource types, we can define events, an event log and its traces in an abstract manner:

Definition 10 *(Event, Event log, Trace). An* event e *is a tuple* $\langle a, ts, id_c, Id_\rho \rangle$*, with an activity name* $a \in \Sigma$*, a timestamp* ts*, a case identifier* id_c *and a set of resource instance identifiers* $Id_\rho \subseteq Id_R$*. Such an event represents that activity* a *occurred at timestamp* ts *for case* id_c *and is executed by resource instances from* Id_ρ *belonging to possibly different resource types.*

An event log L *is a (partially) ordered set of events. These events can be split into traces, defined as projections e.g. on the case identifiers or on the resource identifiers.*

For a process modeled by a Petri net, an activity name corresponds to a transition name or a transition label of the corresponding transition of a (labeled) Petri net. With the projection on case identifiers, we get the events from individual cases, as is mainly used in classical process mining and with projection on resource identifiers, we can get the events from individual resource instances, which are entities in their own right, providing the perspective of a single or multiple resource types.

3 Modeling Resource-Constrained Case Handling Systems

A classical Petri net models a process execution using transition firings and the corresponding changes of markings without making distinction between different cases on which the modeled system works simultaneously. To create a case view, Workflow nets [27] model processes from the perspective of a single case. Systems in which cases share resources need to be modeled in a different way, providing information both about cases and resources. In this paper, we extend the notion of RCWF-nets [31], resource-constrained workflow nets with resource places and id-tokens identifying cases, by loosening some structural restrictions on Petri nets and including information about resource instances working on cases. To achieve that, we make use of ν-Petri nets [22].

3.1 Resource-Constrained Petri Nets

Let R be the set of all resource types. Following the definition of [31], we model each resource type $r \in R$ by a place p_r, where the resources (tokens) are located when they are available. We extend the RCWF-nets definition by adding a place $\bar{p}_r$ for each resource type r. Tokens on $\bar{p}_r$ represent resources working on cases. The structural condition $\mathcal{F}(p_r, t) + \mathcal{F}(\bar{p}_r, t) = 0$ is imposed on the net, which implies that a token can e.g. be moved from p_r to $\bar{p}_r$ to show that the resource gets occupied, moved from $\bar{p}_r$ to p_r to show that the resource becomes available, or there could be tests whether there are free/occupied resources.

We consider *durable* resources only, meaning that resources can neither be created nor destroyed, so in the corresponding net system with initial and final marking m_i and m_f, $m_i(p_r) = m_f(p_r)$ and $m_i(\bar{p}_r) = m_f(\bar{p}_r)$, for any resource type $r \in R$. The net obtained from a resource-constrained net N by removing all resource places P_r together with their incoming and outgoing arcs is called the *production net* of N.

$T_r \subseteq T$ denotes the set of activities in which resource instances of type r are involved. We define $T_r^{in} = p_r^\bullet$ and $T_r^{out} = {}^\bullet p_r$, where resource instances of type r are claimed and released, respectively. Note that both $T_r \setminus (T_r^{in} \cup T_r^{out})$ and $T_r^{in} \cap T_r^{out}$ may be nonempty, since a resource can be claimed by an activity and then released only after executing several other activities, or it can be claimed and immediately released by an activity.

We add modeling restrictions on p_r and $\bar{p}_r$ to exploit structural characteristics of the Petri net later in Sect. 3.3.

Definition 11 *(Resource-constrained net system). Let R be a set of resource types. We define the set of availability resource places $P_r = \{p_r \mid r \in R\}$ and the set of occupancy resource places $\bar{P}_r = \{\bar{p}_r \mid r \in R\}$. A* resource-constrained net system *$SN = (N, m_i, m_f)$ is a regular net system with resource-constrained*

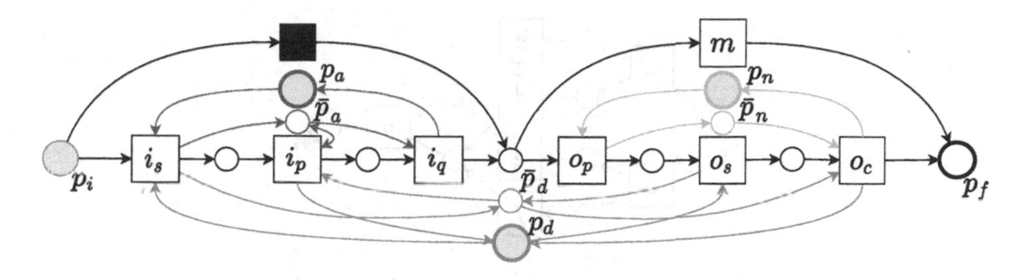

Fig. 1. Running example Petri net.

Petri net $N = \langle P, T, \mathcal{F} \rangle$ *where* $P = P_p \uplus P_r \uplus \bar{P}_r$, *with* P_p *the production places. We have the following modeling restrictions on* p_r *and* $\bar{p}_r$*:*

1. $\forall_{t \in T} [{}^\bullet t(p_r) + {}^\bullet t(\bar{p}_r) = t^\bullet(p_r) + t^\bullet(\bar{p}_r)]$, *i.e.* $\mathcal{F}(p_r, t) + \mathcal{F}(\bar{p}_r, t) = 0$.
2. $m_i(p_r) = m_f(p_r)$ *and* $m_i(\bar{p}_r) = m_f(\bar{p}_r) = 0$*;*

Restriction 1 from Definition 11 enforces the place invariant $(1, 1)$ for each pair of the availability and occupancy resource places p_r and $\bar{p}_r$, which trivially follows from the definition of place invariants. This implies that $m(p_r) + m(\bar{p}_r) = m_i(p_r)$ for any marking m reachable from the initial marking m_i. Restriction 2 requires that all resource tokens are returned to the availability resource place when the net reaches its final marking.

Typically, a variant of the soundness property is imposed on the net system to guarantee that the final marking is reachable from any marking reachable from the initial marking.

3.2 Running Example

As a running example, we use the Petri net representation of a simple process, see Fig. 1. This Petri net models a hospital process in which three types of resources are involved: doctors (modeled with resource places p_d and $\bar{p}_d$), doctor assistants (places p_a and $\bar{p}_a$) and nurses (places p_n and $\bar{p}_n$). Patients undergo two phases of a treatment. The first phase is the intake where a doctor together with an assistant first discuss patient symptoms (transition i_s), after which the doctor provides the plan of approach (transition i_p), and finally the patient asks questions to the assistant (transition i_q). In case of emergency, the whole intake phase can be skipped, which is modeled by the black (silent) transition. The second phase is either medication collection (transition m) or operation. The latter is subdivided in preparation (transition o_p) done by a nurse after which the surgery (transition o_s) and close up (transition o_c) are performed by the nurse and a doctor, ending the process.

An assistant is actively involved in the whole intake phase of the process, which is emphasized by the test arc between the place p_a and transition i_p. Note that the nurse is not involved in the surgery, although it is not released during the entire operation phase.

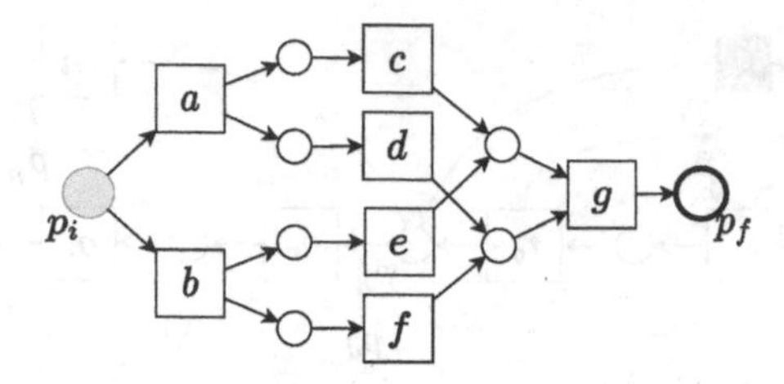

Fig. 2. Example Petri net in need of case isolation.

3.3 ν-Petri Nets

Resource-constrained Petri nets are especially useful when multiple cases are present simultaneously. An example in Fig. 2 shows that distinguishing tokens belonging to different cases is essential for capturing process behavior of simultaneously running cases in a correct way. The shown net does not have the separability property [30]. Trace $\langle a, c, \underline{b}, \underline{f}, g, d, \underline{e}, \underline{g} \rangle$ can be replayed on the Petri net without differentiating between the case ids. However, this trace cannot be formed as an interleaving shuffle of the traces of two separate cases, since each firing of transition g uses tokens belonging to two different cases. Thus we need a mechanism preventing firings of transitions that mix tokens belonging to different resources. Moreover, we also need a mechanism allowing us to keep track of resource instances. In our running example, we need e.g. a possibility to extend the model with a constraint that the doctor who performed the intake is also the doctor who performs the surgery later in the process.

We use ν-Petri nets to provide case and resource isolation. ν-Petri nets, also referred to as Petri nets with names, extend regular Petri nets with the capability of name management. The expressive power of a ν-Petri net strictly surpasses that of Petri nets and they essentially correspond to the minimal object-oriented Petri nets of [11]. In a ν-Petri net, names can be created, communicated and matched which can be used to deal with authentication issues [23], correlation or instance isolation [8]. Name management is formalized by replacing ordinary tokens by distinguishable ones, thus adding color the Petri net.

We first give the definition of regular ν-Petri nets from [22] (see Definition 12), after which we show how we extend the definition to work with resource-constrained Petri nets. Colors are handled by matching variables labeling the arcs of the Petri nets, taken from a fixed set *Var* and a set of special variables $\Upsilon \subset Var$ as defined in Definition 12.

Definition 12 *(ν-Petri net [22]). A ν-Petri net is a tuple $\nu\text{-}N = \langle P, T, \mathcal{F} \rangle$, with a set of places P and a set of transitions T with $P \cap T = \emptyset$, and a flow function $\mathcal{F} : (P \times T) \cup (T \times P) \rightarrow Var^{\oplus}$ such that for every $t \in T$, $\Upsilon \cap pre(t) = \emptyset$ and $post(t) \setminus \Upsilon \subseteq pre(t)$, where $pre(t) = \bigcup_{p \in P} supp(\mathcal{F}(p,t))$ and $post(t) = \bigcup_{p \in P} supp(\mathcal{F}(t,p))$. $\Upsilon \subset Var$ denotes a set of special variables ranged by $\nu, \nu_1, \ldots$ to instantiate fresh names.*

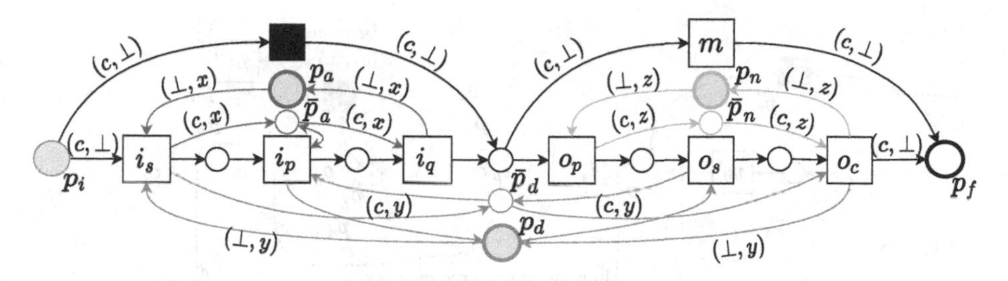

Fig. 3. Process model M: ν-Petri net representation of the running example (note that some arc labels are omitted for clarity).

A marking of ν-N is a function $m : P \to Id^{\oplus}$. $Id(m)$ denotes the set of names in m, i.e. $Id(m) = \bigcup_{p \in P} supp(m(p))$.

A mode μ of a transition t is an injection $\mu : Var(t) \to Id$, that instantiates each variable to an identifier.

For a firing of transition t with mode μ, we write $m \xrightarrow{t_\mu} m'$. t is enabled with mode μ if $\mu(\mathcal{F}(p,t)) \subseteq m(P)$ for all $p \in P$ and $\mu(\nu) \notin Id(m)$ for all $\nu \in \Upsilon \cap Var(t)$. The reached state after the firing of t with mode μ is the marking m', given by:

$$m'(p) = m(p) - \mu(\mathcal{F}(p,t)) + \mu(\mathcal{F}(t,p)) \text{ for all } p \in P \tag{1}$$

ν-Petri nets support instance isolation: we use case ids and resource ids as token colors and require tokens involved in a transition firing to have matching colors. This allows for separating multiple instances simultaneously running in the Petri net. We build on Definition 12 to define resource constrained Petri nets with matching on case instances and resource instances. Instance isolation is achieved by extending the colored tokens to *multi-colored*, for which we have two sets of variables, Var_c and Var_r, for case and resource isolation respectively, instead of the single set Var in ν-Petri nets. This requires modifications in the standard definition of arcs $\mathcal{F}$, marking m and mode μ of the ν-Petri net. The definition of transition firings remains the same.

Definition 13 *(Resource-constrained ν-Petri net). Let $C^{\bot}$ be the set of case ids C extended with ordinary tokens, i.e. $\bullet \in C$, and $R^{\bot}$ be the set of resource ids extended with ordinary tokens.*
A resource-constrained ν-Petri net $N = \langle P, T, \mathcal{F} \rangle$ *is a Petri net system with $\mathcal{F} : (P \times T) \cup (T \times P) \to (Var_c^{\bot} \times Var_r^{\bot})^{\oplus}$, where Var_c denote case variables and Var_r denote resource variable.*

A marking of N is a function $m : P \to (C^{\bot} \times R^{\bot})^{\oplus}$ with case ids C and resources R, which is a mapping from places to multisets of colored tokens.

A mode of a transition t is an injection $\mu : (Var_c^{\bot} \times Var_r^{\bot})(t) \to (C^{\bot} \times R^{\bot})$, that instantiates each variable to an identifier.

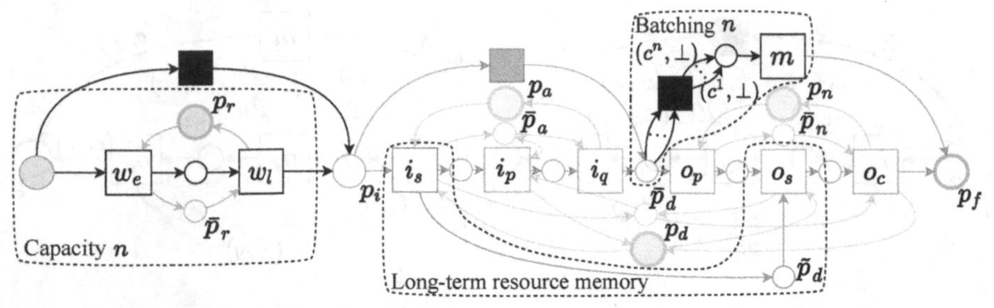

(a) Patterns for capacity, long-term resource memory and batching

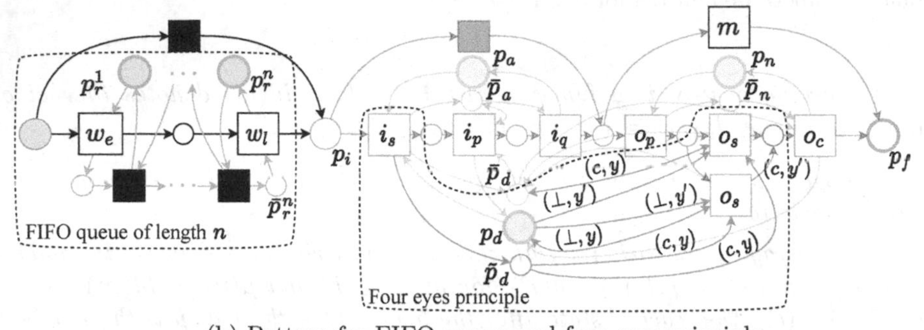

(b) Pattern for FIFO queue and four eyes principle

Fig. 4. Modeling patterns extending the running example showing possible use cases.

The mode determines the case and resource ids of the tokens to consume and produce in a transition firing. Note the role of the occupancy resource places in this definition: they allow to keep track of resources working on individual cases. The place invariant naturally holds for resource-constrained ν-Petri nets as well.

Figure 3 shows the ν-Petri net for our running example.

Note that this example does not include the functionality of new name creation although this could be useful to exploit in some processes. E.g., consider a production process where components are assembled into products. Each component has a unique identifier, of which a subset is merged into a product for which a fresh identifier should be produced, requiring new name creation in the process model.

3.4 Modeling Patterns

Resource-constrained ν-Petri nets open up a number of possibilities in terms of simulating resource-constrained processes, where case and resource isolation are critical for correct simulation.

We illustrate a number of modeling patterns in Fig. 4, building on our running example:

- **Capacity, FIFO queues** A subprocess in Fig. 4a on the left, prior to the intake, models a waiting room with a capacity limited by the initial number

of tokens on the resource place p_r. Figure 4b shows a FIFO (first in first out) version of the waiting room. For the FIFO queue with capacity n, n availability and n occupancy resource places are needed for this pattern to lead patient tokens (with their case ids) through the queue.
- **Long term resource memory**: Place $\tilde{p}_d$ with $\mathcal{F}(i_s, \tilde{p}_d) = \mathcal{F}(\tilde{p}_d, o_s) = (c, y)$ models a long term resource memory with respect to the doctor, ensuring that the resource instance of type doctor that was involved in the intake of a patient is also the same instance that performs the operation for that patient. This construct still allows the resource to be available for other cases (patients) in between the intake and the surgery.
- **Four eyes principle**: Alternatively, as shown in Fig. 4b on the right, $\tilde{p}_d$ can also be used to add the opposite restriction with respect to the doctor resource: the doctor performing the operation (d_2) should be different from the doctor involved in the intake (d_1). This pattern is known as the four eyes principle, meaning that two resources involved in a process should not be equal. Note that the starting transition of the second subprocess should be duplicated, since the intake-doctor d_1, who is not involved in the surgery, could be residing in either p_d or $\tilde{p}_d$.
- **Batching**: Batch processing can be modeled with multiple arrows, like the arrows connected to the silent transition in the net firing before transition m. In this case, a pharmacy may only replenish their inventory for e.g. three orders of medication at once. This pattern is similar to the one in classical Petri nets.

4 Alignments on Resource-Constrained Petri Nets

Several state-of-the art techniques in conformance checking use alignments to relate the recorded executions of a process with a model of this process [1]. A traditional alignment shows how a trace can be replayed on the process model by a sequence of moves representing either a synchronous move, a log move or a model move, denoted as $\binom{a}{a}$, $\binom{a}{\gg}$ and $\binom{\gg}{a}$ respectively. A synchronous move indicates that observed and modeled behavior agree, i.e. the execution of an activity observed in the log can be mimicked by performing this activity in the process model. A log move means that an activity from the log cannot be mimicked in the model, and a model move represents the fact that the model requires an execution of some activity, which is not observed in the log. Log moves and model moves can expose deviations of the real behavior from the model.

In this section we recapitulate the alignment mechanism in its classical form, with computations performed on a case-by-case basis, after which we show its shortcomings when dealing with resource-constrained Petri nets.

4.1 Traditional Case-by-Case Alignments

The foundational work for constructing alignments is presented in [1] and relies on two fundamental concepts: (1) a synchronous product of Petri nets and (2) the marking equation. The synchronous product definition is tuned towards the setting of alignments and it is built for the Petri net model of the process and the trace Petri net (a Petri net representation of the (partially) ordered trace in the event log). The trace Petri net traditionally represents only individual cases from the event log, and consequently does not capture the interdependencies between multiple cases. The case-by-case alignment is then found by a depth-first search on the synchronous product Petri net using the A^* algorithm [1].

A *trace net system* SN_σ of a trace $\sigma = \langle e_1, \ldots, e_n \rangle$ is a net system with the set of transitions $T^l = \{t_i^l \mid e_i \in \sigma\}$, a connection place for every pair of transitions t_i^l and t_{i+1}^l, place p_i being the input place of t_1 and place p_f being the output place of t_n, $m_i = \{p_i\}$ and $m_f = \{p_f\}$.

Given a net system SN modeling the considered process and a trace net system SN_σ modeling a trace from a log, a *synchronous product* ΠSN contains the places and the transitions of SN and SN_σ and additional transitions called synchronous moves: For each pair of transitions $t^m \in T(SN), t^l \in T(SN_\sigma)$ with matching labels $\ell(t^l) = \ell(t^m)$, transition t^s is created with ${}^\bullet t^s = {}^\bullet t^l \cup {}^\bullet t^m$ and $t^{s\bullet} = t^{l\bullet} \cup t^{m\bullet}$. Thus ΠSN contains transitions $T = T^s \cup T^l \cup T^m$, where each $t^s \in T^s$ can be traced back to a pair of a transition $t^l \in T^l$ and a transition $t^m \in T^m$, T^l is the set of transitions of the trace net system and T^m is the set of transitions of the process model SN. While T^s transitions represent synchronous moves in both the trace net and the process model, T^l transitions represent log moves and T^m transitions represent model moves.

The core alignment question is now formalized as follows: given a synchronous product Petri net with a cost function assigning a non-negative cost to each transition firing, find a distributed run from the initial marking to the final marking with the lowest total costs. Synchronous moves have zero costs, since they represent a match between the trace and the model behavior.

Let $\Pi SN = \langle P, T, \mathcal{F} \rangle$ be a synchronous product Petri net with $T = T^s \uplus T^l \uplus T^m$ partitioned into sets of transitions corresponding to synchronous moves, log moves and model moves respectively and let $(\Pi SN, m_i, m_f)$ a corresponding net system. Furthermore let $c : T \rightarrow \mathbb{R}^+$ a cost function.

An *alignment* is a distributed run $\gamma \in \{\gamma \in T^* | (\Pi SN, m_i) \xrightarrow{\gamma} (\Pi SN, m_f)\}$.

An *optimal alignment* is an alignment γ such that $c(\gamma) \leq c(\gamma')$ holds for any alignment γ'.

Optimal alignments can be computed for individual cases in an event log using an A^* based search strategy [1,6,29] where ILP is utilized as a heuristic function, or logic programming [5] is used. Other methods focus on approximations of alignments [26] or provide divide-and-conquer strategies [13]. Although we will not go into the details on the exact workings of these methods, we point out that they all have one fundamental property in common: they all reason over the synchronous product Petri net.

4.2 Unexposed Deviations; the Need for Multi-case and -resource Alignments

When talking about case-by-case alignments with resource-constrained Petri nets, some deviations remain unexposed. Recall that for simulating resource-constrained Petri nets, we need a ν-Petri net representation to correctly isolate the cases and resources. Similar issues emerge when computing alignments case by case using indistinguishable resources. Referring back to the modeling patterns presented in Sect. 3.4, we show some event logs for the extended running example process models from Fig. 4 for which case-by-case alignments fail to expose deviations in resource-constrained Petri nets:

- **Multitasking**: consider the partial event log L_1 given by

  $$L_1 = \langle \dots, \langle i_s, \{d_1, a_1\}\rangle, \langle \underline{i_s}, \{d_1, a_2\}\rangle, \langle i_p, \{d_1, a_1\}\rangle, \langle \underline{i_p}, \{d_1, a_2\}\rangle, \dots \rangle$$

  where the timestamp is abstracted away and the case identifier is denoted by the activity color (and additionally by the bar position). The recorded behavior in L_1 shows that doctor d_1 is multitasking on the intake subprocesses of two patients. The resource-constrained ν-Petri net does not accept this behavior since i_s claims the doctor and the doctor is released again only after i_p is executed. Case-by-case alignments consider every case in isolation and therefore they do not expose any deviations in L_1.
- **Resource switching**: consider the partial event log L_2 given by

  $$L_2 = \langle \dots, \langle i_s, \{d_1, a_1\}\rangle, \langle \underline{i_s}, \{d_2, a_2\}\rangle, \langle i_p, \{d_2, a_1\}\rangle, \langle \underline{i_p}, \{d_1, a_2\}\rangle, \dots \rangle$$

  The behavior recorded in L_2 shows that doctors d_1 and d_2 swapped patients during the intake subprocess, which is not allowed according to the process model where the resources have names (colors).
  Furthermore, consider the partial event log L_3 given by

  $$L_3 = \langle \dots, \langle i_s, \{d_1, a_1\}\rangle, \langle i_p, \{d_1, a_1\}\rangle, \dots, \langle o_s, \{d_2, a_1\}\rangle, \dots \rangle$$

  L_3 shows that doctor d_2 performed an surgery on a patient whose intake was done by doctor d_1, although the long-term resource memory place $\tilde{p}_d$ in the process model implies that the doctor performing the surgery is the same as the one who did the intake.
  These deviations remain undetected by the traditional alignments, computed on classical Petri nets with black tokens.
- **Capacity violations**: consider the partial event log L_4 given by

  $$L_4 = \langle \langle w_e, \{r_1\}\rangle, \langle \underline{w_e}, \{r_1\}\rangle, \langle \overline{w_e}, \{r_1\}\rangle, \langle w_l, \{r_1\}\rangle, \langle \underline{w_l}, \{r_1\}\rangle, \langle \overline{w_l}, \{r_1\}\rangle, \dots \rangle$$

  L_4 shows the behavior from the waiting room subprocess, where a maximum capacity of two patients is in place. Similar to the first example, case-by-case leaves this deviation undetected.

- **Overtaking in FIFO queues**: consider the partial event log L_5 given by

$$L_5 = \langle (w_e, \{r_1\}), (\underline{w_e}, \{r_1\}), (\underline{w_l}, \{r_1\}), (w_l, \{r_1\}), \dots \rangle$$

  The recorded behavior in L_5 does not violate the waiting room's capacity, but the patients leave in a different order than how they arrived, while the process model imposes the FIFO pattern. Such deviations can only be exposed when aligning multiple cases simultaneously.
- **Batching violations**: consider the partial event log L_6 given b

$$L_6 = \langle \dots, (m, \emptyset), (\underline{m}, \emptyset), \dots \rangle$$

  Recall that m is only enabled after e.g. three (with $n = 3$) tokens are in $^{\bullet}o_p$ in order to fire the connected silent transition in the process model. L_6 shows that m occurred for two patients only, deviating from the model. With case-by-case alignments, it is impossible to align m. It also would be impossible in case there were three patients according to the log, since it requires multiple cases being processed simultaneously.

This clearly shows that case-by-case conformance checking is not sufficient and resource identities are essential for detecting deviations from standard resource-related constraints. Therefore, we introduce multi-case and multi-resource alignments using ν-Petri nets that align the *complete* event log to the process model, allowing to expose the deviations listed above.

5 Computing Multi-case and -resource Alignments

Our approach to computing multi-case alignments is based on the traditional alignment-based approach using the synchronous product Petri net. Instead of representing individual cases in the trace Petri net, we capture the complete event log there in order to consider inter-case relations. To retain the case isolation when aligning, we transform the process model given by a resource-constrained Petri net into the resource-constrained ν-Petri net (Definition 13).

As a running example we take the Petri net from Fig. 1 and the event log:

$$L = \langle \boldsymbol{i_s}, \underline{i_s}, \boldsymbol{i_p}, \underline{i_p}, \boldsymbol{o_p}, \boldsymbol{o_s}, \overline{i_s}, \underline{i_q}, \boldsymbol{o_c}, \overline{i_p}, \overline{i_q}, \overline{o_p}, \underline{o_p}, \underline{o_s}, \underline{o_c}, \overline{o_s}, \overline{m} \rangle$$

where the colors (and bar positions) represent the case identifiers. The timestamps and involved resource instances are omitted and made implicit by the ordering and the transitions from the process model respectively. For this example, there is a single resource instance for each resource type.

5.1 Approach

Recall that alignments are computed by taking the synchronous product Petri net constructed from the process model Petri net and a Petri net representation of the trace, called the trace Petri net. With a resource constrained ν-Petri net

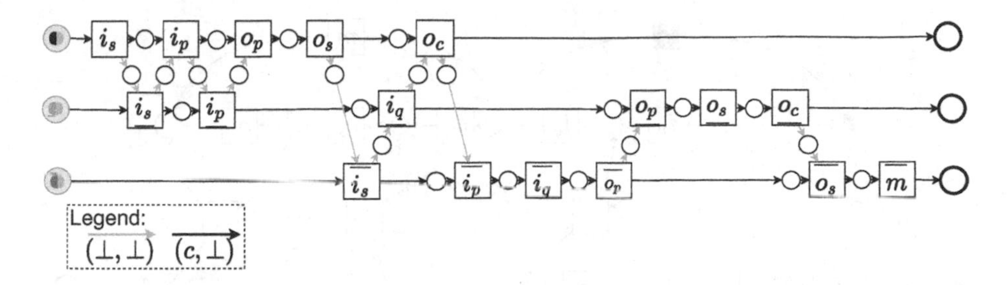

Fig. 5. Running example trace ν-Petri net

as the process model, we have to modify the definition of the trace Petri net that takes into account the case identifiers of the events in the event log. We achieve this by constructing multiple trace Petri nets for each trace in the event log projected on the case identifiers. The cases are then differentiated by turning this into a ν-Petri net with the label $(c, \bot)$ on the arcs. Additionally, we add places between the transitions to enforce the correct ordering as they occurred in the event log, with label $(\bot, \bot)$ on its incoming and outgoing arcs. Formally, this trace ν-Petri net is defined as follows:

Definition 14 *(Trace ν-Petri net). ν-SN = $\langle P, T, \mathcal{F}, \ell \rangle$, a labeled trace ν-Petri net, is constructed from an event log $L = \langle e_i^{id_c,a} \rangle_{1 \leq i \leq n}$, with id_c and a denoting respectively the case identifier and activity of event e_i. For each event $e_i^{id_c,a}$ with $1 \leq i \leq n$, we have a transition $t_i^{id_c}$ with $\ell(t_i) = a$. Places are added between these transitions as follows*[1]*: With i from 1 to n, we add a place $p_i^{id_c}$ between $t_i^{id_c}$ and $\min_{j>i} t_j^{id_{c'}}$ such that $id_{c'} = id_c$ with $\mathcal{F}(t_i^{id_c}, p_i^{id_c}) = \mathcal{F}(p_i^{id_c}, t_j^{id_{c'}}) = (c, \bot)$. Furthermore, we add a place $p_i^{\mathcal{T}}$, enforcing the original ordering, between $t_i^{id_c}$ and $t_{i+1}^{id_{c'}}$ if $id_{c'} \neq id_c$ with $\mathcal{F}(t_i^{id_c}, p_i^{\mathcal{T}}) = \mathcal{F}(p_i^{\mathcal{T}}, t_{i+1}^{id_{c'}}) = (\bot, \bot)$.*

Lastly, initial and final places are added for each case $id_c \in Id_c$: $p_{in}^{id_c}$ to $\min_i t_i^{id_c}$ and $p_{out}^{id_c}$ to $\max_j t_j^c$ with $\mathcal{F}(p_{in}^{id_c}, t_i^{id_c}) = \mathcal{F}(t_j^{id_c}, p_{out}^{id_c}) = (c, \bot)$. The initial and final marking is then defined by $m_i(p_{in}^{id_c}) = m_f(p_{out}^{id_c}) = \{(id_c, \bot)\}$.

The trace ν-Petri net for the running example is shown in Fig. 5. With the redefined trace ν-Petri net we can construct the synchronous product Petri net *ν-ΠSN* consisting of the resource constrained ν-Petri net and the trace ν-Petri net.

Computing the multi-case alignments is now a matter of finding the distributed run in the *ν-ΠSN* for which we can use *any* of the existing methods as described in Sect. 4.1. The optimal alignment is again the one with lowest cost.

Note that while ν-Petri nets are inherently unbounded by generating fresh tokens, we can decide on the tokens to be generated beforehand by preprocessing the event log, and can therefore retain boundedness.

[1] Adding a place p between two transitions t_a and t_b denotes a single arc from t_a to p and a single arc from p to t_b.

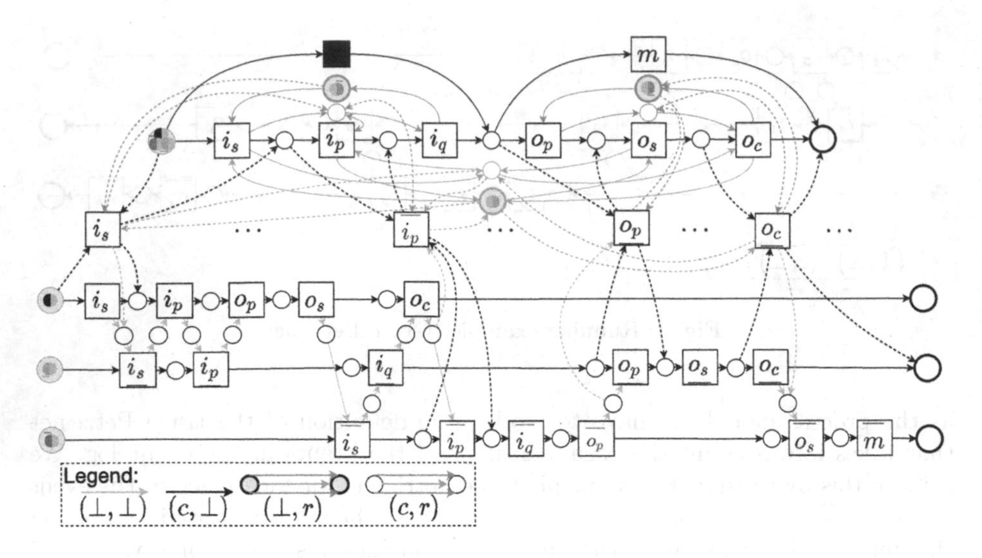

Fig. 6. Running example synchronous product ν-Petri net

For our running example we get the synchronous product ν-Petri net as shown in Fig. 6 and the corresponding optimal alignment is the shortest path through ΠSN and is shown in Table 1. Note that not all synchronous transitions are visualized in the figure for clarity reasons.

Table 1. Alignment from the naive method

$\gamma^{(L)} =$	1		2	3	4	5	6	7	8	9		10	11	12	13	14	15	16	17
L	i_s	$\gg$	i_s	i_p	i_p	o_p	o_s	i_s	i_q	o_c	$\gg$	i_p	i_q	o_p	o_p	o_s	o_c	o_s	m
M	$\gg$	τ	i_s	$\gg$	i_p	o_p	o_s	$\gg$	i_q	o_c	i_s	i_p	i_q	$\gg$	o_p	o_s	o_c	$\gg$	m

5.2 Multi-case and -resource Alignments in Action

With the examples listed in Sect. 4.2 of undetected deviations with traditional alignments, we show here how multi-case and -resource alignments expose them. Note that resource attributes for some event logs are abstracted away when this is implicit from the data (only a single resource was involved). For others, the resource attribute is denoted in the superscript of the event.

- $\gamma^{(L_1)}$ shows an optimal alignment for L_1 computed from the method described above. With multi-case and -resource alignments, it is not possible anymore for all moves to be aligned synchronously because of the doctor's availability during the intake process.

$$\gamma^{(L_1)} = \begin{array}{|c|ccccc|} L_1 & i_s & \gg & \underline{i_s} & i_p & \underline{i_p} \\ \hline M_e & i_s & i_p & \underline{i_s} & \gg & \underline{i_p} \end{array}$$

- $\gamma^{(L_2)}$ shows an optimal alignment for L_2, where we see that i_p and $\underline{i_p}$ should have occurred with doctors d_1 and d_2 respectively according to the model exposing that they have switched positions in the recorded behavior.

$$\gamma^{(L_2)} = \begin{array}{|c|cccccc|} L_2 & i_s^{d_1} & \underline{i_s^{d_2}} & \gg & i_p^{d_2} & \gg & \underline{i_p^{d_1}} \\ \hline M_e & i_s^{d_1} & \underline{i_s^{d_2}} & i_p^{d_1} & \gg & \underline{i_p^{d_2}} & \gg \end{array} \quad \gamma^{(L_3)} = \begin{array}{|c|ccccc|} L_3 & i_s^{d_1} & i_s^{d_2} & \cdots & \gg & o_s^{d_2} \\ \hline M_e & i_s^{d_1} & i_s^{d_2} & \cdots & o_s^{d_1} & \gg \end{array}$$

- $\gamma^{(L_3)}$ shows an optimal alignment for L_3, where we see that the long term resource memory is violated and o_s should have been executed by doctor d_1 instead of d_2.
- $\gamma^{(L_4)}$ shows an optimal alignment for L_4, revealing that w_l should have occurred before $\underline{w_e}$ according to the capacity restriction in the model, i.e. the first patient should have left the waiting room before the third patient entered.

$$\gamma^{(L_4)} = \begin{array}{|c|ccccccc|} L_4 & w_e & \underline{w_e} & \gg & \overline{w_e} & w_l & \underline{w_l} & \overline{w_l} \\ \hline M_e & w_e & \underline{w_e} & w_l & \overline{w_e} & \gg & \underline{w_l} & \overline{w_l} \end{array} \quad \gamma^{(L_5)} = \begin{array}{|c|ccccc|} L_5 & w_e & \underline{w_e} & \gg & \underline{w_l} & w_l \\ \hline M_e & w_e & \underline{w_e} & w_l & \underline{w_l} & \gg \end{array}$$

- $\gamma^{(L_5)}$ shows an optimal alignment for L_5. The model move on w_l shows that the first patient should leave before the second one does, exposing the FIFO violation in the waiting room process.
- $\gamma^{(L_6)}$ shows an optimal alignment for L_6, where the batching restriction is violated. The model moves show that an added third patient should have been included in order to execute m for the three patients.

$$\gamma^{(L_6)} = \begin{array}{|c|ccccc|} L_6 & \overline{\gg} & \overline{\gg} & \overline{\gg} & m & \underline{m} \\ \hline M_e & \overline{\tau} & \overline{\tau} & \overline{m} & m & \underline{m} \end{array}$$

5.3 Relaxing the Synchronous Product Petri Net to Detect Resource-Related Deviations

With the alignments generated as described above it could be difficult to interpret the exposed deviations, especially in terms of the added model moves: was the activity executed but not logged, was it executed by a "wrong" resource and therefore not executable in the model, or was it definitely not executed because no appropriate resource was available?

In this section we show how we can use simple model transformations on the synchronous product Petri net to allow for additional behavior (at some costs), so we can interpret resource-related deviations in more detail. To show these transformations, we use an example subprocess with two activities a and b in which a resource of type r is involved. The corresponding Petri net is shown in Fig. 7, where the transitions $\{\bar{a}, \bar{b}, \tau_1, \tau_2, \tau_3\}$ are added to the model.

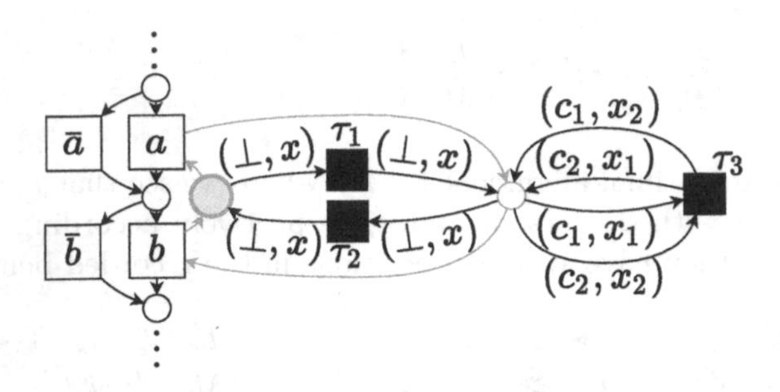

Fig. 7. Resource relaxations for improved alignment interpretability.

Multitasking. From alignment $\gamma^{(L_1)}$ from Sect. 5.2 it is not immediately clear that the resource instance was multitasking. The model transformation in Fig. 7 using silent transitions τ_1 and τ_2 between p_r and $\bar{p}_r$ allows for turning a resource instance from available to occupied and vice versa at any point in time. These silent transitions give additional interpretation to the alignment, showing where the resource might have been released or claimed to work on another case, while it was not allowed by the model. With $c(\tau_1) > 0$ and $c(\tau_2) > 0$, model moves with τ_1 or τ_2, which we denote as *resource moves*, are only selected when necessary.

Model Moves Not Claiming Resources. Transitions $\bar{a}$ and $\bar{b}$ allow for model moves not claiming the resource(s), which we call *control flow moves.* It is important to note that these transitions have no corresponding "resource-free" synchronous move in the synchronous product Petri net ΠSN, since they represent a relaxed version of the model move. Therefore, the cost of a control flow move should be higher than the cost of the corresponding model move (that do claim the resources), i.e. $c(\bar{a}) > c(a)$ and $c(\bar{b}) > c(b)$.

This transformation allows to compute more sensible alignments in cases when a work item is skipped and the resources necessary for it were occupied, making a model move impossible. In such cases, the alignment without control flow moves would fit the model moves in time ranges where the needed resources were available, potentially causing conflicts in earlier or later stages of the process. Therefore, with the additional model moves, the resource claim is bypassed.

τ_1 and τ_2 as introduced above are necessary to avoid deadlocks that can arise in case a control flow move mimics a transition that claims or releases resources.

Resource Switching. From alignments $\gamma^{(L_2)}$ and $\gamma^{(L_3)}$ from Sect. 5.2 it is not immediately clear that the resource instances have switched or the incorrect resource instance is involved. The model transformation in Fig. 7 using τ_3 connected to $\bar{p}_r$ and possibly $\tilde{p}_r$ allows the resources to take over each other's work. For the resource switch from $\gamma^{(L_3)}$, additionally τ_1 and τ_2 are necessary to get d_1 into $\tilde{p}_d$. With $c(\tau_3) > 0$, a model move with τ_3 would only reside in the optimal alignment would it be necessary and it is interpretable showing the resource instance that took over work for a case from a specific resource instance. A model move with τ_3 is also denoted as a resource move.

Resource Type Relaxation. Recall that the resource instances as defined above are strictly typed. In case we want to allow a resource to execute tasks belonging to other resource types (at some cost), e.g. n and a, the following model transformation is sufficient: add a place $p_{n,a}$ with ${}^\bullet p_{n,a} = {}^\bullet p_n \cup {}^\bullet p_a$ and $p_{n,a}^\bullet = p_n^\bullet \cup p_a^\bullet$. A resource instance that could be involved in activities from n and a resides initially in $p_{n,a}$ from where it is able to do both.

Furthermore, when we would want to allow some types, e.g. for type r, to be unnamed, a model transformation making all tokens on place p_r the same color suffices. Note that this was already done for the waiting room resource instances for alignments $\gamma(L_4)$ and $\gamma(L_5)$ from Sect. 5.2.

These, and possible other model transformations could be used to enrich the alignment providing more interpretability.

6 Conclusion

In this paper we proposed a model for processes with shared resources using some features of resource-constrained workflow nets and ν-Petri nets. Our model allows to distinguish both cases and resources. This opens up possibilities in terms of modeling intricate inter-case dependencies and shared resources, including long-term resource memory, while still offering an option to exploit structural properties like well-structuredness of the control flow for e.g. conformance checking.

We showed that traditional alignments for conformance checking fail to detect some deviations that can arise in processes with shared resources. With our extended ν-Petri net representation of the process model, and a newly defined trace ν-Petri net containing the complete event log, we showed that the techniques for computing alignments can be utilized to expose violations on inter-case dependencies and usage of shared resources.

Our proposed extension to ν-Petri nets is a minimal extension that is sufficient for computing alignments on event logs without redundant functionality that other, possibly more sophisticated, extensions may offer as discussed in Sect. 1.

Future Work. Computing alignments on a case-by-case basis is already a complex problem in terms of computational power [6]. In principle, the complexity increases when multiple traces together with resource information are considered. At the same time, resource information available in the log can narrow the actual search space. We plan to look into preprocessing techniques and into structural reductions and decompositions for the Petri net to reduce the search space when computing the alignments.

References

1. Adriansyah, A.: Aligning observed and modeled behavior. Ph.D. thesis, Mathematics and Computer Science (2014)
2. Alizadeh, M., Lu, X., Fahland, D., Zannone, N., van der Aalst, W.M.P.: Linking data and process perspectives for conformance analysis. Comput. Secur. **73**, 172–193 (2018)
3. Barkaoui, K., Petrucci, L.: Structural analysis of workflow nets with shared resources (1998)
4. Berti, A., van der Aalst, W.M.P.: A novel token-based replay technique to speed up conformance checking and process enhancement. Trans. Petri Nets Other Model. Concurr. **15**, 1–26 (2021)
5. Boltenhagen, M., Chatain, T., Carmona, J.: Optimized sat encoding of conformance checking artefacts. Computing **103**(1), 29–50 (2021)
6. Carmona, J., van Dongen, B., Solti, A., Weidlich, M.: Conformance Checking. Springer, Heidelberg (2018)
7. de Leoni, M., van der Aalst, W.M.P.: Aligning event logs and process models for multi-perspective conformance checking: an approach based on integer linear programming. In: Daniel, F., Wang, J., Weber, B. (eds.) BPM 2013. LNCS, vol. 8094, pp. 113–129. Springer, Heidelberg (2013). https://doi.org/10.1007/978-3-642-40176-3_10
8. Decker, G., Weske, M.: Instance isolation analysis for service-oriented architectures. In: 2008 IEEE International Conference on Services Computing, vol. 1, pp. 249–256. IEEE (2008)
9. Fahland, D.: Describing behavior of processes with many-to-many interactions. In: Donatelli, S., Haar, S. (eds.) PETRI NETS 2019. LNCS, vol. 11522, pp. 3–24. Springer, Cham (2019). https://doi.org/10.1007/978-3-030-21571-2_1
10. Ghilardi, S., Gianola, A., Montali, M., Rivkin, A.: Petri nets with parameterised data: modelling and verification (extended version). arXiv preprint arXiv:2006.06630 (2020)
11. Kummer, O.: Undecidability in object-oriented Petri nets. In: Petri Net Newsletter. Citeseer (2000)
12. Lautenbach, K.: Liveness in Petri Nets. Bonn Interner Bericht ISF. Selbstverl, GMD (1975)
13. Lee, W.L.J., Verbeek, H.M.W., Munoz-Gama, J., van der Aalst, W.M.P., Sepúlveda, M.: Replay using recomposition: alignment-based conformance checking in the large. In: BPM (Demos) (2017)
14. Letia, I.A., Goron, A.: Model checking as support for inspecting compliance to rules in flexible processes. J. Vis. Lang. Comput. **28**, 100–121 (2015)
15. Mannhardt, F., De Leoni, M., Reijers, H.A., Van Der Aalst, W.M.P.: Balanced multi-perspective checking of process conformance. Computing **98**(4), 407–437 (2016)
16. Mozafari Mehr, A.S., de Carvalho, R.M., van Dongen, B.: Detecting privacy, data and control-flow deviations in business processes. In: Nurcan, S., Korthaus, A. (eds.) CAiSE 2021. LNBIP, vol. 424, pp. 82–91. Springer, Cham (2021). https://doi.org/10.1007/978-3-030-79108-7_10
17. Montali, M., Rivkin, A.: Model checking Petri nets with names using data-centric dynamic systems. Formal Aspects Comput. **28**(4), 615–641 (2016)

18. Montali, M., Rivkin, A.: DB-Nets: on the marriage of colored Petri nets and relational databases. In: Koutny, M., Kleijn, J., Penczek, W. (eds.) Transactions on Petri Nets and Other Models of Concurrency XII. LNCS, vol. 10470, pp. 91–118. Springer, Heidelberg (2017). https://doi.org/10.1007/978-3-662-55862-1_5
19. Murata, T.: Petri nets: properties, analysis and applications. Proc. IEEE **77**(4), 541–580 (1989)
20. Nielsen, M., Plotkin, G., Winskel, G.: Petri nets, event structures and domains. In: Kahn, G. (ed.) Semantics of Concurrent Computation. LNCS, vol. 70, pp. 266–284. Springer, Heidelberg (1979). https://doi.org/10.1007/BFb0022474
21. Peterson, J.L.: Petri Net Theory and the Modeling of Systems. Prentice Hall PTR (1981)
22. Rosa-Velardo, F., de Frutos-Escrig, D.: Decision problems for Petri nets with names. arXiv preprint arXiv:1011.3964 (2010)
23. Rosa-Velardo, F., de Frutos-Escrig, D., Marroquín-Alonso, O.: On the expressiveness of mobile synchronizing Petri nets. Electron. Notes Theor. Comput. Sci. **180**(1), 77–94 (2007)
24. Rozinat, A., Van der Aalst, W.M.P.: Conformance checking of processes based on monitoring real behavior. Inf. Syst. **33**(1), 64–95 (2008)
25. Taghiabadi, E.R., Gromov, V., Fahland, D., van der Aalst, W.M.P.: Compliance checking of data-aware and resource-aware compliance requirements. In: Meersman, R., et al. (eds.) OTM 2014. LNCS, vol. 8841, pp. 237–257. Springer, Heidelberg (2014). https://doi.org/10.1007/978-3-662-45563-0_14
26. Taymouri, F., Carmona, J.: An evolutionary technique to approximate multiple optimal alignments. In: Weske, M., Montali, M., Weber, I., vom Brocke, J. (eds.) BPM 2018. LNCS, vol. 11080, pp. 215–232. Springer, Cham (2018). https://doi.org/10.1007/978-3-319-98648-7_13
27. Van Der Aalst, W.: Data science in action. In: Van Der Aalst, W. (ed.) Process Mining, pp. 3–23. Springer, Heidelberg (2016). https://doi.org/10.1007/978-3-662-49851-4_1
28. Van der Aalst, W., Adriansyah, A., van Dongen, B.: Replaying history on process models for conformance checking and performance analysis. Wiley Interdisc. Rev. Data Min. Knowl. Discov **2**(2), 182–192 (2012)
29. Dongen, B.F.: Efficiently computing alignments. In: Weske, M., Montali, M., Weber, I., vom Brocke, J. (eds.) BPM 2018. LNCS, vol. 11080, pp. 197–214. Springer, Cham (2018). https://doi.org/10.1007/978-3-319-98648-7_12
30. van Hee, K., Sidorova, N., Voorhoeve, M.: Soundness and separability of workflow nets in the stepwise refinement approach. In: van der Aalst, W.M.P., Best, E. (eds.) ICATPN 2003. LNCS, vol. 2679, pp. 337–356. Springer, Heidelberg (2003). https://doi.org/10.1007/3-540-44919-1_22
31. Van Hee, K., Sidorova, N., Voorhoeve, M.: Resource-constrained workflow nets. Fundamenta Informaticae **71**(2, 3), 243–257 (2006)

From Place Nets to Local Process Models

Viki Peeva(✉), Lisa L. Mannel, and Wil M. P. van der Aalst

RWTH Aachen University, 52062 Aachen, Germany
{peeva,mannel,wvdaalst}@pads.rwth-aachen.de

Abstract. Standard process discovery algorithms find a single process model that describes all traces in the event log from start to end as best as possible. However, when the event log contains highly diverse behavior, they fail to find a suitable model, i.e., a so-called "flower" or "spaghetti" model is returned. In these cases, discovering local process models can provide valuable information about the event log by returning multiple small process models that explain local behavior. In addition to explainability, local process models have also been used for event abstraction, trace clustering, outcome prediction, etc. Existing approaches that discover local process models do not scale well on event logs with many events or activities. Hence, in this paper, we propose a novel approach for discovering local process models composed of so-called place nets, i.e., Petri net places with the corresponding transitions. The place nets may correspond to state- or language-based regions, but do not need to. The goal however is to build multiple models, each explaining parts of the overall behavior. We also introduce different heuristics that measure the model's frequency, simplicity, and precision. The algorithm is scalable on large event logs since it needs only one pass through the event log. We implemented our approach as a ProM plugin and evaluated it on several data sets.

Keywords: Local process models · Process mining · Process discovery

1 Introduction

The main goal of process mining is to help people analyze and improve processes. One subarea of process mining is process discovery which automatically creates process models from available event logs [1]. Process discovery techniques [13, 15,26,27] try to explain and visualize the process from start to end, while other algorithms like sequence and episode mining [19,21] try to mine small patterns that frequently happen in the event log. This paper will focus on local process model discovery which was first introduced in [24] as an individual branch and was positioned between process discovery and pattern mining. Local process models are able to describe complex constructs in contrast to sequences and episodes, but keep the local perspective introduced in pattern mining, which separates them from process discovery. This way, instead of describing a process with one overall model, a set of models is used.

L. Bernardinello and L. Petrucci (Eds.): PETRI NETS 2022, LNCS 13288, pp. 346–368, 2022.
https://doi.org/10.1007/978-3-031-06653-5_18

Some processes we want to analyze are too diverse to have a clear structure. Thus, making it almost impossible to discover an end-to-end model, resulting in a discovery of a so-called "flower" or "spaghetti" models. Hence, one straightforward use-case of local process model discovery is when traditional process discovery approaches fail to produce an understandable model. However, the importance of local process models is not constrained to processes where process discovery fails to produce a good model. Despite the limited number of approaches that offer local process model discovery [2,24], local process models have been used in event abstraction [18], classification of traces [20], clustering of resources [8], as sub-part of end-to-end discovery algorithms [12,17], and in different use-case studies [7,11].

In this paper, we introduce a novel approach for discovering local process models. We are inspired by region-theory discovery algorithms. We assume that the possible regions are already available to us, and instead of building one end-to-end model, we combine the regions in smaller local process models. We accept the regions in the form of place nets, that we can get from any of the existing process discovery approaches. Our proposed algorithm is available as a plugin in ProM[1] [25] (Fig. 1) that allows the input to be defined as a set of place nets or a Petri net. The first notable difference between our approach and the existing ones is that we build the local process models as Petri nets instead of process trees as in [2,24]. This allows us to find constructs like long-term dependencies that are not possible in process trees. The next significant difference is speed and feasibility. We show that in contrast to the existing approaches, we are able to handle event logs with many activities or events and we return results much faster. Previous approaches rely on pruning infrequent local process models early on to gain on speed. Thus, forcing them to return only frequently appearing models. And although we are able to return frequent models, we are not constrained to find only those, since our speed arises from passing the event log only once and not pruning out infrequent patterns. In the future, this would allow for even broader usage and application of local process models in other areas of process analysis. To summarize, our contribution is threefold:

- We introduce an entirely new technique to build local process models that is completely based on Petri nets.
- We offer a technique that is feasible on event logs with many activities or events because it is linear in the size of the event log.
- We do not limit the results to frequent local process models.

We continue the paper by presenting some related work in Sect. 2, and preliminaries in Sect. 3. In Sect. 4, we present the approach for local process model discovery. Section 5 explains our evaluation strategy and the results we get. Section 6 concludes the paper by summarizing and giving an outlook for future work.

[1] The plugin "LocalProcessModelDiscoveryByCombiningPlaces" is available in ProM 6.11 and the Nightly Builds.

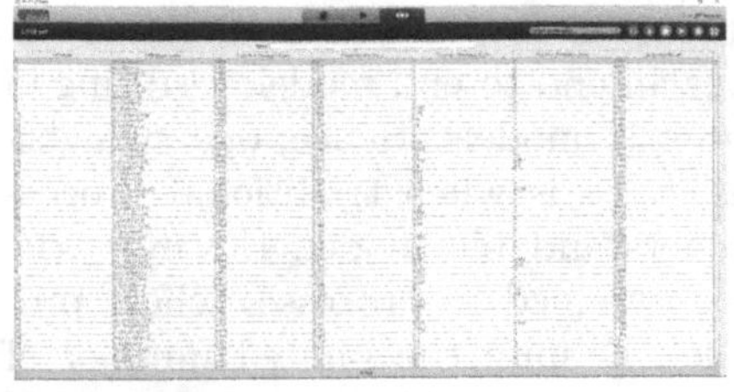

(a) Table with local process models

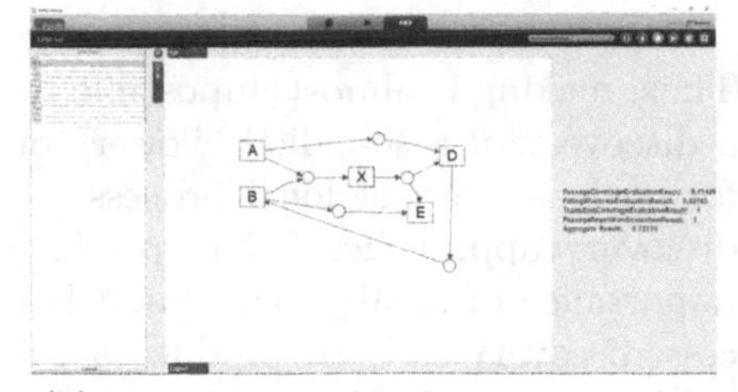

(b) Visualizing a local process model

Fig. 1. Implementation of our approach as a ProM plugin.

2 Related Work

As previously mentioned, local process model discovery is positioned in-between traditional process discovery and episode and sequence mining. Although process discovery approaches [4,6,13,15,26,27] are highly valuable for process analysis, the purpose of local process models is different. Local process models try to explain subsequences of the traces like in episode and sequence mining but can discover much more complex constructs as compared to traditional process discovery. To take advantage of all the different discovery techniques, we can use the output they produce as input for our algorithm. To be flexible regarding the existing and future discovery methods, we only require that the input is a Petri net or a set of place nets no matter how or which algorithm produced them.

To the best of our knowledge, there are two existing techniques for mining local process models from event logs. Both [24] and [2] mine local process models by recursively extending process trees.

The approach in [24] was the first to discover local process models and consists of four main steps. In the first step, for each activity in the event log a process tree containing exactly one leaf node that represents the activity is created. This set is the first set of candidate local process models. In the second step, each of the local process models from the candidate set is evaluated on the event log with different quality metrics, and only a subset of them that satisfy certain thresholds are selected in the third step. In the fourth step, the selected local process models (process trees) are expanded by replacing one of the leaves with each process tree operator (sequence, loop, parallel, and exclusive choice) and adding the replaced leaf as one child and an activity not already present in the process tree as a second child. The expanded local process models become candidates in the next step, and the procedure is repeated until the maximal size of the local process model is achieved or none of the candidates pass the selection phase. Each process tree is evaluated on the entire event log and extended with all activities, making the approach infeasible for event logs with many events or activities.

The approach in [2] was inspired by [24], and also recursively extends process trees. However, they create new process trees by combining two existing process trees, called seeds, that differ only in one leaf node. In the combined process tree this differentiating node is replaced with a process tree operator and the two nodes are added as children to the new operator. In addition to [24] they do not reevaluate the process trees on the entire event log but on projections of the seeds. Additionally, they define compact and maximal process trees and strive to return only such local process models.

Both approaches use monotonicity of model frequency for pruning, but still struggle to return results in reasonable time on mid-sized event logs. To handle this problem, [23] extends the work in [24] by mining local process models for specific subsets of activities decided via heuristics. [22] allows for mining local process models of a specific interest using utility functions, and with [3] the work in [2] is extended to discover patterns that are frequent for a given context. Although not in the focus of this paper, our algorithm can adopt both utility functions and in-context search without impacting our running time significantly.

3 Preliminaries

In this section, we introduce important background information needed for understanding the rest of the paper. We start with some general notations, and we continue with topic-specific definitions.

General. We use sets ($\{a, b, ...\}$), multisets ($[a^2, b, ...]$), sequences ($\langle a, b, ...\rangle$), and tuples ($(a, b, ...)$) as usually defined. Given a set X, X^* represents the set of all sequences over X, and $\mathbb{M}(X)$ is the set of all multisets over X. Given a sequence $\sigma = \langle s_1, s_2, ...s_n\rangle$, we access the i-th element of the sequence with $\sigma[i]$, i.e., $\sigma[i] = s_i$, for $1 \leq i \leq n$. We extend σ with an additional element s_{n+1} by writing $\sigma \cdot s_{n+1}$. We call the sequence σ' a *subsequence* of σ, if and only if $\sigma' = \langle s_l, s_{l+1}, ...s_m\rangle$ and $1 \leq l < m \leq n$ (we write $\sigma' \sqsubseteq \sigma$ or $\sigma' = \sigma[l, m]$ if the indices are known). We call σ' a *relaxed subsequence* (we write $\sigma' \precsim \sigma$) if and only if for some $k \geq 1$ there is $\sigma' = \langle s_{i_1}, s_{i_2}, ...s_{i_k}\rangle$ such that $1 \leq i_1 < i_2 < ... < i_k \leq n$, i.e., we drop any number of elements from σ (at most $n - 1$) and keep the order for the rest. We write $\{\sigma' \text{ op } \sigma\}$ or $[\sigma' \text{ op } \sigma]$ where op $\in \{\sqsubseteq, \precsim, \sqsubseteq_k, \precsim_k\}$, to denote the set or multiset of all sequences σ' that satisfy the given operator in regard to σ. We use $\sqsubseteq_k$ and $\precsim_k$ when we are interested in subsequences respectively relaxed subsequences of a particular length. To recalculate sets or multisets from other sets, multisets or sequences, we use the $\{\cdot\}$ and $[\cdot]$ operators. We use $f(X) = \{f(x)|x \in X\}$ (respectively $f(\sigma) = \langle f(s_1), f(s_2), ...f(s_n)\rangle$) to apply the function f to every element in the set X (the sequence σ) and $f_{\upharpoonright X}$ (respectively $\sigma_{\upharpoonright X}$) to denote the projection of the function f (respectively the sequence σ) on the set X.

Process Mining. The collected data used for process analysis is given in the form of *event logs*. Hence, in Definition 1, we formally define *traces* and *event logs*. Note that although traces are usually defined as sequences of events, in this work, we are interested only in the activity the events represent.

Definition 1. *Given the universe of activities $\mathcal{A}$, we define $\rho \in \mathcal{A}^*$ as a* trace*, and $L \in \mathbb{M}(\mathcal{A}^*)$ as an* event log.

In Definition 2, we define *labeled Petri nets*. Note that a transition $t \in T$ with $l(t) = \tau$ is called silent, and that there may be duplicate transitions $t_1, t_2 \in T$ such that $l(t_1) = l(t_2)$.

Definition 2 (Labeled Petri Net). *A* labeled Petri net *$N = (P, T, F, A, l)$ is a tuple, where P is a set of places and T is a set of transitions such that $P \cap T = \emptyset$. $F \subseteq (P \times T) \cup (T \times P)$ is the flow relation, $A \subseteq \mathcal{A}$ is a set of activities, and $l : T \to A \cup \{\tau\}$ the labeling function.*

Now given a labeled Petri net $N = (P, T, F, A, l)$, for each element $x \in P \cup T$ we define the *preset* of x to be $\bullet x = \{y | (y, x) \in F\}$, and the *postset* of x to be $x\bullet = \{y | (x, y) \in F\}$. We additionally define the set $\overline{N} = \{(t_i, t_o) | \exists_{p \in P}((t_i, t_o) \in \bullet p \times p\bullet)\}$ to denote all pairs of transitions in the net N, that are directly connected via a place. We call each such pair a *passage*.

A labeled Petri net can be in a given state with the help of *markings*. Given a labeled Petri net $N = (P, T, F, A, l)$, we define a *marking* M as $M \in \mathbb{M}(P)$, and with $[\,]$ we denote the *empty marking*. Every element in the marking M represents a *token* in one of the places in P. The state can change by following the *firing rule*. We say that a transition $t \in T$ is *enabled* in the marking M if and only if there is a token in each place in the preset of t, i.e., $\bullet t \subseteq M$ (we write $M[t\rangle$). A transition t can *fire* in marking M if and only if it is *enabled* in M. By firing, the transition changes the marking to $M' = (M \setminus \bullet t) \cup t\bullet$. In this case, we can write $M \xrightarrow{t} M'$. To denote getting from M to M' by firing a sequence of transitions $\sigma = \langle t_1, ..., t_n \rangle \in T^*$ such that $M \xrightarrow{t_1} M_1 \xrightarrow{t_2} M_2 \cdots \xrightarrow{t_n} M'$, we write $M \xrightarrow{\sigma} M'$.

In Definition 3, we define a *union* of two labeled Petri nets and we extend for multiple labeled Petri nets. Then in Definition 4 we define what it means for a labeled Petri net to be *connected*.

Definition 3 (Union of Labeled Petri Nets). *Given two labeled Petri nets $N_1 = (P_1, T_1, F_1, A_1, l_1)$ and $N_2 = (P_2, T_2, F_2, A_2, l_2)$ we define their* union *as $N_1 \cup N_2 = N = (P, T, F, A, l)$ where $P = P_1 \cup P_2$, $T = T_1 \cup T_2$, $F = F_1 \cup F_2$, $A = A_1 \cup A_2$, and $l : T \to A \cup \{\tau, \perp\}$ is the mapping*

$$l(t) = \begin{cases} l_1(t), \text{ if } t \in T_1 \setminus T_2 \\ l_2(t), \text{ if } t \in T_2 \setminus T_1 \\ l_1(t), \text{ if } t \in T_1 \cap T_2 \wedge l_1(t) = l_2(t) \\ \perp, \quad \text{otherwise} \end{cases}.$$

The union is a valid union *if there is no $t \in T$ such that $l(t) = \perp$. We write $\bigcup_{i=1}^{n} N_i = (\cdots((N_1 \cup N_2) \cup N_3) \cdots \cup N_n)$ to denote the* union *of the set of labeled Petri nets $\{N_1, \ldots, N_n\}$.*

Definition 4 (Connected Labeled Petri Net). *A labeled Petri net $N = (P, T, F, A, l)$ is* connected*, if and only if for each two different elements $x, x' \in P \cup T$ there exists a sequence $\langle y_1, \ldots, y_n \rangle$ such that $n \geq 2$, $(y_i, y_{i+1}) \in F$ or $(y_{i+1}, y_i) \in F$ for $1 \leq i < n$ and $y_1 = x$ and $y_n = x'$.*

Local Process Models. Our algorithm discovers a set of *local process models*, that we represent with labeled Petri nets. We discover these *local process models* from an *event log* and a *set of place nets*. To represent our input, in Definition 5 we define a *place net* as a labeled Petri net with only one place.

Definition 5 (Place Net). *A* place net *is a labeled Petri net* $N_p = (\{p\}, T, F, A, l)$, *where* $\{p\}$ *is a set of places containing one place only, and* T *is a set of transitions such that* $P \cap T = \emptyset$. $F \subseteq (\{p\} \times T) \cup (T \times \{p\})$ *is the flow relation,* $A \subseteq \mathcal{A}$ *is a set of activities, and* $l : T \rightarrow A \cup \{\tau\}$ *the labeling function.*

Next, with the help of Definitions 3 and 4, in Definition 6 we define a *local process model* as a *union* of place nets.

Definition 6 (Local Process Model). *Given a set of* place nets $N_{p_i} = (\{p_i\}, T_i, F_i, A_i, l_i)$ *for* $1 \leq i \leq k$, *their union is a* local process model, $LPM = \bigcup_{i=1}^{k} N_{p_i}$, *if and only if* LPM *is a* valid union *and a* connected *labeled Petri net.*

What makes our *local process models* local is the behavior. Therefore, we define the term *locality* or *local distance* to be the maximal length of the trace's subsequences we want our local process models to explain. Since *local process models* are a subset of labeled Petri nets, *markings*, *enabled transitions* and *firing rule*, also hold for them. The opportunity to change states and fire transitions makes it possible *local process models* to describe behavior. Since we want to discover models that explain selected parts of the behavior in an event log, we need to somehow align the two. Thus, in Definition 7 we define how a local process model can replay a sequence of activities. In addition, we want to be able to skip some of the activities during the replay, so we also define *relax replay* (Definition 8).

Definition 7 (Replay). *Given a local process model* $LPM = (P, T, F, A, l)$ *and a sequence of activities* $\rho = \langle a_1, a_2, \dots a_n \rangle$, *we say* LPM replays ρ *if and only if there exists a sequence of transitions* $\sigma = \langle t_1, t_2, \dots, t_m \rangle \in T^*$ *such that* $l(\sigma)_{\restriction A} = \rho$ *and* $[\,] \xrightarrow{\sigma} [\,]$.

Definition 8 (Relaxed Replay). *Given a local process model* $LPM = (P, T, F, A, l)$ *and a sequence of activities* $\rho = \langle a_1, a_2, ... a_n \rangle$, *we say* LPM relax replays ρ *if and only if there exists at least one relaxed subsequence* $\rho' \in [\rho' \precapprox \rho]$ *that* LPM *can* replay.

By defining *replay* and *relax replay* to require starting and ending in an empty marking, makes the subset of *place nets* $N_p = (\{p\}, T, F, A, l)$ for which $\bullet p \subseteq p \bullet$ or $p \bullet \subseteq \bullet p$ unsuitable for our *local process models.* Hence, in the continuation we will discard *place nets* of this type.

In addition, we use *replay* and *relax replay* to define the *language* (Definition 9) and *relaxed language* (Definition 10) for a given local process model LPM.

Definition 9 (Language). *Given a local process model* $LPM = (P, T, F, A, l)$*, we define* $\mathcal{L}(LPM) = \{\rho \in A^* | \exists_{\sigma \in T^*}(l(\sigma)_{\upharpoonright A} = \rho \wedge [\,] \xrightarrow{\sigma} [\,])\}$ *to be the* language *of LPM.*

Definition 10 (Relaxed Language). *Given a local process model* $LPM = (P, T, F, A, l)$*, we define the* relaxed language *of LPM as* $\mathcal{L}_{rlx}(LPM) = \{\rho \in A^* | \exists_{\rho' \in \mathcal{L}(LPM)}(\rho' \sqsubsetapprox \rho)\}$.

We conclude this section, by defining how a local process model can be *compact* in regard to a sequence of activities (Definition 11).

Definition 11 (Compact Local Process Model). *Given a local process model* $LPM = (P, T, F, A, l)$ *and a sequence of activities* $\rho = \langle a_1, a_2, ..., a_n \rangle \in \mathcal{L}_{rlx}(LPM)$*, we say LPM is compact with respect to* ρ *if and only if it holds that* $\exists_{\sigma \in T^*}([\,] \xrightarrow{\sigma} [\,] \wedge l(\sigma)_{\upharpoonright A} \sqsubsetapprox \rho \wedge \forall_{p \in P}(\exists_{t \in \{\sigma\}}(p \in \bullet t \cup t \bullet)))$.

4 Approach

Our algorithm combines place nets into local process models. Hence, as input we require place nets and an event log for which we want to build the local process models. However, for n place nets, there are $2^n - 1$ non-empty candidate local process models. Even if we remove the ones that do not satisfy Definition 6, our search space would still be enormous. Additionally, some of the local process models we build, can be too complicated or not satisfy basic quality expectations. Therefore, we propose a framework with three modules (Fig. 2). Since our search space directly depends on the number of place nets we use, we use the first module for filtering and adapting the place nets to limit their number. However, at the same time the quality of the built local process models directly depends on the quality of the chosen place nets, so we want to choose these wisely. After the place nets are chosen, the second module introduces the main algorithm for building local process models. The goal of the algorithm is to consider different subsets of place nets, construct their union and check whether it can relax replay subsequences of the traces in the given event log. Although we restrict the set of place nets we use, we can still end up with a lot of local process models. Therefore, we also provide a module for evaluating and ranking the found local process models with different metrics. In the following, we introduce each of the modules, with the main focus on the combination algorithm (the second module in Fig. 2).

4.1 Place Net Adaptation and Filtering (PAF)

We use an "oracle" to get the place nets from which we build our local process models. Any algorithm that returns a labeled Petri net or a set of place nets based on an event log can be considered an oracle. The oracle can return many place nets, so for efficiency reasons, we want to limit the number of those we use for building local process models. On the other side, the set of place nets we

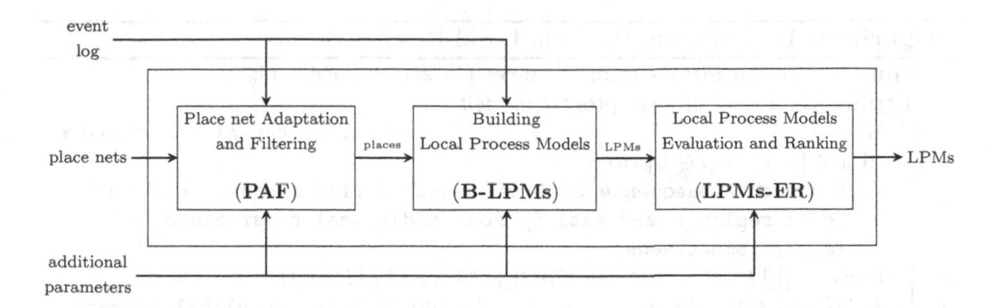

Fig. 2. Top view of our framework for local process model discovery

use restricts our models to a representational bias defined by them. Hence, to promote simplicity and higher relax replay frequency (see Definition 8), we rank the place nets by giving preference to nets with fewer transitions that can relax replay more subsequences. For place nets that rank the same on the previous criteria, we use the lexicographic order of the included transitions. Afterward, we keep the k highest ranking place nets to build local process models, where k is a user-defined parameter. For simplicity in the continuation of the paper, we assume that given the set of place nets P that we return now $\bigcup P$ is a valid union. Otherwise, we keep track of all the subsets where label disagreements exist and do not use multiple contradicting place nets in one local process model.

4.2 Building Local Process Models (B-LPMs)

This module covers the part of the framework that combines place nets into local process models, thus, making it the main contribution of this paper. To explain the approach, we give a high-level pseudo-code in Algorithm 1. There are three main steps that make up the gist of the algorithm, and get us from a set of place nets, to a set of local process models that describe the event log:

1. Focus on locality by iterating all subsequences in the event log of certain length (Line 2).
2. Build local process models for each subsequence separately (Line 3).
3. Store the built local process models in a single structure (Line 4).

The high-level algorithm looks pretty straightforward. However, optimizing the *traversal* of the event log on line 2, the particulars of the *global storage* and how we *create* local process models that relax replay the window, is what makes the algorithm not only feasible but also efficient.

Focus on Locality. We want our local process models to describe what happens within some local distance in the event log. Hence, with a *sliding window* we get subsequences of certain length, that we call *windows*. The sliding window size represents the *locality* we are interested in, and we accept it as an input parameter. We formally define the sliding window in Definition 12.

Algorithm 1: Combining Places in Local Process Models

```
input : L - event log; d - local distance; P - set of place nets;
output: LPM - set of local process models
1 LPM ← [];                                    // initialize the global storage
2 forall w ∈ [ρ' ⊑_d ρ|ρ ∈ L] do
      // for each subsequence of L of length d find subsets of P that
         relax replay w and satisfy some additional constraints AC
         (e.g., compactness)
3     lpms ← {⋃P'|P' ⊆ P ∧ w ∈ L_rlx(⋃P') ∧ AC(⋃P', w)};
4     LPM ← LPM ∪ lpms;                        // add lpms to the global storage
```

Definition 12 (Sliding Window). *Given a trace* $\rho = \langle a_1, a_2, ..., a_n \rangle$ *and locality* $d > 0$, *we define the function* $W_d(i, \rho) = \begin{cases} \rho[i, i+d-1], & \text{if } 1 \leq i \leq n-d+1 \\ \langle\rangle, & \text{otherwise} \end{cases}$ *to be a* sliding window. *Each generated subsequence for a concrete* i *and* ρ *we call a* window.

The sliding window helps us to iterate the event log, and focus on a local level. However, for each window, we need to efficiently and exhaustively (considering our representational bias and limitations) combine places into local process models that can relax replay that window.

Building Local Process Models for One Window. At this point, we have our set of place nets $P = \{N_{p_1}, N_{p_2}, \ldots, N_{p_k}\}$ and a sequence of activities, i.e., our window w. Our goal is to find subsets of P, $P' \subseteq P$, to form local process models, $LPM = \bigcup P'$, that satisfy Definition 6 considering the following constraints:

- LPM can *relax replay* w ($w \in \mathcal{L}_{rlx}(LPM)$)
- LPM is *compact* in regard to w (see Definition 11)

Additionally, we want to be time efficient. Therefore, given a trace $\rho = \langle a_1, a_2, \ldots, a_n \rangle$ we consider that two consecutive windows $W_d(m, \rho)$ and $W_d(m+1, \rho)$, share $d-1$ of their elements. The models found for this overlapping sequence shared by both windows, are the same. Hence, it is important that we do not recalculate these models, which in turn defines the goal to *reuse* local process models shared between consecutive windows.

Idea. The core idea is to create new local process models by extending existing ones with an additional place net such that the relaxed subsequence of the window that they can replay increases in length. We start with the empty local process model that can somehow replay the empty trace, and we want to extend it with carefully selected place nets such that two activities from the window can be replayed. In the next step, we would extend those local process models by adding an additional place net such that the newly created local process models

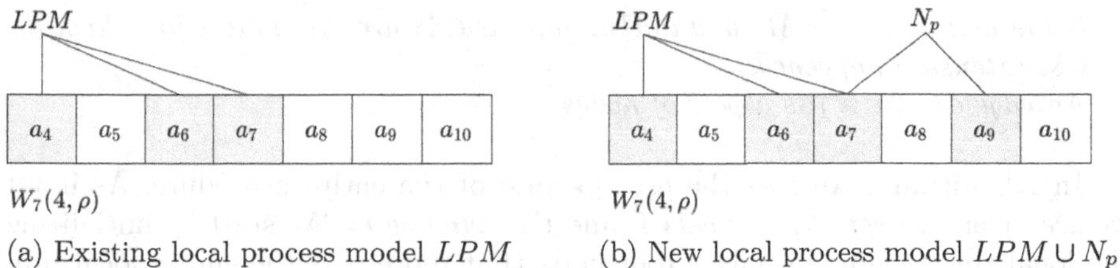

(a) Existing local process model LPM (b) New local process model $LPM \cup N_p$

Fig. 3. Extension of a local process model with a new place net.

can replay three of the activities in the window. We continue as long as there are still unprocessed activities in the window. For example, let us consider the window $W_7(4, \rho)$ in Fig. 3a where ρ is a trace. We have built a local process model LPM such that $\langle a_4, a_6, a_7 \rangle \in \mathcal{L}(LPM)^2$. Since $\langle a_4, a_6, a_7 \rangle \sqsubsetneq W_7(4, \rho)$, $W_7(4, \rho) \in \mathcal{L}_{rlx}(LPM)$. We now want to extend LPM with an additional place net $N_p = (\{p\}, T_p, F_p, A_p, l_p)$. What is specific for N_p is that it should be able to replay a_7 such that a token is put in p, and also replay one of the unprocessed activities (a_8, a_9 or a_{10}) such that the token is removed from p. Hence, the newly built local process model $LPM \cup N_p$ is empty after replaying four activities from the window, and the used firing sequence is an extension of the firing sequence used for replaying $\langle a_4, a_6, a_7 \rangle$ on LPM. In our case the new activity is a_9 and we visualize this in Fig. 3b. To know whether we can extend LPM with N_p we have to check that we do not break the replay of $\langle a_4, a_6 \rangle$. Hence, we need the firing sequence σ for which we replayed $\langle a_4, a_6 \rangle$, to ensure that σ can still fire when the new place net is added. To know where to connect the place net and the local process model such that a_7 can be replayed we need the marking M after firing σ i.e., $[\,] \xrightarrow{\sigma} M$. At the end, we also store the two indices $indIn$ and $indOut$ in the window for which the last extension happened. Note that $\sigma\restriction_A \sqsubsetneq w[1, indIn]$ and $w[1, indOut] \in \mathcal{L}_{rlx}(LPM)$.

Algorithm. We now present an algorithm that builds local process models given a set of place nets P and a window w. We explained that at every step we extend existing local process models with new place nets. To be aware of the extension path from which we got to a particular local process model and how to continue extending it, we organize the local process models in a tree structure that we call *local tree*. Each node in the *local tree* represents a local process model $LPM = (P, T, F, A, l)$ that can relax replay w. There is an edge between two nodes n and n' when the local process model represented by n' was built by extending the local process model in n with an additional place net. We formally define our *local tree* in Definition 13.

Definition 13 (Local Tree). *A local tree $LT = (N, E)$ is a pair, where N is a set of nodes and E a set of edges such that:*

- *A node $n = (LPM, \sigma, M, indIn, indOut)$ is a tuple, where $LPM = (P, T, F, A, l)$ is a local process model, $\sigma \in T^*$ is a sequence of transitions, $M \in \mathbb{M}(P)$*

[2] Note that this doesn't have to be the only one such local process model.

is the marking $[] \xrightarrow{\sigma} M$*, and* $indIn, indOut \in \mathbb{N}$ *are the indices for which the last extension happened.*
- *An edge* $e = (n, n')$ *is a pair of nodes.*

In Algorithm 2 we give the pseudo-code of the entire procedure. As input we are given the *set of place nets* P and the *window* w. We start by initializing the local tree to contain only a root node that represents the empty local process model (line 1). Then we traverse all activity pairs of the window, and for each pair, we extend existing local process models in LT with additional place nets. We get suitable place nets by filtering those that can replay the currently considered two events $w[i]$ and $w[j]$, and suitable nodes by filtering those that contain a local process model in a marking in which $w[i]$ can be replayed (lines 5 and 6). Afterward, we restrict that local process models are extended with a place net only if the place net does not add a new constraint on an already used transition (line 9). Then, we find a common transition of the place net and the local process model that can replay $w[i]$. If there are no such transitions and the node is not the root, the extension can not happen (line 12). If there are multiple such transitions we randomly choose one (line 14). We create a new node n' (line 15) that represents the local process model built by adding the place net N_p to the local process model in the node n, in a marking after replaying $w[i]$. We add the newly created node in the tree and connect it with the node from which it was created (lines 16 and 17). We finish by adding the local process model to the final set if after replay of $w[j]$ we end in the empty marking (lines 18 and 19).

Fulfillment of Constraints and Goals. In the following, we give some intuitions that connect the design of the algorithm to the constraints and the goal. The first constraint is that each returned local process model satisfies Definition 6. In the PAF module we assumed that $\bigcup P$ is a valid union. Hence, the union of any subset $P' \subseteq P$ is also a valid union. That the local process model is connected is satisfied by requiring $T' \neq \emptyset$ when the place net we add is not the first in the local process model (line 12). The constraint that each created local process model can relax replay w is satisfied by combining lines 5, 6, 9 and 18. The filterings of the nodes and place nets, ensure that a local process model is extended with a new place net only when the newly created local process model replays one more activity of the window than its base local process model. In line 9 we make sure we do not break the successful replay of the base local process model, and with line 18 we make sure that there is at least one unprocessed activity in the window, after whose replay the local process model ends in an empty marking. Definition 11 is also satisfied because of the filtering in line 5. A place net is added to a local process model only if a token can be put in the place it represents and removed from it, by replaying two activities. Therefore each place is marked at some point of the replay. Finally, our goal to *reuse* local process models between consecutive windows, is satisfied by the way we organize our nodes in the tree, i.e., how we create edges (line 17). Given two nodes n and n' such that $(n, n') \in LT.E$, we know that $n.LPM.l(n.\sigma) = n'.LPM.l(n'.\sigma)[1, |n'.\sigma| - 1]$. Hence, the most distant ancestor of n' apart from the root, is some node $n*$ that is a child of the

Algorithm 2: Building Local Process Models for a Window

```
input : w - window; P - set of place nets;
output: LPMs - set of local process models
// In the pseudo-code we use a dot notation for accessing elements
   of an object, similar as in object-oriented programming.
1  LT ← (N = {root = (∅, ⟨⟩, [], 0, 0)}, E = ∅);        // initialize the storage
2  d = |w|;                                               // length of the window
3  for j ← 1 to d do
4  |  for i ← 1 to j − 1 do
   |  |  // for each pair of events get suitable place nets and nodes
5  |  |  P' ← {N_p ∈ P | ⟨w[i], w[j]⟩ ∈ 𝓛(N_p)};
6  |  |  N' ← {n ∈ N | n.LPM.l(n.σ)↾_{n.LPM.A} · w[i] ∈ 𝓛(n.LPM)} ∪ {root}
   |  |  // try to extend LPM in each node with each place net
7  |  |  for n = (LPM, σ, M, indIn, indOut) ∈ N' do
8  |  |  |  for N_p = ({p}, T_p, F_p, A_p, l_p) ∈ P' do
9  |  |  |  |  if p• ∩ {σ} ≠ ∅ then
10 |  |  |  |  |  continue;                              // no new constraint
11 |  |  |  |  T' ← {t' ∈ LPM.T ∩ •p | LPM.M[t'⟩ ∧ LPM.l(t') = w[i]};
12 |  |  |  |  if n ≠ root ∧ T' = ∅ then
13 |  |  |  |  |  continue;                              // no common transition
14 |  |  |  |  t ←_R T'                                  // choose any transition
15 |  |  |  |  n' ← (LPM ∪ N_p, σ · t, (M \ •t) ∪ t•, i, j);   // create node
16 |  |  |  |  LT.N ← LT.N ∪ n';                         // add node
17 |  |  |  |  LT.E ← LT.E ∪ (n, n');                    // add edge
   |  |  |  |  // add n'.LPM in final set if w ∈ 𝓛_rlx(n'.LPM)
18 |  |  |  |  if ∃_{t∈n'.LPM.T}(n'.M --t--> [] ∧ n'.LPM.l(t) = w[j]) then
19 |  |  |  |  |  LPMs ← LPMs ∪ {n'.LPM};
20 return LPMs
```

root. Then, $n*.\sigma = \langle n'.\sigma[1]\rangle$. Therefore, if we want to remove all local process models that replay $w[1]$, we just need to remove all children of the root $n*$ for which $n*.LPM.l(n*.\sigma[1]) = w[1]$ (have in mind that $|n*.\sigma| = 1$ for the children of the root).

Example. To clarify how the algorithm works given its input, we additionally provide an example. For simplicity we assume that $t = l(t)$ for each transition. Given the set of place nets P (see Fig. 4a) and the window $w = \langle b, a, x, a, d\rangle$ we build local process models by following Algorithm 2. We first initialize the local tree $LT = (\{root\}, \emptyset)$ and the resulting set $LPMs = \emptyset$. Then we iterate through the window with the indices i and j. We start with $i = 1$ and $j = 2$. Since $w[i] = b$ and $w[j] = a$, we get $P' = \emptyset$ so we continue. For $i = 1$ and $j = 3$ ($w[i] = b$ and $w[j] = x$) we filter $P' = \{N_{p2}\}$, $N' = \{root\}$. Since $p2 \bullet \cap \{root.\sigma\} = \emptyset$ and $T' = \{b\}$, we create the node $n1 = (\{N_{p2}\}, \langle b\rangle, [p2^1], 1, 3)$ and add it as child to the root node ($LT = (\{root, n1\}, \{(root, n1)\})$). Because $[p2^1] \xrightarrow{x} []$ and $w[j] = x$ we add the local process model to the final set ($LPMs = \{\{N_{p2}\}\}$). We skip $i = 2$, $j = 3$ and all pairs for $j = 4$, since $P' = \emptyset$ for them. For $i = 1$

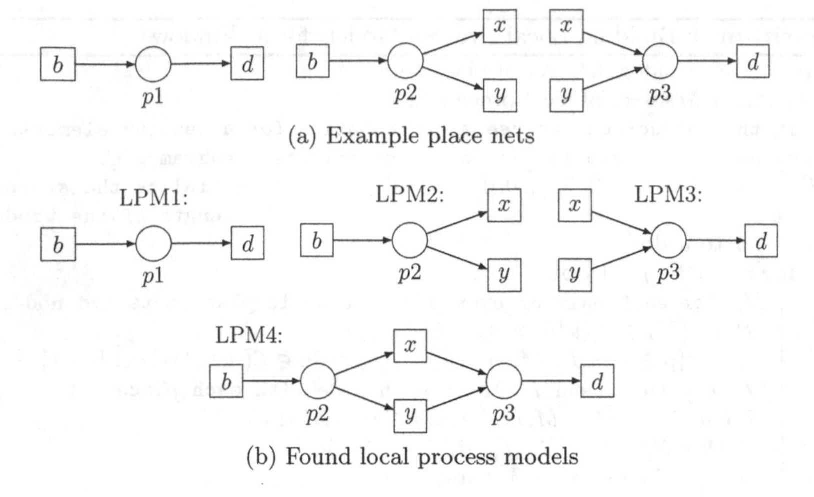

(a) Example place nets

(b) Found local process models

Fig. 4. Place nets and Local Process Models for the example

and $j = 5$ we calculate $P' = \{N_{p1}\}$, $N' = \{root\}$. Given $p1 \bullet \cap \{root.\sigma\} = \emptyset$ and $T' = \{b\}$, we create the node $n2 = (\{N_{p1}\}, \langle b \rangle, [p1^1], 1, 5)$, add it as child to the root node and to the final set ($LT = (\{root, n1, n2\}, \{(root, n1), (root, n2)\})$ and $LPMs = \{\{N_{p2}\}, \{N_{p1}\}\}$). We again skip $i = 2$, $j = 5$ since $P' = \emptyset$. For $i = 3$ and $j = 5$ we calculate $P' = \{N_{p3}\}$, $N' = \{root, n1\}$. For N_{p3} and $root$, $p3 \bullet \cap \{root.\sigma\} = \emptyset$ and $T' = \{x\}$ so we create $n3 = (\{N_{p3}\}, \langle x \rangle, [p3^1], 3, 5)$. For N_{p3} and $n1$, $p3 \bullet \cap \{n1.\sigma\} = \emptyset$ and $T' = \{x\}$ so we create $n4 = (\bigcup\{N_{p2}, N_{p3}\}, \langle b, x \rangle, [p3^1], 3, 5)$. We add $n3$ as child to the root node and $n4$ as child to $n1$. Our local tree now is $LT = (\{root, n1, n2, n3, n4\}, \{(root, n1), (root, n2), (root, n3), (n1, n4)\})$ and final set $LPMs = \{\{N_{p2}\}, \{N_{p1}\}, \{N_{p3}\}, \bigcup\{N_{p2}, N_{p3}\}\}$. We do nothing for $i = 4$ and $j = 5$ since $P' = \emptyset$. The final set $LPMs$ is given in Fig. 4b.

Choice and Concurrency. After processing the window w, the tree contains all local process models LPM for which $w \in \mathcal{L}_{rlx}(LPM)$ and given the used firing sequence σ it holds that $\forall_{t \in \sigma}(M \xrightarrow{t} M' \implies M \cap M' = \emptyset)$, i.e., concurrency is not considered. To build the concurrency constructs, we combine nodes from different branches in the local tree and take the union of the local process models that the nodes contain. The number of transitions that can be concurrent directly depends of the number of nodes we combine. To avoid an explosion of possibilities, the number of concurrent transitions can not be too large since we try all possible node combinations. Because of the place nets, the choice construct is embedded in our input, so no additional processing is needed.

Silent and Duplicate Transitions. The presented algorithm handles the duplicate transitions as all other transitions. However, in the case of silent transitions we convert the set of place nets to a set of paths. Each path is a valid and connected

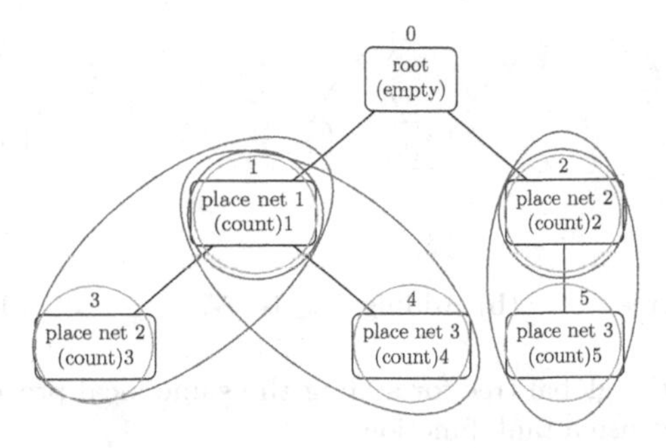

Fig. 5. Global Tree Structure. With red, we denote the place nets in the global tree, blue the local process models, and the count in green is for the number of windows the local process model (starting in the root and ending in that node) can relax replay. (Color figure online)

union of one or multiple place nets connected via silent transitions. Then, on line 5 we check whether the sequence consisted of the two activities, is in the language of the path and the path is compact for the sequence.

Collecting Local Process Models on a Global Level. The local process models we want to store in the global storage are just sets of place nets. Hence, to represent them efficiently, we use a tree structure as shown in Fig. 5. Every node in the tree stores one place net. At the same time each node also represents exactly one local process model by taking the union of the place nets in the path from that node to the root. Hence, in each node we also keep the number of windows the corresponding local process model can relax replay. Any additional information about the local process model that we might want to store in the future, can be stored in the same way as the relax replay count.

Structuring the tree this way we share place nets between the stored local process models. To also make the structure efficiently extendable, we want each path in the tree to represent a unique local process model. Therefore, we introduce a *rank* function. The rank function $rank : P \mapsto \mathbb{N}$ gives priority to each place net which in turn determines the order in which the place nets appear in the tree path representing the local process model. In Fig. 6 we illustrate the problem when a local process model $\bigcup\{N_{p1}, N_{p2}\}$ needs to be added to the tree in Fig. 6a.

After processing each window, we add all discovered local process models to the global storage. At the end, after processing all windows, the tree will contain each local process model we find together with the number of windows each local process model can relax replay.

4.3 Local Process Models Evaluation and Ranking (LPMs-ER)

Our exhaustive search can end up in a large number of local process models. Hence, we need to limit the number of local process models we return and first

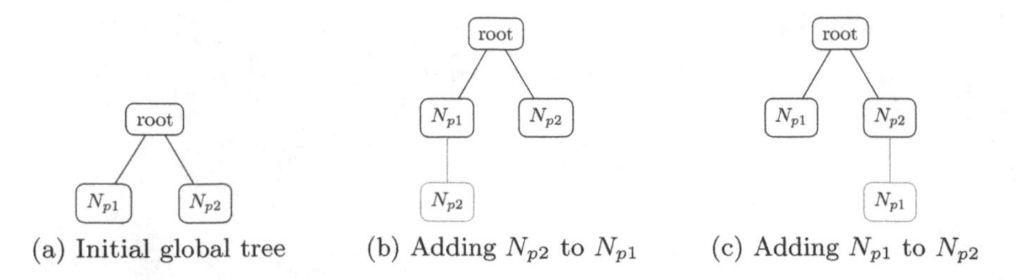

(a) Initial global tree (b) Adding N_{p2} to N_{p1} (c) Adding N_{p1} to N_{p2}

Fig. 6. Difficulty in the global tree for adding the same local process models multiple times when we do not use a rank function.

show the ones we classify as more relevant. One simple restriction is to limit the minimal and maximal number of places and transitions a local process model can have.

To measure the quality of our local process models we propose different heuristics. All metrics are calculated for an event log L, a local distance d, and a local process model $LPM = (P, T, F, A, l)$. With $W^L = [w \in \bigcup_{\rho \in L} [\rho' \sqsubseteq_d \rho]]$ we define the multiset of all windows in L with length d, and with $S^L_{LPM} = [s \in [w' \precsim w] | w \in W^L \wedge s \in \mathcal{L}(LPM)]$ a multiset of the sequences replayed by LPM during relax replay of the windows.

- *Fitting windows evaluation* calculates the fraction of windows a local process model can relax replay (Eq. (1)).

$$fw(LPM, L) = \frac{|\{w \in W^L | w \in \mathcal{L}_{rlx}(LPM)\}|}{|W^L|} \quad (1)$$

  This metric is in a way an adaptation for calculating fitness for the local process models. We never expect one local process model to explain the entire event log, so to make the metric comparable, we compare the values to the best scoring local process model.
- *Passage coverage evaluation* calculates the fraction of the passages used in the relax replay of the fitting windows (Eq. (2)).

$$pc(LPM, L) = \frac{|\{(t_1, t_2) \in \overline{LPM} | \exists_{s \in S^L_{LPM}} (\exists_{i \in \{1, \ldots, |s|-1]\}} (s_i = l(t_1) \wedge s_{i+1} = l(t_2)))\}|}{|\overline{LPM}|} \quad (2)$$

  The values are in the interval $(0, 1]$, where we get 1 when all the local process model passages are used at least once. This metric is similar to precision since lower values mean that the local process model allows more behavior than seen in the event log.
- *Passage repetition evaluation* calculates whether multiple place nets of the local process model contain the same passages (Eq. (3)). We define $\#_{(t_1, t_2)} = |\{p \in P | (t_1, t_2) \in \bullet p \times p \bullet\}$ to be the number of place nets in LPM that have the passage (t_1, t_2).

$$pr(LPM, L) = \frac{|LPM| \cdot |\overline{LPM}| - \sum_{(t_1, t_2) \in \overline{LPM}} \#_{(t_1, t_2)}}{|LPM| \cdot |\overline{LPM}| - |\overline{LPM}|} \quad (3)$$

Table 1. Information about the event logs used in our analysis

Event log alias	Trace variants count	Activities count	Total event count
BPIC2012 [9]	4366	24	182467
BPIC2019 [10]	11973	42	338247
RTFM [14]	231	13	2353
Sepsis [16]	846	16	13775
Artificial Small	2	7	45
Artificial Big	96	13	1624

This metric tries to express the simplicity of the local process model. The value of 1 denotes that each passage is contained by only one place net, and 0 denotes that all passages are contained in all place nets.

- *Transition coverage evaluation* calculates in how many of the relax replayed windows in which a transition t can be used, that transition is actually used during the replay. The average value over all transitions is returned. (Eq. (4)).

$$tc(LPM, L) = \frac{1}{|T|} \cdot \sum_{t \in T} \frac{|[s \in S^L_{LPM} | \exists_{i \in \{1,\dots,|s|\}} (s[i] = l(t))]|}{|[w \in W^L | w \in \mathcal{L}_{rlx}(LPM) \wedge \exists_{i \in \{1,\dots,|w|\}} (w[i] = l(t))]|} \tag{4}$$

The values for the metric are in the interval $(0, 1]$. Low values indicate that we use only a few transitions in our local process model during the relax replay, meaning our model is more complex than necessary.

We finish by ranking the found local process models, by taking the average score of the presented evaluation metrics. The higher the average score, the better the rank of the local process model.

5 Evaluation and Results

In this section, we evaluate our method on real and artificial event logs (see Table 1). We split the evaluation into several parts. We start by discussing how quality is defined and measured for local process models, and the challenges around it. Then, we compare the results of our algorithm with several process discovery approaches and presented related work on a specific event log. Afterward, we present the running time our algorithm has on different event logs, and the effect different parameters have on it. We end the evaluation section by comparing the running time with the running time of existing approaches discussed in related work [2,24]. For all experiments, we use the plugin we implemented in ProM and the eST Miner [15] as a place oracle. To allow for reproducibility of the experiments, we provide the artificial event logs and the sets of place nets we use at https://github.com/VikiPeeva/PlacesAndEventLogs.

5.1 Quality Definition and Challenges

Calculating the quality of local process models is challenging because of all the different ways it can be looked at. From one side we can look at the quality of each individually returned local process model or the quality of all of them as a group. If we use local process model discovery when traditional process discovery fails, the desired result would be a minimal set of local process models that cover the entire event log with as little overlaps between them as possible. This is discussed in [5] where one event log is analyzed by hand and compared to the results from [24]. Both [24] and [2], nor their future work offer this as a possibility and neither our algorithm. However, as discussed in the introduction, that is not the only usage of local process models. If we are interested in what happens when patients are cured, when companies lose money, when employees resign, etc., then we might be interested in finding local process models in regard to some utility functions or different contexts. This is to some degree investigated in [22] and [3] accordingly. Our work, currently does not support this type of local process mining, however, it is orthogonal to the current work, and can be integrated in the algorithm. With the previous information in mind we see how challenging is to give quality comparison on hundreds returned local process models between different approaches, especially when the most straight-forward comparison - event log coverage - is not available for any of them. Hence, for us, the goal was the new approach we propose to be more feasible than the existing ones in regard to running time and number of local process models found, and extendable towards event log coverage and utility mining.

5.2 Discoverability of Constructs

To illustrate the need for local process model discovery, and why the approaches proposed in [24] and [2] are not enough, we give an event log whose traces are generated by repeating the pattern $AXDBXE$. We additionally add noise (from the alphabet l, m, n) between the different occurrences of the pattern and in smaller amount in-between the pattern itself. An example trace in such event log would be $\langle m, A, X, n, D, B, X, E, l\rangle$. We ran the generated event log with α++ miner [26], inductive miner [13], ILP miner [27], the local process model discovery approaches proposed in [24] and [2], and the approach proposed in this paper. We present the results we get in Fig. 7. ILP and α++ miner returned a spaghetti-like models, while the inductive miner returned a model with mostly flowery behavior. In none of these models the pattern is clearly visible. The approaches in [24] and [2] although returning local process models that represent parts of the pattern, are not able to return a local process model that describes the pattern accurately. In contrast, our algorithm finds a local process model that completely describes the pattern (Fig. 1b) in addition to the other local process models that we find. By finding this model we show that we are able to skip in-between noise, and that we can discover constructs like long-term dependencies which the approaches in [2] and [24] cannot because of the representational bias of process trees.

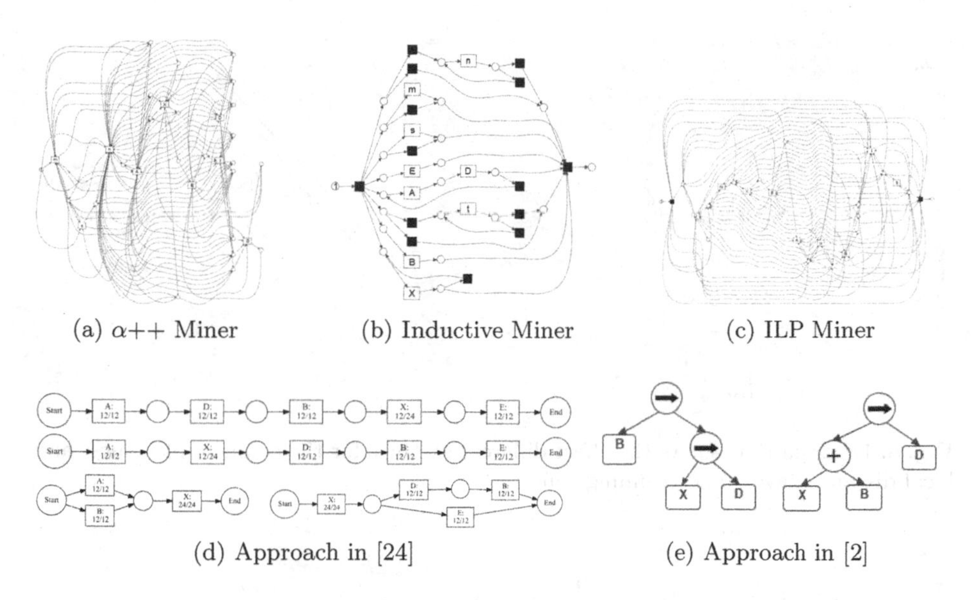

(a) α++ Miner (b) Inductive Miner (c) ILP Miner

(d) Approach in [24] (e) Approach in [2]

Fig. 7. Process models for an event log focusing on the pattern $AXDBXE$.

5.3 Running Time vs Parameters

The main parameters that we can control are the number of place nets we use and the size of the local distance. Other important parameter is the cardinality of the concurrency, i.e., what is the maximal number of transitions we allow to be in a concurrent construct. Hence, we show diagrams to see how these parameters affect the running time of the algorithm.

In Fig. 8, we show the running time for place net counts of [50, 75] and locality of [5, 7, 10, 12]. As expected, given a fixed amount of place nets used, the running time increases as the locality increases, and also the other way around, given a fixed locality, the running time increases as the number of place nets used increases. We can notice that for 50 place nets the algorithm finishes in less than five minutes for all event logs and different localities except for *BPIC2019* and locality 12. However, when considering 75 place nets, for all event logs except *Artificial Small* the limit of ten minutes is reached at locality 12. What is interesting to see is that both *Artificial Big* and *RTFM* have a larger running time for place net count of 75 and localities 5, 7 and 10 than *BPIC2012* and *BPIC2019* although the latter are much larger event logs, both in the number of events and number of activities they contain (see Table 1). This shows the impact the linearity of our algorithm has in regard to the size of the event log, and the importance of how we choose which place nets to use.

Regarding our concurrency cardinality parameter, we see in Fig. 9 that we are able to handle concurrency constructs with 4 transitions for 50 place nets, and 3 transitions for 75 place nets. However, we notice that by adding the possibility for just one more transition, the running time exceeds 10 min.

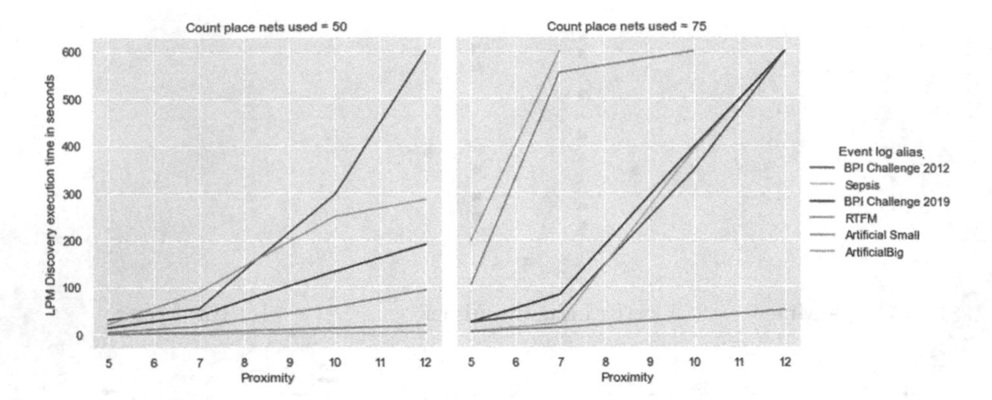

Fig. 8. Diagram that shows the effect different settings for the count of place nets and local distance have on the running time.

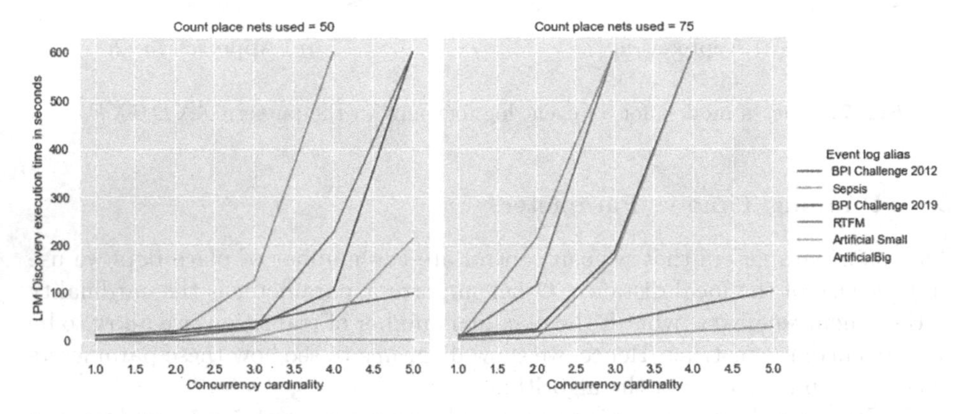

Fig. 9. Diagram that shows how the concurrency cardinality parameter affects the running time.

5.4 Comparison to Other Approaches

In this section, we focus on the comparison of our approach to the ones presented in [24] and [2] in regard to the running time. We run all algorithms on real and artificial event logs with time limit of 10 min on a PC with i7-1.8GHz, 16GB RAM and Windows 10. We use the provided default settings of the plugins where for the approach of Tax et al. the default settings also include the log projections explained in [23]. The only setting we vary for our algorithm is the number of places used (50, 75 and 100). We present the results at Table 2. We see that for the artificial event logs our approach is comparable in the time needed to return results to the one in [24] when we use 50 places. However, when it comes to real event logs, our approach is notably faster than the other two. For example, on the *BPIC2019* event log the other approaches do not return results at all because of memory problems, while we are able to build a large amount of local process

Table 2. Results comparison to [24] and [2]

Event log	Our approach			Approach in [24]		Approach in [2]	
	#places	runtime	#LPMs	runtime	#LPMs	runtime	#LPMs
BPIC2012	50	4s	284	90s	454	out of time	
	75	20s	2473				
	100	**23s**	**6484**				
BPIC2019	50	28s	3190	out of memory		out of memory	
	75	**48s**	**7617**				
	100	out of time	/				
RTFM	50	15s	8967	out of time		out of time	
	75	**368s**	**90862**				
	100	out of time	/				
Sepsis	50	2s	18	56s	4627	125s	375
	75	4s	3384				
	100	**22s**	**14951**				
Artificial Big	50	40s	8979	**70s**	**56110**	out of time	
	75	536s	65383				
	100	out of time	/				
Artificial Small	50	2s	5123	**2s**	**2665**	16s	126
	75	9s	15623				
	100	25s	34844				

models in less than a minute when we use less than 75 places. For the *BPIC2012* event log, [2] needs more than 10 min to return results and [24] investigates 454 candidate local process models in 90 s. This is less than what we can discover and it needs four times more time than our approach. The *Sepsis* and *RTFM* event logs further confirm these results, which shows that our algorithm is able to handle large event logs much better, while returning a large amount of local process models.

6 Conclusion and Outlook

In this paper, we introduced a novel way of discovering local process models. We proposed a first solution to the problem, which can be further investigated and extended. Our first goal was to have an algorithm that can find local process models for large event logs, and we achieved this by building local process models through one pass of the event log. Different quality dimensions that we discussed are returning minimal number of local process models that cover the entire event log or mining using utility functions. These are compelling directions that we plan to investigate as future work. Another point is that we get the place nets from which we build local process models from an oracle which currently is a regular process discovery algorithm. Hence, how to generate place nets valuable for local process model discovery or build the local process models without using place nets is something that warrants further research. The algorithm we propose is able to process large event logs and is flexible to support improvements for the above mentioned topics without destroying the linear complexity on the size of the event log.

Acknowledgments. We thank the Alexander von Humboldt (AvH) Stiftung for supporting our research.

References

1. van der Aalst, W.M.P.: Process Mining - Data Science in Action, 2nd edn. Springer, Cham (2016). https://doi.org/10.1007/978-3-662-49851-4
2. Acheli, M., Grigori, D., Weidlich, M.: Efficient discovery of compact maximal behavioral patterns from event logs. In: Giorgini, P., Weber, B. (eds.) CAiSE 2019. LNCS, vol. 11483, pp. 579–594. Springer, Cham (2019). https://doi.org/10.1007/978-3-030-21290-2_36
3. Acheli, M., Grigori, D., Weidlich, M.: Discovering and analyzing contextual behavioral patterns from event logs. IEEE Trans. Knowl. Data Eng. (2021). https://doi.org/10.1109/TKDE.2021.3077653
4. Bergenthum, R., Desel, J., Lorenz, R., Mauser, S.: Process mining based on regions of languages. In: Alonso, G., Dadam, P., Rosemann, M. (eds.) BPM 2007. LNCS, vol. 4714, pp. 375–383. Springer, Heidelberg (2007). https://doi.org/10.1007/978-3-540-75183-0_27
5. Brunings, M., Fahland, D., van Dongen, B.F.: Defining meaningful local process models. In: van der Aalst, W.M.P., Bergenthum, R., Carmona, J. (eds.) Proceedings of the International Workshop on Algorithms & Theories for the Analysis of Event Data 2020 Satellite event of the 41st International Conference on Application and Theory of Petri Nets and Concurrency Petri Nets 2020, virtual workshop, 24 June 2020. CEUR Workshop Proceedings, vol. 2625, pp. 6–19. CEUR-WS.org (2020)
6. Carmona, J., Cortadella, J., Kishinevsky, M.: A region-based algorithm for discovering Petri nets from event logs. In: Dumas, M., Reichert, M., Shan, M.-C. (eds.) BPM 2008. LNCS, vol. 5240, pp. 358–373. Springer, Heidelberg (2008). https://doi.org/10.1007/978-3-540-85758-7_26
7. Deeva, G., De Weerdt, J.: Understanding automated feedback in learning processes by mining local patterns. In: Daniel, F., Sheng, Q.Z., Motahari, H. (eds.) BPM 2018. LNBIP, vol. 342, pp. 56–68. Springer, Cham (2019). https://doi.org/10.1007/978-3-030-11641-5_5
8. Delcoucq, L., Lecron, F., Fortemps, P., van der Aalst, W.M.P.: Resource-centric process mining: clustering using local process models. In: Hung, C., Cerný, T., Shin, D., Bechini, A. (eds.) SAC '20: The 35th ACM/SIGAPP Symposium on Applied Computing, online event, [Brno, Czech Republic], 30 March–3 April 2020. pp. 45–52. ACM (2020). https://doi.org/10.1145/3341105.3373864
9. van Dongen, B.F.: BPI Challenge 2012 (2012)
10. van Dongen, B.F.: BPI Challenge 2019 (2019)
11. Kirchner, K., Marković, P.: Unveiling hidden patterns in flexible medical treatment processes – a process mining case study. In: Dargam, F., Delias, P., Linden, I., Mareschal, B. (eds.) ICDSST 2018. LNBIP, vol. 313, pp. 169–180. Springer, Cham (2018). https://doi.org/10.1007/978-3-319-90315-6_14

12. Leemans, S.J.J., Tax, N., ter Hofstede, A.H.M.: Indulpet miner: Combining discovery algorithms. In: Panetto, H., Debruyne, C., Proper, H.A., Ardagna, C.A., Roman, D., Meersman, R. (eds.) OTM 2018. LNCS, vol. 11229, pp. 97–115. Springer, Cham (2018). https://doi.org/10.1007/978-3-030-02610-3_6
13. Leemans, S.J.J., Fahland, D., van der Aalst, W.M.P.: Discovering block-structured process models from event logs - a constructive approach. In: Colom, J.-M., Desel, J. (eds.) PETRI NETS 2013. LNCS, vol. 7927, pp. 311–329. Springer, Heidelberg (2013). https://doi.org/10.1007/978-3-642-38697-8_17
14. de Leoni, M., Mannhardt, F.: Road traffic fine management process (2015)
15. Mannel, L.L., van der Aalst, W.M.P.: Finding complex process-structures by exploiting the token-game. In: Donatelli, S., Haar, S. (eds.) PETRI NETS 2019. LNCS, vol. 11522, pp. 258–278. Springer, Cham (2019). https://doi.org/10.1007/978-3-030-21571-2_15
16. Mannhardt, F.: Sepsis Cases - Event Log (2016). https://doi.org/10.4121/uuid:915d2bfb-7e84-49ad-a286-dc35f063a460
17. Mannhardt, F., de Leoni, M., Reijers, H.A., van der Aalst, W.M.P., Toussaint, P.J.: Guided process discovery- a pattern-based approach. Inf. Syst. **76**, 1–18 (2018). https://doi.org/10.1016/j.is.2018.01.009
18. Mannhardt, F., Tax, N.: Unsupervised event abstraction using pattern abstraction and local process models. In: Gulden, J., (eds.) Joint Proceedings of the Radar tracks at the 18th International Working Conference on Business Process Modeling, Development and Support (BPMDS), and the 22nd International Working Conference on Evaluation and Modeling Methods for Systems Analysis and Development(EMMSAD), and the 8th International Workshop on Enterprise Modeling and Information Systems Architectures (EMISA) co-located with the 29th International Conference on Advanced Information Systems Engineering 2017 (CAiSE 2017), Essen, 12–13 June 2017. CEUR Workshop Proceedings, vol. 1859, pp. 55–63. CEUR-WS.org (2017). http://ceur-ws.org/Vol-1859/bpmds-06-paper.pdf
19. Mannila, H., Toivonen, H., Verkamo, A.I.: Discovery of frequent episodes in event sequences. Data Min. Knowl. Disc. **1**(3), 259–289 (1997). https://doi.org/10.1023/A:1009748302351
20. Pijnenborg, P., Verhoeven, R., Firat, M., van Laarhoven, H., Genga, L.: Towards evidence-based analysis of palliative treatments for stomach and esophageal cancer patients: a process mining approach. In: 2021 3rd International Conference on Process Mining (ICPM), pp. 136–143 (2021). https://doi.org/10.1109/ICPM53251.2021.9576880
21. Srikant, R., Agrawal, R.: Mining sequential patterns: generalizations and performance improvements. In: Apers, P., Bouzeghoub, M., Gardarin, G. (eds.) EDBT 1996. LNCS, vol. 1057, pp. 1–17. Springer, Heidelberg (1996). https://doi.org/10.1007/BFb0014140
22. Tax, N., Dalmas, B., Sidorova, N., van der Aalst, W.M.P., Norre, S.: Interest-driven discovery of local process models. Inf. Syst. **77**, 105–117 (2018). https://doi.org/10.1016/j.is.2018.04.006
23. Tax, N., Sidorova, N., van der Aalst, W.M.P., Haakma, R.: Heuristic approaches for generating local process models through log projections. In: 2016 IEEE Symposium Series on Computational Intelligence, SSCI 2016, Athens, 6–9 December 2016, pp. 1–8. IEEE (2016). https://doi.org/10.1109/SSCI.2016.7849948
24. Tax, N., Sidorova, N., Haakma, R., van der Aalst, W.M.P.: Mining local process models. J. Innov. Digit. Ecosyst. **3**(2), 183–196 (2016). https://doi.org/10.1016/j.jides.2016.11.001

25. Verbeek, H.M.W., Buijs, J.C.A.M., van Dongen, B.F., van der Aalst, W.M.P.: Prom 6: the process mining toolkit. In: Rosa, M.L. (ed.) Proceedings of the Business Process Management 2010 Demonstration Track, Hoboken, 14–16 September 2010. CEUR Workshop Proceedings, vol. 615. CEUR-WS.org (2010). http://ceur-ws.org/Vol-615/paper13.pdf
26. Wen, L., van der Aalst, W.M.P., Wang, J., Sun, J.: Mining process models with non-free-choice constructs. Data Min. Knowl. Discov. **15**(2), 145–180 (2007). https://doi.org/10.1007/s10618-007-0065-y
27. van Zelst, S.J., van Dongen, B.F., van der Aalst, W.M.P., Verbeek, H.M.W.: Discovering workflow nets using integer linear programming. Computing **100**(5), 529–556 (2017). https://doi.org/10.1007/s00607-017-0582-5

Data and Process Resonance
Identifier Soundness for Models of Information Systems

Jan Martijn E. M. van der Werf[1(✉)], Andrey Rivkin[2], Artem Polyvyanyy[3], and Marco Montali[2]

[1] Utrecht University, Princetonplein 5, 3584 Utrecht, CC, The Netherlands
j.m.e.m.vanderwerf@uu.nl
[2] Free University of Bozen-Bolzano, piazza Domenicani 3, 39100 Bolzano, Italy
{rivkin,montali}@inf.unibz.it
[3] The University of Melbourne, Parkville, VIC 3010, Australia
artem.polyvyanyy@unimelb.edu.au

Abstract. A model of an information system describes its processes and how these processes manipulate data objects. Object-aware extensions of Petri nets focus on modeling the life-cycle of objects and their interactions. In this paper, we focus on Petri nets with identifiers, where identifiers are used to refer to objects. These objects should "behave" well in the system from inception to termination. We formalize this intuition in the notion of *identifier soundness*, and show that although this property is undecidable in general, useful subclasses exist that guarantee identifier soundness by construction.

Keywords: Information System · Verification · Data and Processes

1 Introduction

Petri nets are widely used to describe distributed systems capable of expanding their resources indefinitely [26]. A *Petri net* describes passive and active components of a system, modeled as places and transitions, respectively. The active components of a Petri net communicate asynchronously with each other via local interfaces. Thus, state changes in a Petri net system have local causes and effects and are modeled as tokens consumed, produced, or transferred by the transitions of the system. A *token* is often used to denote an *object* in the physical world the system manipulates or a *condition* that can cause a state change in the system.

Petri nets with identifiers extend classical Petri nets to provide formal means to relate tokens to objects. Every token in such a Petri net is associated with a vector of identifiers, where each identifier uniquely identifies a data object. Consequently, active components of a Petri net with identifiers model how groups of objects, either envisioned or those existing in the physical world, can be consumed, produced, or transferred by the system.

It is often desirable that modeled systems are correct. Many criteria have been devised for assessing the correctness of systems captured as Petri nets. Those

L. Bernardinello and L. Petrucci (Eds.): PETRI NETS 2022, LNCS 13288, pp. 369–392, 2022.
https://doi.org/10.1007/978-3-031-06653-5_19

criteria target models of systems that use tokens to represent conditions that control state changes. In other words, they can be used to verify the correctness of processes the systems can support and not of the object manipulations carried out within those processes. Such widely-used criteria include boundedness [18], liveness [12], and soundness [1]. The latter one, for instance, ensures that a system modeled as a *workflow net*, a special type of a Petri net used to encode workflow at an organization, has a terminal state that can be distinguished from other states of the modeled system, the system can always reach the terminal state, and every transition of the system can in principle be enabled and, thus, be executed by the system.

Real-world systems, such as information systems [25], are characterized by processes that manipulate objects. For instance, an online retailer system manipulates products, invoices, and customer records. However, correctness criteria that address both aspects, that is, the processes and data, are understood less well. Hence, the paper at hand to address the gap.

In this paper, we propose a correctness criterion for Petri nets with identifiers that combines the checks of the soundness of the system's processes with the soundness of object manipulations within those processes. Intuitively, objects of a specific type are correctly manipulated by the system if every object instance of that type, characterized by a unique identifier, can "leave" the system, that is, a dedicated transition of the system can consume it, and once that happens, no references to that object instance remain in the system. When a system achieves this harmony for its processes and all data object types, we say that the system is *identifier sound*, or, alternatively, that the data and processes of the system are in *resonance*. Specifically, this paper makes these contributions:

- It motivates and defines the notion of *identifier soundness* for checking correctness of data object manipulations in processes of a system; and
- It discusses aspects related to *decidability of identifier soundness* in the general case and for certain restricted, but still useful, classes of systems.

The paper proceeds as follows. The next section introduces concepts and notions required to support subsequent discussions. Section 3 introduces typed Petri nets with identifiers, a model for modeling distributed systems whose state is defined by objects the system manipulates. Section 4 presents the notion of identifier soundness, including a proof that the notion is in general undecidable. Section 5 discusses several classes of systems for which identifier soundness is guaranteed by construction. Finally, the paper concludes with a discussion of related work (Sect. 6) and conclusions (Sect. 7).

2 Preliminaries

Let S and T be sets. The powerset of S is denoted by $\wp(S) = \{S' \mid S' \subseteq S\}$ and $|S|$ denotes the cardinality of S. Given a relation $R \subseteq S \times T$, its range is defined by $\text{RNG}(R) = \{y \in T \mid \exists x \in S : (x, y) \in R\}$. A *multiset* m over S is a mapping of the form $m : S \rightarrow \mathbb{N}$, where $\mathbb{N} = \{0, 1, 2, \ldots\}$ denotes the set of natural numbers. For $s \in S$, $m(s) \in \mathbb{N}$ denotes the number of times s appears in the multiset. We write s^n if $m(s) = n$. For $x \notin S$, we assume $m(x) = 0$. We use $S^{\oplus}$ to denote the set of all finite multisets over S and $\emptyset$ to denote the *empty multiset*. The support of $m \in S^{\oplus}$ is the set of elements that appear in m at least once: $supp(m) = \{s \in S \mid m(s) > 0\}$. Given two multisets m_1 and m_2 over S: *(i)* $m_1 \subseteq m_2$ (resp., $m_1 \subset m_2$) iff $m_1(s) \leq m_2(s)$ (resp., $m_1(s) < m_2(s)$) for each $s \in S$; *(ii)* $(m_1 + m_2)(s) = m_1(s) + m_2(s)$ for each $s \in S$; *(iii)* if $m_1 \subseteq m_2$, $(m_2 - m_1)(s) = m_2(s) - m_1(s)$ for each $s \in S$; and *(iv)* $|m| = \sum_{s \in S} m(s)$. A *sequence* over S of length $n \in \mathbb{N}$ is a function $\sigma : \{1, \ldots, n\} \to S$. If $n > 0$ and $\sigma(i) = a_i$, for $1 \leq i \leq n$, we write $\sigma = \langle a_1, \ldots, a_n \rangle$. The length of a sequence σ is denoted by $|\sigma|$. The sequence of length 0 is called the *empty sequence*, and is denoted by ϵ. The set of all finite sequences over S is denoted by S^*. We write $a \in \sigma$ if there is $1 \leq i \leq |\sigma|$ such that $\sigma(i) = a$. Projection of sequences on a set T is defined inductively by $\epsilon_{|T} = \epsilon$, $(\langle a \rangle \cdot \sigma)_{|T} = \langle a \rangle \cdot \sigma_{|T}$ if $a \in T$ and $\langle a \rangle \cdot \sigma_{|T} = \sigma_{|T}$ otherwise, where $\cdot$ is the sequence concatenation operator. Renaming a sequence with an injective function $r : S \to T$ is defined inductively by $\rho_r(\epsilon) = \epsilon$, and $\rho_r(\langle a \rangle \cdot \sigma) = \langle r(a) \rangle \cdot \rho_r(\sigma)$. Renaming is extended to multisets of sequences as follows: given a multiset $m \in (S^*)^{\oplus}$, we define $\rho_r(m) = \sum_{\sigma \in supp(m)} \sigma(m) \cdot \rho_r(\sigma)$. For example, $\rho_{\{x \mapsto a, y \mapsto b\}}([\langle x, y \rangle^3]) = [\langle a, b \rangle^3]$.

Labeled Transition Systems. To model the behavior of a system, we use *labeled transition systems*. Given a finite set A of (action) labels, a *(labeled) transition system* (LTS) over A is a tuple $\Gamma = (S, A, s_0, \to)$, where S is a (possibly infinite) set of *states*, s_0 is the *initial state* and $\to \subseteq (S \times (A \cup \{\tau\}) \times S)$ is the *transition relation*, where $\tau \notin A$ denotes the silent action [11]. In what follows, we write $s \xrightarrow{a} s'$ for $(s, a, s') \in \to$. Let $r : A \to (A' \cup \{\tau\})$ be an injective, total function. Renaming Γ with r is defined as $\rho_r(\Gamma) = (S, A', s_0, \to')$ with $(s, r(a), s') \in \to'$ iff $(s, a, s') \in \to$. Given a set T, hiding is defined as $\hbar_T(\Gamma) = \rho_h(\Gamma)$ with $h : A \to A \cup \{\tau\}$ such that $h(t) = \tau$ if $t \in T$ and $h(t) = t$ otherwise. Given $a \in A$, $p \overset{a}{\dashrightarrow} q$ denotes a *weak transition relation* that is defined as follows: *(i)* $p \overset{a}{\dashrightarrow} q$ iff $p (\xrightarrow{\tau})^* q_1 \xrightarrow{a} q_2 (\xrightarrow{\tau})^* q$; *(ii)* $p \overset{\tau}{\dashrightarrow} q$ iff $p (\xrightarrow{\tau})^* q$. Here, $(\xrightarrow{\tau})^*$ denotes the reflexive and transitive closure of $\xrightarrow{\tau}$.

Definition 1 (Strong and weak bisimulation). *Let $\Gamma_1 = (S_1, A, s_{01}, \to_1)$ and $\Gamma_2 = (S_2, A, s_{02}, \to_2)$ be two LTSs. A relation $R \subseteq (S_1 \times S_2)$ is called a* strong simulation, *denoted as $\Gamma_1 \prec_R \Gamma_2$, if for every pair $(p, q) \in R$ and $a \in A \cup \{\tau\}$, it holds that if $p \xrightarrow{a}_1 p'$, then there exists $q' \in S_2$ such that $q \xrightarrow{a}_2 q'$ and $(p', q') \in R$. Relation R is a* weak simulation, *denoted by $\Gamma_1 \preccurlyeq_R \Gamma_2$, iff for every pair $(p, q) \in R$ and $a \in A \cup \{\tau\}$ it holds that if $p \xrightarrow{a}_1 p'$, then either $a = \tau$ and $(p', q) \in R$, or there exists $q' \in S_2$ such that $q \overset{a}{\dashrightarrow}_2 q'$ and $(p', q') \in R$.*

R is called a strong (weak) bisimulation, *denoted by $\Gamma_1 \sim_R \Gamma_2$ ($\Gamma_1 \approx_R \Gamma_2$) if both $\Gamma_1 \prec \Gamma_2$ ($\Gamma_1 \preccurlyeq_R \Gamma_2$) and $\Gamma_2 \prec_{R^{-1}} \Gamma_1$ ($\Gamma_2 \preccurlyeq_{R^{-1}} \Gamma_1$). The relation is called* rooted *iff $(s_{01}, s_{02}) \in R$. A rooted relation is indicated with a superscript ^{r}.* ◁

Petri Nets. A weighted Petri net is a 4-tuple (P, T, F, W) where P and T are two disjoint sets of *places* and *transitions*, respectively, $F \subseteq ((P \times T) \cup (T \times P))$

is the *flow relation*, and $W : F \to \mathbb{N}^+$ is a *weight function*. For $x \in P \cup T$, we write ${}^\bullet x = \{y \mid (y, x) \in F\}$ to denote the *preset* of x and $x^\bullet = \{y \mid (x, y) \in F\}$ to denote the *postset* of x. We lift the notation of preset and postset to sets element-wise. If for a Petri net no weight function is defined, we assume $W(f) = 1$ for all $f \in F$. A *marking* of N is a multiset $m \in P^\oplus$, where $m(p)$ denotes the number of *tokens* in place $p \in P$. If $m(p) > 0$, place p is called *marked* in marking m. A *marked Petri net* is a tuple (N, m) with N a weighted Petri net with marking m. A transition $t \in T$ is enabled in (N, m), denoted by $(N, m)[t\rangle$ iff $W((p, t)) \leq m(p)$ for all $p \in {}^\bullet t$. An enabled transition can *fire*, resulting in marking m' iff $m'(p) + W((p, t)) = m(p) + W((t, p))$, for all $p \in P$, and is denoted by $(N, m)[t\rangle(N, m')$. We lift the notation of firings to sequences. A sequence $\sigma \in T^*$ is a *firing sequence* iff $\sigma = \epsilon$, or markings $m_0, \ldots, m_n$ exist such that $(N, m_{i-1})[\sigma(i)\rangle(N, m_i)$ for $1 \leq i \leq |\sigma| = n$, and is denoted by $(N, m_0)[\sigma\rangle(N, m_n)$. If the context is clear, we omit the weighted Petri net N. The set of reachable markings of (N, m) is defined by $\mathcal{R}(N, m) = \{m' \mid \exists \sigma \in T^* : m[\sigma\rangle m'\}$. The semantics of a marked Petri net (N, m) with $N = (P, T, F, W)$ is defined by the LTS $\Gamma_{N,m} = (P^\oplus, T, m_0, \to)$ with $(m, t, m') \in \to$ iff $m[t\rangle m'$.

Workflow Nets. A *workflow net* (WF-net for short) is a tuple $N = (P, T, F, W, in, out)$ such that: *(i)* (P, T, F, W) is a weighted Petri net; *(ii)* $in, out \in P$ are the source and sink place, respectively, with ${}^\bullet in = out^\bullet = \emptyset$; *(iii)* every node in $P \cup T$ is on a directed path from in to out. N is called *k-sound* for some $k \in \mathbb{N}$ iff *(i)* it is proper completing, i.e., for all reachable markings $m \in \mathcal{R}(N, [in^k])$, if $[out^k] \subseteq m$, then $m = [out^k]$; *(ii)* it is weakly terminating, i.e., for any reachable marking $m \in \mathcal{R}(N, [in^k])$, the final marking is reachable, i.e., $[out^k] \in \mathcal{R}(N, m)$; and *(iii)* it is quasi-live, i.e., for all transitions $t \in T$, there is a marking $m \in \mathcal{R}(N, [in])$ such that $m[t\rangle$. The net is called *sound* if it is 1-sound. If it is k-sound for all $k \in \mathbb{N}$, it is called *generalized sound* [15].

3 Typed Petri Nets with Identifiers

Processes and data are highly intertwined: processes manipulate data objects. These manipulations can be complex and involve multiple objects. As an example, consider a retailer shop with three types of objects: *products* that are sold through the shop, and *customers* that can order these products, which is supported through an *order* process. Here, object relations can be many-to-many: e.g., a product can be ordered for many customers and the same customer can order many products. Relations can also be one-to-many, e.g., an order is always for a single customer, but a customer can have many orders. In addition, objects may have their own life cycle, which can be considered to be a process itself. For example, a product may temporarily be unavailable, or customers may be blocked by the shop, disallowing them to order products.

Different approaches have been studied to model and analyse such models that combine objects and processes. For example, data-aware Proclets [7] allow to describe the behavior of individual artifacts and their interactions. Another approach is followed in ν-PN [28], in which tokens can carry a single identifier [27]. These identifiers can be used to reference entities in an information model. However, referencing a fact composed of multiple entities is not possible in ν-PNs. In this paper, we study *typed Petri nets with identifiers* (t-PNIDs), which build upon ν-PNs [28] by extending tokens to carry vectors of identifiers [25,31]. Vectors, represented by multisets, have the advantage that a single token can represent multiple objects or entities at the same time, such as for which customer an order is. Identifiers are typed, i.e., the countable, infinite set of identifiers is partitioned into a set of types, such that each type contains a countable, infinite set of identifiers. Variables are typed as well and can only refer to identifiers of the associated type.

Definition 2 (Identifier Types). *Let $\mathcal{I}$, Λ, and $\mathcal{V}$ denote countable, infinite sets of identifiers, type labels, and variables, respectively. We define:*

- *the* domain assignment *function $I : \Lambda \to \wp(\mathcal{I})$, such that $I(\lambda_1)$ is an infinite set, and $I(\lambda_1) \cap I(\lambda_2) \neq \emptyset$ implies $\lambda_1 = \lambda_2$ for all $\lambda_1, \lambda_2 \in \Lambda$;*
- *the* id typing *function $type_I : \mathcal{I} \to \Lambda$ s.t. if $type_I(\mathtt{id}) = \lambda$, then $\mathtt{id} \in I(\lambda)$;*
- *a* variable typing *function $type_\mathcal{V} : \mathcal{V} \to \Lambda$, prescribing that $x \in \mathcal{V}$ can be substituted only by values from $I(type_\mathcal{V}(x))$.*

When clear from the context, we omit the subscripts of type. $\triangleleft$

In a t-PNID, each place is annotated with a label, called the *place type*. A place type is a vector of types, indicating types of identifier tokens the place can carry. A place with an empty place type, represented by the empty vector, is a classical Petri net place carrying indistinguishable (black) tokens. Each arc is inscribed with a multiset of vectors of identifiers, such that the type of each variable coincides with the place types. This allows to model situations in which a transition may require multiple tokens with different identifiers from the same place.

Definition 3 (Petri net with identifiers). *A* Typed Petri net with identifiers *(t-PNID) N is a tuple (P, T, F, α, β), where:*

- *(P, T, F) is a Petri net;*
- *$\alpha : P \to \Lambda^*$ is the* place typing function*;*
- *$\beta : F \to (\mathcal{V}^*)^\oplus$ defines for each flow a multiset of* variable vectors *such that $\alpha(p) = type(x)$ for any $x \in supp(\beta((p,t)))$ and $type(y) = \alpha(p')$ for any $y \in supp(\beta((t,p')))$ where $t \in T$, $p \in {}^\bullet t$, $p' \in t^\bullet$;*

Figure 1 shows a t-PNID, N_{rs}, of the retailer shop. Each place is colored according to its identifier type. In N_{rs}, places *product* and *unavailable product* are annotated with a vector $\langle product \rangle$, i.e., these places contain tokens that carry only a single identifier of type *product*. Places *customer* and *blocked customer* have type $\langle customer \rangle$. All other places, except for place p, are labeled with type $\langle order \rangle$. Place p maintains the relation between orders and customers, and is typed $\langle order, customer \rangle$, i.e., tokens in this place are identifier vectors of size 2. N_{rs} uses three variables: x for *product*, y for *order* and z for *customer*.

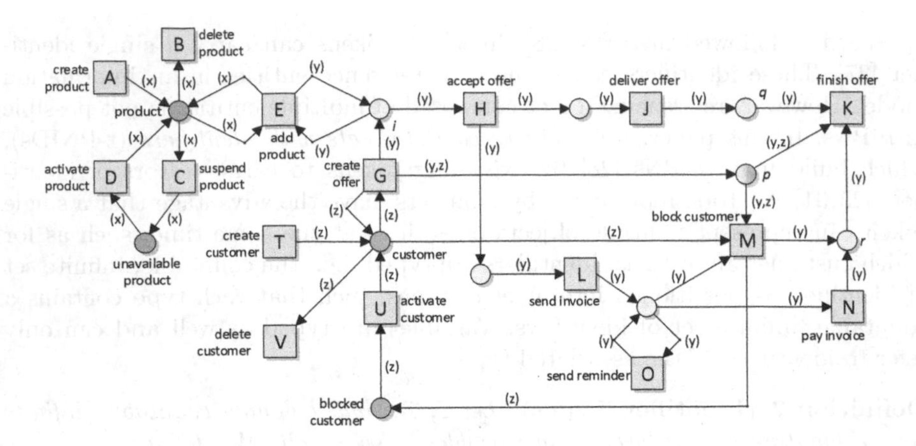

Fig. 1. t-PNID for the retailer shop with types products, customers and orders. Each place is colored according to its type. Place p carries pairs of identifiers: an order and a customer.

A marking of a t-PNID is the configuration of tokens over the set of places. Each token in a place should be of the correct type, i.e., the vector of identifiers carried by a token in a place should match the corresponding place type. All possible vectors of identifiers a place may carry is defined by the set $\mathtt{C}(p)$.

Definition 4 (Marking). *Given a t-PNID $N = (P, T, F, \alpha, \beta)$, and place $p \in P$, its* id set *is $\mathtt{C}(p) = \prod_{1 \leq i \leq |\alpha(p)|} I(\alpha(p)(i))$. A* marking *is a function $m \in \mathbb{M}(M)$, with $\mathbb{M}(M) = P \rightarrow (\Lambda^*)^{\oplus}$, such that $m(p) \in C(p)^{\oplus}$, for each place $p \in P$. The set of identifiers used in M is denoted by $Id(M) = \bigcup_{p \in P} \mathrm{RNG}(supp(M(p)))$. The pair (N, M) is called a* marked t-PNID.

To define the semantics of a t-PNID, the variables need to be valuated with identifiers. In Fig. 1, transition G uses variable y to create an identifier of type *order*, whereas transition K uses the same variable to remove an identifier from the marking.

Definition 5 (Variable sets). *Given a t-PNID $N = (P, T, F, \alpha, \beta)$, $t \in T$ and $\lambda \in \Lambda$, we define the following sets of variables:*

- input variables *as $In(t) = \bigcup_{x \in \beta((p,t)), p \in {}^\bullet t} \mathrm{RNG}(supp(x))$;*
- output variables *as $Out(t) = \bigcup_{x \in \beta((t,p)), p \in t^\bullet} \mathrm{RNG}(supp(x))$;*
- variables *as $Var(t) = In(t) \cup Out(t)$;*
- emitting variables *as $Emit(t) = Out(t) \setminus In(t)$;*
- collecting variables *as $Collect(t) = In(t) \setminus Out(t)$;*
- emitting transitions *as $E_N(\lambda) = \{t \mid \exists x \in Emit(t) \wedge type(x) = \lambda\}$;*
- collecting transitions *as $C_N(\lambda) = \{t \mid \exists x \in Collect(t) \wedge type(x) = \lambda\}$;*
- types in N *as $type(N) = \{\vec{\lambda} \mid \exists p \in P : \vec{\lambda} \in \alpha(p)\}$.* ◁

As customary in colored Petri nets, the firing of a transition requires a *binding* that valuates variables to identifiers. The binding is used to inject new fresh data into the net via variables that emit identifiers, i.e., via variables that appear only on the output arcs of that transition. We require bindings to be an injection, i.e., no two variables within a binding may refer to the same identifier. Note that in this definition, freshness of identifiers is local to the marking, i.e., disappeared identifiers may be reused, as it does not hamper the semantics of the t-PNID. Our semantics allow the use of well-ordered sets of identifiers, such as the natural numbers, as used in [25,27] to ensure that identifiers are globally new. Here we assume local freshness over global freshness.

Definition 6 (Firing rule). *Given a marked t-PNID* (N, M) *with* $N = (P, T, F, \alpha, \beta)$, *a* binding *for transition* $t \in T$ *is an injective function* $\psi : \mathcal{V} \to \mathcal{I}$ *such that* $type(v) = type(\psi(v))$ *and* $\sigma(v) \notin Id(M)$ *iff* $v \in Emit(t)$. *Transition* t *is* enabled *in* (N, M) *under binding* ψ, *denoted by* $(N, M)[t, \psi\rangle$ *iff* $\rho_\psi(\beta(p, t)) \leq M(p)$ *for all* $p \in {}^\bullet t$. *Its firing results in marking* M', *denoted by* $(N, M)[t, \psi\rangle(N, M')$, *such that* $M'(p) + \rho_\psi(\beta(p, t)) = M(p) + \rho_\psi(\beta(t, p))$. ◁

Again, the firing rule is inductively extended to sequences $\eta \in (T \times (\mathcal{V} \to \mathcal{I}))^*$. A marking M' is *reachable* from M if there exists $\eta \in (T \times (\mathcal{V} \to \mathcal{I}))^*$ s.t. $M[\eta\rangle M'$. We denote with $\mathcal{R}(N, M)$ the set of all markings reachable from M for (N, M).

The execution semantics of a t-PNID is defined as an LTS that accounts for all possible executions starting from a given initial marking.

Definition 7. *Given a marked t-PNID* (N, M_0) *with* $N = (P, T, F, \alpha, \beta)$, *its induced transition system is* $\Gamma_{N,M_0} = (\mathbb{M}(N), (T \times (\mathcal{V} \to \mathcal{I})), M_0, \to)$ *with* $M \xrightarrow{(t,\sigma)} M'$ *iff* $M[t, \sigma\rangle M'$.

t-PNIDs are a vector-based extension of ν-PNs. In other words, a ν-PN can be translated into a strongly bisimilar t-PNID with a single type, and all place types are of length of at most 1.

Lemma 1. *For any* ν*-PN there exists a single-typed t-PNID such that the two nets are strongly rooted bisimilar.* ◁

As a result, decidability of reachability for ν-PNs transfers to t-PNIDs [28].

Proposition 1. *Reachability is undecidable for t-PNIDs.* ◁

4 Correctness Criteria for t-PNIDs

Many criteria have been devised for assessing the correctness of systems captured as Petri nets. Traditionally, Petri net-based criteria focus on the correctness of processes the systems can support. Enriching the formalism with ability to capture object manipulation while keeping analyzability is a delicate balancing act. Therefore, object manipulations can only be captured if these are reflected in the token game of the net.

For t-PNIDs, correctness criteria can be categorized as system-level and as object-level. Criteria on the system-level focus on traditional Petri net-based criteria to assess the system as a whole, whereas criteria on object-level address correctness of individual objects represented by identifiers.

4.1 Correctness Criteria on System-Level

System-level properties address the overall behavior of the system. For example, liveness is a typical system-level property. It expresses that any transition is always eventually enabled again. As such, a live system guarantees that its activities cannot eventually become unavailable and never recover again.

Definition 8 (Liveness). *A marked t-PNID* $((P, T, F, \alpha, \beta), M_0)$ *is* live *iff for every marking* $M \in \mathcal{R}(N, M_0)$ *and every transition* $t \in T$*, there exist a marking* $M' \in \mathcal{R}(N, M)$ *and a binding* $\psi : \mathcal{V} \to \mathcal{I}$ *such that* $M'[t, \psi\rangle$. ◁

Boundedness expresses that the reachability graph of a Petri net is finite, i.e., that there are finitely many tokens in the system. Thus, it is a typical system-level property. Many systems have a dynamic number of simultaneously active objects. Designers often do not want to limit themselves on the maximum number of active objects. Consequently, many systems are unbounded by design. Similar to ν-PN, we differentiate between various types of boundedness. *Boundedness* expresses that the number of tokens in any reachable place does not exceed a given bound, whereas *width-boundedness* expresses that the modeled system has a bound on the number of simultaneously active objects. Notice that a width-bounded net may be unbounded if it contains infinitely many tokens referring to finitely many available objects.

Definition 9 (Bounded, width-bounded). *Let* $((P, T, F, \alpha, \beta), M_0)$ *be a marked t-PNID. A place* $p \in P$ *is called:*

- bounded *if there is* $k \in \mathbb{N}$ *such that* $|M(p)| \leq k$ *for all* $M \in \mathcal{R}(N, M_0)$*;*
- width-bounded *if there is* $k \in \mathbb{N}$ *such that* $|Id(M)| \leq k$ *for all* $M \in \mathcal{R}(N, M_0)$*;*

If all places in (N, M_0) *are (width-) bounded,* (N, M_0) *is called (width-) bounded.* ◁

4.2 Correctness Criteria on Object-Level

An object-level property assesses the correctness of individual objects. In t-PNIDs, identifiers can be seen as references to objects: if two tokens carry the same identifier, they refer to the same object. The projection of an identifier on the complete reachability graph of a t-PNID represents the life-cycle of an identifier. Whereas boundedness of a t-PNID implies that states in its reachability graph of the whole system are bounded, *depth-boundedness* expresses that for

Fig. 2. A transition-bordered WF-Net and its closure for soundness [17].

each identifier the number of tokens carrying that identifier is bounded. In other words, if a t-PNID is depth-bounded, the complete system may be unbounded, but the life-cycle of each individual identifier is finite.

Definition 10 (Depth-boundedness). *Let $((P, T, F, \alpha, \beta), M_0)$ be a marked t-PNID. A place $p \in P$ is called* depth-bounded *if there is $k \in \mathbb{N}$ such that $M(p)(\vec{\mathtt{id}}) \leq k$ for all $\vec{\mathtt{id}} \in \mathcal{I}^*$, $M \in \mathcal{R}(N, M_0)$, and $\vec{\mathtt{id}} \in \mathtt{C}(p)$ with $\mathtt{id} \in \vec{\mathtt{id}}$. If all places are depth-bounded, (N, M_0) is called depth-bounded.* ◁

Depth-boundedness is undecidable for ν-PNs [28] and thus also for t-PNIDs.

Proposition 2. *Depth-boundedness is undecidable for t-PNIDs.* ◁

Each type has a life-cycle. Intuitively, an object of a given type "enters" the system via an emitter that creates a unique identifier that refers to the object. The identifier remains in the system, until the object "leaves" the system by firing a collecting transition (that binds to the identifier and consumes it). Hence, once that transition fires, there should be no remaining tokens referring to the removed object. The process of a type is a model that describes all possible paths allowed for a type. It can be represented as a transition-bordered WF-net [17]. Instead of a sink and source place, a transition-bordered WF-net has transitions that represent the start and finish of a process. A transition-bordered WF-net is sound, if its closure is sound. As shown in Fig. 2, its closure is constructed by creating a new source place i s.t. each emitting transition consumes from i, and a new sink place f s.t. each collecting transition produces in f. Consider in t-PNID N_{rs} of Fig 1, identifier type *order*. Its life cycle starts with transition G. Transitions K and V are two transitions that may remove the last reference to an *order*. Soundness of a transition-bordered WF-net would require that firing transition K or transition V would result in the final marking. In the remainder of this section, we develop this intuition into the concept of identifier soundness.

Soundness constitutes three properties: proper completion, weak termination and quasi-liveness. Similarly to [15], we focus on the first two properties. *Proper completion* states that if a marking covers the final marking, it is the final marking. In other words, as soon as a token is produced in the final place, all other places are empty. Following the idea of transition-bordered WF-nets, identifiers should have a similar behavioral property: once an identifier is consumed by a collector, the identifier should be removed from the marking.

Definition 11 (Proper type completion). *Given type* $\lambda \in \Lambda$, *a marked t-PNID* (N, M_0) *is called* proper λ-completing *iff for all* $t \in C_N(\lambda)$, *bindings* $\psi : \mathcal{V} \to \mathcal{I}$ *and markings* $M, M' \in \mathcal{R}(N, M_0)$, *if* $M[t,\psi\rangle M'$, *then for all identifiers* $\mathtt{id} \in \text{RNG}(\psi|_{Collect(t)}) \cap Id(M)$ *and* $type(\mathtt{id}) = \lambda$, *it holds that* $\mathtt{id} \notin Id(M')$.[1] ◁

As an example, consider t-PNID N_{rs} in Fig. 1. For type *customer*, we have $C_{N_{rs}}(customer) = \{K, V\}$. In the current – empty – marking, transition T is enabled with binding $\psi = \{y \mapsto \mathtt{o}, z \mapsto \mathtt{c}\}$, which results in marking M with $M(customer) = [\mathtt{c}]$. Next, transitions G, H, J, L and N can fire, all using the same binding, producing marking M' with $M'(p) = [\mathtt{o}, \mathtt{c}]$, $M'(customer) = [\mathtt{c}]$ and $M'(q) = M'(r) = [\mathtt{c}]$. Hence, transition K is enabled with binding ψ. However, firing K with ψ results in marking M'' with $M''(customer) = [\mathtt{c}]$, while $\psi(z) = \mathtt{c}$. Hence, N_{rs} is not properly *customer*-completing.

Weak termination for a WF-net signifies that from any reachable marking, the final marking can be reached. Translated to identifiers, it should always eventually be possible to remove an identifier from a marking.

Definition 12 (Weak type termination). *Given type* $\lambda \in \Lambda$, *a marked t-PNID* (N, M_0) *is called* weakly λ-terminating *iff for every* $M \in \mathcal{R}(N, M_0)$ *and identifier* $\mathtt{id} \in I(\lambda)$ *such that* $\mathtt{id} \in Id(M)$, *there exists a marking* $M' \in \mathcal{R}(N, M)$ *with* $\mathtt{id} \notin Id(M')$. ◁

Identifier soundness combines the two properties of proper type completion and weak type termination: the former ensures that as soon a collector fires for an identifier, the identifier is removed, whereas the latter ensures that it is always eventually possible to remove that identifier.

Definition 13. *A marked t-PNID* (N, M_0) *is* λ-sound *iff it properly* λ*-completes and weakly* λ*-terminates. It is* identifier sound *iff it is* λ*-sound for every* $\lambda \in type(N)$. ◁

There are two interesting observations that one can make about the identifier soundness property. First, identifier soundness does not imply soundness in the classical sense: any classical net N without types, i.e., $type(N) = \emptyset$, is identifier sound, independently of the properties of N. Second, identifier soundness implies depth-boundedness. In other words, if a t-PNID is identifier sound for all types, it cannot accumulate infinitely many tokens carrying the same identifier.

Lemma 2. *If a t-PNID* (N, M_0) *is identifier sound, then it is depth-bounded.* ◁

Proof. Suppose that (N, M_0) is identifier sound, but not depth-bounded. Then, at least for one place $p \in P$ and identifier $\mathtt{id} \in \mathtt{C}(p)$ of type $\vec{\lambda}$ there exists an infinite sequence of increasing markings M_i, all reachable in (N, M_0), such that $M_i(p)(\mathtt{id}) < M_{i+1}(p)(\mathtt{id})$. Assume $\lambda \in \vec{\lambda}$. As (N, M_0) weakly λ-terminates, there exists $M' \in \mathcal{R}(N, M)$ such that $\mathtt{u} \notin Id(M')$, where $type(\mathtt{u}) = \lambda$. This means that the above sequence cannot exist as all constituents of $\mathtt{id}$ must be eventually removed. Hence, (N, M_0) is depth-bounded. ⊣

As identifier soundness relies on reachability, it is undecidable.

[1] Here, we constrain ψ only to objects of type λ that are only consumed.

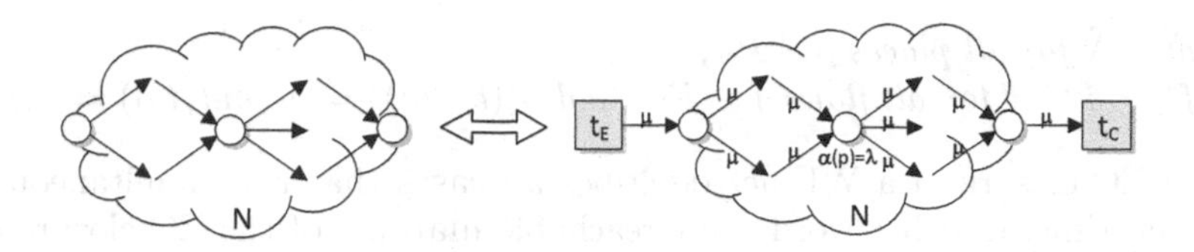

Fig. 3. EC closure of a WF-net N.

Theorem 1. *Identifier soundness is undecidable for t-PNIDs.* ◁

Proof. Let (N, M_0) be a marked t-PNID. By Definition 13, we need to show that it properly completes and weakly terminates. Since the latter requires a reachability test, it is undecidable by Proposition 1. ⊣

The above theorem also naturally follows from the fact that all non-trivial decision problems are undecidable for Petri nets in which tokens carry pairs of data values (taken from unordered domains) and in which element-wise equality comparisons are allowed over such pairs in transition guards [19].

5 Correctness by Construction

As shown in the previous section, identifier soundness is undecidable. However, we are still interested in ensuring correctness criteria over the modeled system. In this section, we propose a structural approach to taming the undecidability and study sub-classes of t-PNIDs that are identifier sound by construction.

5.1 EC-Closed Workflow Nets

WF-nets are widely used to model business processes. The initial place of the WF-net signifies the start of a *case*, the final place represents the goal state, i.e., the process case completion. A firing sequence from initial state to final state represents the activities that are performed for a single case. Thus, a WF-net describes all possible sequences of a single case. Process engines, like Yasper [14] simulate the execution of multiple cases in parallel by coloring the tokens with the case identifier (a similar idea is used for resource-constrained WF-net variants of ν-PNs in [23]). In other words, they label each place with a case type, and inscribe each arc with a variable. To execute it, the WF-net is closed with an emitter and a collector, as shown in Fig. 3. We generalize this idea to any place label, i.e., any finite sequence of types may be used to represent a case.

Definition 14 (EC-Closure). *Given a WF-net N, place type $\vec{\lambda} \in \Lambda^*$ and a variable vector $\vec{v} \in \mathcal{V}^*$ such that $type(\vec{v}) = \vec{\lambda}$. Its* EC-closure *is a t-PNID $\mathcal{W}(N, \lambda, \vec{v}) = (P_N, T_N \cup \{t_E, t_C\}, F_N \cup \{(t_E, in), (out, t_C)\}, \alpha, \beta)$, with:*

- $\alpha(p) = \vec{\lambda}$ *for all places* $p \in P_N$;
- $\beta(f) = \vec{v}^{W(f)}$ *for all flows* $f \in F_N$, *and* $\beta((t_e, in)) = \beta((out, t_c)) = [\vec{v}]$; ◁

The EC-closure of a WF-net describes all cases that run simultaneously at any given time. In other words, any reachable marking of the EC-closure is the "sum" of all simultaneous cases. Lemma 3 formalizes this idea by establishing weak bisimulation between the projection on a single case and the original net.

Lemma 3 (Weak bisimulation for each identifier). *Let N be a WF-net, $\vec{\lambda} \in \Lambda^*$ be a place type and $\vec{v} \in \mathcal{V}^*$ be a variable vector s.t. $type(\vec{v}) = \vec{\lambda}$. Then, for any $\mathtt{id} \in \mathcal{I}^{|\vec{\lambda}|}$, $\rho_r(\Gamma_{\mathcal{W}(N,\vec{\lambda},\vec{v}),\emptyset}) \approx \Gamma_{N,[in]}$ with $r(t,\psi) = r(t)$, if $\psi(\vec{v}) = \mathtt{id}$, and $r((t,\psi)) = \tau$, otherwise.* ◁

Proof. Define $R = \{(M, m) \mid \forall p \in P : M(p)(a) = m(p)\}$. We need to show that R is a weak bisimulation. (⇒) Let M, M' and m be such markings that $(M, m) \in R$ and $M[t,\psi\rangle M'$, with $t \in T$ and $\psi : \mathcal{V} \to \mathcal{I}$. By Definition 14, $\psi(\vec{v}) = u$, for some $u \in \mathcal{I}^{|\vec{\lambda}|}$. From the firing rule, we obtain $M'(p) + [u^{W((p,t))}] = M(p) + [u^{W((t,p))}]$, for any $p \in P$. If $u \neq \mathtt{id}$, then $r(t,\psi) = \tau$, and $(M', m) \in R$. If $u = \mathtt{id}$, there exists such marking m' that $m[t\rangle m'$ (since $m(p) = M(p)(\mathtt{id})$ and thus $m(p) \geq W((p,t)))$ and $m'(p) + W((p,t)) = M(p)(\mathtt{id}) + W((t,p))$. Then, by construction, $m'(p) = M'(p)(id)$ and $(M', m') \in R$.
(⇐) By analogy with the previous argument. ⊣

A natural consequence of this weak bisimulation result is that any EC-closure of a WF-net is identifier sound if and only if the underlying WF-net is sound.

Theorem 2. *Given a WF-Net N, if N is sound, then $\mathcal{W}(N, \vec{\lambda}, \vec{v})$ is identifier sound and live, for any place type $\vec{\lambda} \in \Lambda^*$ and variable vector $\vec{v} \in \mathcal{V}^*$ with $type(\vec{v}) = \vec{\lambda}$.* ◁

Proof. Let $\mathcal{W}(N, \vec{\lambda}, \vec{v}) = (P, T, F, \alpha, \beta)$. By definition of $\mathcal{W}$, $Collect(t) = \emptyset$ for any transition $t \in T \setminus \{t_C\}$. Hence, only transition t_C can remove identifiers, and thus, by construction, $\mathcal{W}$ is properly type completing on all $\lambda \in \vec{\lambda}$.

Next, we need to show that M is weakly type terminating for all types $\lambda \in \vec{\lambda}$. Let $M \in \mathcal{R}(\mathcal{W}, \emptyset)$, with firing sequence $\eta \in (T \times (\mathcal{V} \to \mathcal{I}))^*$, i.e., $M_0[\eta\rangle M$. Let $\mathtt{id} \in \mathtt{C}(p)$ such that $M(p)(\mathtt{id}) > 0$ for some $p \in P$. We then construct a sequence ω by stripping the bindings from ηs.t. it contains only transitions of T. Then, using Lemma 3, we get that $[in][\psi\rangle m$, with $m(p) = M(p)(\mathtt{id})$. Since N is sound, there exists a firing sequence ω' such that $m[\omega'\rangle[out]$. Again by Lemma 3, a firing sequence η' exists such that $M[\eta'\rangle M'$ and $(M', [out]) \in \mathcal{R}(\mathcal{W}, \emptyset)|_{\mathtt{id}}$, where $\mathcal{R}(\mathcal{W}, \emptyset)|_{\mathtt{id}}$ is the set of all reachable markings containing $\mathtt{id}$. Hence, if $M'(p)(\mathtt{id}) > 0$, then $p = out$. Thus, transition t_C is enabled with some binding ψ such that $\psi(\vec{v}) = \mathtt{id}$, and a marking M'' exists such that $M'[t_c, \psi\rangle M''$, which removes all identifiers in $\mathtt{id}$ from M'. Hence, $\mathcal{W}$ is identifier sound.

As transition t_e is always enabled, any transition is live, since N is quasi live. Hence $\mathcal{W}(N, \vec{\lambda}, \vec{v})$ is live. ⊣

5.2 Typed Jackson Nets

A well-studied class of processes that guarantee soundness are block-structured nets. Examples include Process Trees [20], Refined Process Structure Trees [30] and Jackson Nets [13]. Each of the techniques have a set of rules in common from which a class of nets can be constructed that guarantees properties like soundness. In this section, we introduce Typed Jackson Nets (t-JNs), extending the ideas of Jackson Nets [13,17] to t-PNIDs, that guarantee both identifier soundness and liveness. The six reduction rules presented by Murata in [24] form the basis of this class of nets. The rules for t-JNs are depicted in Fig. 4.

Rule 1: **Place Expansion.** The first rule is based on *fusion of a series of places.* As shown in Fig. 4a, a single place p is replaced by two places p_i and p_f that are connected via transition t. All transitions that originally produced in p, produce in p_i in the place expansion, and similarly, the transitions that consumed from place p, now consume from place p_f. In fact, transition t can be seen as a transfer transition: it needs to move tokens from place p_i to place p_f, before the original process can continue. This is also reflected in the labeling of the places: both places have the same place type, and all arcs of transition t are inscripted with $[\vec{\mu}]$, i.e., only consuming and producing a single token in a firing.

Definition 15 (Place expansion). *Let (N, M) be a marked t-PNID with $N = (P, T, F, \alpha, \beta)$, $p \in P$ be a place and $\vec{\mu} \in \mathcal{V}^*$ be a variable vector s.t. $type(\vec{\mu}) = \alpha(p)$. The* place expanded t-PNID *is defined by $R_{p,\vec{\mu}}(N, M) = ((P', T', F', \alpha', \beta'), M')$, where:*

- *$P' = (P \setminus \{p\}) \cup \{p_i, p_f\}$ with $p_i, p_f \notin P$; and $T' = T \cup \{t\}$ with $t \notin T$;*
  $F' = (F \setminus ((\{p\} \times p^\bullet) \cup (^\bullet t \times \{p\})) \cup (^\bullet p \times \{p_i\}) \cup \{(p_i, t), (t, p_f)\} \cup (\{p_f\} \times p^\bullet)$;
- *$\alpha'(q) = \alpha(p)$, if $q \in \{p_i, p_f\}$, and $\alpha'(q) = \alpha(q)$, otherwise.*
- *$\beta'(f) = [\vec{\mu}]$, if $f \in \{(p_i, t), (t, p_f)\}$, $\beta'((u, p_i)) = \beta((u, p))$, if $u \in {}^\bullet p$, $\beta'((p_f, u)) = \beta((p, u))$, if $u \in p^\bullet$, and $\beta'(f) = \beta(f)$, otherwise.*
- *$M'(q) = M(q)$ for all $q \in P \setminus \{p\}$, $M'(p_f) = 0$, and $M'(p_i) = M(p)$.* $\triangleleft$

Inscription $\vec{\mu}$ cannot alter the vector identifier on the tokens, as the type of $\vec{\mu}$ should correspond to both place types $\alpha(p)$ and $\alpha(q)$. Hence, the transition is enabled with the same bindings as any other transition that consumes a token from place p, modulo variable renaming. As such, transition t only "transfers" tokens from place p_i to place p_f. Hence, as the next lemma shows, place expansion yields a weakly bisimilar t-PNID.

Lemma 4. *Let (N, M_0) be a marked t-PNID with $N = (P, T, F, \alpha, \beta)$, $p \in P$ be a place to expand and $\vec{\mu} \in \mathcal{V}^*$ be a variable vector. Then $\Gamma_{N,M_0} \approx^r \text{Ⱨ}_{\{t\}}(\Gamma_{R_{p,\vec{\mu}}(N,M_0)})$, with transition t added by $R_{p,\vec{\mu}}$.* $\triangleleft$

Proof. Let $(N', M_0') = R_{p,\mu}(N, M_0)$. We define $Q \subseteq \mathbb{M}(N) \times \mathbb{M}(N')$ such that $M(q) = M'(q)$ for all places $q \in P \setminus \{p\}$ and $M'(p_i) + M'(p_f) = M(p)$. Then $(M_0, M_0') \in Q$, hence the relation is rooted.

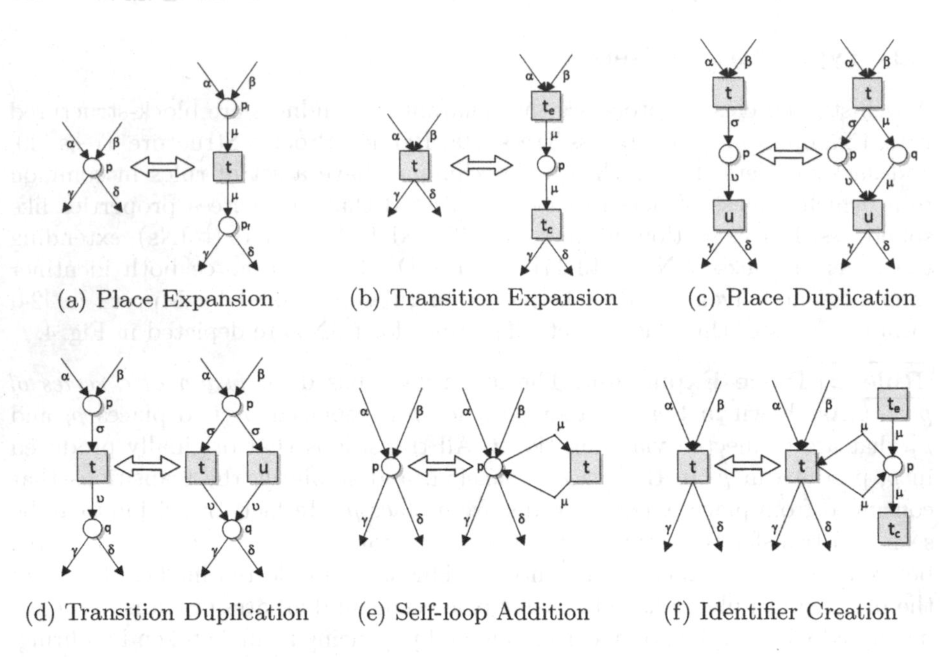

(a) Place Expansion (b) Transition Expansion (c) Place Duplication

(d) Transition Duplication (e) Self-loop Addition (f) Identifier Creation

Fig. 4. Construction rules of the typed Jackson Nets.

($\Rightarrow$) Let $(M, M') \in Q$ and $M[u, \psi\rangle \bar{M}$, where u is as in Definition 15. We need to show that there exists marking $\bar{M}'$ such that $M' \overset{(t,\psi)}{\dashrightarrow} \bar{M}'$ and $(\bar{M}, \bar{M}') \in Q$.

Suppose $p \notin {}^\bullet u$. Then $M'(q) = M(q)$ and $M(q) \geq \rho_\psi(\beta((p, u)))$ (note that $\rho_\psi(\beta((p, u))) = \rho_\psi(\beta'((p, u))))$. By the firing rule, a marking $\bar{M}'$ exists with $M'[u, \psi\rangle \bar{M}'$, $\bar{M}(q) = \bar{M}'(q)$ for all $q \in P'$. Thus, $(\bar{M}, \bar{M}') \in Q$. Suppose $p \in {}^\bullet u$. Then $\rho_\psi(\beta((p_f, u))) \leq M(p) = M'(p_i) + M'(p_f)$. If $\rho_\psi(\beta(p_f, u))) \leq M'(p_f)$, then transition u is enabled, and a marking $\bar{M}'$ exists with $M'[u, \psi\rangle \bar{M}'$ and $(M', \bar{M}') \in Q$. Otherwise, $\rho_\psi(\beta(p_f, u))) \leq M'(p_i)$. Construct a binding ψ' by letting $\psi'(\mu(i)) = \psi(\beta(p, u)(i))$, for all $1 \leq i \leq |\mu|$. Then, $\rho_{\psi'}(\mu) = \rho_\psi(\beta(p, u))$, and transition t is enabled with binding ψ'. Hence, a marking exists M'' with $M'[t, \psi'\rangle M''$ and $\rho_\psi(\beta((p', u))) \leq M''(p')$. Then $(M, M'') \in Q$ and t is labeled τ in $\hat{\mathrm{H}}_{\{t\}}(R_{(p,\vec{\mu})}(N))$, and the first case applies on M''. In all cases, $M' \overset{(t,\psi)}{\dashrightarrow} \bar{M}'$.
($\Leftarrow$) By analogy with the previous argument.

Rule 2: **Transition Expansion.** The second rule is transition expansion, which corresponds to Murata's *fusion of series transitions.* As shown in Fig. 4b, transition t is divided into two transitions, t_c that consumes the tokens, and a second transition t_p that produces the tokens. The two transitions are connected with a single, fresh place p. This place can have any type, as long as it does not hamper firing the post transition t_p, i.e., place p should ensure that all variables consumed by t_c, and that are required by t_e are passed. Transition t_c is allowed to emit new identifiers, as long as these are not already produced by t_p.

Definition 16 (Transition expansion). *Let (N, M) be a marked t-PNID with $N = (P, T, F, \alpha, \beta)$, let $t \in T$, and let $\lambda \subseteq \Lambda^*$ and $\mu \in (\mathcal{V} \setminus Emit(t))^*$ such that $type(x) \in \lambda$ and $x \in \mu$, for all $x \in In(t)$, and $type(\mu) = \lambda$. The* transition expanded t-PNID *is defined by $R_{t,\lambda,\mu}(N, M) = ((P', T', F', \alpha', \beta'), M)$, where:*

- $P' = P \cup \{p\}$ *with* $p \notin P$; *and* $T' = (T \setminus \{t\}) \cup \{t_e, t_c\}$ *with* $t_e, t_c \notin T$;
- $F' = (F \setminus ((^\bullet t \times \{t\}) \cup (\{t\} \times t^\bullet))) \cup (^\bullet t \times \{t_e\}) \cup \{(t_e, p), (p, t_c)\} \cup (\{t_c\} \times t^\bullet)$;
- $\alpha'(p) = \lambda$ *and* $\alpha'(q) = \alpha(q)$ *for all* $q \in P$;
- $\beta'(f) = [\mu]$ *if* $f \in \{(t_e, p), (p, t_c)\}$, $\beta'((q, t_e)) = \beta((q, t))$ *for* $q \in {^\bullet t}$, $\beta'((t_c, q)) = \beta((t, q))$ *for* $q \in t^\bullet$, *and* $\beta'(f) = \beta(f)$ *otherwise.* ◁

Transition t_e is allowed to introduce new variables, but key is that inscription μ contains all input variables of transition t. Consequently, μ encodes the binding of transition t. We use this to prove weak bisimulation between a t-PNID and it transition expanded net. The idea behind the simulation relation Q is that the firing of t_e is postponed until t_c fires. In other words, Q encodes that tokens remain in place q until transition t_c fires.

Lemma 5. *Given marked t-PNID (N, M_0) with $N = (P, T, F, \alpha, \beta)$, transition $t \in T$, $\lambda \in \Lambda^*$ and $\mu \in \mathcal{V}^*$. Let t_e, t_c be the transitions added by the expansion. Then $\Gamma_{N,M_0} \approx^r \rho_r(\Gamma_{R_{t,\lambda,\mu}(N,M_0)})$ with $r = \{(t_e, \tau), (t_c, t)\}$.* ◁

Proof. Let $N' = R_{t,\lambda,\mu}(N)$. Define relation $Q \subseteq \mathbb{M}(N) \times \mathbb{M}(N')$ such that $M(q) = M'(q)$ for all places $q \in P \setminus {^\bullet t}$ and $M(q) = M'(q) + \sum_{b \in supp(M'(p))} M'(p)(b) \cdot \rho_{\mu(b)}\beta((q, t))$, where $\mu(b)$ is a shorthand for the binding $\psi : \mathcal{V} \to \mathcal{I}$ with $\psi(x) = b(i)$ iff $\mu(i) = x$ for all $1 \leq i \leq |\mu|$. Then $(M_0, M_0) \in Q$.
($\Rightarrow$) Follows directly from the firing rule, and the construction of μ.
($\Leftarrow$) Let $(M, M') \in Q$ and $M'[u, \psi\rangle \bar{M}'$. We need to show a marking $\bar{M}$ exists such that $M \overset{(t,\psi)}{\dashrightarrow} \bar{M}$ and $(\bar{M}, \bar{M}') \in Q$. If $t_e \neq u \neq t_c$, the statement holds by definition of the firing rule. Suppose $u = t_e$, i.e., $r(u) = \tau$. Hence, we need to show that $(M, \bar{M}') \in Q$. Let $q \in {^\bullet t}$. Since $(M, M') \in Q$, we have $M(q) = M'(q) + \sum_{b \in supp(M'(p))} M'(p)(b) \cdot \rho_{\mu(b)}\beta((q, t))$. By the firing rule, we have $\bar{M}'(p) = M'(p) + [\rho_\psi(\mu)]$ and $M'(q) = \bar{M}'(q) + \rho_\psi(\beta((q, t)))$. By construction, ρ_ψ and $\rho_{\mu([\rho_\psi(\mu)])}$ are identical functions. Rewriting gives $M(q) = \bar{M}'(q) + \sum_{b \in supp(\bar{M}'(p))} M'(p)(b) \cdot \rho_{\mu(b)}\beta((q, t))$, and thus $(M, \bar{M}') \in Q$.

Suppose $u = t_c$, i.e., $r(u) = t$ and $[\rho_\psi(\mu)] \leq M'(p)$. Let $q \in {^\bullet t}$. Then $M(q) = M'(q) + \sum_{b \in supp(M'(p))} M'(p)(b) \cdot \rho_{\mu(b)}\beta((q, t))$. Since $\bar{M}'(p) + [\rho_\psi(\mu)] = M'(p)$ and $\rho_\psi(\beta((q, u))) = \rho_{\mu([\rho_\psi(\mu)])}(\beta((q, u)))$, we obtain $M(q) = M'(q) + \left(\sum_{b \in supp(\bar{M}'(p))} \bar{M}'(p)(b) \cdot \rho_{\mu(b)}\beta((q, t))\right) + \rho_\psi\beta((q, t))$. Hence, a marking $\bar{M}$ exists such that $M[t, \psi\rangle \bar{M}$. Rewriting gives $(\bar{M}, \bar{M}') \in Q$. ⊣

Rule 3: Place Duplication. Whereas the previous two rules only introduced ways to create sequences, the third rule introduces parallelism by duplicating a place, as shown in Fig. 4c. It is based on the *fusion of parallel transitions* reduction rule of Murata. For t-PNIDs, duplicating a place has an additional advantage: as all information required for passing the identifiers is already guaranteed, the duplicated place can have any place type.

Definition 17 (Duplicate place). *Let (N, M) be a marked t-PNID with $N = (P, T, F, \alpha, \beta)$, let $p \in P$, such that $M(p) = \emptyset$, and some transitions $t, u \in T$ exist with ${}^\bullet p = \{t\}$ and $p^\bullet = \{u\}$. Let $\lambda \in \Lambda^*$ and $\mu \in (\mathcal{V} \setminus \mathit{Emit}(u))^*$ such that $\mathit{type}(\mu) = \lambda$. Its* duplicated place t-PNID *is defined by $D_{p,\lambda,\mu}(N, M) = ((P', T, F', \alpha', \beta'), M)$, where:*

- $P' = P \cup \{q\}$, *with* $q \notin P$, *and* $F' = F \cup \{(t, q), (q, u)\}$;
- $\alpha' = \alpha \cup \{q \mapsto \lambda\}$ *and* $\beta' = \beta \cup \{(t, q) \mapsto [\mu], (q, u) \mapsto [\mu]\}$. ◁

As the duplicated place cannot hamper the firing of any transition, all behavior is preserved by a strong bisimulation on the identity mapping.

Lemma 6. *Given a marked t-PNID (N, M_0) with $N = (P, T, F, \alpha, \beta)$, place $p \in P$, $\lambda \in \Lambda^*$ and $\mu \in \mathcal{V}^*$. Then $\Gamma_{N,M_0} \sim^r \Gamma_{D_{p,\lambda,\mu}(N,M_0)}$.* ◁

Proof. Let $(N', M'_0) = D_{p,\lambda,\mu}(N)$. Define relation $Q \subseteq \mathbb{M}(N) \times \mathbb{M}(N')$ such that $(M, M') \in Q$ iff $M(p) = M'(p)$ for all places $p \in P$. The bisimulation relation trivially follows from the firing rule. ⊣

Rule 4: **Transition Duplication.** As already recognized by Berthelot [6], if two transitions have an identical preset and postset, one of these transitions can be removed while preserving liveness and boundedness. Murata's fusion of parallel places is a special case of this rule, requiring that the preset and postset are singletons. For t-JNs, this results in the duplicate transition rule: any transition may be duplicated, as shown in Fig. 4d.

Definition 18 (Duplicate place). *Let (N, M) be a marked t-PNID with $N = (P, T, F, \alpha, \beta)$, and let $t \in T$ such that some places $p, q \in P$ exist with ${}^\bullet t = \{p\}$ and $t^\bullet = \{q\}$. Its* duplicated transition t-PNID *is defined by $D_t(N, M) = ((P, T', F', \alpha, \beta'), M)$, where:*

- $T' = T \cup \{u\}$, *with* $t' \notin T$, *and* $F' = F \cup \{(p, u), (u, q)\}$;
- $\beta'((p, u)) = \beta((p, t))$, $\beta((u, q)) = \beta((t, q))$ *and* $\beta'(f) = \beta(f)$ *for all* $f \in F$. ◁

As the above rule only duplicates $t \in T$, the identity relation on markings is a strong rooted bisimulation. The proof is straightforward from the definition.

Lemma 7. *Given a marked t-PNID (N, M_0) with $N = (P, T, F, \alpha, \beta)$, and transition $t \in T$. Then $\Gamma_{N,M_0} \sim^r \rho_{\{(u,t)\}}(\Gamma_{D_t(N,M_0)})$.* ◁

Proof. Let $(N', M'_0) = D_t(N)$. Define relation $Q \subseteq \mathbb{M}(N) \times \mathbb{M}(N')$ such that $(M, M') \in Q$ iff $M(p) = M'(p)$ for all places $p \in P$. The bisimulation relation trivially follows from the firing rule. ⊣

Rule 5: **Adding Identity Transitions.** In [6], Berthelot classified a transition t with an identical preset and postset, i.e., ${}^\bullet t = t^\bullet$ as irrelevant, as its firing does not change the marking. The reduction rule *elimination of self-loop transitions* is a special case, as Murata required these sets to be singletons. For t-JNs, adding a self-loop transition is the fifth rule, as shown in Fig. 4e.

Definition 19 (Self-loop addition). *Let $(N, M$ be a marked t-PNID with $N = (P, T, F, \alpha, \beta)$, and let $p \in P$. Its* Self-loop Added t-PNID *is defined by $A_p(N, M) = ((P, T', F', \alpha, \beta'), M)$, where:*

- *$T' = T \cup \{t\}$, with $t \notin T$, and $F' = F \cup \{(p, t), (t, p)\}$;*
- *$\beta'((p, t')) = \beta'((p, t')) = [\vec{\mu}]$ with $\vec{\mu} \in \mathcal{V}^*$ such that $type(\mu) = \alpha(p)$, and $\beta'(f) = \beta(f)$ otherwise.* ◁

Similar to the duplicate transition rule, the self-loop addition rule does not introduce new behavior, except for silent self-loops. Hence, the identity relation on markings is a weak rooted bisimulation.

Lemma 8. *Given a marked t-PNID (N, M_0) with $N = (P, T, F, \alpha, \beta)$, and place $p \in P$. Then $\Gamma_{N,M_0} \approx^r \hat{\mathrm{H}}_{\{t\}}(\Gamma_{A_p(N,M_0)})$ with t the added self-loop transition.* ◁

Proof. Let $(N', M_0') = A_p(N, M_0)$. Define relation $Q \subseteq \mathbb{M}(N) \times \mathbb{M}(N')$ such that $(M, M') \in Q$ iff $M(p) = M'(p)$ for all places $p \in P$. The bisimulation relation trivially follows from the firing rule. ⊣

Rule 6: **Identifier Introduction.** The first five rules preserve the criteria of block-structured WF-nets. Murata's *elimination of self-loop places* states that adding or removing a marked place with identical preset and postset does preserve liveness and boundedness. This rule is often used to introduce a fixed resource to a net, i.e., the number of resources is determined in the initial marking. Instead, identifier introduction adds dynamic resources, as shown in Fig. 4f: transition t_e emits new identifiers as its inscription uses only "new" variables (i.e., those that have not been used in the net), and place p works like a storage of the available resources, which can be removed by firing transition t_c.

Definition 20 (Identifier Introduction). *Let (N, M) be a marked t-PNID with $N = (P, T, F, \alpha, \beta)$, let $t \in T$, let $\vec{\lambda} \in (\Lambda \setminus type(N))^*$ and $\vec{\mu} \in \mathcal{V}^*$ such that $type(\vec{\mu}) = \vec{\lambda}$. The* Identifier introducing t-PNID *is defined by $A_{t,\vec{\lambda},\vec{\mu}}(N, M) = ((P', T', F', \alpha', \beta'), M)$, where:*

- *$P' = P' \cup \{p\}$ and $T' = T \cup \{t_e, t_c\}$, for $p \notin P$ and $t_e, t_c \notin T$, and $F' = F \cup \{(p, t), (t, p), (t_e, p), (p, t_c)\}$;*
- *$\alpha' = \alpha \cup \{p \mapsto \vec{\lambda}\}$ and $\beta' = \beta \cup \{(p, t) \mapsto [\vec{\mu}], (t, p) \mapsto [\vec{\mu}], (t_e, p) \mapsto [\vec{\mu}], (p, t_c) \mapsto [\vec{\mu}]\}$;* ◁

Lemma 9. *Given a marked t-PNID (N, M_0) with $N = (P, T, F, \alpha, \beta)$, transition $t \in T$, $\vec{\lambda} \in \Lambda^*$ and $\vec{\mu} \in (\mathcal{V} \setminus Var(t))^*$. Then $\Gamma_{N,M_0} \approx^r \hat{\mathrm{H}}_{\{t_e,t_c\}}(\Gamma_{A_{(p)}(N,M_0)})$ with t_e, t_c the added transitions.* ◁

Proof. Let $N' = (P', T', F', \alpha', \beta')$. Define $Q \subseteq \mathbb{M}(N) \times \mathbb{M}(N')$ such that $(M, M') \in Q$ iff $M(p) = M'(p)$ for all $p \in P$.
($\Rightarrow$) Suppose $M[u, \psi\rangle \bar{M}'$. If $t \neq u$, the statement directly follows from the firing rule. If $t = u$, then a marking M'' and binding ψ' exists such that $M'[t_e, \psi'\rangle M''$. Then $M'(p) > \emptyset$, $(M, M'') \in Q$, and $M''[t, \psi\rangle$. Hence, markings $\bar{M}''$ and $\bar{M}'$ exist such that $M''[t, \psi\rangle \bar{M}''[t_c, \psi'\rangle \bar{M}'$, and $(M', \bar{M}''), (M', \bar{M}') \in Q$.
($\Leftarrow$) Follows directly from the firing rule. ⊣

As shown in [28], unbounded places are width-bounded, i.e., can contain an infinite number of identifiers, or depth-bounded, i.e., for each identifier, the number of tokens carrying that identifier is bounded, or both. The place added by the identifier creation rule is by definition width-unbounded, as it has an empty preset. However, it is identifier sound, and thus depth-bounded, as shown in the next lemma.

Lemma 10. *Given a marked t-PNID (N, M) with $N = (P, T, F, \alpha, \beta)$. Then $A_{t,\vec{\lambda},\vec{\mu}}(N, M)$ is identifier sound iff (N, M) is identifier sound.* ◁

Proof. Let $(N', M') = A_{t,\vec{\lambda},\vec{\mu}}(N, M)$, and let $p \in P' \setminus P$. Let $\vec{\lambda} \in type(N')$. If $\vec{\lambda} \in type(N)$, it is $\vec{\lambda}$-sound by Lemma 9. Suppose $\vec{\lambda} \notin type(N)$. Then $E_{N'}(\lambda) = \{t_e\}$ and $C_{N'}(\lambda) = \{t_c\}$. Let $\mathtt{a} \in Id(M)$ such that $type(\mathtt{a}) = \vec{\lambda}$. Then an $\vec{id} \in \mathtt{C}(p)$ exists with $\mathtt{a} \in \vec{id}$ and $M(p)(\vec{id}) = 1$. Hence, a binding ψ exists such that $\psi(\vec{\mu}) = \vec{id}$, and transition t_c is enabled with ψ. Let M' be a marking such that $M[t_c, \psi\rangle M'$. Then $\mathtt{a} \notin Id(M')$, which proves the statement. ⊣

Any net that can be reduced to a net with a single transition using these rules is called a typed Jackson Net (t-JN).

Definition 21. *The class of* typed Jackson Nets $\mathcal{T}$ *is inductively defined by:*

- $((\emptyset, \{t\}, \emptyset, \emptyset, \emptyset), \emptyset) \in \mathcal{T}$;
- *if* $(N, M) \in \mathcal{T}$, *then* $R_{p,\vec{\mu}}(N, M) \in \mathcal{T}$;
- *if* $(N, M) \in \mathcal{T}$, *then* $R_{t,\vec{\lambda},\vec{\mu}}(N, M) \in \mathcal{T}$;
- *if* $(N, M) \in \mathcal{T}$, *then* $D_{p,\vec{\lambda},\vec{\mu}}(N, M_0) \in \mathcal{T}$;
- *if* $(N, M) \in \mathcal{T}$, *then* $D_t(N, M) \in \mathcal{T}$;
- *if* $(N, M) \in \mathcal{T}$, *then* $A_p(N, M) \in \mathcal{T}$;
- *if* $(N, M) \in \mathcal{T}$, *then* $A_{t,\vec{\lambda},\vec{\mu}}(N, M) \in \mathcal{T}$. ◁

As any t-JN reduces to a single transition, and each construction rule goes hand in hand with a bisimulation relation, any liveness property is preserved. Consequently, any t-JN is identifier sound and live.

Theorem 3. *Any typed Jackson Net is identifier sound and live.* ◁

Proof. We prove the statement by induction on the structure of t-JNs. The statement holds trivially for the initial net, $((\emptyset, \{t\}, \emptyset, \emptyset, \emptyset), \emptyset)$. Suppose $(N, M) \in \mathcal{T}$. Then for each of the rules, the statement follows directly from the respective bisimulation relations of Lemma 4–9, and the result of Lemma 10. ⊣

To solve the problem of the running example, several solutions exist. One solution is shown in Fig. 5. In this example, the net is a t-JN: starting from transition G, a self loop is added (transition G). The transition is then expanded with transition Z. Place p is then duplicated to create the subnet R. In this way, subnet R only knows of type *offer*, as the connection with *customer* is stored in place p. Consequently, the net is identifier sound, and live.

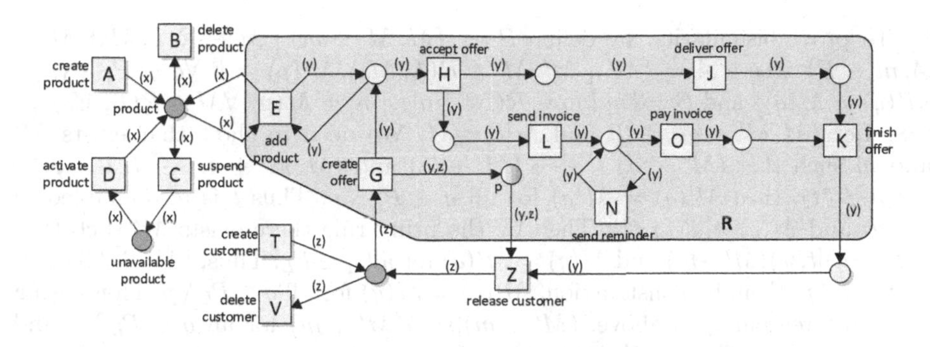

Fig. 5. The example of the retailer shop as a typed Jackson Net.

5.3 Workflow Refinement

A well-known refinement rule is workflow refinement [15]. In a WF-net, any place may be refined with a generalized sound WF-net. If the original net is sound, then the refined net is sound as well. In this section, we present a similar refinement rule. Given a t-PNID, any place may be refined by a generalized sound WF-net. In the refinement, each place is labeled with the place type of the refined place, and all arcs in the WF-net are inscribed with the same variable vector.

Definition 22 (Workflow refinement). *Let* $L = (P_L, T_L, F_L, \alpha_L, \beta_L)$, *be a t-PNID,* $p \in P_L$ *a place, and* $N = (P_N, T_N, F_N, W_N, in_N, out_N)$ *a WF-net.* Workflow refinement *is defined by* $L \oplus_p N = (P, T, F, \alpha, \beta)$, *where:*

- $P = (P_L \setminus \{p\}) \cup P_N$ *and* $T = T_L \cup T_N$;
- $F = (F_N \cap ((P \times T) \cup (T \times P))) \cup F_L \cup \{(t, in_N) \mid t \in {}^\bullet p\} \cup \{(out, t) \mid t \in p^\bullet\}$;
- $\alpha(q) = \alpha_N(q)$ *for* $q \in P_L \setminus \{p\}$, *and* $\alpha(q) = \alpha(p)$ *for* $q \in P_N$;
- $\beta(f) = \beta_L(f)$ *for* $f \in F_L$, $\beta(f) = [\vec{\mu}]^{(W(f))}$*for* $f \in F_N$ *and* $type(\vec{\mu}) = \alpha(p)$, $\beta((t, in)) = \beta((t, p))$ *for* $t \in {}^\bullet t$ *and* $\beta((out, t)) = \beta((p, t))$ *for* $t \in t^\bullet$. ◁

Generalized soundness of a WF-Net ensures that any number of tokens in the initial place are "transferred" to the final place. As shown in Sect. 5.1, the EC-closure of a sound WF-net is identifier sound and live. A similar approach is taken to show that the refinement is weakly bisimilar to the original net. Analogously to [15], the bisimulation relation is the identity relation, except for place p. The relation maps all possible token configurations of place p to any reachable marking in the WF-net, given p's token configuration.

Lemma 11. *Let* $L = (P_L, T_L, F_L, \alpha_L, \beta_L)$ *be a t-PNID with initial marking* M_0, *let* $p \in P_L$ *be a place s.t.* $M_0(p) = \emptyset$, *and let* $N = (P_N, T_N, F_N, W_N, in_N, out_N)$ *be a WF-net. If* N *is generalized sound, then* $\Gamma_L \approx^r \text{Ħ}_{T_N}(\Gamma_{L \oplus_p N})$. ◁

Proof. For simplicity, we start by defining a type extension of N as a t-PNID $N' = (P_N, T_N, F_N, \alpha, \beta)$, where $type(\vec{v}) = \vec{\lambda}$, $\alpha(p) = \vec{\lambda}$ for all places $p \in P_N$, and $\beta(f) = \vec{v}^{W(f)}$ for all $f \in F_N$, and $\beta((t_e, in)) = \beta((out, t_c)) = [\vec{v}]$.

To prove bisimilarity, we define $R = \{(M, M' + m) \mid M \in \mathcal{R}(L, M_0), M' \in \mathcal{A}, m \in \mathcal{B}\}$ where $\mathcal{A} = \{M' \mid M', M \in \mathcal{R}(L, M_0), M'(p) = \emptyset, \forall q \in P_L \setminus \{p\} : M'(q) = M(q)\}$ and $\mathcal{B} = \{m \mid m \in \mathcal{R}(N', [in]), [in] = M(p), \forall M \in \mathcal{R}(L, M_0)\}$.
($\Rightarrow$) Let $(M, M' + m) \in R$ and $M[t, \psi\rangle \bar{M}$. We need to show that exists $\bar{M}'$ and $\bar{m}$ such that $(M' + m) \overset{(t,\psi)}{\dashrightarrow} (\bar{M}' + \bar{m})$ and $(\bar{M}, \bar{M}' + \bar{m}) \in R$. If $t \notin {}^\bullet p$ (or $p \notin {}^\bullet t$), then $M'(q) = M(q)$ for all $q \in P_L \setminus p$. Thus t is also enabled in $M'(q)$ and $M(q) \geq \beta((q, t))$. Then by the firing rule there exists $\bar{M}'$ such that $(M' + m)[t, \psi\rangle(\bar{M}' + m)$ and $\bar{M}(q) = \bar{M}'(q)$ for all $q \in P_L$. Thus, $(M', \bar{M}' + m) \in R$. If $t \in {}^\bullet p$, then, by construction, $M'(q) = M(q)$ for all $q \in P_L \setminus p$. Thus, using the same reasoning as above, $(M' + m)[t, \psi\rangle(\bar{M}' + \bar{m})$ for all $q \in P_L \setminus p$ and $\bar{m}(in) = M'(p)$. Thus, $(M', \bar{M}' + \bar{m}) \in R$.

Now, assume that $p \in {}^\bullet t$ and $\rho_\psi(\beta((p, t))) = \texttt{id}$. Given that N is generalized sound and by applying Lemma 3, there exists a firing sequence η for N' that carries identifier id to *out*. This means that, by construction, $M'(q) = M(q)$, for all $q \in P_L$, and $m(out)(\texttt{id}) = M(p)(\texttt{id})$. Hence, t is enabled in $(M' + m)$ under binding ψ' that differs from ψ everywhere but on place *out*. By the firing rule, there exists $(\bar{M}' + \bar{m})$ s.t. $(M' + m)[t, \psi'\rangle(\bar{M}' + \bar{m})$ and $(M, (\bar{M}' + \bar{m})) \in R$.
($\Leftarrow$) By analogy with the previous argument.

As a consequence of the bisimulation relation, the refinement is identifier sound and live if the original net is identifier sound.

Theorem 4. *Let (L, M) be a marked t-PNID and N be a generalized sound WF net. Then (L, M) is identifier sound and live iff $(L \oplus N, M)$ is identifier sound and live.* ◁

The refinement rule allows to combine the approaches discussed in this section. For example, a designer can first design a net using the construction rules of Sect. 5.2, and then design generalized WF-nets for specific places. In this way, the construction rules and refinement rules ensure that the designer can model systems where data and processes are in resonance.

6 Related Work

This work belongs to the line of research that aims at augmenting pure control-flow description of processes with data, and study formal properties of the resulting, integrated models. When doing so, it becomes natural to move from case-centric process models whose analysis focuses on the evolution of a single instance in isolation, to so-called *object-centric process models* where multiple related instances of the same or different processes co-evolve. This is relevant for process modeling, analysis, and mining [2].

Different approaches to capture the control-flow backbone of object-oriented processes have been studied in literature, including declarative [4] and database-centric models [22]. In this work, we follow the Petri net tradition, which comes with three different strategies to tackle object-centric processes.

A first strategy is to represent objects implicitly. The most prominent example in this vein is constituted by proclets [7]. Here, each object type comes with a Petri net specifying its life cycle. Special ports, annotated with multiplicity constraints, are used to express generation and synchronization points in the process. Correctness analysis of proclets is an open research topic.

A second strategy is to represent objects explicitly. Models adopting this strategy are typically extensions of ν-PNs [28], building on their ability to generate (fresh) object identifiers and express guarded transitions relating multiple objects at once. The ISML approach [25] equips Petri nets with identifiers (PNIDs) [16] with the ability of manipulating populations of objects defining the extensional level of an ORM data model. For such models, correctness properties are assessed by imposing that the overall set of object identifiers is finite, and fixed a-priori. Catalog-nets [10] extend PNIDs with the ability of querying a read-only database containing background information. Decidability and other meta-properties, as well as actual algorithms for verification based on SMT model-checking, are given for safety properties, whereas (data-aware) soundness can only be assessed for state-bounded systems [5,22].

The third, final strategy for modeling object-centric processes with Petri nets is to rely on models that highlight how multiple objects of different types may flow through shared transitions, without considering object identifier values. This approach is followed in [3], where object-centric nets are extracted from event logs, where logged events might come with sets of object identifiers. Soundness for this model is studied in [21].

The approach studied in this paper focuses on the essence of Petri net-based object-centric processes adopting the explicit approach, that is, grounded on PNIDs. We provide, for the first time, a notion of *identifier soundness* that conceptually captures the intended evolution of objects within a net, show that such a property is undecidable to check in general, and provide a pattern-based construction technique that guarantees to produce identifier-sound models.

7 Conclusions

Achieving harmony in models that describe how processes data objects manipulate is challenging. In this paper, we use typed Petri nets with Identifiers (t-PNIDs) to model these complex interactions of multiple objects, referred through their identifiers. We propose identifier soundness as a correctness criterion that conceptually captures the expected evolution of each object. Identifier soundness is, in general, undecidable for t-PNIDs. For two subclasses, we show that identifier soundness is guaranteed and that the overall model remains live.

Many systems allow for a dynamic number of simultaneously active objects. In theory, this number can be infinite, and thus such models become width-unbounded. However, for many systems, there is a natural upper bound, which can be either assumed or guaranteed with different modeling techniques (such as multiplicity upper bounds on objects [22] or resources [23,29]. One can extend t-PNIDs by enriching objects with attributes over different data types, similar as it is done in data modeling and knowledge graphs. This calls for combining the techniques studied in this paper with data abstraction techniques used to deal with numerical data types, possibly equipped with arithmetics [8,9].

We plan to provide tool support for designers of such systems. Although many correctness criteria are undecidable, this does not mean designers should be left in the dark. Since the ISM-suite [31] already allows to model t-PNIDs, we intend to work on extending it with verification techniques to support the modeler in designing systems where processes and data are in resonance.

References

1. van der Aalst, W.M.P.: Verification of workflow nets. In: Azéma, P., Balbo, G. (eds.) ICATPN 1997. LNCS, vol. 1248, pp. 407–426. Springer, Heidelberg (1997). https://doi.org/10.1007/3-540-63139-9_48
2. van der Aalst, W.M.P.: Object-centric process mining: dealing with divergence and convergence in event data. In: Ölveczky, P.C., Salaün, G. (eds.) SEFM 2019. LNCS, vol. 11724, pp. 3–25. Springer, Cham (2019). https://doi.org/10.1007/978-3-030-30446-1_1
3. van der Aalst, W.M.P., Berti, A.: Discovering object-centric petri nets. Fundam. Informaticae **175**(1–4), 1–40 (2020)
4. Artale, A., Kovtunova, A., Montali, M., van der Aalst, W.M.P.: Modeling and reasoning over declarative data-aware processes with object-centric behavioral constraints. In: Hildebrandt, T., van Dongen, B.F., Röglinger, M., Mendling, J. (eds.) BPM 2019. LNCS, vol. 11675, pp. 139–156. Springer, Cham (2019). https://doi.org/10.1007/978-3-030-26619-6_11
5. Bagheri Hariri, B., Calvanese, D., Montali, M., Deutsch, A.: State-boundedness in data-aware dynamic systems. In: Proceedings of KR 2014. AAAI Press (2014)
6. Berthelot, G.: Verification de Reseaux de Petri. Ph.D. thesis, Université Pierre et Marie Curie (Paris) (1978)
7. Fahland, Dirk: Describing behavior of processes with many-to-many interactions. In: Donatelli, Susanna, Haar, Stefan (eds.) PETRI NETS 2019. LNCS, vol. 11522, pp. 3–24. Springer, Cham (2019). https://doi.org/10.1007/978-3-030-21571-2_1
8. Felli, P., de Leoni, M., Montali, M.: Soundness verification of decision-aware process models with variable-to-variable conditions. In: Proceedings of ACSD 2019, pp. 82–91. IEEE (2019)
9. Felli, P., Montali, M., Winkler, S.: Linear-time verification of data-aware dynamic systems with arithmetic. In: Proceedings of AAAI 2022. AAAI Press (2022)
10. Ghilardi, S., Gianola, A., Montali, M., Rivkin, A.: Petri nets with parameterised data. In: Fahland, D., Ghidini, C., Becker, J., Dumas, M. (eds.) BPM 2020. LNCS, vol. 12168, pp. 55–74. Springer, Cham (2020). https://doi.org/10.1007/978-3-030-58666-9_4
11. Glabbeek, R.: The linear time - branching time spectrum II: the semantics of sequential systems with silent moves. In: CONCUR 1993. LNCS, vol. 715, pp. 66–81. Springer (1993). https://doi.org/10.1007/3-540-57208-2_6
12. Hack, M.: The recursive equivalence of the reachability problem and the liveness problem for Petri nets and vector addition systems. In: Proceedings of SWAT 1974. pp. 156–164. IEEE Computer Society (1974)

13. van Hee, K.M., Hidders, J., Houben, G.J., Paredaens, J., Thiran, P.: On the relationship between workflow models and document types. Inf. Syst. **34**(1), 178–208 (2009)
14. van Hee, K.M., Oanea, O., Post, R., Somers, L.J., van der Werf, J.M.E.M.: Yasper: a tool for workflow modeling and analysis. In: Proceedings of ACSD 2006, pp. 279–282. IEEE (2006)
15. van Hee, K., Sidorova, N., Voorhoeve, M.: Soundness and separability of workflow nets in the stepwise refinement approach. In: van der Aalst, W.M.P., Best, E. (eds.) ICATPN 2003. LNCS, vol. 2679, pp. 337–356. Springer, Heidelberg (2003). https://doi.org/10.1007/3-540-44919-1_22
16. van Hee, K.M., Sidorova, N., Voorhoeve, M., van der Werf, J.M.E.M.: Generation of database transactions with petri nets. Fundam. Inform. **93**(1–3), 171–184 (2009)
17. van Hee, K.M., Sidorova, N., van der Werf, J.M.: Business process modeling using petri nets. In: Jensen, K., van der Aalst, W.M.P., Balbo, G., Koutny, M., Wolf, K. (eds.) Transactions on Petri Nets and Other Models of Concurrency VII. LNCS, vol. 7480, pp. 116–161. Springer, Heidelberg (2013). https://doi.org/10.1007/978-3-642-38143-0_4
18. Karp, R.M., Miller, R.E.: Parallel program schemata. J. Comput. Syst. Sci. **3**(2), 147–195 (1969)
19. Lasota, S.: Decidability border for petri nets with data: WQO dichotomy conjecture. In: Kordon, F., Moldt, D. (eds.) PETRI NETS 2016. LNCS, vol. 9698, pp. 20–36. Springer, Cham (2016). https://doi.org/10.1007/978-3-319-39086-4_3
20. Leemans, S.J.J., Fahland, D., van der Aalst, W.M.P.: Discovering block-structured process models from event logs - a constructive approach. In: Colom, J.-M., Desel, J. (eds.) PETRI NETS 2013. LNCS, vol. 7927, pp. 311–329. Springer, Heidelberg (2013). https://doi.org/10.1007/978-3-642-38697-8_17
21. Lomazova, I.A., Mitsyuk, A.A., Rivkin, A.: Soundness in object-centric workflow Petri nets. CoRR abs/2112.14994 (2021)
22. Montali, M., Calvanese, D.: Soundness of data-aware, case-centric processes. Int. J. Softw. Tools for Technol. Transf. **18**(5), 535–558 (2016). https://doi.org/10.1007/s10009-016-0417-2
23. Montali, M., Rivkin, A.: Model checking Petri nets with names using data-centric dynamic systems. Formal Aspects Comput. **28**(4), 615–641 (2016). https://doi.org/10.1007/s00165-016-0370-6
24. Murata, T.: Petri nets: properties, analysis and applications. Proc. IEEE **77**(4), 541–580 (1989)
25. Polyvyanyy, A., van der Werf, J.M.E.M., Overbeek, S., Brouwers, R.: Information systems modeling: language, verification, and tool support. In: Giorgini, P., Weber, B. (eds.) CAiSE 2019. LNCS, vol. 11483, pp. 194–212. Springer, Cham (2019). https://doi.org/10.1007/978-3-030-21290-2_13
26. Reisig, W.: Understanding Petri Nets - Modeling Techniques, Analysis Methods. Case Studies. Springer, Berlin Heidelberg (2013). https://doi.org/10.1007/978-3-642-33278-4
27. Rosa-Velardo, F., Alonso, O.M., de Frutos-Escrig, D.: Mobile synchronizing petri nets: a choreographic approach for coordination in Ubiquitous systems. Electron. Notes Theor. Comput. Sci. **150**(1), 103–126 (2006)
28. Rosa-Velardo, F., de Frutos-Escrig, D.: Decidability and complexity of Petri nets with unordered data. Theor. Comput. Sci. **412**(34), 4439–4451 (2011)
29. Sidorova, N., Stahl, C.: Soundness for resource-constrained workflow nets is decidable. IEEE Trans. Syst. Man Cybern. Syst. **43**(3), 724–729 (2013)

30. Weidlich, M., Polyvyanyy, A., Mendling, J., Weske, M.: Causal behavioural profiles - efficient computation, applications, and evaluation. Fundam. Inform. **113**(3–4), 399–435 (2011)
31. van der Werf, J.M.E.M., Polyvyanyy, A.: The information systems modeling suite. In: Janicki, R., Sidorova, N., Chatain, T. (eds.) PETRI NETS 2020. LNCS, vol. 12152, pp. 414–425. Springer, Cham (2020). https://doi.org/10.1007/978-3-030-51831-8_22

Author Index

Printed in the United States
by Baker & Taylor Publisher Services